Ant Ecology

Ant Ecology

EDITED BY

Lori Lach, Catherine L. Parr, and Kirsti L. Abbott

OXFORD
UNIVERSITY PRESS

OXFORD
UNIVERSITY PRESS

Great Clarendon Street, Oxford OX2 6DP
Oxford University Press is a department of the University of Oxford.
It furthers the University's objective of excellence in research, scholarship,
and education by publishing worldwide in
Oxford New York
Auckland Cape Town Dar es Salaam Hong Kong Karachi
Kuala Lumpur Madrid Melbourne Mexico City Nairobi
New Delhi Shanghai Taipei Toronto
With offices in
Argentina Austria Brazil Chile Czech Republic France Greece
Guatemala Hungary Italy Japan South Korea Poland Portugal
Singapore Switzerland Thailand Turkey Ukraine Vietnam

Published in the United States
by Oxford University Press Inc., New York

Reprinted 2010

ISBN 978-0-19-954463-9

Printed and bound in Great Britain by CPI Antony Rowe,
Chippenham and Eastbourne

Contents

Foreword

Edward O. Wilson

This book devoted to the many dimensions of ant ecology has been delivered at the right time. The number of biologists who include ants in their research, especially in ecology and sociobiology, and thus the overall capability of research, is rising swiftly. So is the number of myrmecologists, those who study ants as their primary objective. So numerous have these researchers become, and such is the high overall quality of their work, that myrmecology is poised to take its place among such more immediately recognizable taxon-defined disciplines as ornithology, herpetology, and nematology. Myrmecology can be properly regarded as a part of entomology, but is no longer subordinate to it.

When I began my own studies on ants as a teenager, in 1946, there were fewer than a dozen myrmecologists in the United States actively publishing, not including narrowly focused economic entomologists. There were two in South America, and several more each in Europe, Australia, and Asia. Today the number worldwide is in the hundreds, and rising steeply. As the surviving myrmecologist with the longest continuous track record of research (65 years), I have recently, at last stopped trying to keep up with all of the literature. If I and a few other old timers can be said to have been carrying the torch of myrmecology, I am happy to have it wrested from our grasp and hurried forward.

There are compelling additional reasons why the discipline should continue to grow and take its place among the prominent biological sciences. Ants make up most of the insect biomass, and they weigh more than all the land vertebrates combined save human beings. In part they have accomplished this feat by elaborate symbioses, including, variously among species sapsucker herds, ant-plant associations, arboreal ant gardens, elaiosome-mediated seed dispersal, and fungus gardens. In some species, most notably the *Oecophylla* weaver ants and leafcutter garden ants, we find the most elaborate nonhuman systems of communication and division of labor. Their systems are rivaled only by those of the apine bees and macrotermitine mound-building termites.

The communication systems have shown us to what degree it is possible for pheromone communication to evolve, and what its ultimate limitations may be, on this or perhaps any other planet. By studying self-organization as simple colonies evolve into superorganisms, myrmecologists have made important advances in defining the process of group selection. They have disclosed some of the true nature of emergent traits during the emergence of new levels of biological organizations.

Yet while the scientific natural history of ants has grown to maturity during the past two centuries of research, telling us much about basic habits and the life cycles of hundreds of species, and while the past half century has added a great deal of information about how ant colonies are put together, myrmecologists have only begun to explore the ecology of ants. We understand little of the environmental factors that shaped the social adaptations of these insects, how assemblages of species have evolved as an evolutionary product. We have only begun to explore the full impact of ants on the natural ecosystems of the planet and those they share with humanity.

The authors of the present volume have made an important contribution by summarizing much of what we have learned about the ecology of ants and by suggesting the shape of what is to come.

Preface

From scorching, barren deserts to humid tropical forests, from deep in the soil to high in the tree canopies, ants are everywhere! Their near-ubiquitous occurrence on every continent except Antarctica, combined with their enormous abundance and high diversity make ants deserved of special attention.

Ants are one of the few insect groups that can be reliably identified to family by just about anyone, regardless of age or background. They are the wise and hard-working creatures of biblical and fable lore, the endearing underdogs of Hollywood animation to some (and exaggerated villains of B-grade films to others), the unwitting victims of children with magnifying glasses, and the unwanted guests of picnics. Politicians, economists, and traffic planners are among the non-biologists who have mined ant society structure for potential application to human behaviour. Understanding of ant behaviour and collective intelligence has contributed to advances in robot development, computer science, telecommunication networks, and the stock market.

To myrmecologists – those who study ants – ants are the 'premier soil turners, channelers of energy, and dominatrices of the insect fauna' (Hölldobler and Wilson 1990). Indeed, it would be difficult to overstate the importance of ants in the functioning of terrestrial ecosystems. Estimated to number between 25,000 and 30,000 species, currently just more than 12,500 ant species are described, accounting for less than 1% of all described insect species (Bolton *et al.* 2006; May 1988). Despite their relatively small contribution to overall global biodiversity, they are omnipresent in virtually every terrestrial habitat. The estimated 10,000 trillion individual ants alive at any one time weigh about as much as all human beings combined (Hölldobler and Wilson 1994). Sustaining and sheltering their sheer numbers dictates that ants engage in a variety of ecological roles: competitors, predators, prey, scavengers, mutualists, gardeners, and soil engineers.

In their need for food and shelter, they are like any other organism on the planet. But as eusocial organisms, ants have evolved to partition reproduction and resource acquisition among different individuals. This division of labour has dramatic consequences for the ecology of ants. With the exception of colony-founding events, queens stay in protected nest enclaves with the sole purpose of producing eggs. The workers are responsible for foraging, maintaining and defending the colony, and only very rarely reproduce. Since a single worker is only one of many that undertakes these tasks and does not represent a reproductive unit, its survival is not integral to the longevity of the colony. These observations were once thought 'fatal to' the theory of natural selection (Darwin 1859); how could worker ants evolve if they are incapable of reproducing? Recasting ants as 'superorganisms', and framing their social organization within the context of kin selection, where natural selection acts on the colony, and workers maximize colony efficiency in the absence of 'interindividual conflict for reproductive privilege', resolves this natural selection conundrum and goes a long way towards explaining why eusocial insects have been so successful: organized groups outcompete individuals, and larger groups outcompete smaller ones of the same species (Hölldobler and Wilson 2008).

Thus, in ecology the importance of ants is reflected by their ubiquity and the great number of interactions in which they are capable of participating within an ecological community. As such, the study of ants has led to significant advances in our understanding of insect evolution, global diversity patterns, competitive interactions, mutualisms, ecosystem responses to change, and biological invasions. But ants are also important to study and understand because they are different; their status as superorganisms places them at a level of organization between individuals and ecosystems (Hölldobler and Wilson 2008). Their social structure provides a rich ground for exploring how division of labour affects the acquisition of resources, foraging and defensive behaviours, and coevolution with the flora and fauna with which they interact. In turn, how elements of their social structure, such as colony founding, caste differentiation, and nestmate recognition, are influenced by their environment deserves investigation.

Why *Ant Ecology?*

Several excellent texts have described the social organization and evolution of social insect societies (e.g., Bourke and Franks 1995; Crozier and Pamilo 1996; Gordon 1999; Hölldobler and Wilson 1990). Our purpose in compiling this book was fourfold: to complement and build on these fundamental works, to highlight the contributions of myrmecology to ecology more broadly, to synthesize the current state of knowledge, and to add to the growing body of work that seeks to promote interest in insects both among ecologists and in the world of conservation. We also seek to inspire current and future myrmecologists to seize the opportunities presented by the gaps in research that are identified throughout the book. We hope that this volume will appeal to community and behavioural ecologists, population biologists, macroecologists, evolutionary biologists, as well as those involved with conservation and natural resource management.

Ecology is not a linearly structured science; it is as complex and interconnected as the world that it seeks to understand. Ant ecology is no different. Thus, although we have organized the book into four parts: Global Ant Diversity and Conservation, Community Dynamics, Population Ecology, and Invasive Ants, the boundaries between them blur and blend. Each section begins with a brief introduction that identifies common themes and defines terms applicable to the subsequent four chapters. The first three sections provide a firm foundation in ant ecology, while the fourth applies this foundation to the problem of ant invasions. Interspersed throughout the book are short boxes that further explain important techniques, terms, or methods, or highlight an interesting discovery, debate, or application relevant to the chapter. A notable strength of the book is that it draws on the knowledge and experience of so many myrmecologists and ecologists; 53 authors and 55 reviewers from around the world have contributed their ideas, time, and energy to the pages that follow.

Acknowledgements

Ant Ecology was conceived in boundless enthusiasm at the seaside café Chocolate Fish in Wellington, New Zealand, on 30 August 2006. Since then many individuals have helped and supported us through the process of turning our ambitious proposal into this tangible volume.

We are extremely grateful to Ian Sherman, Helen Eaton, and Elmandi Du Toit from Oxford University Press for providing guidance, advice, and feedback at crucial times and for always being so encouraging of this project. Thank you for putting this book on your portfolio for 2009.

We are most indebted to our authors, who joined us in this journey and remained committed and patient with us throughout. Edited volumes such as this benefit from the variety of perspectives brought to the project by each author, and the creativity of each contributor in describing his or her own part of this amazing scientific discipline. Many authors contributed to the terms and definitions in the glossary, and we thank them for their diligence.

The dazzling array of weird and wonderful ants and their fascinating biologies are brought to life through images supplied by Alex Wild. We sincerely thank you for such superb photographs. We thank Andrew Mercer for providing the domain (www.funkyant.com) that hosted all the *Ant Ecology* emails, and Pat Lach for her incredible patience and hard work in compiling the reference list. We are very grateful to Adam Beaumont, Dan Borg, Stephanie Chapple, Natalie Funtera, Jeremy Gibson, Courtney Johnson, Emily McGuire, Mike Parr, and Amber Tritt who put in some long hours proofreading, commenting on chapters and text boxes, and providing us with non-myrmecologists' perspectives on it all.

Each chapter and box was improved with comments from multiple peer reviews. We are indebted to the following colleagues for their constructive feedback: Donat Agosti, Craig Allen, Leeanne Alonso, Inge Armbrecht, Alan Andersen, Nico Blüthgen, Barry Bolton, Koos Boomsma, Raphael Boulay, Seán Brady, Chris Burwell, Steven Chown, Cameron Currie, Robert Cowie, Jacques Delabie, Claire Detrain, Anna Dornhaus, Bart Drees, Karl Evans, Don Feener, Patricia Folgarait, Deborah Gordon, Aaron Gove, Wulfila Gronenberg, Richard Harris, Jürgen Heinze, Joan Herbers, Bob Johnson, Adam Kay, David Keith, Laurent Keller, Paul Krushelnycky, Phil Lester, Jack Longino, Jonathan Majer, Melodie McGeoch, Lloyd Morrison, Tim New, Dennis O 'Dowd, Paulo Oliveira, Jes Pedersen, Stacy Philpott, Scott Powell, Tom Richardson, Alexei Rowles, Ted Schultz, Steve Shattuck, Dan Simberloff, Roy Snelling, Andy Suarez, Melissa Thomas, Chad Tillberg, John Vandermeer, Heraldo Vasconcelos, Darren Ward, Phil Ward, Seike Yamane, and Alan Yen, as well as those who preferred to remain anonymous. We are also grateful to Alan Andersen, Mike Kaspari, Dennis O 'Dowd, and multiple anonymous reviewers of our initial proposal for their ideas and encouragement.

Lori thanks Richard Hobbs and Viki Cramer for advice on book editing and all members of the Hobbs lab for their support. She is ever grateful to Manfred, Sydney, and Mom for the many manifestations of their encouragement, support, and inspiration. Lori was partially funded by the Australian Research Council.

Kate is extremely grateful to Richard for his patience during the many evenings at the computer and for nourishing, supporting, and inspiring her on this long endeavour. The Trapnell Fund provided funds to Kate.

Kirsti wishes to thank Katie Barnett and Charlotte Malycon for discussions on the etiquette and technicalities of publishing. Kirsti could not have been a part of the project if it were not for the support of Jeremy, Toni, Carol, and Mum, creating time where there was none, and allowing her the time away from parenting when it was most needed.

List of Contributors

Kirsti L. Abbott Faculty of Science, Monash University 3800, VIC, Australia.
Email: Kirsti.Abbott@sci.monash.edu.au

Leeanne E. Alonso Rapid Assessment Program, Conservation International, 2011 Crystal Drive, Suite 500, Arlington, VA 22202, USA.
Email: l.alonso@conservation.org

Alan N. Andersen Tropical Ecosystems Research Centre, CSIRO Sustainable Ecosystems, PMB 44, Winnellie 0822, NT, Australia.
Email: Alan.Andersen@csiro.au

Inge Armbrecht Universidad del Valle, Departamento de Biología, Apartado Aéreo 25360, Cali, Colombia.
Email: inge@univalle.edu.co

Andrew J. Beattie Department of Biological Sciences, Macquarie University, 2109, NSW, Australia.
Email: abeattie@bio.mq.edu.au

Brandon T. Bestelmeyer USDA-ARS Jornada Experimental Range, New Mexico State University, Las Cruces, NM 88003, USA.
Email: bbestelm@nmsu.edu

Nico Blüthgen Department of Animal Ecology & Tropical Biology, University of Würzburg, Biozentrum, Am Hubland, 97074 Würzburg, Germany.
Email: bluethgen@biozentrum.uni-wuerzburg.de

Keith Christian School of Environmental and Life Sciences, Charles Darwin University, Darwin 0909, NT, Australia.
Email: keith.christian@cdu.edu.au

Ross H. Crozier School of Marine and Tropical Biology, James Cook University, DB23, Townsville 4811, QLD, Australia.
Email: Ross.Crozier@jcu.edu.au

Peter Davis Department of Agriculture and Food Western Australia, South Perth 6151, WA, Australia.
Email: pdavis@agric.wa.gov.au

Patrizia d'Ettorre Centre for Social Evolution, Department of Biology, Universitesparken 15, DK-2100, Copenhagen, Denmark.
Email: PDettorre@bi.ku.dk

Anna Dornhaus Department of Ecology and Evolutionary Biology, University of Arizona, P.O. Box 210088, Tucson, AZ 85721, USA.
Email: dornhaus@email.arizona.edu

Robert R. Dunn Department of Biology, North Carolina State University, Raleigh, NC 27695–7617, USA.
Email: Rob_Dunn@ncsu.edu

Christine Errard Domaine des Blaines F-36350 La Pérouille, France.
Email: christine.errard@wanadoo.fr

Heike Feldhaar Behavioural Biology, University of Osnabrück. Barbarastr. 11 49076 Osnabrück, Germany.
Email: feldhaar@biologie.uni-osnabrueck.de

Brian L. Fisher Entomology, California Academy of Sciences, 55 Music Concourse Drive, San Francisco, CA 94118, USA.
Email: bfisher@calacademy.org

Heloise Gibb Department of Zoology, La Trobe University, Bundoora 3086, VIC, Australia.
Email: h.gibb@latrobe.edu.au

Peter T. Green Department of Botany, La Trobe University, Bundoora 3086, VIC, Australia.
Email: p.green@latrobe.edu.au

Benoit Guénard Department of Biology. North Carolina State University, Raleigh. NC 27695–7617, USA.
Email: zeroben@gmail.com

Benjamin D. Hoffmann Tropical Ecosystems Research Centre, CSIRO Sustainable Ecosystems, PMB 44, Winnellie 0822, NT, Australia.
Email: Ben.Hoffmann@csiro.au

David A. Holway Ecology, Behaviour & Evolution, University of California San Diego, Division of Biological Sciences, 9500 Gilman Drive, La Jolla, CA 92093–0116, USA.
Email: dholway@biomail.ucsd.edu

Linda M. Hooper-Bùi Department of Entomology, Louisiana State University, Baton Rouge, LA 70803, USA.
Email: LHooper@agcenter.lsu.edu

Paul D. Krushelnycky U.S. Geological Survey, Pacific Island Ecosystems Research Center, Haleakala Field Station P.O. Box 369, Makawao, HI 96768 USA.
And
Department of Plant and Environmental Protection Sciences, 3050 Maile Way, Gilmore 310, University of Hawai'i, Honolulu, HI 96822, USA.
Email: pauldk@hawai'i.edu

Lori Lach School of Plant Biology, M084
The University of Western Australia
35 Stirling Highway, Crawley, WA 6009, Australia.
Email: lorilach@cyllene.uwa.edu.au

Edward G. LeBrun Brackenridge Field Laboratory, University of Texas at Austin, Austin, TX 78703, USA.
Email: elebrun@mail.texas.edu

Alain Lenoir IRBI, UMR CNRS 6035, Institut de Recherche sur la Biologie de l 'Insecte, Université François Rabelais, Tours, France.
Email: alain.lenoir@univ-tours.fr

Terrence P. McGlynn Department of Biology, California State University Dominguez Hills, 1000 E. Victoria St., Carson, CA 90747, USA.
Email: terry.mcglynn@gmail.com

Mathieu Molet Laboratoire Ecologie et Evolution, CNRS UMR 7625, Université Pierre-et-Marie Curie, 7 quai Saint Bernard, 75005 Paris, France.
Email: mathieu.molet@snv.jussieu.fr

Kailen Mooney Department of Ecology & Evolutionary Biology, University of California, 321 Steinhaus Hall, Irvine, CA 92697–2525, USA.
Email: mooneyk@uci.edu

Joshua Ness Biology Department, Skidmore College, 815 North Broadway, Saratoga Springs, NY 12866, USA.
Email: jness@skidmore.edu

Simon O 'Connor Ministry of Agriculture & Forestry, Biosecurity New Zealand, PO Box 2526, Wellington, New Zealand.
Email: simon.oconnor@maf.govt.nz

Dennis J. O 'Dowd Australian Centre for Biodiversity, School of Biological Sciences, Monash University 3800, VIC, Australia.
Email: dennis.odowd@sci.monash.edu.au

Catherine L. Parr Environmental Change Institute, School of Geography and the Environment, University of Oxford, South Parks Road, Oxford, OX1 3QY, UK.
Email: kate.parr@ouce.ox.ac.uk

Christian Peeters Laboratoire Ecologie et Evolution, CNRS UMR 7625, Université Pierre-et-Marie Curie, 7 quai Saint Bernard, 75005 Paris, France.
Email: christian.peeters@snv.jussieu.fr

Renkang Peng School of Environmental and Life Sciences, Charles Darwin University, Darwin 0909, NT, Australia.
Email: renkang.peng@cdu.edu.au

Ivette Perfecto School of Natural Resources and Environment, University of Michigan, MI 48109–1041, USA.
Email: perfecto@umich.edu

Stacy M. Philpott Department of Environmental Sciences, University of Toledo, Toledo, OH 43606–3390, USA.
Email: stacy.philpott@utoledo.edu

Scott Powell Department of Biology, University of Missouri–St Louis, One University Boulevard, Saint Louis, MO 63121–4499, USA.
And
Universidade Federal de Uberlândia, Brazil, Campus Umauarama Bloco 2D-sala 28, 38400–902 Uberlandia MG, Brazil.
Email: scottpowell@mac.com

Leticia Rios-Casanova Facultad de Estudios Superiores-Iztacala (UBIPRO). Universidad Nacional Autónoma de México Av. de los Barrios 1, Tlalnepantla 54090, Estado de México, México.
Email: leticiarc@campus.iztacala.unam.mx

Simon Robson School of Marine & Tropical Biology, James Cook University, Townsville 4811, QLD, Australia.
Email: simon.robson@jcu.edu.au

Nathan J. Sanders Department of Ecology & Evolutionary Biology, 569 Dabney Hall, University of Tennessee, Knoxville, TN 37996, USA.
Email: nsanders@utk.edu

Birgit C. Schlick-Steiner School of Marine and Tropical Biology, James Cook University, DB23, Townsville 4811 QLD, Australia.
And
Institute of Ecology, University of Innsbruck, Technikerstr. 25, A-6020 Innsbruck, Austria.
Email: birgit.schlick-steiner@uibk.ac.at

Ted R. Schultz Department of Entomology, Smithsonian Institution, P.O. Box 37012, NHB, CE516, MRC 188, Washington, D.C. 20013–7012, USA.
Email: schultzt@si.edu

Joseph C. Spagna William Paterson University of New Jersey, 300 Pompton Rd., Wayne, NJ 07470, USA.
Email: spagnaj@wpunj.edu

Florian M. Steiner School of Marine and Tropical Biology, James Cook University, Townsville 4811 QLD, Australia.
And
Institute of Ecology, University of Innsbruck, Technikerstr. 25, A-6020 Innsbruck, Austria.
Email: florian.m.steiner@uibk.ac.at

Andrew V. Suarez Department of Entomology and Department of Animal Biology, School of Integrative Biology, University of Illinois at Urbana-Champaign, 505 S. Goodwin Ave., Urbana, IL 61801, USA.
Email: avsuarez@life.uiuc.edu

Robert W. Taylor 26 Ambalindum Drive, Hawker 2614, ACT, Australia.
Email: bob.taylor@homemail.com.au

Chadwick V. Tillberg Department of Biology, Linfield College, 900 S.E. Baker St., McMinnville, OR 97128, USA.
Email: ctillbe@linfield.edu

Neil D. Tsutsui Department of Environmental Science, Policy & Management, University of California-Berkeley, 137 Mulford Hall, MC3114, Berkeley, CA 94720–3114, USA.
Email: ntsutsui@nature.berkeley.edu

Philip S. Ward Department of Entomology and Center for Population Biology, University of California at Davis, Davis, CA 95616, USA.
Email: psward@ucdavis.edu

Michael D. Weiser Department of Biology, North Carolina State University, Raleigh, NC 27695, USA.
Email: mdweiser@ncsu.edu

Grant P. Weston Ministry of Agriculture and Forestry, Biosecurity New Zealand, Auckland – Biosecurity Centre, PO Box 53–066, Auckland, 2150, New Zealand.
Email: westong@maf.govt.nz

Stephen P. Yanoviak Department of Biology, University of Arkansas at Little Rock, 2801 S. University Ave., Little Rock, AR 72204–1099, USA.
Email: spyanoviak@ualr.edu

List of Plates

List of Abbreviations

AFLP	Amplified Fragment Length Polymorphism
ANeT	Asian Ant Network
BAP	Biodiversity Action Plan
BDFFP	Biological Dynamics Forest Fragment Project
CHC	Cuticular Hydrocarbon
COI	Cytochrome c Oxidase I, also Cytochrome Oxidase I
CR	Critically Endangered
DCF	Dependent Colony Foundation
DIVA	Dispersal-Vicariance Analysis
DD	Data Deficient
EFN	Extrafloral Nectary
EPBC Act	Environment Protection and Biodiversity Conservation Act
EN	Endangered
EX	Extinct
EXW	Extinct in the Wild
FB	Food Bodies
GALS	Giant African Land Snail
GBIF	Global Biodiversity Information Facility
CBA	Cost–Benefit Analysis
GIS	Geographic Information System
IBA	Important Bird Areas
ICF	Independent Colony Foundation
ICZN	International Code of Zoological Nomenclature
IGR	Insect Growth Regulator
ISPM	International Standards for Phytosanitary Measure
IPM	Integrated Pest Management
JH	Juvenile Hormone
KBA	Key Biodiversity Areas
LC	Least Concern
LLAMA	Leaf Litter Arthropods of Mesoamerica
GCM	Geometric Constraints Model
mtDNA	mitochondrial DNA
MTE	Metabolic Theory of Ecology
Mya	Million years ago
NT	Near Threatened
NPP	Net Primary Productivity
SISG	Social Insects Specialist Group
SSC	Species Survival Commission
SRLI	Sampled Red List Index
USDA	United States Department of Agriculture
VU	Vulnerable
YCA	Yellow Crazy Ant

PART I

GLOBAL ANT DIVERSITY AND CONSERVATION

Explaining the origin and distribution of biodiversity has been a long-standing goal of ecology. As with other taxa, ant diversity varies among continents and biogeographic regions. Part I provides the evolutionary, biogeographic, and macroecological perspectives necessary to better understand global and regional patterns in ant diversity. This section addresses questions such as how many species and major lineages of ants are there? What is their evolutionary history? How are different taxonomic groups distributed globally? What processes are important in determining diversity in space and time? And, how can we apply this understanding to the conservation of ants?

Chapter 1 unravels the evolutionary history of ants using recent morphological evidence, molecular phylogenetic studies, and new fossil discoveries. Ward reports on the substantial progress that has been made in recent years in identifying major clades of ants and clarifying phylogenetic relationships. Currently there are more than 12,500 described species of ants in 290 genera belonging to 21 extant subfamilies. At the species level, although cataloguing and understanding diversity presents many challenges, advancements have been made with a relatively stable generic classification, several global species revisions, and new technological developments including imaging, Web-based databases, and DNA bar-coding.

Chapter 2 builds on themes presented in the first chapter to focus on biogeographic patterns of global ant diversity. The present-day distribution of ants reflects the combined influence of geography, geology, and climate on the origin, diversification, and spread of lineages. Gondwanaland fragments of South America, Africa, and Australia have the highest percentage of endemic genera, while species-level hotspot areas include lowland tropical regions. Using phylogenetic-based biogeography, Fisher presents three case studies that demonstrate how information on geography and climatic events can be inferred. Available information on ant extinctions and the palaeogeographic distribution of ant fossils are reviewed.

Taking both global and regional perspectives, Chapter 3 describes diversity patterns and examines the underlying causes of these gradients. Whether for species richness, range size, or body size, ant diversity patterns generally reflect diversity patterns of other terrestrial fauna and flora. Dunn and colleagues summarize patterns of ant diversity along latitudinal and elevational diversity gradients for multiple spatial and taxonomic scales. For each gradient, potential mechanisms are explored. Knowledge about the causes and consequences of ant diversity gradients is likely to provide crucial information for improving our understanding of the effects of climate change on ant diversity.

Given the threats posed by extensive habitat modification and loss globally, especially in tropical areas that harbour exceptional levels of ant diversity and endemism, it is essential that ants are included in conservation efforts. In Chapter 4, Alonso reviews the principal threats, approaches, and challenges to ant conservation and the current status of ant conservation, concluding that attention should be focused on hotspots of richness, endemism, and islands. This chapter provides a clear call to action for all myrmecologists and lists key actions for improved ant conservation in the future.

Chapter 1

Taxonomy, Phylogenetics, and Evolution

Philip S. Ward

1.1 Introduction

Since their origin about 120 million years ago (Mya), ants have evolved to become the most species-rich and ecologically diverse group of social insects (Grimaldi and Engel 2005; Hölldobler and Wilson 1990). Currently there are about 12,500 described species of extant ants (Bolton *et al.* 2006), but this statistic is undermined by several sources of uncertainty. On the one hand, as taxonomic knowledge improves, some of these species names will prove to be redundant (synonyms of older names). On the other hand, it is also evident that there are many ant species remaining to be discovered and/or formally described. In recent taxonomic monographs the number of new synonyms is typically less than the number of new species, sometimes by a wide margin (e.g. Bolton 2000, 2007, Wilson 2003), suggesting that the total diversity of ants could well exceed 25,000 species.

Ant systematics is concerned with delimiting species and understanding the phylogenetic relationships among them. Demarcation of ant species typically entails detailed morphological scrutiny of the worker caste – supplemented by examination of queens and males, if available – with the aim of discovering phenotypic gaps that indicate the existence of distinct evolutionary lineages (Ward 2001). Phylogenetics involves application of various methods of inference in an attempt to estimate the historical relationships among taxa. In recent years there has been a surge of interest in ant phylogeny, with DNA sequences supplanting morphology as the principal source of evidence. Such molecular studies promise to provide robust phylogenies that will be of great benefit to ant ecologists and behaviourists. This phylogenetic knowledge is also leading to an improved higher classification of ants, one that reflects the main features of evolutionary history. Species-level taxonomy has advanced more fitfully than ant phylogenetics, however, and a great deal remains to be accomplished before most ant species are well characterized (Ward 2007c).

This chapter summarizes recent progress in ant phylogeny and provides an outline of the higher classification of ants that is consistent with this new knowledge. The major lineages of ants are identified and features of their biology are discussed. The state of species-level taxonomy is evaluated, and resources that are available to ecologists for the identification of ant species and genera are documented. The emphasis is on extant taxa, with occasional reference to the fossil record where relevant to the discussion. Ant biogeography is considered in Chapter 2 of this volume by Fisher.

1.2 Phylogeny: inferring the evolutionary history of ants

Our knowledge of ant phylogeny has improved remarkably over the last decade. Several factors have contributed to this: a new comprehensive classification of ants based on reevaluation of the morphological evidence (Bolton 2003), a series of molecular phylogenetic studies employing multiple nuclear genes (Brady 2003; Brady *et al.* 2006; Moreau *et al.* 2006; Rabeling *et al.* 2008; Schultz and

Brady 2008a; Ward and Downie 2005), and new fossil discoveries (Dlussky *et al.* 2004; Engel and Grimaldi 2005; Grimaldi and Agosti 2000; Nel *et al.* 2004; Perrichot *et al.* 2008a; Perrichot *et al.* 2008b). As a result we are now at a stage where the broad outlines of ant evolutionary history are becoming apparent. This situation was preceded by an earlier period of uncertainty and disagreement over the relationships among the main lineages of ants, and a lack of consensus on the delimitation of these lineages (Baroni Urbani *et al.* 1992; Brown 1954; Grimaldi *et al.* 1997; Taylor 1978; Wilson *et al.* 1967). In retrospect it seems that these difficulties arose because of insufficient appreciation of the potential for extensive morphological convergence, especially in the worker caste, from which most evidence about ant phylogeny had been taken.

Twenty-one extant subfamilies of ants are currently recognized, and for most of these there is compelling evidence of monophyly (Rabeling *et al.* 2008; Ward 2007c). From molecular phylogenetic studies it is now evident that most of these subfamilies belong to a well-supported group, known as the formicoid clade (Figure 1.1). This includes three large ant subfamilies, Dolichoderinae, Formicinae, and Myrmicinae, as well as army ants and relatives

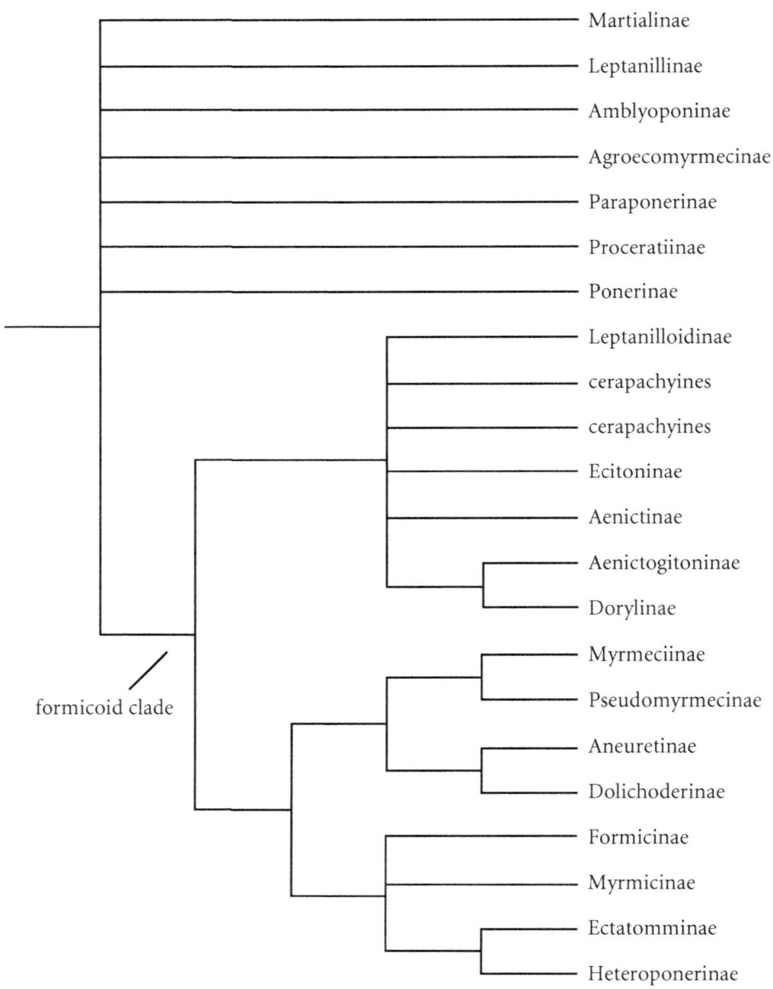

Figure 1.1 Summary of well supported relationships among the extant ant subfamilies. Modified from Ward (2007c).

(dorylomorphs), bulldog ants (Myrmeciinae), big-eyed arboreal ants (Pseudomyrmecinae), and a scattering of other smaller groups. No single morphological feature has been discovered that distinguishes formicoids from other ants but they emerge as a solid group in all molecular phylogenetic analyses (Brady *et al.* 2006; Moreau *et al.* 2006; Ohnishi *et al.* 2004; Ouellette *et al.* 2006; Saux *et al.* 2004; Ward and Brady 2003; Ward and Downie 2005).

The evolutionary history of formicoid ants is becoming increasingly well known. At the base of the formicoid tree, as sister to the others, is the well-supported dorylomorph clade, comprising army ants, cerapachyines, and leptanilloidines. Within this group, the subfamily Cerapachyinae is not monophyletic, however, and the question of army ant monophyly remains an open one (cf. Brady 2003), although the enigmatic subfamily Aenictogitoninae, known until recently only from isolated males, is now clearly shown to be sister to army ants in the genus *Dorylus* (Brady *et al.* 2006). Additional research is needed to resolve relationships among the cerapachyine lineages and to clarify the history of army ant evolution within the dorylomorph clade.

Among the remaining formicoids there is strong support for five major clades: (*a*) myrmeciomorphs, consisting of Myrmeciinae and Pseudomyrmecinae; (*b*) dolichoderomorphs (Dolichoderinae and Aneuretinae); (*c*) ectaheteromorphs (Ectatomminae and Heteroponerinae); (*d*) Myrmicinae; and (*e*) Formicinae. Inferred relationships among these groups are shown in Figure 1.1. The last three clades are treated as a trichotomy because current molecular data cannot clearly resolve relationships among them.

Outside the formicoids we can recognize two major groups of ants: (*a*) highly modified subterranean species belonging to the subfamily Leptanillinae, and (*b*) five 'poneroid' subfamilies (Agroecomyrmecinae, Amblyoponinae, Paraponerinae, Ponerinae, and Proceratiinae). The relationship of poneroids and leptanillines to one another and to the formicoids remains uncertain. In some molecular phylogenetic analyses, Leptanillinae is recovered as sister to all other ants, with poneroids forming a clade that is sister to the formicoids (Brady *et al.* 2006; Moreau *et al.* 2006; Saux *et al.* 2004), but this result may be an artefact of long-branch attraction (Bergsten 2005; Holland *et al.* 2003) between leptanillines and aculeate wasp out-groups. When Brady *et al.* (2006) performed an unrooted analysis on a seven-gene data set, excluding aculeate wasp out-groups but including representatives of all ant subfamilies, they obtained a strongly supported bipartition between formicoids and all other ants. However, the poneroids could not be made monophyletic under any possible rooting of this tree, because Leptanillinae was nested *within* the poneroids close to Agroecomyrmecinae and Amblyoponinae. The root of the ant tree could still be on the leptanilline branch, but this would imply that poneroids are paraphyletic. Under the likelihood-based Shimodaira–Hasegawa test, Brady *et al.* (2006) could not reject alternate placements of the root within poneroids or on the bipartition separating poneroids plus leptanillines from formicoids.

The recent discovery of a bizarre new ant species from the Amazon rainforest has generated further insight and intrigue into questions of early ant evolution (Rabeling *et al.* 2008). Given the name *Martialis heureka*, this species has a pale, blind worker caste, and a suite of unusual morphological features (Figure 1.2). Based on Bayesian analyses of three nuclear genes (18S rDNA, 28S rDNA, and elongation factor 1-alpha F2), *Martialis* appears to be sister to all other extant ants, with Leptanillinae as sister to the remaining ants (Rabeling *et al.* 2008). It remains to be seen whether the inclusion of additional genes will strengthen support for this result. In any event, the prevalence of hypogaeic habits in *Martialis*, Leptanillinae, and poneroids suggests that either the ancestral ant was subterranean and cryptic (contrary to the impression given by early ant fossils – discussed later) or that the surviving members of these old lineages have retreated to subterranean habitats, and convergently lost sight and pigment.

It is now clear that some features of worker ant morphology have indeed undergone greater convergence than previously realized. For example, the subfamily Agroecomyrmecinae, represented by the extant genus *Tatuidris*, had been placed in or near the Myrmicinae (Bolton 2003), on the basis of their common possession of

Figure 1.2 The recent discovery of *Martialis heureka* in the Amazon rainforest near Manaus, Brazil, has shed light on the general pattern of ant evolution and radiation. The species is in a monotypic genus representing a new subfamily (Martialinae) that appears to be a sister lineage to all other living ants. (Photo: Rabeling *et al.* 2008)

certain features of abdominal morphology, including the presence of a postpetiole and tergosternal fusion of the petiole sclerites. Yet we now know that these two groups are only distantly related, with the Agroecomyrmecinae falling outside the formicoid clade. Within the formicoids there is an intermingling of taxa with both highly derived social behaviour and morphology as well as those with more generalized (ancestral) habits and appearance, suggesting that the characteristics associated with the derived taxa (such as trophallaxis, complex chemical communication, marked worker–queen differentiation, and worker caste polymorphism) arose independently multiple times.

The fossil record provides an additional source of information about the evolutionary history of ants, particularly concerning their first appearance and the timeline of their diversification. When considering the contribution of fossils, it is useful to make a distinction between crown group and stem group taxa (Magallón 2004). Crown group ants are the clade composed of the most recent common ancestor of all extant ants and their descendants. Stem group ants are more inclusive, containing all organisms more closely related to ants than to any other extant taxa. We can employ the term 'stem ants' to refer to extinct taxa that are outside the crown group but that are inferred to be more closely related to ants than to any other living aculeate wasps. Comparable stem and crown group distinctions can be made at any level in the taxonomic hierarchy.

By this criterion the ant-like Cretaceous fossils assigned to the subfamilies Sphecomyrminae and Armaniinae (the latter is sometimes treated as its own family, Armaniidae) can be regarded as stem ants (Ward 2007c). They have been recorded from Cretaceous deposits ranging in age from about 80 to 100 Mya, mostly from the northern hemisphere (Perrichot *et al.* 2008a). Interestingly, a few putative crown group ants are known from the same time period, but assignment of most of these to extant subfamilies is problematic. An exception is the remarkable fossil *Kyromyrma neffi* from New Jersey amber (Turonian, 90 Mya). This taxon belongs to the Formicinae (Grimaldi and Agosti 2000) and establishes a minimum age of 90 million years for stem group formicines. Earlier in the Cretaceous there are fossil aculeate wasps, going as far back as 140 Mya (Grimaldi and Engel 2005), but no ants (Perrichot *et al.* 2008a).

Divergence date estimates based on molecular data that incorporate the full range of information from the hymenopteran fossil record suggest that crown group ants arose about 115–135 Mya (Brady *et al.* 2006). This is consistent with the fossil record (Grimaldi and Engel 2005) but contradicts other molecular studies that inferred a Jurassic origin for ants (Crozier *et al.* 1997; Moreau *et al.* 2006).

Ants are scarce as Cretaceous fossils and increase markedly in abundance in Tertiary deposits (Grimaldi and Agosti 2000), making up 5% of all insects in Baltic amber (Eocene), 20% in Florissant shales (early Oligocene), and 36% in Dominican amber (Dlussky and Rasnitsyn 2003). The increasing ascendance of ants in the Paleogene has been attributed to codiversification with angiosperms (Moreau *et al.* 2006; Wilson and Hölldobler 2005), with the success of ants being linked to the development of angiosperm forests with a complex leaf litter layer, and to the tending of honeydew-producing hemipterans. This is an intriguing idea, but a rigorous test remains elusive.

1.3 Higher classification: a survey of the major lineages

Table 1.1 provides an outline of the higher classification of ants that is consistent with the new findings about phylogeny. In this Section I briefly review the biology and distribution of these groups at the subfamily level. Brown (2000) produced a useful compendium of similar information, arranged by genus.

1.3.1 Martialinae

This new subfamily was erected for a remarkable ant species discovered recently in lowland Amazon rainforest near Manaus, Brazil (Rabeling *et al.* 2008). *Martialis heureka* is known from a single stray worker collected on the rainforest floor at dusk. Judging from its pale cuticle, absence of eyes, and small size (~2.5 mm in length) *Martialis* is likely to be a hypogaeic species, foraging in concealed locations. The enlarged fore legs, bulbous head, and elongate, slender mandibles suggest specialized predatory habits. Great interest awaits the discovery of additional specimens of *Martialis* and the elucidation of its biology.

1.3.2 Leptanillinae

Leptanillines are small to minute, pale, blind ants, known only from tropical and warm temperate regions of the Old World. Colonies of these ants are apparently restricted to subterranean habitats, and the most frequent indication of their presence is the occurrence of males in light traps and pan traps (Robertson 2000). Little is known about the biology of most leptanillines, but Masuko (1990) documented nomadic, army ant-like behaviour in *Leptanilla japonica*, whose workers prey on geophilomorph centipedes. He also discovered that the adult ants feed on larval haemolymph from a special gland located on the fourth abdominal segment (Masuko 1989). Queens of Leptanillini are wingless and dichthadiiform, while those of Anomalomyrmini are deciduously winged (Baroni Urbani and de Andrade 2006; Bolton 2003).

1.3.3 Poneroids

Poneroids are a heterogeneous assemblage of ants that represents either a grade or a clade at the base of the ant tree (discussed earlier). One of the more unusual groups in this assemblage is the subfamily Agroecomyrmecinae, consisting of an extant genus of Neotropical ants (*Tatuidris*), whose small cryptic workers are associated with rainforest leaf litter, and two fossil genera, described from Colorado Florissant shales and Baltic amber, respectively (Bolton 2003; Brown and Kempf 1968). Nothing is known about the biology of these ants. Based on the specialized mandibles and well-developed sting, Brown and Kempf (1968) speculated that the ants attack 'some active or slippery live arthropod prey in the soil or soil cover'. The unique species *Paraponera clavata*, the sole living representative of the subfamily Paraponerinae (Bolton 2003), emerges in some molecular analyses as sister to *Tatuidris*. *P. clavata* is a large aggressive ant that inhabits Neotropical rainforests. Workers are generalist predators and scavengers, and also collect substantial quantities of extrafloral nectar (Young and Hermann 1980).

The Amblyoponinae (~100 described species) comprise several genera of ants with distinctive morphology and specialized predatory behaviour (Brown 1960). The workers are small-eyed and generally cryptic in their foraging habits. Recorded prey items include geophilomorph centipedes, beetle larvae, and diplurans. Workers have been shown to consume larval haemolymph by puncturing the integument at specific locations (Masuko 1986; Wild 2005). Members of the genus *Onychomyrmex* exhibit nomadic behaviour and synchronized brood cycles in the manner of army ants (Miyata *et al.* 2003). Amblyoponines are found in all biogeographic regions, but they are most common in forested habitats. They are often treated as exemplars of 'primitive' ant behaviour (Thorne and Traniello 2003), although some of their traits are highly derived.

The subfamily Proceratiinae (~120 species) is another group of specialized predators with hypogaeic nesting and foraging habits. The eyes of the

Table 1.1. A higher classification of extant ants (Hymenoptera: Formicidae) consistent with recent molecular phylogenetic findings. The table lists currently valid names of subfamilies (-inae) and tribes (-ini). Extinct taxa are excluded. Groups that are known or suspected to be non-monophyletic—as they are currently defined—are marked with an asterisk. A few of these taxa can be readily transformed into monophyletic groups (e.g., Ponerini becomes a clade when Thaumatomyrmecini is subsumed within it) but others represent more problematic situations. Based on Engel and Grimaldi (2005), Bolton *et al.* (2006), Brady *et al.* (2006), Ward (2007b), Rabeling *et al.* (2008) and Ward *et al.* (2009).

Martialine clade		
	Martialinae	
Leptanilline clade		
	Leptanillinae	
		Anomalomyrmini
		Leptanillini
***Poneroids**		
	Agroecomyrmecinae	
	*Amblyoponinae	
	Paraponerinae	
	Ponerinae	
		Platythyreini
		*Ponerini
		Thaumatomyrmecini
	Proceratiinae	
		Probolomyrmecini
		Proceratiini
Formicoid clade: dorylomorphs		
	Aenictinae	
	Aenictogitoninae	
	*Cerapachyinae	
		Acanthostichini
		*Cerapachyini
		Cylindromyrmecini
	Dorylinae	
	Ecitoninae	
		Cheliomyrmecini
		*Ecitonini
	Leptanilloidinae	
Formicoid clade: myrmeciomorphs		
	Myrmeciinae	
		Myrmeciini
		Prionomyrmecini
	Pseudomyrmecinae	
Formicoid clade: dolichoderomorphs		
	Aneuretinae	
	Dolichoderinae	
		Bothriomyrmecini
		Dolichoderini
		Leptomyrmecini
		Tapinomini
Formicoid clade: ectaheteromorphs		
	Ectatomminae	
		*Ectatommini
		Typhlomyrmecini
	Heteroponerinae	
Formicoid clade: Formicinae		
	Formicinae	
		Camponotini
		Dimorphomyrmecini
		Formicini
		Gigantiopini
		*Lasiini
		Melophorini
		Myrmecorhynchini
		Myrmoteratini
		Notostigmatini
		Oecophyllini
		*Plagiolepidini
Formicoid clade: Myrmicinae		
	Myrmicinae	
		Adelomyrmecini
		Ankylomyrmini
		Attini
		Basicerotini
		Blepharidattini
		Cataulacini
		Cephalotini
		Crematogastrini
		*Dacetini
		*Formicoxenini
		Lenomyrmecini
		Liomyrmecini
		Melissotarsini
		Meranoplini
		Metaponini
		Myrmecinini
		*Myrmicini
		Paratopulini
		Phalacromyrmecini
		*Pheidolini
		*Solenopsidini
		Stegomyrmecini
		*Stenammini
		Tetramoriini

workers are reduced or absent. These ants are widespread but infrequently encountered in tropical and warm temperate regions, nesting in the ground or in rotten wood. Species of *Proceratium* and *Discothyrea* have been recorded preying on eggs of spiders and other arthropods (Brown 1980).

The largest and most diverse of the poneroid subfamilies is the Ponerinae, with more than 1,100 described species (Bolton *et al.* 2006). These are predacious ants, and include both large-eyed epigaeic (above-ground foraging) species as well as small-eyed hypogaeic taxa. Ponerines are widespread in warm temperate and tropical habitats, becoming especially abundant in the wet tropics. While most species appear to be rather generalized in their choice of prey, some have become specialized hunters of particular kinds of arthropods such as termites, millipedes, and isopods (Hölldobler and Wilson 1990). A few ponerine species have acquired group-raiding and nomadic behaviour (Maschwitz *et al.* 1989).

1.3.4 Dorylomorphs

This is a very distinctive clade of mostly tropical ants, that includes not only army ants (Aenictinae, Aenictogitoninae, Dorylinae, and Ecitoninae) but also the core group (cerapachyines) from which they evidently evolved. The army ants possess a unique suite of characteristics: mass foraging, nomadism, and highly modified (dichthadiiform) queens (Brady 2003). The more common species have been the subject of extensive scientific study (summary in Gotwald 1995). Much less attention has been focused on the biology of cerapachyine ants. Most *Cerapachys* species appear to be brood predators of other ants, while species of *Acanthostichus* and *Cylindromyrmex* are reported to prey on termites (Brown 1975). Also included in the dorylomorph clade is the subfamily Leptanilloidinae, a small group of blind, subterranean ants confined to the New World tropics (Brandão *et al.* 1999; Ward 2007b).

1.3.5 Myrmeciomorphs

The myrmeciomorphs comprise two groups of active, large-eyed stinging ants: the Myrmeciinae (~100 species) and the Pseudomyrmecinae (~300 species). There are two extant genera of myrmeciines, *Myrmecia* and *Nothomyrmecia*, both confined to the Australian region. The fossil record reveals that myrmeciines were formerly present in South America and the northern hemisphere (Archibald *et al.* 2006; Ward and Brady 2003). The living species of Myrmeciinae are generalist predators and scavengers and they nest mostly in the ground. In contrast, ants in the subfamily Pseudomyrmecinae are almost exclusively arboreal, nesting in dead twigs and, in a minority of cases, in live plant domatia (Ward 1991). Pseudomyrmecines are widespread in tropical Africa and Australasia, but reach their highest diversity in the New World tropics (Ward and Downie 2005).

1.3.6 Dolichoderomorphs

The subfamily Dolichoderinae is one of the 'big three' among the formicoids, with about 880 described species and a worldwide distribution. These ants are generalist scavengers and predators, and avid consumers of hemipteran honeydew. In many ant communities – but especially in Australia – they are among the most numerically and behaviourally dominant ants (Andersen 1995). Dolichoderines have produced several prominent invasive species, most notably the Argentine ant (*Linepithema humile*) which has severely disrupted native ant communities in most places where it has been introduced (Holway *et al.* 2002a; see also Part IV). The sister taxon to Dolichoderinae is the Aneuretinae, represented by a single living species confined to the wet forests of Sri Lanka (Wilson *et al.* 1956). Aneuretines were more widespread and diverse in the past, being known from Paleogene fossils in Europe, Asia, and North America (Dlussky and Rasnitsyn 2003).

1.3.7 Ectaheteromorphs

The two subfamilies, Ectatomminae (~260 species) and Heteroponerinae (22 species), that make up this group were previously placed in the Ponerinae (*sensu lato*). Most ectaheteromorphs appear to be generalist predators and scavengers but some species in the genus *Gnamptogenys* have become specialized predators on millipedes, beetles, and other ants (Lattke 1995). Ectatomminae and Heteroponerinae are widespread and mostly tropical or subtropical in distribution but they are absent from the Afrotropical region.

1.3.8 Myrmicinae

This is the largest of all ant subfamilies, with more than 6,700 described species and many others awaiting description. Myrmicines encompass an extraordinary range of foraging behaviours, nesting habits, and colony structure. They include omnivores, generalized predators and scavengers, specialist predators, seed harvesters, primitive fungus-growers, and leaf-cutting ants (Hölldobler and Wilson 1990). They are found on all major land masses and in essentially all habitats occupied by ants. Myrmicines well known to the public include fire ants (*Solenopsis* spp.), leaf-cutting ants (*Atta, Acromyrmex*), and seed harvesters (*Pogonomyrmex, Messor*).

1.3.9 Formicinae

The subfamily Formicinae is another cosmopolitan group, with about 3,600 described species. Formicines are very diverse in nesting and feeding habits, although in comparison to myrmicines they show less of a tendency towards reduction in size and occupation of concealed microhabitats. Most species appear to be generalist predators and scavengers; specialized predation is rare (Hölldobler and Wilson 1990). A few formicine lineages are predominantly hypogaeic and have developed trophobiotic relationships with subterranean root-feeding hemipterans. Familiar formicines include carpenter and sugar ants (*Camponotus*), wood ants and congeners (*Formica*), honeypot ants (*Myrmecocystus*), and weaver ants (*Oecophylla*).

1.4 Species-level taxonomy and identification

For most ant ecologists the aspect of ant systematics of greatest practical importance is the availability of tools for species identification. This in turn depends on the quality of species-level taxonomic research that has been carried out. Progress in ant species delimitation has been mixed, however, and overshadowed in recent years by advances in ant phylogeny. We are a long way from having the ability to recognize most species of ants, especially in the species-rich tropics. This inability to employ species names of universal applicability (Box 1.1) results in a continuing impediment to studies of ant biology.

The rate of description of new ant species was quite modest until the middle of the nineteenth century (Ward 2007c). Then the pace picked up as ant specialists appeared on the scene, and descriptive activity reached a peak in the early twentieth century when myrmecologists such as Emery, Forel, Santschi, and Wheeler engaged in prolific naming of numerous ant species, subspecies, and 'varieties'. Unfortunately many of these named taxa were poorly characterized, described in isolation from related species, and not incorporated into a more comprehensive taxonomic framework. In the last 60 years infraspecific taxa have been abandoned in ant taxonomy, and a 'population perspective' of species-level variation has been gradually adopted. But the earlier legacy of uncertain and dubious names is still with us and retards progress, especially in large widespread genera such as *Pheidole, Camponotus*, and *Crematogaster*.

There are also features intrinsic to ants that make species delimitation challenging (see Box 1.2). The worker caste is the most readily available form and the one on which most ant taxonomy is based. Yet workers often exhibit less pronounced differences among species than the sexual queens and males. Interspecific differences can also be obscured by worker caste polymorphism within species. Recent reports of cryptic species of ants (Pusch *et al.* 2006; Schlick-Steiner *et al.* 2006b), clonal reproduction (Foucaud *et al.* 2007; Kobayashi *et al.* 2008; Pearcy *et al.* 2004), and hybridization (Feldhaar *et al.* 2008; Helms-Cahan *et al.* 2002; Schwander *et al.* 2007) add further complications to the task of inferring species boundaries in ants.

Despite this, the situation for identification of ant species continues to improve. The generic classification of ants is relatively stable, and keys for identification of ant genera are widely available. Bolton's guide (1994) to ant genera of the world is the gold standard. There are also up to date

Box 1.1 Applications of taxonomy: why should we name ants?
Philip S. Ward

Scientific names are labels applied to taxa to aid in communication. Under the conventions of the current (fourth) edition of the *International Code of Zoological Nomenclature* (ICZN) (International Commission on Zoological Nomenclature 1999), animal species have compound names, comprising the genus name (first letter capitalized) and the species name (all lower case), with the name of the author, the person who described the species, as an optional third component. The year of publication of the original species description may also be inserted after the author's name. If the author's name is placed in parentheses, this signifies that he or she originally described the species in a genus different from the one in which it is currently placed. For example, *Lasius flavus* (Linnaeus, 1758) was originally described by Linnaeus in the genus *Formica*, and later transferred to *Lasius*.

Such generic transfers occur because taxonomists refine their concepts of the limits of a given genus or other higher taxon (tribe, subfamily, etc.). For example, the 17 species of ants described by Linnaeus (1758) were all placed in the genus *Formica*, but they are now allocated to 11 different genera in four subfamilies (Ward 2007c). The higher-level classification of ants has gradually stabilized, however, to the point where about 290 extant genera are currently recognized (Bolton *et al.* 2006), and this number is not likely to change radically.

The ICZN specifies rules for determining whether a scientific name is available (properly published) and valid (considered the correct name for a taxon). If two or more different names have been proposed for what is considered to be the same species then, in general, the oldest available name becomes the valid name, while the younger name is relegated to the status of junior synonym (an invalid name). Junior synonyms are common in ant taxonomy, largely due to a spate of species descriptions that occurred in the late nineteenth and early twentieth centuries, often unaccompanied by clear diagnostic features. At that time, names were also commonly applied to intraspecific forms (subspecies, varieties), a practice now largely abandoned in ant taxonomy.

An ant species by any other name would smell as sweet (or foul), but sharing information about the species is greatly facilitated by having a universal label. The scientific name of a species provides a link to information about phylogenetic relationships, distribution, ecology, behaviour, and other aspects of the species' biology. When non-systematists study ants there is sometimes a tendency – encouraged by situations where taxonomic knowledge is incomplete or inaccessible – to identify specimens to genus only and then assign morphospecies tags (*Pheidole* sp. A., *Pheidole* sp. B., etc.). This precludes linkage to other studies of the same species identified with different monikers, and as a result the development of global knowledge about the species is inhibited. In some instances there may be no other option, but it is then especially important to deposit voucher specimens in a public institution so that species identities can be checked by future investigators. This is true even for specimens identified to species – the identifications might be in error, or concepts of species limits might change in the future.

At a more applied level, having a unique and universal label for each ant species enhances our ability to use these organisms in biodiversity assessment and conservation planning. In particular, if we wish to evaluate not just local richness (alpha-diversity) but also species turnover (beta-diversity), or if we seek to understand the phylogenetic heritage of ant communities, then accurate identification of species assumes critical importance.

Box 1.2 How to identify ants
Brian L. Fisher

There is a great satisfaction and reward in establishing the identity of an ant, especially if it is one you encountered in the field. After spending days or weeks collecting and preparing a series of specimens, there is nothing like that first look under the scope. What do these ants have to tell us about their species, their environment, and their place on the planet?

To answer these questions, you must first identify the genus and species of the ant. In theory, identification should be a straightforward process of comparing characteristics in published keys. In practice, however, the technical background required and the paucity of taxonomic references represent considerable challenges. Ants may be among the most dominant and important groups of terrestrial organisms, but we have a long way to go before taxonomic knowledge and identification tools for this group matches those of birds and butterflies.

a

b

Figure 1.2.1 Views from above and side of a point-mounted ant specimen showing position of ant on point and labels on pin. The specimen here is *Ierataner alluaudi* from Madagascar. (Photos: www.AntWeb.org and Erin Prado)

Like any task worth doing, ant identification demands some preparation. First, collect the specimens into 95% EtOH. Because of intraspecific variation, be sure to obtain an adequate series of workers and soldier castes and, if present, queens and males. For each collection, record at least minimal locality and collection data, including locality name, latitude and longitude (using the decimal degrees format to an accuracy of five decimal places), elevation, date, habitat, method, and collector.

Learn to be skilled at point-mounting (Fig. 1.2.1), and practise it regularly. Properly label specimens, and always deposit voucher specimens at a museum of your choice. Learn ant morphology (Fig. 1.2.2). The glossary in Bolton (1994) provides a solid foundation in morphological terminology for ants, and Richards (1977) provides the same for the broader Hymenoptera. With this background and preparation, both online and printed resources (e.g. Bolton 1994; Bolton 2003; www.AntWeb.org) can help to distinguish the worker castes of the more than 280 described ant genera. With experience, identification to a genus is possible in the field using a 20x hand lens.

continues

Box 1.2 continued

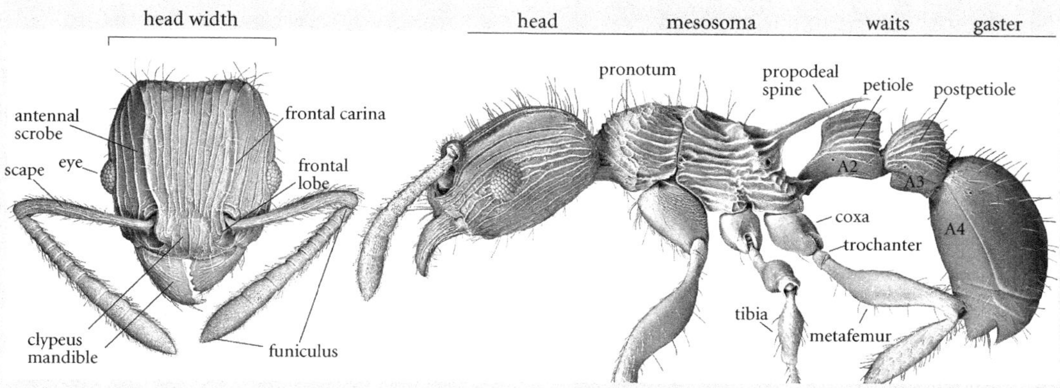

Figure 1.2.2 Common morphological features used to identify ant genera and species. Measurements are often necessary to distinguish similar species. Abdominal segments A2-A4 are indicated. Shown here is a worker of *Tetramorium* from Madagascar.

Not all genera, however, are well defined or identified, even based on the worker caste (Brady *et al.* 2006; Ward 2007c), and keys to genera based on males are lacking for most regions (Yoshimura and Fisher 2007).

Molecular analyses and faunal inventories will allow a more comprehensive diagnosis of clades (i.e. genera) based on characters from workers, queens, and males, and will remain an active and productive area of research over the next decade.

The next step in the enterprise, species-level identification, is often fraught with frustration and failure. The odds are stacked against you, since as many as half of all ant species have yet to be described. Furthermore, keys for many groups are non-existent, and published species descriptions are not much help since many species were described in isolation without comparisons to similar taxa (Ward 2007c). Meanwhile, differences among polymorphic worker and soldier castes within a species may obscure the subtle differences that exist between species. Many characters are used to identify ants at the species level, including hairs (location, number, length, shape, direction), scape (shape, length), mandible (shape, length, dentition), sculpturing, spines (location, length), head (width, length), and eye (location, size). Complicating the process further is the fact that population differentiation and speciation are ongoing, meaning that intermediates and hybrids are to be expected occasionally. The bottom line is that species-level taxonomy in ants is difficult.

The good news is that several changes underway are facilitating the practice of good taxonomy and the publication of user friendly keys: (*a*) Taxonomic resources are increasingly available in digital format. (*b*) Large-scale inventories are capturing specimens across their entire range of variation. (*c*) DNA techniques are facilitating the linkage of sexes and castes, highlighting taxa requiring further study (Fisher and Smith 2008). (*d*) The identification of monophyletic species groups, especially in hyperdiverse genera such as *Pheidole*, *Monomorium*, *Solenopsis*, *Crematogaster*, and *Camponotus*, provide smaller taxonomic units for taxonomic study (divide and conquer). (*e*) The use of matrix-based online interactive keys such as Lucidcentral.org provide flexibility in character selection during identification and are updatable as new species are discovered.

Box 1.2 continued

For many of us, the delimitation of higher ant taxa and species cannot happen fast enough. In the meantime, a taxonomist's best options are to consult the literature (Bolton 2003, and see Table 1.2) and scan the Web for regional keys and images. In many cases an expert may be needed to confirm your identifications. Your chances of having an expert look at your material increases if you first mount, label, and sort specimens to morphospecies, labelling them with your tentative identifications.

regional keys to ant genera of Central Europe (Seifert 2007), Japan (Imai *et al*. 2003), Taiwan (Lin and Wu 2003), Borneo (Hashimoto 2003), North America (Fisher and Cover 2007), the Neotropics (Palacio and Fernández 2003), and Australia (Shattuck 1999). Within the last three decades, several large ant genera and numerous smaller ones have received the benefit of global species-level revisions or comprehensive treatment over large geographic regions. For some countries in Europe and Asia, there are well illustrated identification guides for all ant species (e.g. Imai *et al*. 2003; Seifert 2007). Identification resources and high-quality images of ants are becoming increasingly available on the internet, through such sources as AntWeb (www.AntWeb.org), Antbase (www.antbase.org), Australian Ants Online (http://anic.ento.csiro.au/ants/), Ants of Costa Rica (http://academic.evergreen.edu/projects/ants/AntsofCostaRica.html) and Japanese Ant Image Database (http://ant.edb.miyakyo-u.ac.jp/E).

The literature concerned with species identification of ants is quite scattered, but Brown (2000) and Bolton (2003) provide helpful documentation. For identification of ants from specific regions or countries, researchers are advised to first consult the list in Bolton (2003: 78–9). Under the treatment of individual subfamilies and genera Bolton (2003) also indicates what species keys, if any, are available. More recent publications that are not cited in Bolton (2003) are listed in Table 1.2.

Table 1.2 Recently published ant species identification keys not cited in Bolton (2003).

Taxon	Region/Country	Reference(s)
Formicidae	Central Europe	Seifert (2007)
Formicidae	Serbia and Montenegro	Petrov (2006)
Formicidae	North Korea	Radchenko (2005a)
Formicidae	Micronesia	Clouse (2007)
Formicidae	Socotra Archipelago	Collingwood *et al*. (2004)
Poneroids, ectaheteromorphs	Colombia	Jiménez *et al*. (2008)
Acropyga	Worldwide	LaPolla (2004); LaPolla and Fisher (2005)
Adelomyrmex	Worldwide	Fernández (2003)
Aenictus	Australia	Shattuck (2008b)
Allomerus	Neotropical	Fernández (2007a)
Anochetus	Malagasy	Fisher and Smith (2008)
Aphaenogaster	Australia	Shattuck (2008a)
Axinidris	Afrotropical	Snelling (2007)
Azteca	Costa Rica	Longino (2007)
Basiceros	Neotropical	Feitosa *et al*. (2007)
Camponotus	Australia	McArthur (2007)
Camponotus	Poland	Suchocka *et al*. (2008)
Camponotus (*Karavaievia*)	Southeast Asia	Dumpert *et al*. (2006)

Camponotus (*festinatus* group)	United States	Snelling (2006)
Carebara	New World	Fernández (2004, 2006)
Carebara	China	Xu (2003)
Centromyrmex	Afrotropical	Bolton and Fisher (2008a)
Crematogaster	Morocco	Cagniant (2005)
Crematogaster (Physocrema)	Indochinese Peninsula	Hosoishi and Ogata (2008)
Diacamma	Australia	Shattuck and Barnett (2006)
Eurhopalothrix	Neotropical	Ketterl *et al.* (2004)
Forelius (part)	New World	Guerrero and Fernández (2008)
Formica (*pallidefulva* group)	Nearctic	Trager *et al.* (2007)
Gnamptogenys	Oriental and Australian	Lattke (2004)
Gnamptogenys	New World	Lattke *et al.* (2007)
Lachnomyrmex	Neotropical	Feitosa and Brandão (2008)
Lasius (*Dendrolasius*)	East Palearctic	Radchenko (2005b)
Leptothorax	East Palearctic	Radchenko (2004)
Linepithema	New World	Wild (2007b)
Lordomyrma	Fiji	Sarnat (2006)
Mayriella	Oriental and Australian	Shattuck (2007); Shattuck and Barnett (2007)
Meranoplus (part)	Australia	Andersen (2006)
Meranoplus (*diversus* group)	Australia	Schödl (2007)
Metapone	Madagascar	Alpert (2007)
Monomorium	Madagascar	Heterick (2006)
Monomorium	Neotropical	Fernández (2007b)
Mycetarotes	Neotropical	Mayhé-Nunes and Brandão (2006)
Mycocepurus	North and Central America	Mackay *et al.* (2004)
Myrmelachista	Costa Rica	Longino (2006)
Myrmica (social parasites)	Palearctic	Radchenko and Elmes (2003)
Myrmica	Korea	Lyu (2006)
Mystrium	Indo-Australian	Bihn and Verhaagh (2007)
Neivamyrmex	United States	Snelling and Snelling (2007)
Nesomyrmex	Southern Africa	Mbanyana and Robertson (2008)
Odontomachus	Malagasy	Fisher and Smith (2008)
Oxyepoecus (*vezenyii* group)	Neotropical	Albuquerque and Brandão (2004)
Perissomyrmex	Worldwide	Zhou and Huang (2006)
Perissomyrmex	Worldwide	Ogata and Okido (2007)
Peronomyrmex	Australia	Shattuck (2006)
Pheidologeton	China	Zhou *et al.* (2006)
Pheidole	Northern Vietnam	Eguchi (2008)
Pheidole (*roosevelti* group)	Fiji	Sarnat (2008)
Phrynoponera	Afrotropical	Bolton and Fisher (2008b)
Polyrhachis	India	Karmaly (2004)
Polyrhachis	Sulawesi	Kohout (2008b)
Polyrhachis (*Aulacomyrma*)	Melanesia	Kohout (2007b)
Polyrhachis (*Campomyrma*)	Borneo	Kohout (2007a)
Polyrhachis (*Cyrtomyrma*)	Borneo, Melanesia, Australia	Kohout (2006)
Polyrhachis (flavoflagellata-grp)	Southeast Asia	Kohout (2008a)
Prionopelta	Indo-Pacific	Shattuck (2008c)
Pristomyrmex	Philippines	Zettel (2006)
Probolomyrmex	Oriental	Eguchi *et al.* (2006)
Proceratium	Worldwide	Baroni Urbani and de Andrade (2003)

(*Continued*)

Table 1.2 Continued

Taxon	Region/Country	Reference(s)
Pyramica	China	Xu and Zhou (2004)
Rogeria	Guyana	Lapolla and Sosa-Calvo (2006)
Solenopsis	Galápagos Islands	Pacheco *et al.* (2007)
Stegomyrmex	Neotropical	Feitosa *et al.* (2008)
Strumigenys	China	Zhou and Xu (2003)
Technomyrmex	Worldwide	Bolton (2007)
Technomyrmex	New World	Fernández and Guerrero (2008)
Temnothorax	southeast United States	Deyrup and Cover (2004)
Temnothorax	East Palearctic	Radchenko (2004)
Tetramorium (*chefketi* complex)	Palearctic	Csosz *et al.* (2007)
Tetramorium (*tortuosum* group)	México	Vásquez-Bolaños (2007)
Tetraponera	China	Xu and Chai (2004)
Tetraponera (*ambigua* group)	Afrotropical	Ward (2006)
Trachymyrmex (part)	New World	Mayhé-Nunes and Brandão (2005, 2007)
Trachymyrmex	Nearctic	Rabeling *et al.* (2007)
Wasmannia	New World	Longino and Fernández (2007)

1.5 Future directions in species delimitation

Effective species-level taxonomy requires the accumulation and comparison of large population samples and an intelligent consideration of the factors generating intra- and interspecific variation. By its very nature this work is laborious, and requires sustained dedication of time and energy. New developments in technology can be employed to facilitate many aspects of the process, however. Improvements in imaging (e.g. the Automontage system) allow colour illustrations of unprecedented quality to become a standard feature in taxonomic revisions. We can envisage a time in the near future when colour images of most of the world's described ants, including type specimens, will be available online, along with the relevant taxonomic literature. For working taxonomists this will largely obviate the need for expensive museum visits to Europe, where many of the type specimens of the numerous poorly characterized ant taxa have been deposited. Electronic capture of specimen data and measurements will continue to facilitate the analysis and dissemination of this information. Interactive keys such as Lucid (www.lucidcentral.org) offer increased flexibility for species identification and more efficient handling and sharing of character state information. For ant ecologists seeking more user-friendly identification tools such interactive keys present an attractive alternative to conventional dichotomous keys.

Nevertheless in some taxonomically challenging situations it may be difficult to clearly demarcate ant species on the basis of morphology alone. If we consider ant species to be evolutionarily independent lineages that are reproductively isolated from one another, then genetic data should be informative about species boundaries. DNA barcoding, using a 658 bp fragment of the mitochondrial gene cytochrome c oxidase I (COI), has been touted as an effective tool both for species identification (Hebert *et al.* 2003) and for species discovery and delimitation (Smith *et al.* 2006). This gene – alone or in combination with COII – has proven to be informative about species boundaries in *Cardiocondyla* (Heinze *et al.* 2005), *Tetramorium* (Schlick-Steiner *et al.* 2006b), and in various groups of Malagasy ants (Smith *et al.* 2005). There are other instances in which COI has performed less effectively in the discrimination of ant species, for example in the genera *Anochetus* (Fisher and Smith 2008), *Solenopsis* (Ross and Shoemaker 2005), *Cataglyphis* (Knaden *et al.* 2005), and *Linepithema* (Wild 2009). In the last three cases nuclear gene markers showed greater species fidelity than mitochondrial DNA (mtDNA), a result that could be attributed to the greater tendency of organelle DNA to leak cross species boundaries (Hudson and Coyne 2002). This argues, of course, for the use of multiple lines of evidence when evaluating species limits. Under the most favourable conditions one can anticipate a process of reciprocal illumination, where inferences from morphology guide initial taxon sampling and then subsequent results from DNA sequencing lead to reevaluation of morphological differences, which in turn generates further exploration of genetic and phenotypic variation.

Another potential complication with the use of mtDNA markers is posed by nuclear pseudogenes of mitochondrial origin, or numts (Bensasson *et al.* 2001), which compete with the targeted mitochondrial gene(s) during amplification. Recently numts have been documented in *Atta cephalotes*, in which they were found in all 67 individual ants examined (Martins *et al.* 2007) and they are known or suspected to occur in a variety of other ant taxa (Kronauer *et al.* 2007b; Ward unpubl. data). Numts are abundant in the genome of the honeybee, *Apis mellifera* (Pamilo *et al.* 2007). Thus, when sequencing mitochondrial genes in ants care is needed to ensure that the protocols for extraction and amplification yield the targeted mtDNA genes rather than pseudogenes.

Although inferences about species boundaries are strengthened by the use of multiple independent markers, we are hampered by a dearth of nuclear genes with appropriate levels of variability for evaluating relationships among closely related populations and species of ants. Other methods for capturing genetic differences among populations, such as the use of amplified fragment length polymorphisms (AFLPs) (Vos *et al.* 1995) or microsatellites (e.g. Goodisman and Hahn 2005) may have greater utility. Genomic approaches to species delimitation (Shaffer and Thomson 2007) also show promise. In applying new molecular techniques to species-level taxonomy it is important to avoid falling into the trap of typological thinking – if there is an overarching lesson to be learned from evolutionary biology it is that species are usually variable entities, and that speciation is a gradual and complicated process.

1.6 Summary

As the most species-rich and biologically eclectic of all social insects, ants present considerable challenges to the process of cataloguing and understanding their remarkable diversity. Substantial progress has been made in recent years, however, in identifying the major clades of ants and clarifying their evolutionary history. We now have a higher classification of ants in which most of the subfamilies and tribes appear to be monophyletic and well diagnosed, with a few notable exceptions. Uncertainty persists regarding the phylogenetic relationships among old lineages at the base of the ant tree and concerning the time frame of ant evolution. The species-level taxonomy of ants has advanced more fitfully, and ant ecologists have an extensive but far-from-complete set of resources for identifying ant species. Ongoing and sustained effort is needed in the area of species discovery and delimitation, and in the delivery of this information to potential users.

Acknowledgements

I thank Barry Bolton, Steve Shattuck, and Ted Schultz for comments on an earlier draft of this chapter. The support of the University of California (Davis) and the US National Science Foundation (grants DEB-0344731 and EF-0431330) is also gratefully acknowledged.

Chapter 2

Biogeography

Brian L. Fisher

2.1 Introduction

Understanding the processes that have allowed ants to spread into and dominate so many different habitats is an active area of research involving analysis of their current distribution as well as historical and geographical factors that affect dispersal and radiation (e.g. Brady *et al.* 2006; Moreau *et al.* 2006). In this chapter, I examine global diversity patterns for present-day and fossil taxa, analyse taxonomic case histories using the genus *Crematogaster* and the subfamilies Pseudomyrmecinae and Dolichoderinae as examples, describe 'hotspots' of world ant diversity, and discuss how islands' species distributions serve as a model system for understanding the biotic evolution in a region and ant biogeography in general.

2.2 Global biogeographic patterns

If you want to travel the world to encounter as many ant species and genera as possible, where would you go? If you could only visit a few places, which biogeographic regions would you choose? Studies of global ant distribution patterns are providing answers to these types of questions.

There are currently a total of 290 extant ant genera (Appendix 1) and over 12,500 described extant species (Bolton *et al.* 2006; see also Chapter 1). Given the high rate of new species descriptions (Ward 2007c) and the large number of undescribed species in collections, the total number of ant species (described and undescribed) may be as high as 30,000. However, species are not randomly or uniformly distributed across the earth. Geography, geology, and climate all play a role in the diversification and spread of lineages. These factors explain how and why species and genera have assembled in a given region, and why endemic taxa are clustered in particular areas. As will be discussed in Section 2.6, species distributions, especially on oceanic islands, may also reflect an element of chance — the rare and fortuitous dispersal of a species from a source population.

On a global scale, ant fauna can be divided into biogeographic regions that contain endemic and closely related taxa and, at their boundaries, show rapid turnover of species (Figure 2.1). Deciding on the number of regions and their boundaries is arbitrary and open to debate (Cox 2001; Morrone 2002). Early researchers of birds and mammals defined zoogeographic regions somewhat subjectively, based on their intuition about how to interpret geographic patterns (e.g. Wallace 1855). They observed that the range boundaries of species and genera are generally coincident within regions. Today, more rigorous approaches to the characterization and interpretation of biogeographic history are possible based on more detailed information on the distribution of species and their relationships. However, there is still debate as to the best approach to establish the boundaries of regions (Morrone 2002). Finer subdivisions may contain more information, but are less useful as a general reference system.

The classical biogeographic partitions of Pielou (1979) are based mostly on vertebrates and plants and include seven regions: Nearctic, Neotropic, Palaearctic, Afrotropic, Indomalaya, Australasia, and

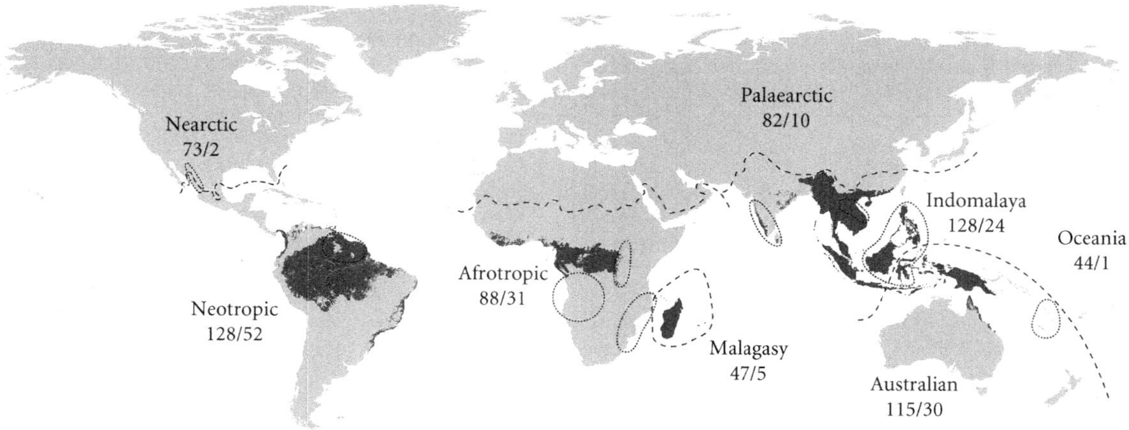

Figure 2.1 Biogeographic regions (delimited by long-dashed lines) and number of native ant genera and endemics. Biogeographic regions based on the classical regions of Pielou (1979) and Olson *et al.* (2001). Areas depicted in dark grey show exceptional diversity. Outlined areas (short-dashed lines) remain in need of exploration.

Oceania (excluding Antarctica, where no ants are found). Ant biogeographers such as Brown (1973) and Bolton (1994, 1995b) further divide them. Brown designates Madagascar as a distinct region from the Afrotropic, based on high levels of species endemism (>95%) (Fisher 2003). Bolton separates out Australia on the same grounds. In both cases, these divisions provide a practical advantage of generic lists and keys for regional analysis (Bolton 1994).

Defining regional boundaries is another challenge. Transitions between the historically isolated Nearctic and Neotropic faunas and the Palaearctic, Indomalaya, and Australian regions occur over a wide area, making it difficult to draw a definitive line between them. Most biogeographers use Wallace's Line, which runs just east of Bali, Borneo, and the Philippines, as the boundary between the Indomalaya and Australian regions (Figure 2.1). Bolton (1995b), in contrast, deviates from this conventional boundary and instead keeps the islands of southeast Asia (Malay Archipelago) together in the same region (Indo-Australian). While Bolton is correct that Wallace's Line is not a striking boundary for ant genera, it has been shown to be important at the species level. For example, Ward (2001) found that most of the 33 *Tetraponera* species do not cross Wallace's Line: 23 species were restricted to the Indomalaya region, while 7 species were confined to Australia, New Guinea, and adjacent islands. Only three species ranged through both the Indomalaya and Australasian realms.

With the above-mentioned caveats in mind, I evaluate the distribution of genera of living ants for the biogeographic regions outlined by classical biogeographers (Olson *et al.* 2001; Pielou 1979), with the addition of the Malagasy region, and compared them to Bolton's regions (1995b) (Table 2.1). A biogeographic summary based on species, rather than genera, is limited by our incomplete knowledge of species distributions (Dunn *et al.* 2007d) and a great number of undescribed species. However, Bolton's taxonomic catalogue (Bolton *et al.* 2006) does provide the country of origin for the type specimen(s) of each species. Based on these data, the rank of biogeographic regions in relation to the number of described species from each of the regions is shown in Table 2.1.

By all measures, the Neotropic is a regional hotspot for diversity, with the highest number of lineages (genera) and species, and the greatest number of endemic genera (Table 2.1). Not surprisingly, the larger, more isolated remnants of Gondwanaland (the Neotropic, Afrotropic, and Australia) show the greatest endemicity (Bolton 1995b). Overall, over half (53%) of all 290 genera, are restricted to one of the eight classical biogeographic regions.

Table 2.1 Comparison of the ranking of biogeographic regions based on number of genera, percentage of endemic genera, and complementarity, which maximizes the accumulation of the greatest number of genera, and number of described species between the classical biogeographic regions of Pielou (1979) and Olson *et al.* (2001) (classical) and those defined in Bolton (1995b). Number of plant species is based on Kier *et al.* (2005) and Qian and Ricklefs (2008).

Number of genera		Endemic genera (%)		Complementarity		Number of described species		Plant species richness
Classical	**Bolton**	**Classical**	**Bolton**	**Classical**	**Bolton**	**Classical**	**Bolton**	**Classical**
NEO (128)	NEO (128)	NEO (41)	NEO (41)	NEO	NEO	NEO	NEO	NEO
IND (128)	INA (122)	AFR (35)	AFR (35)	IND	INA	IND	INA	IND
AUS (115)	ORI (112)	AUS (26)	AUS (21)	AFR	AFR	AUS	AFR	AFR
AFR (88)	AUS (99)	IND (19)	INA (12)	AUS	AUS	AFR	PAL	AUS
PAL (82)	AFR (88)	PAL (12)	PAL (12)	PAL	PAL	PAL	AUS	PAL
NEA (73)	PAL (82)	MAL (11)	MAL (11)	MAL	ORI	NEA	ORI	NEA
MAL (47)	NEA (73)	NEA (3)	ORI (6)	NEA	MAL	MAL	NEA	MAL
OCE (44)	MAL (47)	OCE (2)	NEA (3)	OCE	NEA	OCE	MAL	OCE

Note: NEO = Neotropic, IND = Indo-malaya, INA = Indo-Australian, AUS = Australian (classical) Australasian (Bolton), AFR = Afrotropic, PAL = Palaearctic, NEA = Nearctic, MAL = Malagasy, OCE = Oceania, ORI = Oriental

The Malagasy region, a less isolated Gondwanaland remnant, still shows a remarkable degree of diversity relative to its small area, with more endemic genera than Oceania. Although the Palaearctic encompasses more than twice the land area of the Neotropic region, its diversity is just two-thirds that of the Neotropics, a reflection of its colder, drier, high latitude climate. Climatic conditions comparable to the Palaearctic are found in the Nearctic region, which is home to a similar degree of diversity when its relative isolation and smaller landmass is considered. Differences between faunas grow progressively less distinct as one moves between the Palaearctic and Indomalaya, and the Australasian regions.

Overall, however, patterns of global ant diversity adhere closely to patterns for other terrestrial fauna and flora, with the most diverse communities found in lowland tropical regions. Species richness accounts of plants (Kier *et al.* 2005 and references therein; Qian and Ricklefs 2008) rank the biogeographic regions similarly, with one notable difference. Ant species richness is relatively greater in the Australasian region than that of plants, suggesting more ant than plant diversity in the drier areas of the continent.

Gondwanaland endemicity is particularly striking in the Afrotropic and Neotropic regions. These two regions show no overlap among native ant species (Brown 1973). Of the 128 genera in the Neotropics, only 36 are found in the Afrotropics. However, this is unsurprising given the long, 100-million-year period of separation between Africa and South America (Ali and Aitchison 2008) and the ages of extant ant lineages, which generally emerged long after the breakup of Gondwanaland (Brady *et al.* 2006). These results suggest that it is unlikely that extant genera in South America were present when Africa and South America were connected (see Section 2.3).

While the era of ant exploration is clearly far from over, with many regions likely to yield treasures of undescribed genera and species (see circled areas in Figure 2.1), the overall species and genera richness patterns described here are likely to hold. Our growing understanding of ant relationships will further improve the analysis of biogeographic patterns. Phylogenetic studies, when combined with divergence times estimates, permit analyses of the origin, dispersal, radiation, and spread of taxa across regions. Section 2.4 includes three case histories where phylogeny was included in a historical biogeographic study.

2.3 Palaeogeographical distribution of fossil ants

Understanding how extant ant lineages arrived at their current distribution patterns requires a study of their origins and distribution history. This section examines what we know about ant history based on the fossil record (see also Chapter 1).

2.3.1 Geography

With almost 50% of extant genera restricted to just one of the biogeographic regions, were early ant ancestors equally isolated on different land masses? For early ants, where was the hotspot of diversity as compared to today's tropical lowland forest? Unfortunately, the geographic origins and patterns of early ants are somewhat obscure. The scarcity of early ant fossils challenges our ability to compare historical and current patterns.

The oldest known fossil ants are from French and Burmese ambers in the early-to-mid-Cretaceous period (Figure 2.2; see also Chapter 1). These are surprisingly rich, including at least seven distinct genera. The palaeoenvironment of the French amber is estimated to have been a subtropical rainforest (Perrichot *et al.* 2008a). Thus, these early ants already seem to exhibit a preference for moist and hot places. The contemporaneous occurrence of these genera implies that by the Albian (~105 Mya), ants had already significantly diverged and were widespread with multiple lineages co-occurring on the same continent. Unfortunately, our picture of ant evolution before the Albian is blank and lacks a single ant fossil. Because they were already diverse by the Albian, I share the view of Perrichot *et al.* (2008a) that eventually fossils will be found earlier in the Cretaceous.

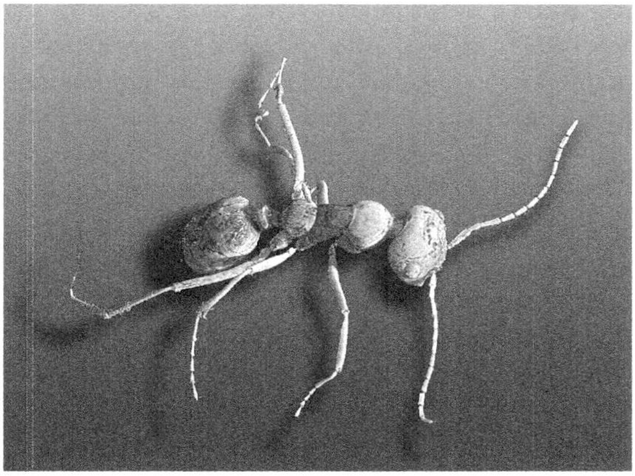

Figure 2.2 *Sphecomyrma* sp., Sphecomyrminae, from early Cenomanian amber of La Buzinie, Charente, SW France. Three dimensional virtual reconstruction in phase contrast synchrotron microtomography. (Image: Lak [CNRS/ESRF] / Tafforeau [ESRF] / Perrichot [Kansas Univ.] – ANR AMBRACE.)

The distribution of Cretaceous specimens demonstrates that ants had spread across much of Laurasia (today's northern hemisphere continents) early on in their evolution (Perrichot *et al.* 2008a). We cannot say much about their spread through Gondwana. Early ants are conspicuously absent from Gondwanan fossil deposits from the early to mid-Cretaceous. The first accurate record is a diverse set of Formicidae from Botswana dating slightly later from the Turonian (~93 Mya). The absence of Gondwanan deposits before the Turonian may reflect the limited extent of early ant habitat as well as the chance nature of locating fossils.

The findings from fossil taxa combined with phylogenetic divergence data suggest that the distribution of extant genera was not driven by Gondwanan vicariance events. The dating studies show that most subfamilies originated (stem group) after the breakup of Gondwana and in the late Cretaceous, and followed by within-subfamily diversification in the Palaeogene (Brady *et al.* 2006; Ward 2007c). These dating estimates imply that during the breakup of Gondwana (~100 Mya), the ant genera now found in South America, Africa, and Madagascar were not yet present. Thus, the current distribution of the army ants *Dorylus* (Africa) and *Eciton* (New World) cannot be a consequence of the breakup (cf. Brady 2003; Brady *et al.* 2006). Nor can the Gondwanaland distribution of one of the early branching lineages of extant ants, *Amblyopone* and *Mystrium*, found in just a handful of pockets around the world, be attributed to the breakup. Instead, the ages of these four lineages imply that their current distribution is a product of dispersal, radiation, and spread of taxa across these isolated biogeographic regions.

The overall results are that the modern ant collector is dealing with a fauna that arose less than 50 or 60 Mya. For example, one peculiarity that emerged in the early Eocene (~50–55 Mya) was the giant ants (*Formicium giganteum*), subfamily Formiciinae (Lutz 1986, 1990, 1992). Ant collectors of today can only dream about what it might be like to collect these extinct lineages. The common use of a pooter (aspirator) would not have been advised to gather these ants. Workers are not yet known, but full-bodied queens have been found in Germany and Tennessee (USA) and males of one species in Germany. These giants were likely carnivorous and grew up to 5.5 cm, with 13 cm wingspans that were larger than those of some modern hummingbirds.

2.3.2 Geographic patterns of ant extinction

The fossil record has provided evidence that genera and subfamilies with a modern restricted distribution may represent the survivors of a lineage that at early times was more widespread. For example, the sole surviving representative of the subfamily Aneuretinae, *Aneuretus simoni*, is found exclusively in central Sri Lankan rainforest and is the sister group of the Dolichoderinae (Brady *et al.* 2006). During the Mesozoic and early Palaeogene, aneuretines were distributed widely in North America and Eurasia (Dlussky and Rasnitsyn 2003; Engel and Grimaldi 2005). What led to the extinction of other aneuretines? Engel and Grimaldi (2005) propose the Eocene–Oligocene (~35 Mya) climatic shift that altered biogeography of numerous insect lineages (Grimaldi and Engel 2005). It is unclear how *A. simoni* was able to survive the factors underlying the extinction of its relatives.

The Myrmeciinae were also much more diverse historically than their modern distribution would indicate. Present-day native Myrmeciinae are restricted to Australia and New Caledonia. Though fossil records for the subfamily are restricted to the Eocene, the subfamily included a number of genera distributed throughout the world, including fossils from North America, South America, and Eurasia (Archibald *et al.* 2006; Dlussky and Rasnitsyn 2003; Ward and Brady 2003).

Genera also show patterns of extant lineages occupying restricted ranges compared to their ancestors. *Leptomyrmex* is today found only in New Guinea (and nearby islands), eastern Australia, and New Caledonia, but traces of one fossil species were found also in Central America (Dominican amber) (Baroni Urbani and Wilson 1987).

We understand very little about why some representatives of lineages survive while others do not (e.g. *Nothomyrmecia*, the only living representative of the ancient lineage Prionomyrmecini; see Box 2.1). For some taxa, at least, nesting site appears to have played a role. The proposed earliest branching lineages of extant ants include the subfamilies Leptanillinae and Martialinae (Brady *et al.* 2006; Rabeling *et al.* 2008). In both these subfamilies, extant species are thought to forage and nest underground. It is possible that the taxa that have persisted today were exclusively subterranean. Their underground habitat could have provided protection from competitors, climatic shifts, or other environmental changes that drove their relatives to extinction (Rabeling *et al.* 2008).

2.4 Phylogenetic-based biogeography

Analyses of phylogenetic relationships among ants can yield far more than just lineage information. Considered together with habitat requirements and mutualistic relationships, they can shed considerable light on the regional history of climatic, tectonic, and other geographic shifts. I have examined three case studies: the genus *Crematogaster*, where historical analysis sheds light on geographic and climactic events; and the subfamilies Pseudomyrmecinae and Dolichoderinae, where phylogenetic patterns correlate with geography.

2.4.1 *Crematogaster*

One example of historical biogeography is the phylogenetic study of mutualistic myrmicine ants of the genus *Crematogaster* in Sundaland, southeast Asia (Quek *et al.* 2007). Sundaland is an extension of the continental shelf of southeast Asia that

Box 2.1 The remarkable rediscovery of the Dinosaur Ant, *Nothomyrmecia macrops*
Robert W. Taylor

The 'Dinosaur Ant', *Nothomyrmecia macrops*, is considered to be perhaps the most archaic living formicid. It is related to the Australian bulldog ants (*Myrmecia*), to the Baltic amber *Prionomyrmex*, and the Argentinian fossil *Ameghinoa* (Ward and Brady 2003). The significance of this morphologically 'primitive' ant was recognized by its describer John Clark (1934). Brown and Wilson (1959b) reviewed its known history, making the prediction that such a pale-coloured, large-eyed creature must be nocturnal.

The first two known worker specimens were collected in 1931, almost certainly on the remote 120 km bush track between Balladonia Station and Mount Ragged in southeast Western Australia. Several specifically targeted expeditions subsequently failed to rediscover the species, which to date has not been collected again in Western Australia, but is now known to range from Poochera in South Australia, southeastwards into the Eyre Peninsula, and west towards the Nullarbor Plain. *Nothomyrmecia* was finally rediscovered in 1977, almost 1,200 km to the east of its original collection, near the hamlet of Poochera, South Australia; a Mecca for myrmecologists, and a place now targeted by many ecotourists.

The rediscovery of *N. macrops* is a tale of unexpected triumph. Five team members, including Don Colless, Murray Upton, John Lawrence, John Feehan, and myself, set out to search the distant Mount Ragged track in Western Australia for *Nothomyrmecia*, in a last-ditch Australian attempt to find the ant, following word that a well-financed expedition (his third) was being planned by the noted American myrmecologist William L. Brown. Two days westwards from Canberra, we were delayed at Wudduna, South Australia, for vehicle repairs. I distributed colour slides of the *Nothomyrmecia* types to the group, anticipating the coming rediscovery, which unbeknown to us was fatefully then only hours away. Later, while refuelling at Poochera, we decided to make camp nearby, still many hours short of the originally projected campsite that night.

Colless valiantly proposed collecting in the camp area, to which everyone else reluctantly agreed. I left the caravan last to meet an icy south-west wind inauspicious for ant activity. After about five fruitless and begrudged minutes, I moved back towards the warmth when my headlamp caught a *Eucalyptus* trunk about 15 m from the caravan. There, on the tree trunk was a spotlighted *Nothomyrmecia* worker! The amazing serendipity of the night was complete. I rushed to the caravan where a light sheet was in operation and famously proclaimed: 'The bloody bastard's here.' We collected more workers from the same tree, and yet some more the following morning from the ground nearby. Later, following an unsuccessful search at Mount Ragged, I flew to Canberra from Perth, and was back at Poochera by mid-November with then Sydney University student Phil Ward to collect the first live colonies (Taylor 1978). The rest is history.

Nothomyrmecia has the usual 'formicid' attributes (Figures 2.1.1 and 4.2), including metapleural glands, dealation by recently mated queens; an apterous, mesosomally reduced worker caste, which is a generation younger than the colony queen; elbowed antennae; a petiolate waist; and a non-cellular nest in which eggs, larvae, and pupae are not segregated in individual cells. Its 'primitive' features include the powerful (and painful) sting, the low dimorphism between queens and workers, and the presence of worker ocelli and pupal cocoons. Specialized features are the obligate nocturnal foraging activity, the peculiarly reduced wings of virgin queens, and the ventral rather than dorsal abdominal stridulatory organ, a structure almost unique among the Hymenoptera. The diploid chromosome number $2n=94$ is the second highest known for any non-polyploid animal (Imai *et al.* 1990).

continues

Box 2.1 continued

Figure 2.1.1 *Nothomyrmecia macrops* queens, worker (lower left), and pupae. (Photo: Robert W. Taylor)

Founding queens cohabit in groups of up to four in nests excavated in the soil. They forage like workers during this period, and are reduced by aggression to one when the first daughter workers appear. Nests extend nearly a metre below ground as colonies grow to contain up to 200 workers. Lone foragers gather insect prey on trees near their nests, and individually return to the same tree, night after night. The contents of waste middens accumulated deep in the nests consist largely of hemipteran and dipteran remains, with very few beetle or lepidopteran fragments. Proteinaceous food is supplemented by sugary liquids, including honeydew deposits (hemipteran excretions). Navigation involves exceptional visual acuity using the tree canopy pattern against the night sky, and possibly also polarized-light sky patterns, as a map. All foragers depart nests within the hour following nightfall. Successful huntresses return during the night, while those without prey return in numbers at dawn. Researcher Birgit Greiner has commented that their eyes are so strongly dark-adapted that they are essentially blind in daylight.

includes Borneo, Malaya, and Sumatra. During the Pliocene (~1.8–5 Mya), climate fluctuations caused wet periods to alternate with regimes of cooler and drier weather. During the same era, rising sea levels alternately inundated and reconnected the Sunda Shelf landmasses. Phylogenetic studies of *Crematogaster* have yielded insights into the climatic and geographical changes that accompanied these events.

One clade of *Crematogaster* ants, the subgenus *Decacrema*, evolved an extremely close relationship with trees of the *Macaranga* genus in Sundaland. The ants live exclusively in the hollow stems or domatia of the trees and consume food bodies in the leaves. In exchange, the colony defends the plants against encroachment by other animals and vines. The trees themselves are restricted to areas of continuously wet rainforest, and cannot withstand drought or seasonality. For this reason, the evolutionary relationships among *Crematogaster* ants can serve as a surrogate index of climate change in the Sunda Shelf.

Molecular phylogenetic studies of the *Decacrema* ant complex indicate that of the three locations in Sundaland, Borneo contains by far the greatest number of lineages, suggesting it is the home of the ancestral species. Chloroplast DNA studies point to a similar origin for *Macaranga* trees. Meanwhile, the highest lineage diversity of ants on all three islands is found on mountaintops. This finding indicates that the ants in high-elevation rainforests enjoyed moist conditions throughout the Pleistocene, allowing them to spread and diversify without interruption. By contrast, cooling and drying climate shifts shrank the rainforest cover on lower elevation slopes, and reduced ant diversity.

The relationships among Sumatra, Malaya, and Borneo ant lineages have also suggested times when these areas were connected. The relative ages of Sumatra and Malaya ant lineages that are most closely related to Borneo lineages likely reflect periods of low sea level when land bridges connected some areas but not others. Meanwhile, lineages with constricted ranges or patchy distributions among the three sites likely reflect past dramatic range reductions that severed shelf connections and turned these areas into refugia for rainforest and ants alike.

2.4.2 Pseudomyrmecinae

The ant subfamily Pseudomyrmecinae comprises big-eyed arboreal ants that are widespread in tropical and subtropical regions throughout the world and number about 300 species. Most of these species colonise dead twigs, stems, and branches, although about 40 species have obligate mutualistic relationships with domatia-bearing plants. In their study of the subfamily, Ward and Downie (2005) used a combination of molecular data and morphology to investigate the biogeography and biological evolution of the Pseudomyrmecinae.

The current distribution of these ants suggests that they originated in a portion of Gondwanaland during the mid-Cretaceous. Molecular genetics point to an exceptionally long stem lineage that was initially marked by limited diversification. The phylogenetic analyses of Ward and Downie (2005) indicate an origin in the Old World Tropics (paraphyletic *Tetraponera*) followed by dispersal to the New World Tropics and subsequent diversification (*Pseudomyrmex*). Therefore, much of this species divergence took place after the continents had broken up and reached their current locations. Using results that show *Tetraponera* as a paraphyletic grade at the base of the Pseudomyrmecinae, Ward and Downie (2005) proposed that the ancestral area for the genus is Indo-Australia and not Africa as proposed by Ward (2001).

The current pattern of pseudomyrmecine diversity resembles geographic trends seen in other taxa inhabiting both Neotropical and Palaeotropical forests. The greatest number of species (200+) is found in the Neotropics (Table 2.1). This region includes a wide variety of habitats due to active mountain building and other geographic characteristics, which may explain this proliferation of species. The greater diversity of the region's habitats, combined with a relatively consistent climate and large area, may have provided conditions ideal for diversification. The lowest number of Pseudomyrmecinae species, 25, is found in Africa, a landmass that has experienced high rates of extinction due to large climate shifts, and where tropical forests cover a smaller area at relatively high elevations (~500 m above sea level).

2.4.3 Dolichoderinae

The subfamily Dolichoderinae is a cosmopolitan group of ants known for using chemical defences and sheer numbers to dominate ant communities. The 840-plus species in the group include several of the world's most successful invasive ants, including the Argentine ant (*Linepithema humile*), the ghost ant (*Tapinoma melanocephalum*), and white-footed ants (*Technomyrmex albipes, T. difficilis,* and *T. vitiensis*). Fossil records suggest that the dolichoderines declined in the northern hemisphere starting in the late Eocene, although their abundance and diversity have remained strong in the southern hemisphere, especially in Australia.

Combining both fossil and molecular data in a dispersal-vicariance analysis (DIVA), P.S. Ward and colleagues (unpublished) address the historical biogeography and diversification of the group. Their work indicates that the crown group Dolichoderinae arose in the Palaeocene (~65 Mya) and was preceded by ~30 million years of stem lineage evolution (and presumed extinction).

Their work had identified four main clades within the subfamily Dolichoderinae. Based on the DIVA, the crown group Tapinomini, the sister group of all other extant dolichoderines, arose in the Afrotropics < 60 Mya. The sister to the remaining dolichoderines, Bothriomyrmecini is estimated to have their crown group origin in the Indomalaya region. Both Tapinomini and Bothriomyrmecini have remained diverse in the Palaeotropics, but a few representatives have colonised the Nearctic and Neotropical regions (e.g. species in the genera *Bothriomyrmex, Technomyrmex, Liometopum,* and *Tapinoma*). The genus *Dolichoderus*

(=tribe Dolichoderini) was not evaluated in the analysis but the crown group is currently widespread, being absent only from the Afrotropics. The remaining lineage, tribe Anonychomyrmini, originated and diversified in the Neotropics into hundreds of species that now include groups in North America (*Forelius, Dorymyrmex*) and multiple dispersal events from South America to Australia during the mid-Tertiary. One of these dispersals, by the common ancestor of *Linepithema* and *Iridomyrmex*, led to a spectacular radiation that has produced several of Australia's most dominant ant species (Andersen 1995). Thus, the arrival, diversification, and dominance of dolichoderines in the Australian region occurred later than in other parts of the world. Interestingly, the close relationship of the dolichoderine fauna in Australia to *Linepithema* may explain the limited invasion by the human-dispersed Argentine ant (*L. humile*) over the last 100 years.

2.5 Hotspots: ants are more diverse in lowland, low-latitude forest

The world's most diverse ant communities tend to reside in low-elevation, low-latitude forests. In general, there is a strong latitudinal gradient in species richness, with tropical latitudes containing far more species than temperate zones (Ward 2000). Possible factors driving this pattern, discussed in detail in Chapter 3, include differences in temperature and the faster pace of species diversification in the tropics (Allen *et al.* 2006; Kaspari 2004). The warmth and higher predation rates of lowland tropical areas are correlated with a reduction in colony mass and an increase in ant abundance (Kaspari 2004).

The large size of lowland tropical forests further bolsters ant species richness in this biome (Rosenzweig 1995). Both the Amazon of South America and the Congo Basin rainforests of Central Africa are distributed in relatively large, unbroken blocks inhabited by relatively widespread species. Even the island of Madagascar contains a strip of eastern wet tropical forest that stretches for nearly 1,500 km from north to south. The continuity of these habitats helps sustain high levels of species diversity.

In general, ants have difficulty in tolerating cold and wet climates. In tropical regions, species diversity drops off in montane forest (Brown 1973; Fisher 1999b; Kaspari *et al.* 2004; Malsch *et al.* 2008). Ants are absent above about 2,300 m in all closed-canopy broadleaf forests, even those located in the tropics. However, they can be found at altitudes over 3,500 m in the open ground of the Andes or Himalayas (Brown 1973). Kaspari *et al.* (2000a, 2004) discuss the role of temperature in global ant patterns. But few studies address the factors behind the steep decline of ant species richness with increasing elevation and the general restriction of ants to relatively low altitudes in the tropics. Malsch *et al.* (2008) studied the biotic and abiotic factors in parallel among ground and lower vegetation ant communities along an elevational gradient. The study site consisted of evergreen tropical rainforest on Malaysia's Mount Kinabalu. They demonstrated that the steep decline in ant species richness with increasing altitude was correlated with several factors: (*a*) temperature decrease; (*b*) high humidity (comprising the relative humidity of the air, fog, rain, and waterlogging); (*c*) scarcity of nesting space; and (*d*) scarcity of nutritional resources. Overall, they found temperature to be the fundamental factor modulating other abiotic and biotic resources that determine this pattern. Ground temperature within closed-canopy forests is more likely to drop below the threshold necessary for ants to forage or develop efficiently (Brown 1973), reinforcing the idea that lowland tropical forests foster the most ideal conditions for ants.

Patterns of richness along elevational gradients are now of particular interest in light of climate changes (Deutsch *et al.* 2008). Janzen (1967b) proposed that tropical mountain passes are more effective barriers to dispersal than temperate-zone passes of equivalent altitude. He argued that because annual variation in ambient temperature at any site in the tropics is relatively low, it not only reduces seasonal overlaps in temperature between low- and high-altitude areas, but also selects for narrow temperature tolerances. As a result, tropical lowland organisms experience mountain passes as higher, more insurmountable barriers to dispersal than more temperate-zone species. This tendency in turn favours smaller species distributions such as those seen among tropical ants, and an increase in species turnover in ant assemblages along elevational gradients.

Climate change will favour organisms that can quickly acclimate, adapt, disperse, or change their behaviour (Deutsch *et al.* 2008). As Janzen suggested, the greatest biological diversity occurs in the tropics where change (e.g. rapid adaptation to climate change) is the hardest. Unlike deforestation, which is obvious and often noisy, climate change may drive tropical insects into silent extinction. But for ants that dominate the lowland forest, such as army ants, there is another point to consider. Once limited in elevation by wet and cold, these predatory insects will move to now warmer and less-cloudy higher elevations. At loftier elevations, they will encounter and threaten many groups such as beetles in the families Carabidae and Staphylinidae that are unaccustomed to competition with ants. To explore the potential impacts of climate change, ant communities along elevation gradients, especially at the cloud forest transition, should be monitored.

2.6 Islands

Islands offer a particularly clear lens through which ant biogeography can be viewed. The early studies of ants on islands (Caribbean, Melanesia, and Polynesia) by Wilson were of particular influence in the development of island biogeographic theory (MacArthur and Wilson 1967). Worldwide, more than half of the estimated 30,000 species of ants remain undescribed, a clear impediment in the study of biogeographic patterns. However, on the other hand, islands are much smaller in area and harbour fewer species than continents, making exhaustive inventories of their ant species possible. Careful study of this more limited species assemblage, combined with an array of islands differing in age, size, and isolation, can shed light on processes that affect ant composition, dispersal, extinction, and radiation. This natural laboratory, however, has been damaged and continues to be at risk. Increased habitat fragmentation, and the accelerated pace of ant species introductions, threaten endemic island ecosystems worldwide (Abbott 2005, 2006; Fisher 2005; Lach 2008b; O'Dowd *et al.* 2003; Underwood and Fisher 2006). Lowland tropical island faunas are especially susceptible to introduced ants; ant faunas that have been eradicated or severely reduced will complicate the analysis of biogeographic patterns.

2.6.1 Chance dispersal

The composition of the ant fauna on any particular island typically reflects the age, size, and relative isolation of the island (Figure 2.3). Ants often reach oceanic islands via accidental 'sweepstake routes' (Wilson 1988). As a result, even neighbouring islands tend to have unique assemblages of ant species. Ants in the sweepstakes face a low success rate, but those that successfully establish themselves enjoy a huge potential pay-off. Species that gain a foothold on a large island have the opportunity to radiate and fill many empty ecological niches.

Ants can arrive at islands via any of four common dispersal routes. A newly inseminated winged queen might be blown across the open ocean to distant shores. An entire ant colony might raft to an island inside a rotten log. Land bridges to other continents can be exposed during periods of low sea level. Lastly, ants are well adapted for transport by unwitting humans (Holway *et al.* 2002a; Wetterer 2006; Wilson 2005).

An island's size and geography determines much about its ant diversity (Figure 2.3). It is the primary factor driving whether or not dispersing ants can land and establish a foothold. Larger islands offer bigger targets for wayfaring ants to hit and more diverse habitats to occupy. The world's three largest tropical islands, New Guinea, Borneo, and Madagascar, have more endemic ant genera and species than any other islands on earth (Fisher 2009). While most island ants tend to originate from adjacent continents, prevailing winds and currents will also affect the sources of colonisation. Though Madagascar is much closer to Africa, a few of its ant lineages are related to taxa from Asia, where prevailing currents originate (Fisher 2000, 2003). Island age, too, plays a role in colonisation, as older islands offer ants more time to arrive and colonise. Moreover, the existing ant community, vegetation, and habitat determine whether new ants can survive and/or proliferate.

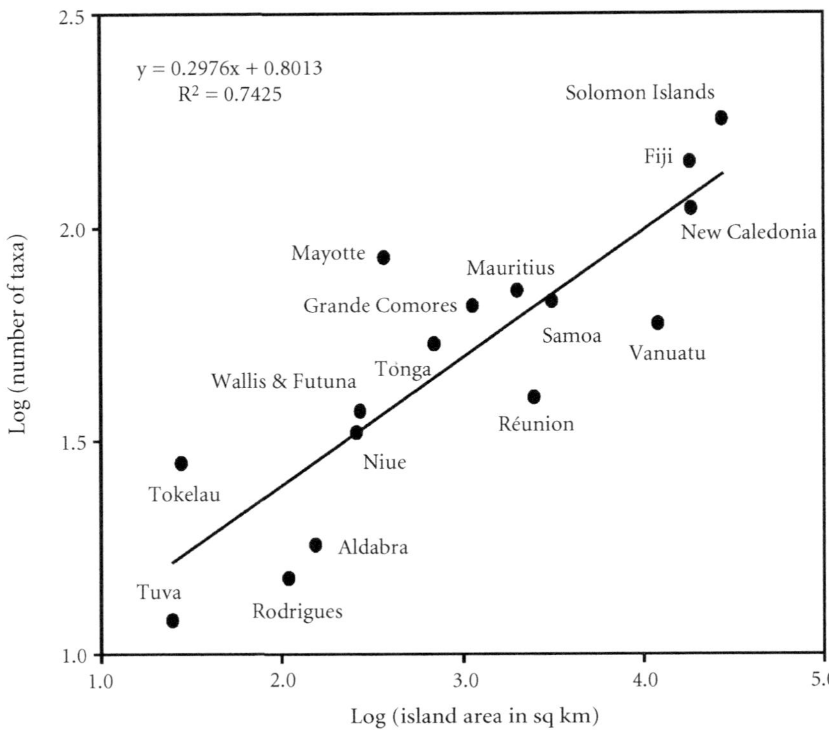

Figure 2.3 Log-linear regression analysis of ant species richness versus area of islands in the Pacific Ocean and south-west Indian Ocean. Though the size of an island is important, variation in species richness also reflects the age and relative isolation of the island. For example, Mayotte's proximity to Madagascar and great age (7.7–15 million years, Nougier *et al.* 1986) may explain the high number of species on the island. Data from Abbott *et al.* (2006), Ward and Wetterer (2006), and www.AntWeb.org.

The sweepstakes model of colonisation is exemplified by the pattern of ant diversity across the Antilles. This New World archipelago arcs across the Caribbean in a chain of more than 7,000 islands (Wilson 1988). While the smaller islands have fewer endemic species, those islands farthest from the mainland have fewer ant genera. In fact, few to no endemics live on Caribbean islands under 1,000 km^2, with one exception. Trinidad, located just 7 miles from mainland Venezuela, has an ant fauna characteristic of South America. Its species assemblage includes 17 genera widespread on the continent but absent from the rest of the Antilles.

Hawai'i offers a far more extreme example of biogeographic forces at work. A young and extremely isolated island chain, it is one of the few places on earth that lacks native ants. Since the arrival of humans, however, >50 ant species have been established. Many of these are extremely invasive and have devastated the islands' native insect faunas (Krushelnycky *et al.* 2005b).

Approximately midway between Hawai'i and New Zealand is the Pacific island nation of Tokelau. Though Tokelau is located at the hypothesized limit of native ants in the Pacific, a surprising number of tramp ants have assembled in a very small area. Tokelau consists of three isolated low-lying oceanic atolls which comprise 11 km^2 of terrestrial habitat, making it the nation with the world's smallest land area. Intensive sampling on the atolls recorded 28 ant species, with perhaps no natives or endemics, but a recently assembled community of human-dispersed tramps (Abbott *et al.* 2006).

At the opposite end of the age and diversity spectrum lies Madagascar, a very old island long isolated in the south-west Indian Ocean. Above 95% of its more than 1,000 ant species are endemic to the island, having arrived from Asia and Africa after Madagascar had been split off from Gondwanaland over 120 Mya (Fisher 2003).

2.6.2 Radiation

The relative paucity of insect species and the availability of empty niches on islands tend to encourage adaptive radiation among new arrivals (Zimmerman 1970). Larger islands may contain more diverse habitats and more niches to fill, encouraging the evolution of more endemic species. By the same token, older islands afford established species some additional time to diverge. For example, Cuba and Hispaniola, both relatively large islands, provided ideal platforms for endemic radiations of the genus *Temnothorax*. *Temnothorax* now constitute more than 25% of the ant fauna in Cuba alone and occupy habitats ranging from soil to limestone crevices and epiphytic plants (Wilson 1988). The biological diversity of this group is comparable to the range usually seen in several genera.

On Madagascar, ant genera (*Camponotus, Cerapachys, Hypoponera, Pheidole, Strumigenys*, and *Tetramorium*) demonstrate high levels of radiation (Fisher 2003). The morphological and niche diversity represented within *Cerapachys* alone is stunning, with some species having developed characteristics more typical of African army ants.

The composition of ants on an island at the time of arrival of a new species likely influences radiation as well. The lack of dominant mainland ants (e.g. army ants) on Cuba, Hispaniola, and Madagascar may have helped new species persist and radiate.

2.6.3 Taxon cycle

Based on studies of ants on the islands of Melanesia, Wilson (1959, 1961) proposed that species pass through 'taxon cycles', phases of expansion, and contraction in distribution accompanied by habitat shifts. He observed that expanding taxa tended to be recent arrivals that occupy coastline habitats. Wilson suggested that subsequent arrivals push species that arrived earlier farther inland and higher in altitude. As a result, older and endemic species are more likely to have fragmented ranges that consist of interior, montane habitats.

Because the taxon cycle is an historical model, an assessment of the model requires phylogenetics-based biogeographic methods to reconstruct the past history of events. Based on a phylogeographic analysis, the taxon cycle model has been supported in some studies, for instance of birds in the Lesser Antilles (Ricklefs and Bermingham 2002). No such study has been conducted for ants.

Though phylogenetic studies were not conducted, Fisher and Smith (2008) document an interesting pattern in the genera *Anochetus* and *Odontomachus* on the island of Madagascar that could be evaluated in the context of a taxon cycle model. In both genera, one or two species are restricted to higher elevation fragments, while another one or two species are widespread across lowland habitats. In both cases, the widespread species belong to groups found in Africa, while the restricted species are most similar to groups found only in Asia. An historical study is needed to evaluate if the African species-group taxa colonised after the Asian species-group taxa. If so, the first colonists of the lowlands may have been gradually pushed up into montane forest by new incursions of African species.

2.6.4 Turnover

The composition of ant species can vary considerably across an island's history. The primary forces that affect island biogeography — size, isolation, and habitats — also exert great influence on species turnover through time. Because islands are small and more prone to climate and colonisation shifts, species turnover among island ants can be surprisingly rapid. On Hispaniola, amber fossils indicate that 20 Mya, the island's ant fauna was closely related to the continental fauna of México (Wilson 1988). During this time, Hispaniola and its Greater Antilles neighbours were all located much closer to the mainland. But of the 38 genera and subgenera found in Dominican amber, only 22 persist today on Hispaniola. The farther the island travelled from the mainland, the more taxa were lost. Far from sources of new ants, few species arrived. Highly specialized species or those less able to establish themselves on new ground were the most likely to disappear. Volcanism, climate shifts, inundations, and other large-scale changes have caused similar effects on species turnover on other islands.

2.7 Future directions

Lack of a well-resolved phylogeny for many ant clades together with taxonomic uncertainties at the species level have limited the progress of understanding ant biogeography. With only an estimated 50% of ant species described, there is still a great need for species exploration and description. The recent discovery of *Martialis* demonstrated that new discoveries can shed light on the general pattern of ant evolution and radiation (Rabeling *et al.* 2008). Biogeographic studies will further benefit from these new species discoveries. Most importantly, studies that incorporate molecular phylogenies with divergence times estimated using previously established calibration points from fossil taxa will be necessary to understand the origin, dispersal, radiation, and spread of taxa across isolated biogeographic regions. Island systems, such as the southwest Indian Ocean islands, offer a model for exploring biogeographic questions. Because islands are smaller and often show a simplified ecology consisting of fewer species whose arrival can be dated, some questions can be easier to address than in larger, more complex, continental ecosystems. These questions include: (*a*) How many colonisation events occurred for each genus? (*b*) Did islands serve as stepping stones in dispersal? (*c*) What was the time frame for dispersal events?

The growing availability of standardized, geotagged data on ant distributions gathered from around the world (i.e. www.AntWeb.org) combined with enhanced geographic tools (e.g. Google Earth) will facilitate exploring fundamental questions regarding the distribution and history of ants on this planet. Geographic tools help visualize the role of topography, moisture, vegetation, and other environmental layers on species differentiation. The next technological challenge will be to juxtapose evolutionary relationships and distribution information atop geographic data. Such an online visualization tool will help reveal relationships among speciation and geographic barriers, connections to environmental conditions, and shifts in species over time.

The historical study of species distribution and how species have changed over time will also become increasing important as we try to understand how species will respond to climate change. We lack answers to simple questions about how the biota will respond to these new climate regimes. We do not understand how fast animals are changing their ranges, where they are moving, or which components of ecological communities (e.g. terrestrial versus arboreal arthropods) are most vulnerable to extinction. Answers to these questions are necessary for formulating adaptation strategies to minimize the impacts of global climate destabilization. One approach to predicting the impact of climate change is to evaluate how communities have changed in the past. Knowing how communities changed during past climatic shifts may be our best hope in mitigating current changes.

2.8 Summary

The present-day distribution of ants reflects the influence of geography, geology, and climate on the origin, diversification, and spread of a lineage. Though the process is complex, often difficult to reconstruct for a given taxon, and limited by the high number of undescribed taxa, two important overall patterns emerge: taxa are neither randomly nor uniformly distributed across the earth, and endemic taxa are clustered in particular regions. The greatest diversity is found in the tropics and the Gondwanaland fragments of South America, Africa, and Australia, which have the highest percentage of endemic genera, and where remarkable hotspots are found in moist lowland and low-latitude forests.

Approaches that combine exhaustive inventories, taxonomic revisions, and phylogenetics will enable a more rigorous approach to the study of biogeography. A newer approach, incorporating fossil records into studies of molecular divergence, shows promise for clarifying the ancient and relatively rapid origins of ant genera.

The rapid rate of ant species discovery continues to add nuance and critical missing links to the ant family tree.

Islands offer excellent model systems to explore outstanding questions of ant biogeography. One system with great potential includes the southwest Indian Ocean islands of Comoros, Madagascar, the Mascarenes, and the Seychelles. This region is diverse in origin, represented by coralline, volcanic, and Gondwanaland fragments. Ranging in age from 15,000 to 120 million years, the islands vary widely in size, degree of isolation, and habitat types. Such historical and geographic diversity makes these islands an ideal place to explore the relative impact of biogeographical factors on species diversity.

Acknowledgements

I thank Phil Ward, Vincent Perrichot, Benoit Guénard, and Bruce Archibald for discussions on ant biogeography. V. Perrichot kindly provided the synchrotron image of the newly discovered *Sphecomyrma* specimen. This study was partially supported by the National Science Foundation under Grant No DEB-0344731 and EF-0431330.

Appendix

The distribution of ant genera across eight biogeographic regions. The table is a compilation of many published works, especially Bolton (1995), Brown (1973), Fisher (1997), and museum records. The biogeographic regions are the same as those defined by classical biogeographers (Olson *et al.* 2001; Pielou 1979). For each genus, "0" indicates absence and "1" indicates presence in that biogeographical region. Genera known only from probable tramp or introduced species in a given region are noted in [] but not included in the totals. I have not recorded all tramp species in all biogeographic regions, but have made efforts to note the most common recorded. NEA: Nearctic; NEO: Neotropical; AFR: Afrotropic; MAL: Malagasy; PAL: Palearctic; IND: Indomalaya; AUS: Australian; OCE: Oceania. It should be noted that because genera are constantly being revised, changes to the listed genera are likely in the future.

Genus	Subfamily	NEA	NEO	AFR	MAL	PAL	IND	AUS	OCE
Acanthognathus	Myrmicinae	0	1	0	0	0	0	0	0
Acanthomyrmex	Myrmicinae	0	0	0	0	0	1	1	0
Acanthoponera	Heteroponerinae	0	1	0	0	0	0	0	0
Acanthostichus	Cerapachyinae	1	1	0	0	0	0	0	0
Acromyrmex	Myrmicinae	1	1	0	0	0	0	0	0
Acropyga	Formicinae	1	1	1	1	1	1	1	1
Adelomyrmex	Myrmicinae	0	1	0	0	0	0	1	1
Adetomyrma	Amblyoponinae	0	0	0	1	0	0	0	0
Adlerzia	Myrmicinae	0	0	0	0	0	0	1	0
Aenictogiton	Aenictogitoninae	0	0	1	0	0	0	0	0
Aenictus	Aenictinae	0	0	1	0	1	1	1	0
Agraulomyrmex	Formicinae	0	0	1	0	0	0	0	0
Alloformica	Formicinae	0	0	0	0	1	0	0	0
Allomerus	Myrmicinae	0	1	0	0	0	0	0	0
Amblyopone	Amblyoponinae	1	1	1	1	1	1	1	[1]
Ancyridris	Myrmicinae	0	0	0	0	0	0	1	0
Anergates	Myrmicinae	1	0	0	0	1	0	0	0
Aneuretus	Aneuretinae	0	0	0	0	0	1	0	0
Anillidris	Dolichoderinae	0	1	0	0	0	0	0	0
Anillomyrma	Myrmicinae	0	0	0	0	0	1	1	0
Anisopheidole	Myrmicinae	0	0	0	0	0	0	1	0
Ankylomyrma	Myrmicinae	0	0	1	0	0	0	0	0
Anochetus	Ponerinae	1	1	1	1	1	1	1	1
Anomalomyrma	Leptanillinae	0	0	0	0	1	1	0	0
Anonychomyrma	Dolichoderinae	0	0	0	0	0	1	1	0
Anoplolepis	Formicinae	0	1	1	[1]	1	[1]	[1]	[1]
Aphaenogaster	Myrmicinae	1	1	0	1	1	1	1	1
Aphomomyrmex	Formicinae	0	0	1	0	0	0	0	0
Apomyrma	Amblyoponinae	0	0	1	0	0	0	0	0
Apterostigma	Myrmicinae	0	1	0	0	0	0	0	0
Aptinoma	Dolichoderinae	0	0	0	1	0	0	0	0
Arnoldius	Dolichoderinae	0	0	0	0	1	1	1	0
Asphinctanilloides	Leptanilloidinae	0	1	0	0	0	0	0	0
Asphinctopone	Ponerinae	0	0	1	0	0	0	0	0
Atopomyrmex	Myrmicinae	0	0	1	0	0	0	0	0
Atta	Myrmicinae	1	1	0	0	0	0	0	0
Aulacopone	Heteroponerinae	0	0	0	0	1	0	0	0
Axinidris	Dolichoderinae	0	0	1	0	0	0	0	0
Azteca	Dolichoderinae	0	1	0	0	0	0	0	0
Bajcaridris	Formicinae	0	0	0	0	1	0	0	0
Bannapone	Amblyoponinae	0	0	0	0	0	1	0	0
Baracidris	Myrmicinae	0	0	1	0	0	0	0	0

Bariamyrma	Myrmicinae	0	1	0	0	0	0	0	0
Basiceros	Myrmicinae	0	1	0	0	0	0	0	0
Belonopelta	Ponerinae	0	1	0	0	0	0	0	0
Blepharidatta	Myrmicinae	0	1	0	0	0	0	0	0
Boloponera	Ponerinae	0	0	1	0	0	0	0	0
Bondroitia	Myrmicinae	0	0	1	0	0	0	0	0
Bothriomyrmex	Dolichoderinae	0	1	0	0	1	1	0	0
Brachymyrmex	Formicinae	1	1	0	[1]	1	0	0	[1]
Bregmatomyrma	Formicinae	0	0	0	0	0	1	0	0
Calomyrmex	Formicinae	0	0	0	0	0	0	1	0
Calyptomyrmex	Myrmicinae	0	0	1	1	0	1	1	1
Camponotus	Formicinae	1	1	1	1	1	1	1	1
Cardiocondyla	Myrmicinae	1	1	1	1	1	1	1	1
Carebara	Myrmicinae	1	1	1	1	1	1	1	1
Carebarella	Myrmicinae	0	1	0	0	0	0	0	0
Cataglyphis	Formicinae	0	0	1	0	1	1	0	0
Cataulacus	Myrmicinae	0	0	1	1	0	1	1	0
Centromyrmex	Ponerinae	0	1	1	0	0	1	0	0
Cephalotes	Myrmicinae	1	1	0	0	0	0	0	0
Cerapachys	Cerapachyinae	1	1	1	1	1	1	1	1
Chalepoxenus	Myrmicinae	0	0	0	0	1	1	0	0
Cheliomyrmex	Ecitoninae	0	1	0	0	0	0	0	0
Chimaeridris	Myrmicinae	0	0	0	0	0	1	1	0
Chronoxenus	Dolichoderinae	0	0	0	0	0	1	0	0
Cladomyrma	Formicinae	0	0	0	0	0	1	0	0
Colobostruma	Myrmicinae	0	0	0	0	0	0	1	0
Concoctio	Amblyoponinae	0	0	1	0	0	0	0	0
Condylodon	Incertae sedis	0	1	0	0	0	0	0	0
Crematogaster	Myrmicinae	1	1	1	1	1	1	1	1
Cryptomyrmex	Myrmicinae	0	1	0	0	0	0	0	0
Cryptopone	Ponerinae	1	1	1	0	1	1	1	1
Cylindromyrmex	Cerapachyinae	0	1	0	0	0	0	0	0
Cyphoidris	Myrmicinae	0	0	1	0	0	0	0	0
Cyphomyrmex	Myrmicinae	1	1	0	0	0	0	0	0
Dacatria	Myrmicinae	0	0	0	0	1	1	0	0
Dacetinops	Myrmicinae	0	0	0	0	0	1	1	0
Daceton	Myrmicinae	0	1	0	0	0	0	0	0
Decamorium	Myrmicinae	0	0	1	0	0	0	0	0
Diacamma	Ponerinae	0	0	0	0	0	1	1	0
Dicroaspis	Myrmicinae	0	0	1	0	0	0	0	0
Dilobocondyla	Myrmicinae	0	0	0	0	0	1	1	0
Dinoponera	Ponerinae	0	1	0	0	0	0	0	0
Diplomorium	Myrmicinae	0	0	1	0	0	0	0	0
Discothyrea	Proceratiinae	1	1	1	1	1	1	1	1
Doleromyrma	Dolichoderinae	0	0	0	0	0	0	1	0
Dolichoderus	Dolichoderinae	1	1	0	0	1	1	1	0
Dolioponera	Ponerinae	0	0	1	0	0	0	0	0
Dolopomyrmex	Myrmicinae	1	0	0	0	0	0	0	0
Dorylus	Dorylinae	0	0	1	0	0	1	1	0
Dorymyrmex	Dolichoderinae	1	1	0	0	0	0	0	0
Echinopla	Formicinae	0	0	0	0	0	1	1	0

continued

Genus	Subfamily	NEA	NEO	AFR	MAL	PAL	IND	AUS	OCE
Eciton	Ecitoninae	0	1	0	0	0	0	0	0
Ecphorella	Dolichoderinae	0	0	1	0	0	0	0	0
Ectatomma	Ectatomminae	0	1	0	0	0	0	0	0
Emeryopone	Ponerinae	0	0	0	0	1	1	0	0
Epopostruma	Myrmicinae	0	0	0	0	0	0	1	0
Euprenolepis	Formicinae	0	1	0	0	0	1	1	0
Eurhopalothrix	Myrmicinae	1	1	0	0	0	1	1	1
Eutetramorium	Myrmicinae	0	0	0	1	0	0	0	0
Feroponera	Ponerinae	0	0	1	0	0	0	0	0
Forelius	Dolichoderinae	1	1	0	0	0	0	0	0
Forelophilus	Formicinae	0	0	0	0	0	1	0	0
Formica	Formicinae	1	1	0	0	1	1	0	0
Formicoxenus	Myrmicinae	1	1	0	0	1	0	0	0
Froggattella	Dolichoderinae	0	0	0	0	0	0	1	0
Gauromyrmex	Myrmicinae	0	0	0	0	0	1	0	1
Gesomyrmex	Formicinae	0	0	0	0	0	1	0	0
Gigantiops	Formicinae	0	1	0	0	0	0	0	0
Gnamptogenys	Ectatomminae	1	1	0	0	1	1	1	1
Goniomma	Myrmicinae	0	0	0	0	1	0	0	0
Gracilidris	Dolichoderinae	0	1	0	0	0	0	0	0
Harpagoxenus	Myrmicinae	1	0	0	0	1	0	0	0
Harpegnathos	Ponerinae	0	0	0	0	0	1	0	0
Heteroponera	Heteroponerinae	0	1	0	0	0	0	1	0
Huberia	Myrmicinae	0	0	0	0	0	0	1	0
Hylomyrma	Myrmicinae	0	1	0	0	0	0	0	0
Hypoponera	Ponerinae	1	1	1	1	1	1	1	1
Indomyrma	Myrmicinae	0	0	0	0	0	1	0	0
Iridomyrmex	Dolichoderinae	0	0	0	0	1	1	1	1
Ishakidris	Myrmicinae	0	0	0	0	0	1	0	0
Kartidris	Myrmicinae	0	0	0	0	0	1	0	0
Labidus	Ecitoninae	1	1	0	0	0	0	0	0
Lachnomyrmex	Myrmicinae	0	1	0	0	0	0	0	0
Lasiomyrma	Myrmicinae	0	0	0	0	0	1	0	0
Lasiophanes	Formicinae	0	1	0	0	0	0	0	0
Lasius	Formicinae	1	1	0	0	1	1	0	0
Lenomyrmex	Myrmicinae	0	1	0	0	0	0	0	0
Lepisiota	Formicinae	0	0	1	1	1	1	0	1
Leptanilla	Leptanillinae	0	0	1	0	1	1	1	0
Leptanilloides	Leptanilloidinae	0	1	0	0	0	0	0	0
Leptogenys	Ponerinae	1	1	1	1	1	1	1	1
Leptomyrmex	Dolichoderinae	0	0	0	0	0	0	1	0
Leptothorax	Myrmicinae	1	1	1	1	1	1	0	0
Linepithema	Dolichoderinae	1	1	0	0	[1]	[1]	[1]	[1]
Liometopum	Dolichoderinae	1	1	0	0	1	1	0	0
Liomyrmex	Myrmicinae	0	0	0	0	0	1	1	0
Loboponera	Ponerinae	0	0	1	0	0	0	0	0
Lophomyrmex	Myrmicinae	0	0	0	0	0	1	0	0
Lordomyrma	Myrmicinae	0	0	0	0	1	1	1	1
Loweriella	Dolichoderinae	0	0	0	0	0	1	0	0
Machomyrma	Myrmicinae	0	0	0	0	0	0	1	0

Manica	Myrmicinae	1	0	0	0	1	0	0	0
Martialis	Martialinae	0	1	0	0	0	0	0	0
Mayriella	Myrmicinae	0	0	0	0	0	1	1	0
Megalomyrmex	Myrmicinae	0	1	0	0	0	0	0	0
Melissotarsus	Myrmicinae	0	0	1	1	0	0	0	0
Melophorus	Formicinae	0	0	0	0	0	0	1	0
Meranoplus	Myrmicinae	0	0	1	1	0	1	1	0
Mesostruma	Myrmicinae	0	0	0	0	0	0	1	0
Messor	Myrmicinae	1	1	1	0	1	1	0	0
Metapone	Myrmicinae	0	0	1	1	0	1	1	1
Microdaceton	Myrmicinae	0	0	1	0	0	0	0	0
Monomorium	Myrmicinae	1	1	1	1	1	1	1	1
Mycetagroicus	Myrmicinae	0	1	0	0	0	0	0	0
Mycetarotes	Myrmicinae	0	1	0	0	0	0	0	0
Mycetophylax	Myrmicinae	0	1	0	0	0	0	0	0
Mycetosoritis	Myrmicinae	1	1	0	0	0	0	0	0
Mycocepurus	Myrmicinae	0	1	0	0	0	0	0	0
Myopias	Ponerinae	0	0	0	0	0	1	1	0
Myopopone	Amblyoponinae	0	0	0	0	0	1	1	0
Myrcidris	Pseudomyrmecinae	0	1	0	0	0	0	0	0
Myrmecia	Myrmeciinae	0	0	0	0	0	0	1	0
Myrmecina	Myrmicinae	1	1	0	0	1	1	1	1
Myrmecocystus	Formicinae	1	1	0	0	0	0	0	0
Myrmecorhynchus	Formicinae	0	0	0	0	0	0	1	0
Myrmelachista	Formicinae	0	1	0	0	0	0	0	0
Myrmica	Myrmicinae	1	1	0	0	1	1	0	0
Myrmicaria	Myrmicinae	0	0	1	0	0	1	0	0
Myrmicocrypta	Myrmicinae	0	1	0	0	0	0	0	0
Myrmoteras	Formicinae	0	0	0	0	0	1	1	0
Myrmoxenus	Myrmicinae	0	0	0	0	1	0	0	0
Mystrium	Amblyoponinae	0	0	1	1	0	1	1	0
Nebothriomyrmex	Dolichoderinae	0	0	0	0	0	0	1	0
Neivamyrmex	Ecitoninae	1	1	0	0	0	0	0	0
Nesomyrmex	Myrmicinae	1	1	1	1	0	1	0	0
Nomamyrmex	Ecitoninae	1	1	0	0	0	0	0	0
Noonilla	Incertae sedis	0	0	0	0	0	1	0	0
Nothomyrmecia	Myrmeciinae	0	0	0	0	0	0	1	0
Notoncus	Formicinae	0	0	0	0	0	0	1	0
Notostigma	Formicinae	0	0	0	0	0	0	1	0
Ochetellus	Dolichoderinae	1	0	0	[1]	1	1	1	1
Ochetomyrmex	Myrmicinae	0	1	0	0	0	0	0	0
Octostruma	Myrmicinae	0	1	0	0	0	0	0	0
Ocymyrmex	Myrmicinae	0	0	1	0	0	0	0	0
Odontomachus	Ponerinae	1	1	1	1	1	1	1	1
Odontoponera	Ponerinae	0	0	0	0	1	1	0	0
Oecophylla	Formicinae	0	0	1	0	0	1	1	0
Onychomyrmex	Amblyoponinae	0	0	0	0	0	0	1	0
Opamyrma	Amblyoponinae	0	0	0	0	0	1	0	0
Opisthopsis	Formicinae	0	0	0	0	0	0	1	0
Orectognathus	Myrmicinae	0	0	0	0	0	0	1	0
Overbeckia	Formicinae	0	0	0	0	0	1	0	0

continued

Genus	Subfamily	NEA	NEO	AFR	MAL	PAL	IND	AUS	OCE
Oxyepoecus	Myrmicinae	0	1	0	0	0	0	0	0
Oxyopomyrmex	Myrmicinae	0	0	0	0	1	0	0	0
Pachycondyla	Ponerinae	1	1	1	1	1	1	1	1
Papyrius	Dolichoderinae	0	0	0	0	0	0	1	0
Paraponera	Paraponerinae	0	1	0	0	0	0	0	0
Paraprionopelta	Amblyoponinae	0	1	0	0	0	0	0	0
Paratopula	Myrmicinae	0	0	0	0	0	1	1	0
Paratrechina	Formicinae	1	1	1	1	1	1	1	1
Parvimyrma	Myrmicinae	0	0	0	0	0	1	0	0
Perissomyrmex	Myrmicinae	0	1	0	0	1	1	0	0
Peronomyrmex	Myrmicinae	0	0	0	0	0	0	1	0
Petalomyrmex	Formicinae	0	0	1	0	0	0	0	0
Phalacromyrmex	Myrmicinae	0	1	0	0	0	0	0	0
Phasmomyrmex	Formicinae	0	0	1	0	0	0	0	0
Phaulomyrma	Leptanillinae	0	0	0	0	0	1	0	0
Pheidole	Myrmicinae	1	1	1	1	1	1	1	1
Pheidologeton	Myrmicinae	0	0	1	0	1	1	1	0
Philidris	Dolichoderinae	0	0	0	0	0	1	1	1
Phrynoponera	Ponerinae	0	0	1	0	0	0	0	0
Pilotrochus	Myrmicinae	0	0	0	1	0	0	0	0
Plagiolepis	Formicinae	1	1	1	1	1	1	1	[1]
Platythyrea	Ponerinae	1	1	1	1	0	1	1	[1]
Plectroctena	Ponerinae	0	0	1	0	0	0	0	0
Podomyrma	Myrmicinae	0	0	0	0	0	1	1	0
Poecilomyrma	Myrmicinae	0	0	0	0	0	0	0	1
Pogonomyrmex	Myrmicinae	1	1	0	0	0	0	0	0
Polyergus	Formicinae	1	1	0	0	1	0	0	0
Polyrhachis	Formicinae	0	0	1	0	1	1	1	1
Ponera	Ponerinae	1	1	1	[1]	1	1	1	1
Prenolepis	Formicinae	1	1	0	0	1	1	1	0
Prionopelta	Amblyoponinae	1	1	1	1	0	1	1	1
Pristomyrmex	Myrmicinae	0	0	1	1	1	1	1	1
Proatta	Myrmicinae	0	0	0	0	0	1	0	0
Probolomyrmex	Proceratiinae	0	1	1	0	1	1	1	0
Proceratium	Proceratiinae	1	1	1	1	1	1	1	1
Procryptocerus	Myrmicinae	0	1	0	0	0	0	0	0
Proformica	Formicinae	0	0	0	0	1	0	0	0
Prolasius	Formicinae	0	0	0	0	0	0	1	0
Promyopias	Ponerinae	0	0	1	0	0	0	0	0
Protalaridris	Myrmicinae	0	1	0	0	0	0	0	0
Protanilla	Leptanillinae	0	0	0	0	1	1	0	0
Protomognathus	Myrmicinae	1	0	0	0	0	0	0	0
Psalidomyrmex	Ponerinae	0	0	1	0	0	0	0	0
Pseudoatta	Myrmicinae	0	1	0	0	0	0	0	0
Pseudolasius	Formicinae	0	0	1	1	1	1	1	0
Pseudomyrmex	Pseudomyrmecinae	1	1	0	0	0	0	0	[1]
Pseudonotoncus	Formicinae	0	0	0	0	0	0	1	0
Pyramica	Myrmicinae	1	1	1	1	1	1	1	1
Ravavy	Dolichoderinae	0	0	0	1	0	0	0	0
Recurvidris	Myrmicinae	0	0	0	0	1	1	1	0

Rhopalomastix	Myrmicinae	0	0	0	0	0	1	1	0
Rhopalothrix	Myrmicinae	0	1	0	0	0	0	1	0
Rhoptromyrmex	Myrmicinae	0	0	1	0	1	1	1	0
Rhytidoponera	Ectatomminae	0	0	0	0	0	0	1	0
Rogeria	Myrmicinae	1	1	0	0	0	0	1	1
Romblonella	Myrmicinae	0	0	0	0	0	1	1	1
Rossomyrmex	Formicinae	0	0	0	0	1	0	0	0
Rostromyrmex	Myrmicinae	0	0	0	0	0	1	0	0
Rotastruma	Myrmicinae	0	0	0	0	0	1	0	0
Santschiella	Formicinae	0	0	1	0	0	0	0	0
Secostruma	Myrmicinae	0	0	0	0	0	1	1	0
Sericomyrmex	Myrmicinae	0	1	0	0	0	0	0	0
Simopelta	Ponerinae	0	1	0	0	0	0	0	0
Simopone	Cerapachyinae	0	0	1	1	0	1	1	0
Solenopsis	Myrmicinae	1	1	1	1	1	1	1	1
Sphinctomyrmex	Cerapachyinae	0	1	1	0	0	1	1	0
Stegomyrmex	Myrmicinae	0	1	0	0	0	0	0	0
Stenamma	Myrmicinae	1	1	0	0	1	1	0	0
Stereomyrmex	Myrmicinae	0	0	0	0	0	1	1	0
Stigmacros	Formicinae	0	0	0	0	0	0	1	0
Streblognathus	Ponerinae	0	0	1	0	0	0	0	0
Strongylognathus	Myrmicinae	0	0	0	0	1	0	0	0
Strumigenys	Myrmicinae	1	1	1	1	1	1	1	1
Talaridris	Myrmicinae	0	1	0	0	0	0	0	0
Tapinolepis	Formicinae	0	0	1	0	0	0	0	0
Tapinoma	Dolichoderinae	1	1	1	1	1	1	1	1
Tatuidris	Agroecomyrmecinae	0	1	0	0	0	0	0	0
Technomyrmex	Dolichoderinae	1	1	1	1	1	1	1	1
Teleutomyrmex	Myrmicinae	0	0	0	0	1	0	0	0
Temnothorax	Myrmicinae	1	1	1	1	1	1	0	0
Terataner	Myrmicinae	0	0	1	1	0	0	0	0
Teratomyrmex	Formicinae	0	0	0	0	0	0	1	0
Tetheamyrma	Myrmicinae	0	0	0	0	0	1	0	0
Tetramorium	Myrmicinae	1	1	1	1	1	1	1	1
Tetraponera	Pseudomyrmecinae	0	0	1	1	1	1	1	0
Thaumatomyrmex	Ponerinae	0	1	0	0	0	0	0	0
Trachymyrmex	Myrmicinae	1	1	0	0	0	0	0	0
Tranopelta	Myrmicinae	0	1	0	0	0	0	0	0
Turneria	Dolichoderinae	0	0	0	0	0	0	1	0
Typhlomyrmex	Ectatomminae	1	1	0	0	0	0	0	0
Tyrannomyrmex	Myrmicinae	0	0	0	0	0	1	0	0
Vollenhovia	Myrmicinae	1	0	0	0	1	1	1	1
Vombisidris	Myrmicinae	0	0	0	0	0	1	1	0
Wasmannia	Myrmicinae	1	1	0	0	0	0	1	[1]
Xenomyrmex	Myrmicinae	1	1	0	0	0	1	0	0
Yavnella	Leptanillinae	0	0	0	0	1	1	0	0
	total	73	128	88	47	82	128	115	44
	endemic	2	52	31	5	10	24	30	1

Chapter 3

Geographic Gradients

Robert R. Dunn, Benoit Guénard, Michael D. Weiser, and Nathan J. Sanders

3.1 Introduction

Linnaeus, upon travelling north from southern Sweden, was struck by the differences between that colder land and his homeland. The differences in the composition of species begged explanation. Ever since (and probably before), naturalists, ecologists, and biogeographers have sought to understand why species and the communities they make up differ from place to place. Why does body size, both within and among species, vary geographically? Where are there more species and why? The ease of posing such questions has attracted generations of biologists – each lured by the possibility of finding elegant answers to superficially simple problems. But to date, consensus explanations remain elusive. And so how should we move forward? Major headway, we suspect, will come not from the search for general explanations of ecological gradients, but instead through the careful study of a few focal taxa and the factors that influence their distribution along gradients. The most obvious focal taxa are likely to be vertebrate groups (e.g. birds and mammals). But because most animal species are insects, having even a single well-studied insect taxon whose ecological gradients are well understood would be disproportionately valuable. To date, the best candidate for such intensive focus is ants. They are relatively easy to sample (see Box 3.1), well known, ecologically important and have been the subject of gradient studies for nearly 50 years.

In this chapter, we review our current understanding of gradients in ant diversity, body size, range size, and other life-history traits. We examine the underlying causes of these gradients and discuss their consequences for some of the processes ants mediate, such as predation and seed dispersal. We start with the patterns themselves by reviewing latitudinal and elevational gradients in ant diversity at multiple spatial and taxonomic scales. Then, we move on to two other common gradients in ants – gradients in body size and range size. For each gradient, we explore some of the potential underlying mechanisms.

3.2. Empirical patterns along gradients

3.2.1 Species diversity and latitude

Since as early as the work of Forster and von Humboldt in the early 19th century (see Hawkins 2001), scientists have been aware of latitudinal gradients in diversity, particularly for vertebrates and plants (Brown and Lomolino 1998; Rosenzweig 1995; Willig and Lyons 1998; Willig *et al*. 2003). Generally, diversity declines with latitude and ants are no exception. Kusnezov (1957) was the first to attempt to quantify the latitudinal gradient in ant diversity by compiling total species lists for regions (e.g. Alaska, Iowa, Utah, Cuba, Trinidad, parts of Argentina, and Brazil). His finding that ant species richness decreased with latitude was among the first clear demonstrations of a latitudinal diversity gradient in insects. In the 1970s, similar results were reported based on data compiled from the ants of the British Isles and northern Europe – a decline in species diversity with latitude (Baroni-Urbani and Collingwood 1976, 1977). Cushman *et al*. (1993) followed these earlier leads and compiled species lists and museum data to examine whether there was a latitudinal gradient in ant diversity in

Box 3.1 Field techniques for sampling ants
Brandon Bestelmeyer and Leticia Rios Casanova

Ants occur in most environments and ecologists ask a diverse array of questions involving ants. Thus, a key consideration in ant studies is to match the environment and question (and associated environmental variables) to the ant sampling technique. Since each technique has distinct limitations, using a complementary suite of techniques is often beneficial (Agosti and Alonso 2000). For further details, see Bestelmeyer *et al.* (2000) and newer applications.

There are two general questions that an investigator should consider in choosing one or more techniques to sample ants: (*a*) What distinct microhabitats or sampling strata are available or of interest in a study area? And (*b*) What is the ant variable of interest? From top to bottom, distinct sampling strata include: (*a*) large trees including emergents (>5 m tall), (b) small trees and/or *shrubs* (0.5–5 m), (c) small shrubs and/or herbs (<0.5 m), (d) rotting wood, (e) leaf litter and/or humus, (f) ground surface, and (g) subsurface soil. There are three general types of variables used to represent ant species and communities: (*a*) *abundance*, some measure of the number of individuals of a species within samples, or frequency of occurrence across samples; (*b*) *richness*, number of species in a sample, also presence or absence; and (*c*) *behaviour*, observations of species interactions or foraging that can be used to quantify behavioural dominance. Certain techniques also estimate forager or colony density. A subjective ranking of the value of technique types for sampling different strata, with respect to the type of data best produced by that technique, is indicated in Table 3.1.1 (see also Bestelmeyer *et al.* 2000). Brief descriptions of the classes of techniques follow.

Canopy fogging samples ants from the canopy of large trees. Fogging involves four steps that include (*a*) identification of tree species to be sampled and replication, (*b*) placement of collection funnels under trees (with a standard size, number, and arrangement), (*c*) fogging of the tree for a fixed period (e.g. 4 min) using an insecticide (often pyrethrum), and (*d*) collection of specimens that fall from the tree after a fixed time period (e.g. 90 min). Only ants active on the tree surface are collected using this technique. Ant behaviour (e.g. ants living in epiphytes) and the time period at which fogging takes place will influence the likelihood of ant species' records in samples (Schulz and Wagner 2002).

Beating samples ants from large to small woody plants within reach of poles. Ants dislodged from vegetation struck by the pole fall onto a sheet (e.g. 2 × 2 m) where they are collected (e.g. Majer *et al.* 1994). Beating can also occur in conjunction with the use of direct searching (see later) for ant species inhabiting bark and twigs that may be missed by beating. As with canopy fogging, ant behaviour and time period will influence the collection of particular species.

Sweeping and vacuuming are typically used to estimate abundance from low woody and herbaceous strata. Sweeping involves moving a sweep net through vegetation to dislodge and collect ants in the net. The number of sweeps, extent sampled, and pattern of sampling is standardized for comparison (e.g. Andersen *et al.* 2007). Plant substrates can also be vacuumed. Philpott *et al.* (2006) used a garden vacuum to suck arthropods into plastic bags where they were killed with ethyl acetate. Samples were compared using the number of individuals per gram of foliage. The ground surface can also be vacuumed to provide a density value for high-density foragers (Abbott 2006). There are similar considerations regarding ant behaviour and period of sampling as discussed earlier.

Bait sampling encompasses a diverse array of techniques used to evaluate the presence and behaviour of active ant foragers on the ground surface, below ground, or in vegetation. Baits are protein-, fat-, or carbohydrate-rich food substances that attract foraging ants to points where they are observed and/or collected. Baits are used to indicate which ant species are present, especially where other methods cannot be used (e.g. pitfall traps in rocky terrain). Changes in the abundance and species of ants at baits as well as interactions over time can be used to determine dominance (Andersen 1992;

continues

Table 3.1.1 A qualitative ranking of the utility of different methods for sampling ants in different strata (from + to +++) and the types of data provided by each method other than species presence or absence (that is common to all methods).

Method	Canopy fogging	Beating	Sweeping/ vacuum	Bait sampling	Pitfall trapping	Quadrat count	Colony sampling	Direct sampling	Litter extraction	Soil sampling
Data type	Abundance, richness	Abundance, richness	Abundance, richness	Behaviour	Abundance, richness	Abundance, richness behaviour	Abundance (colonies)	Behaviour	Abundance, richness	Abundance, richness
Stratum										
Large tree	+++	+[a]		+++			+[a]	++		
Small tree/ shrub		+++		+++	+			++		
Small shrub/ herbs		++	+++	+				++		
Rotting wood							++	+++		
Litter/humus					+		++	+	+++	
Ground surface			+	+++	+++	+++	+++	+	+	
Belowground soil				++				++	+	+++

[a]If using a canopy crane.

continues

Box 3.1 continued

see Chapter 5). Ant contributions to ecosystem processes, such as myrmechochory, can be measured using seed baits. Finally, baiting is an excellent tool for measuring behavioural attributes. Differences in behavioural dominance, foraging behaviour, and preferences for food items will often result in an incomplete representation of the larger ant community.

Pitfall trapping is the most common method for sampling ground-active ants, although traps may also be used on woody strata. Open containers are placed flush with the surface. Ants fall into them and are usually killed and preserved in a liquid such as propylene glycol. Trap diameter and trapping duration (e.g. 1–4 days) need to be considered when using this technique. Trapping is relatively simple, integrates the sample across daily periods used by different ant species, and is perhaps the most commonly used method to evaluate ant community structure across broad areas and in situations where leaf litter is not deep. Differences in ant movement rates among habitats can bias the data (Melbourne 1999).

Quadrat sampling can be used for sampling surface-active ants, typically in open habitats. Behavioural observations can also be gathered, such as food items collected by the ants. A quadrat is used to delineate the observation area. In cases where activity is very high, the quadrat may be a small (e.g. 10 × 10 cm) card. Species of ants seen inside or entering the quadrat over a fixed time interval are counted and/or collected, and identified either in the field or back in the laboratory. Quadrats provide data similar to those of pitfall traps without trap-related biases. The method requires great skill in identification when ant activity is high and some species may be impossible to distinguish in the field. Observations may need replication at different times of day.

Colony sampling enumerates ant colonies in a defined area, providing a measure of abundance or density based not on individual ants but on colonies as units. Ants with characteristic nest structures facilitate direct enumeration and even mapping (Schooley and Wiens 2003). In habitats where most nests are inconspicuous, examination of soil under rocks, litter, nuts, and twigs or branches is necessary. Multiple nest entrances may overestimate colonies, especially when ants are polydomous. This method can provide a population-oriented perspective on ant communities and a distinct community characterization when compared to pitfall trapping (Schlick-Steiner *et al.* 2006a).

Direct searching records the presence of ant species inhabiting a habitat element or area. Different microhabitats or habitat elements (e.g. trees, foliage, and especially inside rotting wood) are systematically searched and ant foragers (or entire colonies) are collected. Ant behaviour within microhabitats (e.g. at extrafloral nectaries) can also be directly observed and quantified (Oliveira *et al.* 1999). Collections can be performed for fixed time periods to aid standardized comparisons of richness or composition. Nonetheless, standardized comparison across investigators can be difficult due to differences in ability. Direct sampling permits a complete list of the ant fauna in relatively little time by experienced collectors, but abundance is difficult to estimate. It is often used as supplementary technique or in cases where other methods are impractical.

Litter extraction measures abundance and richness of ants inhabiting a volume of leaf litter. A quantity of moist leaf litter is collected and placed in an extraction apparatus. Ants migrate from the litter and fall into a collecting receptacle. The migration behaviour of ants can be caused by disturbance (Winkler sack) or changes in microclimate (Berlese funnel). Litter techniques are needed for complete and cost-effective community characterization where litter horizons are deep, as in many tropical forests wherein the bulk of non-arboreal species inhabit the leaf-litter column (Fisher 1999b).

Soil sampling for subterranean ants is accomplished by excavating and searching a volume of soil for ants. Alternatively, chambered, tubular soil probes can be installed into the soil using an auger wherein baits placed in the chambers attract ants into the chambers to measure presence (Ryder Wilkie *et al.* 2007). This technique allows ants to be associated with particular soil horizons or depth increments in a cost-effective way.

Europe and found that species diversity declined with latitude.

The advantage of such large-grain, large-scale, geographic approaches is that they consider the spatial grain at which speciation and often extinction occur. The disadvantage of these scales and grains is that data on species are necessarily crude because it is easier to collect all the species in a metre square or even a hectare than in an entire geopolitical region. There are very few regions on earth where the species-level faunas of ants or any other insect taxon are completely known. To avoid the difficulties associated with generating species lists at large spatial grains, Jeanne (1979) collected ants at 10 sites, from Minnesota to Brazil. He was the first to show that the number of ant species in local communities increased towards the tropics, just as earlier authors had observed at larger sampling grains. Similarly, Kaspari *et al.* (2003) compared 49 New World sites that span similar conditions to those considered by Kusnezov (1957) (tropical forest to arctic), but with a bias towards the northern hemisphere. At both the plot (m^2) and transect (30 samples from a 330 m transect) grains, species diversity decreased with increasing latitude (Kaspari *et al.* 2003). In a related study – based on leaf-litter ant samples from 96 sites from Australia, Madagascar, North America, and South America – Kaspari and colleagues also found a strong latitudinal gradient in ant species diversity (Kaspari *et al.* 2004).

In sum, as for nearly all major groups of organisms (Hillebrand 2004; Willig *et al.* 2003), ants exhibit a strong latitudinal gradient in diversity. However, broad similarity in patterns of diversity among grains does not preclude more subtle differences in the form (e.g. quadratic versus linear) or parameterization (magnitude of coefficients) of the relationships between latitude and diversity (Levin 1992). Similarly, the relationship between diversity and latitude differs between hemispheres for some other taxa, with more species for a given latitude in the southern hemisphere (Chown *et al.* 2004). Such a possibility remains poorly explored for ants (but see Dunn *et al.* 2009). Further, patterns of ant diversity with latitude may differ at different scales, or depending on how diversity is measured (although see Kaspari *et al.* 2003). Understanding differences in diversity gradients among grains, and, in particular, patterns of beta diversity (turnover in species composition among focal sampling units, such as plots or grid cells) along gradients represent interesting areas for future work. Where it has been studied in other taxa, beta diversity has been shown to decline with distance from the equator (Qian 2008; Qian *et al.* 2009; Soininen *et al.* 2007; Stevens and Willig 2002), although sometimes only weakly (Gaston *et al.* 2007). Similar variation in beta diversity with latitude (and elevation) might be predicted for ants (Soininen *et al.* 2007).

3.2.2 Elevational gradients

Elevational gradients have long been seen as analogous to latitudinal gradients (e.g. von Humboldt 1808), because many of the processes that shape latitudinal gradients may also operate along elevational gradients. For example, temperature declines with distance from the equator and might decline similarly with elevation. Such environmental gradients can be logistically easier to study on elevational gradients than on latitudinal gradients because a much greater range of conditions can be encountered in a much smaller distance. While mean annual temperature decreases, on average, approximately 0.7°C/100 km of latitude, it decreases with elevation much more steeply, at approximately 0.6°C/100 m elevation (Colwell *et al.* 2008). It is not surprising then that patterns of ant diversity along elevational gradients often mirror the larger-scale latitudinal patterns. Ant species diversity typically declines with elevation (e.g. Atkin and Proctor 1988; Brown 1973; Cole 1940; Collins 1980; Janzen 1973; Janzen *et al.* 1976; Weber 1943) as it does with latitude, though with some interesting exceptions.

Most studies of ants along elevational gradients have been in the tropics. Several of these studies found no species at the highest elevations surveyed (e.g. 3,380 m in Costa Rica, Janzen 1973; 3,200 m in Sudan, Weber 1943). While these studies were modest in their sampling, they revealed that richness generally declines with elevation, a pattern that has been observed elsewhere. Species diversity of leaf-litter ants in a Malaysian rainforest decreased exponentially with increasing elevation (500–2,600 m; Brühl *et al.* 1999). Similarly, monotonic declines in leaf-litter ant species richness with increasing elevation have been found in Panamanian rainforest (Olson 1994), Costa Rica (30–2,900 m; D. Olson personal communication; Colwell *et al.* 2008), Madagascar (Fisher 1996), and Tanzania (Robertson 2002). Finally, in the only study to consider correlates of canopy ant diversity, ant diversity decreased linearly with elevation (Majer *et al.* 2001).

A handful of other studies of tropical elevation gradients have detected peaks in species richness at mid-elevations (Fisher 1999a; Olson 1994; Samson *et al.* 1997). In two of these cases, it has been suggested that the drop in diversity at the lowest elevations may be due to sampling incompleteness at low elevations (e.g. Olson 1994; Samson *et al.* 1997) or higher rates of disturbance at lower elevations making habitat less suitable. In the third study, the lowest elevation sampled was relatively high (430 m; Fisher 1999a), leaving open the possibility that the lowest elevations might still be the most diverse.

Just as for tropical regions, temperate regions also often show negative relationships between various measures of ant diversity and elevation, whether for eastern USA (Cole 1940; Lessard *et al.* 2007; Sanders *et al.* 2007b), Austria (Glaser 2006), or Japan (Ito *et al.* 1998). Three exceptions to this pattern are gradients for which low elevations are dominated by (or at least include) arid biomes (Botes *et al.* 2006; Sanders 2002; Sanders *et al.* 2003b). In each of these cases, ant diversity exhibits more of a mid-elevational peak, an observation to which we will return to later. These exceptions notwithstanding, the relatively consistent linear relationship between any of the variety of measures of ant diversity and elevation contrasts with results for the elevational diversity gradient literature in general (Rahbek 2005; Rahbek *et al.* 2007), where a large percentage of invertebrates and other taxa show peaks of diversity at mid-elevations.

3.3 Explanations for latitudinal and elevational gradients in ant diversity

Patterns of diversity of ants along elevational and latitudinal gradients are both strong and relatively consistent. The key question then is why do these gradients exist? What causes the decline in ant diversity with elevation and latitude? There are no fewer than 30 hypotheses to explain patterns of diversity in space (Clarke and Gaston 2006; Evans *et al.* 2005; Willig *et al.* 2003) and additional hypotheses continue to be put forth (surveyed in Dunn 2008b). Fortunately, a smaller number of hypotheses have been the focus of recent research (Willig *et al.* 2003). We focus on six of those hypotheses here, but acknowledge that essentially none of the more than 30 explanations for large-scale patterns of diversity have been conclusively rejected (Dunn 2008b). Three of the six hypotheses on which we will focus are related to speciation and extinction differences among regions: the hypotheses of temperature-dependent kinetics (Allen *et al.* 2002; Rohde 1999); effective evolutionary age (Pianka 1966); and geographic area (Rosenzweig 1995). One hypothesis relates most directly to differences in extinction rates among regions: species-energy theory (Srivastava and Lawton 1998). Finally, two hypotheses, the geometric constraints models (GCMs) (Colwell *et al.* 2004) and niche conservatism models (Wiens and Graham 2005) relate to the consequences of random places of geographic ranges (GCMs) and random or biased placement of species' environmental niches in space (niche conservatism models), respectively. Here we briefly review some of the theory relevant to gradients in diversity, and highlight the extent to which each theory has or has not been tested for ants.

3.3.1 Speciation and extinction rate differences

Several theories predict differences in rates of diversification among regions due to differential rates of speciation or extinction, differences in the time for speciation, and the influences of area on speciation and extinction. We consider each of these bodies of theory in turn.

Speciation rates: temperature-dependent kinetics hypotheses

Temperature-dependent kinetics hypotheses posit that, all else being equal, speciation rates are faster in warmer places than in colder places due to the effects of higher metabolism on mutation rates (Allen *et al.* 2002; Rohde 1999). Evidence for at least one prediction of the temperature-dependent kinetics hypothesis appears to be growing. Several taxa, including butterflies (Cardillo 1999), birds (Cardillo 1999; Cardillo *et al.* 2005; Ricklefs 2006), primates (Bohm and Mayhew 2005), and foraminifera (Buzas *et al.* 2002) have been shown to have higher net diversification rates in tropical latitudes. However, it is unclear whether the magnitude of the difference is sufficient to explain contemporary patterns of diversity (Evans and Gaston 2005). In addition, the differences in net diversification among regions could reflect either differences in speciation rates or extinction rates. Rates of molecular evolution do not appear to differ with latitude (Bromham and Cardillo 2003), but would be expected to if differences in diversification rates with latitude were due solely to differences in the rate of speciation. The all-encompassing Metabolic Theory of Ecology (MTE), like the original temperature-dependent kinetics hypothesis (Allen *et al.* 2002; Rohde 1999), predicts speciation rates should be a positive function of temperature. However, the MTE predicts not only that temperature should influence diversity via its effects on speciation, but also that the effect should be so great and universal that there is a globally consistent slope of the relationship between temperature and diversity, independent of the study taxon or scale of study (Allen *et al.* 2002; Brown *et al.* 2004). This more extreme formulation of the temperature-dependent kinetics hypothesis has not been supported for ants (whose empirical slopes deviate from MTE predictions (Hawkins *et al.* 2007; Sanders *et al.* 2007b), or other taxa more generally, where slopes vary wildly among domains and taxa (Hawkins *et al.* 2007).

Extinction rates

Biodiversity theories typically focus on how differences in speciation rates in space affect net diversification and ultimate diversity gradients, but extinction rates may also vary in space. For example, the increased rate of net diversification in the tropics sometimes observed for birds and mammals may be due to the decreased extinction rates in the tropics rather than increased speciation rates (Weir and Schluter 2007). Speciation rates (as opposed to diversification rates) as determined by recent cladogenesis events, were found to be higher in cold, high-latitude regions suggesting that higher net diversification in the lower latitudes must be due to lower extinction rates (Weir and Schluter 2007), as has been observed from marine bivalve fossil data (Jablonski *et al.* 2006). Very little is known about extinction rates in ants (see Chapter 2), particularly with regard to how extinction rates vary in space. It is, known, however, that a number of ant taxa such as lineages of Myrmeciinae, *Oecophylla*, and other genera have gone extinct from northern habitats of the northern hemisphere (Archibald *et al.* 2006) in line with the idea that extinction rates are higher in cool climates. However, Dunn *et al.* (2009) have recently argued that these extinctions may be more strongly linked to the change in temperature in the northern hemisphere since the Eocene, rather than current lower temperatures per se.

The effective evolutionary age hypothesis

One way that diversity might vary among regions independent of diversification rates is if the time for diversification has differed between those same regions. The concept of effective evolutionary time posits that diversity should be greatest where the time for speciation has been longest, that is in the regions that have been least climatically stable over millions of years (Pianka 1966). While many of those scientists actively studying diversity gradients believe effective evolutionary time to be important (surveyed in Dunn 2008b), it is a difficult model to study empirically and perhaps for that reason has been less thoroughly explored than have other hypotheses, whether for ants or any other taxon. The biggest difficulty is estimating the age of different biomes or climatic conditions, and while palaeoclimate data continue to be gathered, it is likely to be difficult always to estimate the age of different biomes with any accuracy. Perhaps the best (albeit still anecdotal) evidence that the age of biomes, or at least their relative stability, is important to contemporary diversity comes from a recent study by Dunn *et al.* (2009). Dunn *et al.* (2009) found that even after accounting for climate that local ant communities were more diverse in the northern than in the southern hemisphere. This difference is in line with what might be expected if the greater climate change in the northern hemisphere since the Eocene was associated with higher extinction rates.

The geographic area hypothesis

In reality, the diversity of a given region or site is undoubtedly a reflection of speciation rates, extinction rates, and the time between major pulses of extinction. In practice, however, these aspects of diversity tend to be treated separately. The area hypothesis is one of few theories that considers the effects of both speciation and extinction (though it ignores evolutionary age). It argues that speciation rate increases with the area of continents because larger continents are more likely to be subdivided by random processes or dispersal barriers (e.g. rivers and mountains; Rosenzweig 1995; Rosenzweig and Sandlin 1997). In addition, species inhabiting larger habitats and biomes are predicted to have the opportunity of having larger spatial extents for populations and ranges, and hence lower per species rates of extinctions. However, empirical evidence to date for a positive relationship between the spatial extent of species' geographic ranges and speciation rate is limited. Some evidence points toward a unimodal relationship between range size and speciation probability (Gaston 1996; Gaston and Blackburn 1997), though the shape of the relationship undoubtedly depends on the size of the domain being considered.

As has been pointed out by many authors, rewinding the evolutionary tape is impossible, and no experiments could confirm or reject any of these hypotheses about diversification, speciation, or extinction, at least at global scales. Furthermore, the key independent variables pertaining to different theories are correlated such that tropical forests, for example, are hot, have high Net Primary Productivity (NPP), are large and relatively old, and so would be predicted to be diverse under essentially every theory. Given that elevational gradients may mirror latitudinal gradients, carrying out experiments aimed at elucidating the underlying mechanisms and predictions of some of these evolutionary models may be tenable. Yet, it is reasonable to question whether short-term responses of species or taxa to experiments along elevational gradients are sufficient tests of theories meant to apply over global spatial scales and temporal scales of millions of years. Arguably, the theories most testable using experiments along elevational gradients are those related to energy and coexistence, an area of theory broadly referred to as species-energy theory, to which we now turn.

3.3.2 Species-energy theory

In the quest for mechanistic links between climatic conditions and patterns of local (or regional) biodiversity, many recent studies on a variety of taxa have tested a body of theory that is sometimes called species-energy theory (see Clarke and Gaston 2006). Species-energy theory is based on the premise that large-scale patterns of biodiversity occur because diversity tracks patterns in the density of individual organisms and patterns in the density of organisms track the energy available per unit area per unit time (typically g C/area/time) (Hutchinson 1959; Pianka 1966; Wright 1983). Theory posits that where more energy is available there are lower local extinction rates because more individuals of all species can be sustained per unit area by the greater flux of resources. Lower local extinction rates are predicted to, in turn, lead to lower rates of local extinction and hence higher diversity.

If energy availability determines local ant diversity, then NPP, or some proxy of NPP, should be the best predictor of local ant diversity. However, access to resources may be mediated for ants by more than simply the total energy available. The energy available to ants is affected by any variables that affect when and how long ants forage. Because ants are thermophilic, temperature (along with other variables) should limit the resource availability above and beyond the limits imposed by NPP. Depending on the relative importance of temperature and NPP, and whether the region being considered is temperature limited (with temperature likely to limit ant fitness, foraging, and abundance at lower temperatures than those at which it affects plants) or precipitation limited (with precipitation likely to limit plants before ants), the local relationship between ant diversity and climate might be complex and vary among regions (Kaspari *et al.* 2003).

Several studies have now found patterns broadly reconcilable with some form of species-energy theory. Ant species diversity was positively correlated with light availability, length of the growing season, and July temperatures (which are all correlated with each other) in the British Isles and northern Europe (Baroni-Urbani and Collingwood 1976, 1977). Kaspari *et al.*'s studies (2000, 2003) of New World ant communities revealed that ant diversity patterns at smaller grains (plot and transect) were best explained by differences in the density of ant colonies. At the largest grain (with diversity measured as Fisher's α, which accounts for differences in abundances), temperature and ecoclimatic area were the two best predictor variables. Similarly, Sanders *et al.* (2007b) in the Great Smoky Mountains National Park in the eastern United States showed that ant diversity at small grains (m^2) was best explained by abundance and temperature. At progressively larger scales, temperature was a progressively better predictor of diversity than was abundance. Most recently, Dunn *et al.* (2009) found that temperature was the single strongest predictor of local ant diversity when more than 1,000 samples from all over the globe were considered.

However, even if temperature and energy are consistently good predictors of local ant diversity, the relationship remains correlative (and could, for example, also support a link between temperature and speciation and extinction). There was initial enthusiasm for a 'more individuals' mechanism linking energy availability to species diversity (Kaspari *et al.* 2000b), but that enthusiasm has waned of late (Clarke and Gaston 2006) for a variety of reasons. First, it remains unclear why more energy should lead to more individuals of different species rather than just larger population sizes of the species already there. Second, the 'more individuals' hypothesis predicts that diversity increases as a function of increases in the total number of individuals summed across species. In general, energy and diversity seem better correlated than do energy and the density of individuals, the opposite of what would be predicted were energy to directly affect individual density (e.g. Clarke and Gaston 2006). For ants, the links between ant density (individuals per area) and energy are difficult to explore because it is not obvious whether the variable that should be considered as a test of theoretical predictions is nest density or individual density. However, in the one case where the density of ant nests was compared to NPP (Kaspari *et al.* 2003), ground-foraging density was well correlated with NPP as well as temperature and temperature seasonality. Experiments are sorely needed that examine the mechanisms linking climatic conditions and energy availability to ant diversity at a variety of scales. But the good news is that species-energy theory is, unlike other diversity theories, actually amenable to small-scale experimental tests.

3.3.3 Mid-domain null models

Geometric constraint models

Latitudinal and any other large-scale spatial patterns of diversity are ultimately a result of the overlap in the distribution of species' ranges. As a consequence, some have argued that it is not only interesting, but necessary to examine the pattern of diversity that would result were species' ranges to be randomly arranged on the domain being studied. What would happen, for example, if a lineage started in some random position on a continent and then diversified, with new species no more likely to occupy any one part of the continent than any other?

Models that simulate the expected pattern of species diversity in space where species' ranges to be arranged at random are usually called Geometric Constraints Null Models (GCMs) (see Colwell *et al.* 2004). The random process simulated in most of the many null models employed to date is the random arrangement of geographic or elevational ranges along the study gradient (Colwell and Hurtt 1994; Colwell and Lees 2000; Colwell *et al.* 2004). These models have also been generalized to other domains, such as the distribution of riparian plants along rivers (Dunn *et al.* 2006), phenologies of plant species through time (Morales *et al.* 2005) and most recently the seasonal activity patterns (Dunn *et al.* 2007c) and reproductive phenologies of ants (Dunn *et al.* 2007b). Just as a pile of pancakes is almost always higher in the middle of a plate as a consequence of the overlap of pancakes, GCM randomizations almost always produce a hump-shaped pattern in diversity, with the highest number of species at the middle of the domain (i.e. the 'mid-domain effect'; Colwell *et al.* 2004), whether it is one-dimensional spatial domain (examples in

Colwell *et al.* 2004), a temporal domain (examples in Dunn *et al.* 2007b,d]), or a two-dimensional domain (examples in Colwell *et al.* 2004). Thus, humped-shaped patterns of species diversity (e.g. the latitudinal gradient in species diversity) are predicted by GCMs in the absence of spatial variation in climate, biogeographic history, etc. When the domain considered is an elevational gradient, GCMs also predict hump-shaped patterns, with highest species diversity at mid-elevations.

To date, there have been only a handful of tests of the utility of GCMs as explanations of ant diversity along latitudinal or elevational gradients. GCMs alone accounted for 13% of the variation in ant species diversity along a regional elevational gradient in Colorado, 91% of the variation in Nevada, and 37% of the variation in Utah (Sanders 2002). In contrast, there was no support for GCMs in a study of leaf-litter forest ants along an elevational gradient in the southern Appalachians (Sanders *et al.* 2007b), nor was there for continental-scale patterns of ant diversity in North America (Kaspari *et al.* 2004). Thus, of the elevational studies of ant diversity to date, the only datasets that seem to coincide well with null model predictions are those from areas where low-elevation conditions are hot and exceedingly dry, and high-elevation conditions are, like everywhere, colder than low elevations. We suspect that ant diversity patterns will tend to consistently differ from null expectations of GCMs except in those cases where climatic predictions and null models overlap. Such cases represent the coincidence of null model predictions and climatic effects.

Independent of the question of whether ant diversity patterns deviate from the expectations of GCMs is the question of whether GCM null models represent a useful approach to considering patterns of diversity. GCMs have been strongly criticized on the grounds that they are biologically unrealistic and make inappropriate assumptions (Hawkins and Diniz 2002; Hawkins *et al.* 2005; Zapata *et al.* 2003, 2005). Disagreements involving the specifics of GCM simulations and their biological plausibility may, we suspect, be resolved by altering aspects of the models themselves, in particular by considering GCM models as diversification models in which speciation and extinction are considered to be random with respect to geography and history. More realistic models can be created by adding additional realism where useful, whether that realism is more sophisticated models of speciation or aspects of phylogenetic niche conservatism.

Niche conservatism models

Phylogenetic niche conservatism is the idea that species traits evolve relatively slowly, such that related species are more similar to each other than are any two random species. Such similarity includes not only morphological and behavioural traits but also traits related to environmental tolerance and hence the conditions in which a species can live and its geographic range. As such, one can imagine that each lineage begins in some point of origin with a given set of climatic conditions. Over time, new species will arise that will tend to have similar climatic preferences to the original species. Slowly, climatic preferences (niches) will shift, but initially the centre of diversity for the lineage in question will be centred near the niche preferences of the first species. Given enough time, lineages will spread from their original conditions and begin to fill the domain being studied (as they do niche conservatism models will begin to look very similar to geometric constraints models).

Some assumptions of niche conservatism models can be tested. For example, one can examine whether niches tend to be conserved by plotting measures of niche space on a phylogenetic tree. However, niche conservatism models differ (along with GCMs, to some extent) from other diversity models in making no single prediction of what the 'correlates of diversity' should be. Instead, niche conservatism models have lent themselves to simulation and in particular to GCM-like models of diversity patterns. Recent work has explored the patterns of diversity resulting simply from assuming that lineages exhibit niche conservatism (Rahbek *et al.* 2007; Rangel *et al.* 2007). These new models are essentially GCM models into which more biology has been added and such models may represent an important new step forward in considering patterns of diversity. While these simulation approaches seem a productive avenue for exploring the consequences of niche conservatism, they may also offer an important means to better understand the predictions and assumptions of diversity theories more generally.

3.4 Diversity of higher taxonomic levels and phylogenetic diversity: an example

To date, studies of ant diversity patterns have focused on species diversity. Ultimately, our understanding of ant diversity will likely depend upon additional insights about diversity at higher taxonomic and phylogenetic levels. What are the spatial patterns of species, genus, and subfamily diversity and how are they related? Do similar processes explain patterns at different taxonomic scales? Do phylogenetically independent clades track environmental variables similarly? Answers to these questions may help us understand not only patterns of diversity, but also the evolutionary processes underlying the patterns.

The study of diversity patterns of higher taxa, such as genera, has a long history in the palaeontological literature where complete sampling and identification of species are problematic (reviewed in Willig *et al.* 2003). Generic and higher taxon diversity has also been useful for diverse modern groups such as plants (Qian 1998) and copepods (Woodd-Walker *et al.* 2002) and, here, ants. Although it is possible to compare the diversity of sampled plots across biomes, knowing the diversity for larger grain sizes for comparisons that include the tropics remain a formidable challenge, because a large proportion of ant species remain undescribed. However, genera are relatively well described, even in areas such as the tropics where study is difficult and diversity high. Although new genera remain to be found (see, for example the recent discovery of a new ant subfamily and genus, Rabeling *et al.* 2008), the discovery of a new ant genus is now relatively rare. Furthermore, new genera are likely to be geographically rare and consequently have little influence on overall diversity patterns. Therefore, if ant species diversity is correlated in space with generic diversity, as is the case for many other taxa (e.g. trees; Enquist *et al.* 2002), we may be able to use generic diversity as a proxy for species diversity in examining truly global patterns of regional diversity and endemicity. See Box 3.2 for a description of ant databases that may facilitate such analyses.

A strong correlation between generic and species diversity bodes well for using patterns of generic diversity as a proxy for patterns of species diversity. In the Great Smoky Mountains in the eastern United States, genus and species diversities were highly correlated ($R^2 = 0.89$, Ordinary Least Squares Regression) (data re-analyzed from Sanders *et al.* 2007b). A similarly strong relationship exists between species diversity and generic diversity in well-studied political regions in North America (exclusive of México) and Europe (Guénard *et al.*, unpublished data; see Figure 3.1). Even if species diversity and generic diversity are not always well-correlated (as is likely to be the case in Australia, where generic diversity is highest in the tropics, but species diversity appears highest in drier regions; Shattuck 1999), the genus may represent an interesting taxonomic level to explore in its own right.

One of us (B. Guénard) has compiled distribution data on ant genera for the better-studied countries and political regions of the world (described in more detail in Figure 3.1). This work enables an exploration of whether patterns of diversity differ for ants

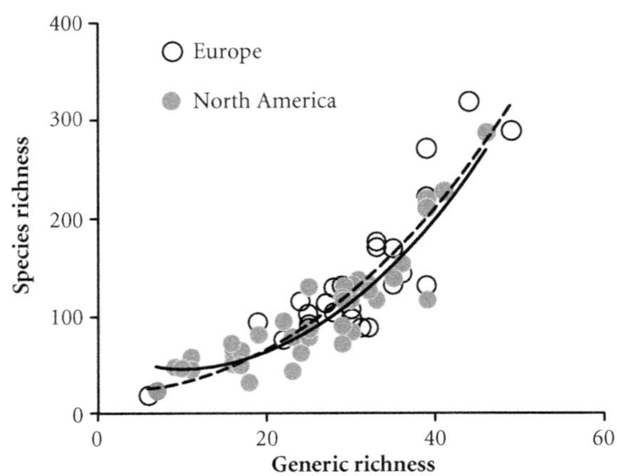

Figure 3.1 Genus and species diversity of ants for North American (exclusive of México-grey circles and solid line) and European political regions (white circles and dashed black line). Each point is a political region (e.g., country, state). $R^2 = 0.854$ ($y = 0.1757x^2 - 3.7254x + 66.93$) for an exponential function for North America. $R^2 = 0.803$ ($y = 0.135x^2 - 0.776x + 26.18$) for an exponential function for Europe.

Box 3.2 The emergence and utility of online global ant databases
Nathan J. Sanders

Most of what is known about global patterns of diversity is based on birds, plants, and to a lesser extent, reptiles and butterflies. However, it is little appreciated that ants were among the first taxa to be examined through a macroecological lens. Wheeler (1910) wrote: 'The great importance of ants in the study of geographical distribution has not been overlooked by students of this fascinating subject', and Gregg (1963) discussed the 'macroecology' of Colorado ants. Despite this head-start in macroecological studies, quantitative studies of global patterns of ant diversity are still in their infancy. Kusnezov (1957), Jeanne (1979), Kaspari *et al.* (2000b), and Ward (2000) were pioneers in the study of global ant diversity. With the exception of Kusnezov, each relied on his own field collected data. However, ecologists working on other taxa often take advantage of massive databases that are, for the most part, freely available online. Could ant ecologists do the same? The answer seems to be yes, as a number of exciting databases provide freely available data on the distribution and diversity of ants.

diversity from over 3,000 ant assemblages around the globe (Figure 3.2.1; www.antmacroecology.org). In addition, the database also includes whether the assemblage has been invaded by exotic ants, whether it has been disturbed (either naturally or otherwise), and the type of habitat the assemblage is situated in. It is possible the Global Ant Community Database of Dunn and colleagues is the largest of its kind, for any taxon.

The Global Biodiversity Information Facility (GBIF) (www.gbif.org) is intended to be a georeferenced database for the diversity of all life on the planet. There are at least 187,401 records of ants (locations at which a species has been collected) for species in >280 genera. To my knowledge, only one study on ants (Geraghty *et al.* 2007) has used information from GBIF. Though the distribution of the data is spatially and taxonomically patchy, it is possible to estimate the northern and southern latitudinal extent of species ranges, especially in well-sampled areas, as was done by Geraghty *et al.* (2007) in their study on the relationships among body size, colony size, and range size in ants of eastern North America.

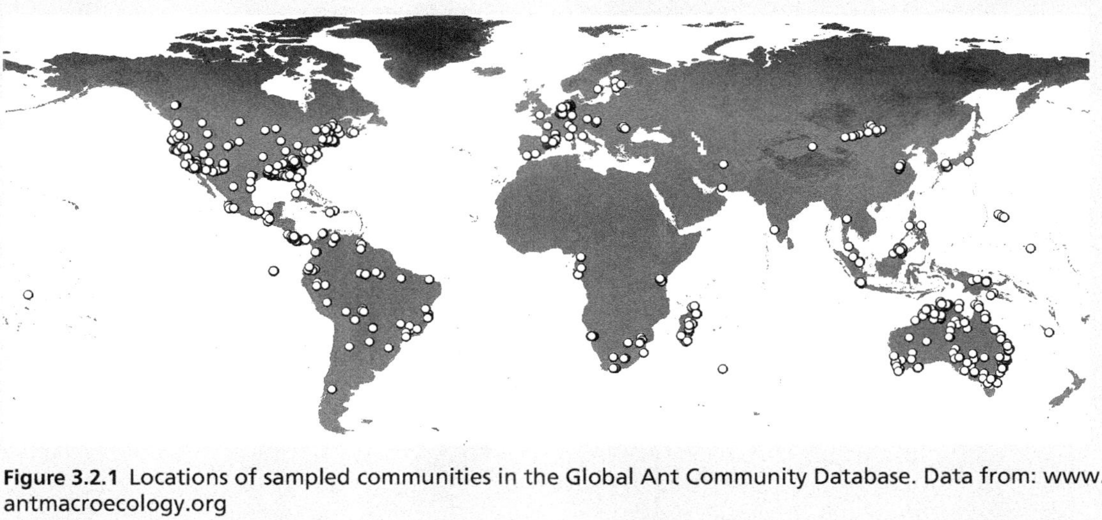

Figure 3.2.1 Locations of sampled communities in the Global Ant Community Database. Data from: www.antmacroecology.org

The Global Ant Community Database that Dunn and colleagues have compiled will no doubt be useful (see Chapter 3). It is a growing database that includes information on ant bio-

The web site AntWeb (www.AntWeb.org) 'provides tools for exploring the diversity and identification of ants' with the ultimate aim of describing and providing a high resolution

continues

Box 3.2 continued

photo and distribution map of every ant species. This web site contains a pioneering online database linking actual specimen localities with a Google Earth ™map so that one can easily visualize the distribution of a particular ant species. Moreover, antweb.org allows one to see detailed information on the species occurring in particular geopolitical units, such as the ants of Madagascar or the ants of Mississippi.

The www.antbase.org is a warehouse of information about ant systematics. Some of the key goals of antbase.org are to provide up-to-date information about ant systematics and provide access to all of the literature on ant systematics. An interesting component of antbase.org is that it provides an up-to-the-minute estimate of the total number of recorded ant species (12,591 as of 17 September 2009).

Very recently, www.plazi.org has been launched as follow-up of antbase.org. It allows for searching full text for published ant descriptions. Furthermore, distribution data from original publications now directly feed into GBIF, and thus can be harvested for global biodiversity studies.

At more regional scales, Klingenberg and Verhaagh (2005) provide a valuable compilation of a growing number of web sites covering mainly regional-scale fauna studies and species lists. With more databases on the horizon, they finally quantify what Wheeler (1910) and Gregg (1963) knew so long ago — ants are an ideal taxon with which to document and understand global patterns of biodiversity and macroecological processes.

relative to other taxa, and also whether patterns of ant diversity along these large gradients differ among taxa. Ant generic diversity decreases in diversity with latitude (see Figure 3.2), just as does species diversity in local plots. Generic diversity is highest in the broad latitudinal band of the tropics in the Americas, in Africa and Europe, and in Australasia and is higher in the southern than the northern hemisphere for comparable latitudes. There is a dip in diversity in North Africa commensurate with what might be

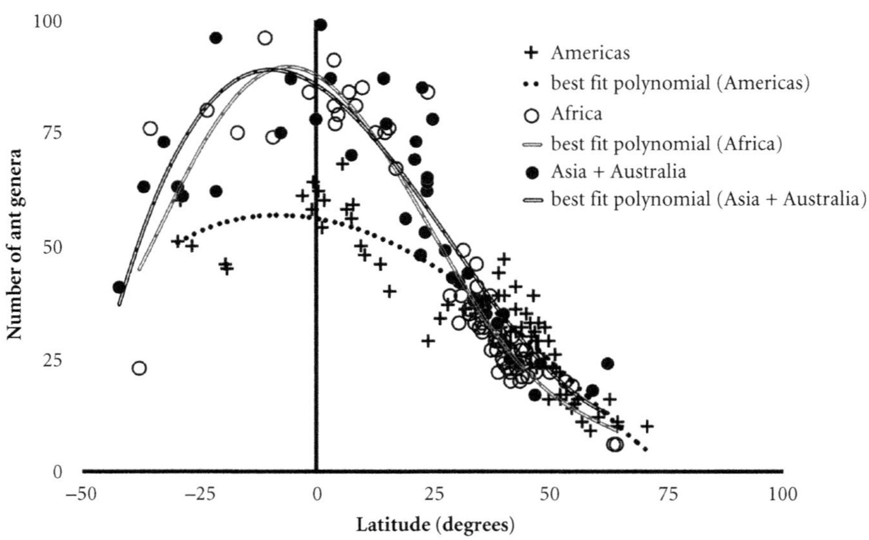

Figure 3.2 Generic diversity of ants in each political region in the Americas, Asia + Australia, and Africa plotted against latitude. Lines show best-fit polynomials.

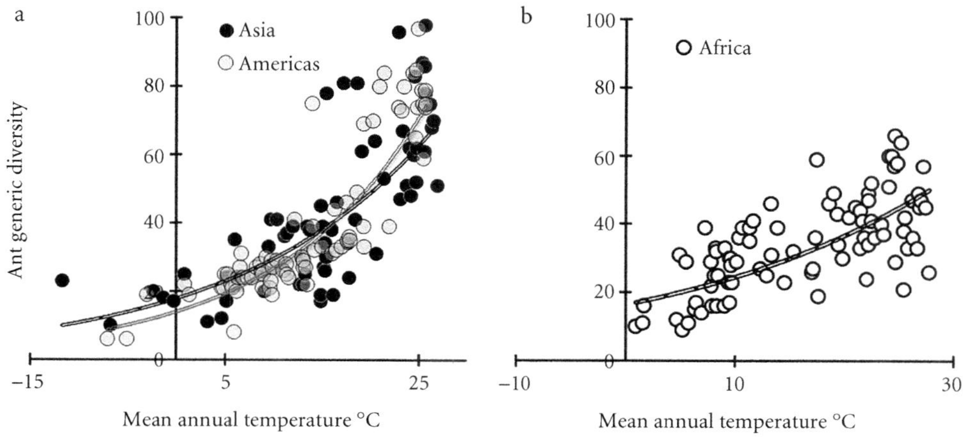

Figure 3.3 Generic diversity of ants in each political region in (a) the Americas, Asia + Australia, and (b) Africa plotted against mean annual temperature (°C). Lines show best-fit polynomials. Note the relatively lower diversity for a given temperature in Africa than in Asia or the Americas. Temperature data were derived using ArcView ver.9.1 to extract mean monthly temperature for each point occurring within the boundaries of each political entity at 0.5° degree resolution.

expected, given the extreme climatic conditions for that region (see Figure 3.2). The overall patterns are nearly identical when differences in area are corrected for by plotting the residuals of the area × genus diversity regression against latitude instead of raw genus diversity (see Figure 3.2). The decline in diversity with latitude appears asymmetrical, as has been shown for some other taxa (reviewed in Chown *et al.* 2004) and as was also recently shown for local ant diversity (Dunn *et al.* 2009), where diversity is lower in the northern than in the southern hemisphere.

When generic diversity is regressed on mean annual temperature, most of the global variation in generic diversity is accounted for (see Figure 3.3). While the relationship between temperature and diversity is very similar for the Americas and Asia, high-temperature sites are less diverse in Africa than in other regions (see Figure 3.3). In part, this might be due to the relative dryness of Africa. However, it appears even tropical countries with tropical forest in Africa are less diverse than their counterparts in Asia or the Americas. Diversity is also lower in Africa relative to the Americas and Asia in amphibians (Buckley and Jetz 2007), and birds (C. Rahbek, personal communication), but not for plants (Kreft and Jetz 2007), even after accounting for differences in climate.

Plotting the most diverse higher groups of ants separately (Myrmicinae, Formicinae, and Poneroids) shows that these groups differ both in their absolute diversity (as would be expected) and also in their patterns of diversity relative to temperature. The subfamily Myrmicinae contains the highest number of genera, and drives the overall pattern of generic diversity (as it probably does in most studies of ant species diversity along gradients). The Poneroids, in which most species forage at higher trophic levels (see Figure 3.4) show more peaked patterns of diversity with latitude and represent a much higher proportion of generic diversity at higher temperatures. The most generalist ant groups considered (Formicinae in Figure 3.4, and Dolichoderinae, not shown) show weak and more inconsistent relationships between diversity and temperature. Regardless of the cause of the differences among taxa, it is clear that although temperature seems to be consistently important, its effect is not identical among taxa. From the perspective of naturalists, this is perhaps an obvious result, but from the perspective of theory, this difference follows from several, but not all hypotheses, meant to explain global patterns of diversity. Under metabolic theory, for example, the effects of temperature on kinetics and ultimately speciation are not expected to differ among ectothermic taxa or trophic levels.

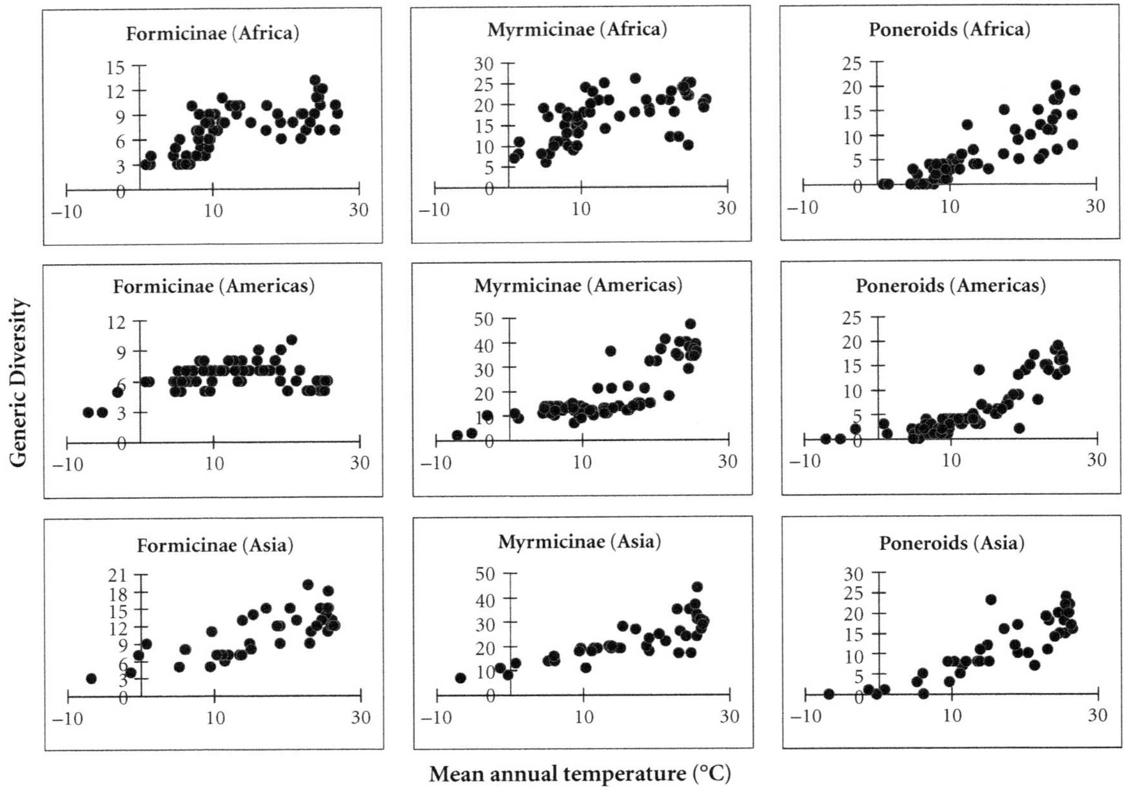

Figure 3.4 Relationship between generic ant diversity and mean annual temperature for three bioregions of the world, by the major higher ant groups (Formicinae, Myrmicinae, and Poneroid). Each point represents the generic ant diversity in some political unit (state, province, or country). Scale bars differ among panels.

3.5 Patterns of range size

Ultimately, one would hope to be able to predict not just patterns of species diversity, but also patterns of endemicity and range size. With respect to conservation, it is often not species diversity per se we care about, but rather the diversity of rare, evolutionarily unique, or regionally restricted species (see Chapter 4).

To date, studies of range size and endemicity patterns in ants have focused on analyses of Rapoport's rule. Rapoport's rule asserts that range size decreases with elevation (Stevens 1992) and latitude (Stevens 1989). Because conditions at high latitudes and/or elevations are more seasonal than those at lower latitudes and/or elevations, Stevens (1989) predicted that species at high latitudes and/or elevations would have large ranges as a consequence of their necessarily broader environmental tolerances. Support for the rule as an explanation for such patterns is mixed at best (Colwell and Hurtt 1994; Gaston *et al.* 1998; Rohde 1996; Taylor and Gaines 1999).

Two studies of ants have considered patterns in geographic range size among species and both have been limited to parts of North America. In both the Great Smoky Mountains National Park (Geraghty *et al.* 2007), and in Colorado, Nevada, and Utah, ant species at lower elevations tended to have smaller ranges than those species at higher elevations (species with larger ranges also have higher elevational range midpoints; Sanders 2002) in accordance with Rapoport's rule predictions. Neither study explicitly considered whether Rapoport's rule would explain patterns of diversity, and too few studies have examined the causes of variation in the range sizes of ant species for us to say whether the results are general. In the Great Smoky Mountains study (Geraghty *et al.* 2007), the size and position of elevational and latitudinal ranges were correlated (e.g. species with big elevational ranges had big latitudinal ranges). While the idea that latitudinal and elevational ranges should be correlated makes intuitive sense, it has seldom been tested.

An alternative approach to considering patterns of range size and rarity is to explicitly model the correlates of diversity of small-ranged or rare species. For other taxa, such as birds (Jetz and Rahbek 2001) the correlates of diversity for narrow- and wide-ranging species are very different. Because widespread species drive large-scale patterns of diversity, simply considering total diversity can mask patterns of rare and/or small-ranged species diversity, unless the two are considered separately. Any of a variety of processes might lead to differences in the diversity patterns of wide- and narrow-ranged species. Recent work (Dynesius and Jansson 2000; Jansson and Dynesius 2002) suggests climatic history, for example, may have a much stronger impact on the distribution of rare species than it does on overall patterns of diversity. We know of only a handful of studies of patterns of diversity in rare ant or small-ranged species. Diversity of rare ant species in Great Smoky Mountains National Park in the eastern United States was concentrated in the warmest sites, just as for species diversity overall (Lessard *et al.* 2007). Similarly, along an elevational gradient in Costa Rica, nearly all small-ranged ant species were found at low elevations (Colwell *et al.* 2008). In the Philippines, only 2 of the 77 species encountered were found at either of the two high elevations sampled (1,550 m and 1,750 m) and only one of those species was restricted to those elevations (Samson *et al.* 1997). In Madagascar, a high proportion of the species found at the highest elevation sites (1,800–2,000 m) was found only at those sites (50% compared to 24% at the lowest elevation site), but the absolute number of species restricted to the highest elevation was still small (10 species of 273 total; Fisher 1996). In general, rare and narrow-ranged ant species seem much more likely to be at low than at high elevations. These patterns may not hold, however, for other regions and deserve better testing.

One final approach to considering Rapoport's rule, or more generally, patterns in species distributions, would be to directly compare the niches, and in particular one aspect of the niche, the climatic tolerance of species with latitude or elevation. The idea that niche sizes might be smaller in the tropics pre-dates Rapoport's rule by many years (Klopfer and Macarthur 1961) but remains untested. Are, for example, species in tropical forests constrained to live in a narrower range of temperatures than are species in temperate forests? For the moment, this question remains unresolved in general, not just with regard to ants.

3.6 Patterns of life history and morphology

There is a long history of the study not only of variation along gradients in diversity but also in traits and their distribution. To date, the limited work on variation in ant life history has focused on colony and body size, but we also consider here other traits, such as the prevalence of social parasitism, that might vary along gradients.

One of the best-explored patterns in a life-history trait along gradients is Bergmann's rule (James 1970), which stems from the observation that body size tends to increase with latitude, and by analogy, elevation. Bergmann's rule was first applied to variation in body size among species along gradients, but has also been considered within species. Just as for the latitudinal diversity gradient, in practice Bergmann's rule refers primarily to the pattern of body size along gradients (larger body size at higher latitudes and elevations), rather than to a particular mechanism underlying that pattern. The first question we consider is whether ants show the pattern referred to as Bergmann's rule. Unlike for solitary animals, for ants and other social insects, body size can vary along gradients either as a function of the body size of individual (e.g. a worker ant) or as a function of the 'body size' of colonies, where colonies are considered to be the scale at which investment decisions are made.

3.6.1 Colony size

Analyses of relationships between colony size and latitude reveal mixed results. In the Americas, ground-dwelling tropical ant species have smaller colonies than do temperate ant species (Kaspari and Vargo 1994). Kaspari and Vargo (1994) argue that larger colony size in colder environments facilitates overwintering ability because larger colonies are thought to have a greater ability to withstand attrition due to winter mortality. However, their results are reconcilable with alternative explanations. Perhaps, for example, ant species in the tropics are more likely than temperate ant species to occupy the litter, and for any of a variety of reasons, leaf-litter lifestyle necessitates smaller colonies. Large colonies are at least not a requirement of cold climates since among those species that inhabit the coldest realms (Francoeur 1997), colonies often include no more than tens of individuals. If larger colonies are advantageous as buffers to winter mortality or are in any other way more advantageous at higher latitudes and elevations, then one would expect species found in colder climates would have larger colonies, but also that within species, populations from colder climates would have larger colonies. However, there is no relationship between latitude and colony size in the holarctic ant, *Leptothorax acervorum* (Heinze *et al.* 2003), and work on a different *Leptothorax* species has shown no relationship between winter mortality and colony size (Herbers and Johnson 2007).

With results limited to one study, the relationship between elevation and colony size remains ambiguous. Geraghty *et al.* (2007) found no correlation between colony size and elevation among ant species in the Great Smoky Mountains National Park. A strong effect of elevation on colony size might not be expected within temperate regions, if the primary driver of the latitudinal colony size gradient is variation in the proportion of species living in the litter. An alternative explanation for the discrepancy between elevational and latitudinal studies has to do with differences in species turnover. In the Great Smoky Mountains, for example, the pattern of species distribution and diversity with elevation is nested (Lessard *et al.* 2007) such that the species found at high elevation are, for the most part, a cold-tolerant subset of those at low elevations. In contrast, as one goes north, even from the Great Smoky Mountains, one encounters a new set of species. Diversity is not typically nested with latitude. Thus, while the northern latitude endemic species may adapt to local conditions, the species found at high elevations in the Great Smoky Mountains are also found at low elevations and so likely have selection for one colony size at high elevations and another colony size at low elevations. Whether these distinctions generalize to other regions is an open and interesting question.

3.6.2 Body size

In addition to variation in colony size, the size of individual ants might also be expected to decrease with increasing temperature along the climatic gradients for which elevation and latitude are proxies. Just as colony survival may depend on the adaptiveness of a particular nest size for a particular climate, survival of individual ants may also vary with climate as a function of body size, with consequence for spatial patterns in body size along gradients. Results for body size, like for colony size, suggest ant species tend to be larger at higher latitudes in the northern hemisphere (Cushman *et al.* 1993). This pattern is mirrored by variation in individual body size within species. For example, colonies of *Leptothorax acervorum* at higher latitudes had larger individuals than those at lower latitudes (Heinze *et al.* 1998). Just as for colony size, a variety of mechanisms related to phylogenetic history, migration ability, and starvation resistance might explain this pattern. An interesting next step would be to conduct laboratory studies on the relationship between body size and survivorship for a large number of species.

3.6.3 Other ant-related life-history patterns

Reproductive strategies can vary with latitude and elevation. Such variation is perhaps best characterized for marine invertebrates (Rohde 1999), and birds (Lack 1947; McNamara *et al.* 2008) in which both species and individuals at higher latitudes produce fewer offspring than those at lower latitudes (McNamara *et al.* 2008; Rohde 1999). For birds, it has been suggested that smaller clutch sizes (with more investment in each offspring) result from the greater seasonality (and hence urgency for rapid development) in high latitudes. We know of no work comparing egg production by queen ants along gradients, but the study of reproductive investment along gradients in general would be an interesting area of inquiry.

However, apparent empirical spatial patterns in rates of social parasitism (any of a variety of systems whereby individuals of one ant species take advantage of colonies of another ant species during some part of their life cycle) have long attracted attention. Most of the more than 300 social parasite species (Buschinger 1990, 1991) known today are found under temperate climates and at high elevations. Many of these social parasites, especially among inquilines, have been described from extreme environments such as mountainous or arid regions. This observation has led to the idea that low temperatures found in higher altitude or latitude could favour the emergence of social parasitism, perhaps because just as has been suggested for benthic invertebrates (Rohde 1999), mortality associated with independent colony foundation is greater at those latitudes. Some have argued that social parasitism may be far more common in the tropics than is now appreciated (Feitosa and Brandão 2008; Wilson 1984), but the overall pattern in social parasitism seems unlikely to change.

As latitude increases, plants flower later, are reproductively active for a shorter period of time, and are older at the time of first reproduction (Olsson and Agren 2002). Similar patterns in reproduction might be expected for ants along elevational or latitudinal gradients. There is some suggestion that in tropical forests reproductive flights of ants are more continuous than they are in temperate forest ecosystems (Dunn *et al.* 2007b; Kaspari *et al.* 2001a,b). Further, within temperate regions, reproductive flights are more continuous at lower elevations than at higher elevations (Dunn *et al.* 2007b). These two patterns suggest that there may be general patterns in reproduction along environmental gradients, but we have barely begun to explore. A study comparing reproductive strategies of a suite of ant species along an elevational or latitudinal gradient would contribute greatly not just to our understanding of ants, but also to the responses of species to gradients more generally.

3.7 Consequences of gradients in ant communities

It has been suggested that rates and types of interactions might differ with elevational and latitudinal gradients. Biotic interactions have been speculated to be more important in low latitudes and abiotic drivers more important in high latitudes. In the broader literature, support for such gradients is strong in the few cases where they have been examined (reviewed in Rohde 1999). Perhaps the strongest evidence for gradients in biotic interactions in terrestrial organisms comes from ants, where predation rates and interactions with some mutualists appear to vary along gradients. Predation rates in forests, particularly in the forest canopy (Jeanne 1979) decrease with latitude (Jeanne 1979; Novotny *et al.* 2006), possibly due to concomitant decreases in ant diversity with latitude. Figure 3.5 suggests such a relationship for ground-foraging ants in forests.

Seed dispersal of myrmecochorous plants also changes along environmental gradients across which ant communities vary (Gove *et al.* 2007; Zelikova *et al.* 2008). Finally, the decline in the

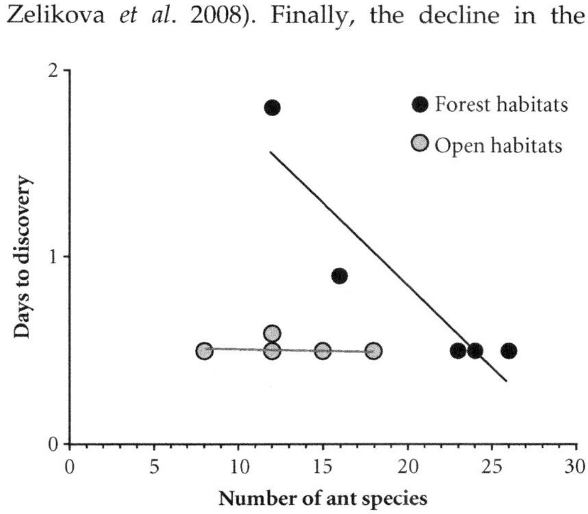

Figure 3.5 Days to discovery of baits (larval wasps) by ants from samples along a latitudinal gradient. Figure derived from data in Jeanne (1979). Each point represents a site where baits were observed.

proportion of treehopper species dependent on ants with elevation is thought to be due to the decline in the diversity and abundance of ants with elevation (Olmstead and Wood 1990). Other ant mutualists might be expected to show similar patterns.

3.8 Climate change and gradients of diversity

What can theory and empirical patterns tell us about the potential for change in patterns with climate change? Given that ant communities and species are influenced strongly by climate, and in particular temperature, ants may be particularly predisposed to respond directly to the temperature changes associated with climate change. However, to date, studies that focus on the relationship between climate change and ant communities are scarce and restricted almost exclusively to the future distributions of invasive species (Levia and Frost 2004; Morrison *et al.* 2005; Chapter 13). No studies have yet examined whether historical shifts have occurred in ant distributions with climate change (see reviews for other taxa in Hughes 2000; Parmesan and Yohe 2003), though the long history of ant collecting in many of the regions that have experienced the greatest climate change makes ants a good candidate for such work. However, our knowledge about the patterns in ant communities along gradients allows us to make strong *a priori* predictions as to what might be expected in ant communities as climate changes. For example, to the extent that ant diversity, abundance, and ecological roles all increase with temperature, in many regions increases in temperature with climate change seem likely to increase the local diversity, abundance, and relative importance of ants.

That the abundance of some ants will increase in a warmer world seems very likely. Less clear is which ant species and lineages will become more abundant and widespread. For example, in regions like southwestern Australia and South Africa, which currently have relatively wet and cool winters but are surrounded by larger more arid habitats, the regional species pool may account for many of the 'new' colonists under warmer conditions. Conversely, for other regions, particularly islands and mainland habitats in which there are no adjacent species pools corresponding to future climatic conditions, source pools are less clear and may be dominated by invasive and introduced species (Colwell *et al.* 2008). The tip of Florida, in the United States, is a small patch of very isolated subtropical habitat, in which reside no fewer than 50 introduced ant species (e.g. Deyrup 2003). With climate change, subtropical conditions are predicted to expand through the south-eastern United States and the introduced species at the tip of Florida may spread. If invasive species are able to colonize new, warmer habitats faster than migrating native species they may preempt the arrival of natives or, at the very least, reduce their ability to track shifting climates. Thus, although our most general prediction is that higher temperatures will increase ant abundance, diversity, and impacts, locally there are regions where such changes will be pre-empted or altered by invasive species.

Nailing down which species will be negatively affected by to climate change is difficult. Studies of vertebrates suggest that endemic species found mostly in cool climates (at high elevations and latitudes) are likely to be most affected by climatic warming (McDonald and Brown 1992). However, for ants, at least in temperate systems, there are few high-elevation endemic species (see the earlier discussion; Lessard *et al.* 2007). Even in tropical systems, a smaller percentage of ants than other taxa seem confined to high elevations (comparisons in Colwell *et al.* 2008; Samson *et al.* 1997). If the elevational gradients studied to date are indicative, ant populations may be less at risk of extinction at high elevations than is the case for other taxa. Detailed studies of not just the diversity of ants along elevational and latitudinal gradients, but also the patterns of range size would help to understand the relative risk of high-elevation ants. Low-elevation tropical species also may be at risk due to climate change because as tropical habitats warm, such species will find themselves in much smaller areas of habitable forest (mid-elevations occupy a smaller area than do low elevations, in general, since area declines with elevation) (Colwell *et al.* 2008). Key to this prediction is the idea that tropical species, at least on average, are poorly able to tolerate conditions warmer than those they currently inhabit. Physiological tests suggest that tropical insects do tend to have narrower thermal tolerances than do temperate species (Deutsch *et al.* 2008), though this pattern deserves much better exploration.

As species distributions shift with climate change, trait groups should also shift. If high-latitude and high-elevation species tend to have large bodies and large colonies, the first species to be affected by climate change can be predicted to have the same traits (see Bergmann's rule in Section 3.6). Similarly social parasites, because of their cool-climate bias and relative rarity may also be at disproportionate risk. The ecological consequences of ants in communities may also shift. At high elevations and latitudes, expansion of ant populations and communities may dramatically affect predation and other processes mediated by ants, in general accelerating them. At low elevations, the effect might be more idiosyncratic.

Ultimately, understanding the effects of climate change on ants will be an important test for our understanding of the relationship between ants and environmental gradients more generally. The consequences of gradients in environmental conditions have long been interesting to ecologists in general and ant ecologists in particular. The practical value of such understanding has, however, been limited. Climate change represents an opportunity in which macroecological relationships can prove useful to conservation and application. It may be that in testing the utility of our current understanding of gradients and ants, we also come a great deal further in understanding the robustness of that understanding.

3.9 Future directions

In traditional models of science, progress ultimately relies on rejecting hypotheses, and the hypothesis with the most support might be arrived at through winnowing of possibilities. To date, few or perhaps no serious explanations for gradients in diversity or other attributes of ecological communities have been completely rejected. The reasons for these difficulties are straightforward. Most studies of patterns in community composition have been correlative. Most of the existing explanations for patterns of diversity, for example, make similar predictions with regard to spatial patterns of diversity – that diversity should be high or highest in the tropics, or that diversity should be positively correlated with temperature. As such, rejecting hypotheses may depend on testing secondary predictions of the hypotheses. Unfortunately, for all but a few of the hypotheses, secondary predictions are poorly resolved or variable. For example, one version of the temperature-dependent kinetics hypothesis predicts a precise slope for the relationship between species richness and temperature (Brown *et al.* 2004). The universality of that slope has been rejected (Hawkins *et al.* 2007; Sanders *et al.* 2007b), but there remain other versions of the temperature-dependent kinetics hypothesis that do not assume a universal slope. So even when secondary predictions can be clearly rejected, hypotheses do not disappear, they just change.

We propose that there are three key approaches to better understanding not just the patterns, but the causes to gradients in animal communities, whether for ants or any other taxon. First, we need a better understanding of how patterns of composition of communities have changed through time and how patterns of diversification and, separately, speciation and extinction, vary in space. Well-sampled, dated phylogenies will help both of these endeavours, particularly if they can be coupled with analyses of changes in the ant fossil record. A second important approach will be to simulate patterns of diversification and the diversification of traits under the assumptions made by different hypotheses (Rangel *et al.* 2007). Such models help to understand what patterns of diversity and community structure are conceivable given different hypotheses and may make clear that some hypotheses are unable to produce observed patterns of diversity given realistic parameters. In addition, the process of simulating diversification helps to make obvious which hypotheses do and, in the far more common scenario, do not, make explicit predictions about extinction rates, speciation rates, and rates of dispersal. Finally, we suggest that, under some circumstances, experimental manipulations of potential driving variables might help elucidate the underlying causes of some diversity gradients.

3.10 Summary

Generally, ant diversity tends to decline with latitude and elevation. To date, results for latitudinal studies, elevational studies, and an ongoing global study of genera can be reconciled with the notion that temperature limits resource availability, access to resources, and ultimately, diversity of ants. However, critical mechanistic links remain missing, including tests of the relationship between abundance and extinction. In addition, energy variables might also have direct effects on speciation and hence regional species pools, but this possibility remains untested. Because the recent continental-scale studies of ant species diversity have focused on relatively small sampling grains (with the exception of the generic analysis herein), the importance of area as a determinant of large-scale diversity patterns also remains poorly tested. Work on elevational gradients suggests area may indeed have strong direct (Sanders 2002) and indirect (Romdal and Grytnes 2007) effects on ant diversity (though the effect on generic diversity at big spatial grains is minimal). In general, and perhaps more so than other taxa, we find ant diversity patterns deviate from the expectations of geometric constraints models due to aspects of climate associated with energy. Additionally, broad-scale patterns in body size and colony size, and their underlying causes, are in need of more attention.

Understanding the causes of those patterns is more difficult whether for ants or for other taxa. Phylogenetic work and experiments may help us to better understand causal mechanisms. In the meantime, observed patterns in ant communities allow us to make clear predictions regarding the consequences of climate change for ants. In many areas, ants are likely to form a larger part of the overall fauna in terms of biomass. Native ant species may also become more locally diverse, though this may be pre-empted by invasive species. Ants appear likely (because of their thermophilic bias in distribution) to be less negatively influenced by climate change than other taxa.

Acknowledgements

Thanks to Sean Menke, Leanne Alonso, J. P. Lessard, Matt Fitzpatrick, and Monica Sanchez for helpful advice. Two anonymous (and diligent) reviewers helped to provide insights and additional references. RRD, NJS, and MW were supported by a DOE-NICCR grant, DOE-PER grant DE-FG02-08ER64510 and funding from Discover Life. Benoit Guénard was supported by a fellowship from the Keck Center for Behavioral Biology at NCSU.

Chapter 4

Ant Conservation: Current Status and a Call to Action

Leeanne E. Alonso

"People need insects to survive, but insects do not need us. If all humankind were to disappear tomorrow, it is unlikely that a single insect species would go extinct, except three forms of human body and head lice.... But if insects were to vanish, the terrestrial environment would soon collapse into chaos".

—E. O. Wilson (2006)

4.1 Introduction

Ants are one of the most ubiquitous, widespread, and abundant groups of animals on earth. Ant biomass is one of the highest of any taxonomic group and their diversity rivals many other insect groups, with over 12,500 described species and many more yet to be discovered and described (Chapter 1). Ants play critical roles in every terrestrial ecosystem: recycling nutrients, dispersing seeds, engaging in mutualistic associations with other organisms (Chapter 6), and serving as predators (Philpott and Armbrecht 2006) and scavengers. Their nests and underground activities have such broad effects on other organisms that they often act as ecosystem engineers (Folgarait 1998). Many ant species, particularly army ants (e.g. *Eciton* and *Dorylus*), can be considered top predators because they exert a significant impact on other arthropod populations (O'Donnell *et al.* 2007).

Many of the ecological roles that ants fill are directly or indirectly beneficial to humans, including natural pest control (e.g. crop herbivores: Perfecto 1991; Philpott and Armbrecht 2006), soil aeration (e.g. Gabet *et al.* 2003), and nutrient cycling (e.g. Wagner *et al.* 2004). The economic value of ecological services provided by insects in the United States, to which ants contribute a large proportion, is estimated at over $50 billion per year (Losey and Vaughan 2006). Thus, economically and ecologically the conservation of ants and the services they provide are of high importance.

This chapter reviews the current status of ant conservation as well as the threats and challenges. As a call to action to the myrmecological and conservation communities, I identify priority areas and species needing conservation, then outline strategies for the conservation of ants and the inclusion of Formicidae into broader biodiversity planning. The chapter concludes with an outline of the key actions needed to improve ant conservation in the future.

4.2 Threats to ants and challenges to ant conservation

Like most taxa, many ant species and populations face a range of threats to their continued survival (see Section 4.4 for characteristics of vulnerability to threats). The most immediate and widespread threat comes from the loss, disturbance, or alteration of habitat. Land-use changes associated with farming, mining, livestock, and urban growth all displace key habitat for many ant species (Chapter 8). Fragmentation studies have revealed that ant species richness and genetic diversity can be affected even in large forest patches of 40 km^2 (Bickel *et al.* 2006; Brühl *et al.* 2003). Nomadic ant species such as army ants in the Neotropics and driver ants in Africa and Asia need large expanses of habitat to find enough food to feed their exceptionally large colonies (Gotwald 1995). Likewise, deforestation and forest fragmentation can cause local extinctions of the Neotropical swarm-raiding army ant, *Eciton burchelli*, and other army ants (Boswell *et al.* 1998; Kumar and O'Donnell 2009) and roads have serious impacts on leaf-cutting (*Atta* spp.) colony dynamics (Vasconcelos *et al.* 2006).

Global climate change is likely already affecting the distribution of many ant species. For example, Colwell *et al.* (2008) predict that as many as 80% of the ant species of a lowland rainforest could decline or disappear from the lowlands due to upslope range shifts and lowland extinctions (biotic attrition) resulting from the increased temperatures. While species range shifts at higher latitudes may be compensated for by species from lower latitudes as the climate warms, there are no species to replace the lowland tropical species.

Little is known about modern extinction rates of ants and other insects, although given their predominance, insects will probably make up most of the extinctions over the next few years (Dunn 2005). In addition, based on what we know of insect extinctions so far, insects may be going extinct in ways that differ from other taxa, particularly through extinction of narrow habitat specialists (Section 4.3.2) and coextinction of insects closely aligned with other species (Section 4.4.2) (Dunn 2005; Koh *et al.* 2004).

Invasive ant species that out-compete native ant species for food and other resources, or kill them directly (Chapter 15) also threaten native ant species, especially on islands and in degraded habitats. Global climate change and the increasing international trade in pet ants, particularly in Europe, are likely to further spread invasive ants (R. Dunn, personal communication; Buschinger 2004; Chapter 13). Finally, agricultural and urban pesticides often target ants directly, and non-targeted applications can also kill ants and their colonies.

Given the scale and magnitude of the threats, our ability to conserve intact ant assemblages is a formidable task. Limited funding is just one of the many impediments to the conservation of native ants. A lack of information on ant species distributions (particularly for tropical regions) makes identifying rare and threatened species very difficult. Although North American and European ant distributions are relatively well known, we currently have only a general idea of global ant species distributions because new species are still being found and described (see Chapters 1 and 2), and their biology understood. Moreover, ants are small and easily overlooked by both the general public and conservationists, and are often perceived as pest organisms rather than in need of conserving. Their presence inside houses and in gardens can be a nuisance and people assume they are doing damage. While there are a few ant species that have become widespread invasive pests (see Part IV), most described ant species are unobtrusive and beneficial to natural ecosystems and humans. Finally, much conservation action is largely based on the assumption that other taxa, such as plants, birds, or mammals, can serve as surrogates for the conservation needs of invertebrates and other lesser-known taxa (Gardner *et al.* 2008; Rodrigues and Brooks 2007). However, few studies or analyses of surrogacy have included ants; those that have generally indicate that ant diversity patterns and responses of ants to disturbance are not the same as that of most 'umbrella taxa' (Alonso 2000). Ant species richness and distribution generally correlate best with other terrestrial, ground-dwelling invertebrates (Alonso 2000), but these taxa are also not usually included in conservation planning.

4.3 Where to conserve? Identifying priority areas

Given the substantial financial and practical constraints of conservation, conserving all species everywhere is not a viable option. One of the most widely applied strategies in global biodiversity conservation is to prioritize efforts in areas with high biodiversity and/or high endemism that are also highly threatened. In this way, conservation of a number of species from many taxonomic groups is achieved simultaneously. Such global approaches include Conservation International's Biodiversity Hotspot approach (areas with high levels of species diversity and endemism and over

86% of habitat already lost (Mittermeier *et al.* 2004; Myers *et al.* 2000b) and World Wildlife Fund's Ecoregions (Olson *et al.* 2001). An alternative approach is to focus efforts on areas with large concentrations of threatened species (based on the IUCN Red List; IUCN 2008). For example, BirdLife International (2000, 2008) has identified Important Bird Areas (IBA) throughout the world; these are sites that harbour a high number of bird species including a key number of threatened bird species. Within broader Hotspots and Ecoregions, Conservation International and partners now identify Key Biodiversity Areas (KBA) based on the presence of threatened species of all taxa evaluated on the IUCN Red List (for criteria see Eken *et al.* 2004; Langhammer *et al.* 2007).

4.3.1 Hotspots of ant richness and endemism

Current data on described ant species indicate tropical bioregions are 'hotspots' of ant richness and endemism (Chapters 2 and 3; see also Ward 2000). This pattern is similar to that found for many other taxa, and overlaps broadly with global multi-taxa terrestrial biodiversity hotspots, most of which are tropical (Mittermeier *et al.* 2004). The Neotropical, Indomalayan, and Afrotropical bioregions rank highest in terms of number of ant genera per bioregion, while the Neotropical, Afrotropical, and Australian bioregions have the highest percentage of their ant genera endemic to the region (Chapter 2). While these patterns reflect what we currently know about generic ant distributions and are likely to be upheld with future data additions (Chapter 2), there may be some sampling bias. A large database of over 225 ant diversity studies published since 1987 reveals that more diversity studies have been conducted in the Neotropical and Australian bioregions than in Asia and Africa, thus leaving substantial gaps in ant distribution data (Dunn *et al.* 2007d).

Recent mapping of global ant generic richness by B. Guénard and colleagues confirm that, in terms of number of ant genera, Oceania, India, southeast Asia, northern Australia, Central America, Brazil, and the northern part of South America are the most diverse (Chapter 3). Africa is not as diverse as these areas but does have many genera endemic to the continent, and recent studies from countries including Cameroon and Gabon are increasing the number of genera recorded. Unfortunately, many of the places that are potentially very diverse are also the least explored and studied, such as most of Africa and India, east Asia (Myanmar, Laos, and Cambodia), northeastern Brazil, northern Argentina, the Yucatán Peninsula, and the northern part of México.

Within a bioregion, ant species richness tends to be higher at lower elevations than at higher elevations, regardless of latitude (Johnson and Ward 2002; Lessard *et al.* 2007; Chapter 3). In many tropical regions, ant species generally do not occur above 2,000 m asl in closed canopy forests (Ward 2000). Mid-elevational peaks in species richness around 500 m have been documented (Fisher 1998; Olson 1994; Ward 2000; Chapters 2 and 3). Preliminary analyses from a global database of ant biodiversity studies indicate that ant diversity is highest in biomes with high temperatures (Dunn *et al.* 2007d; Chapter 3). The degree of endemism within lowland areas is not well known (except perhaps for islands, see Section 4.2.3), but it is likely that higher elevation mountains have higher endemism for ants than lowlands due to their isolation and distinctive microclimates (J. Longino, personal communication).

Given that habitat conversion and other threats are high in the lowland tropics (Hansen *et al.* 2008), these bioregions should be considered a priority for ant conservation. However, the composition of ant genera and subfamilies differs significantly among biogeographic regions (Ward 2000); in fact, over half of all 290 genera are restricted to one of the eight bioregions (Chapter 2). Thus, some ant conservation is warranted in each bioregion for taxa not found elsewhere. In addition, many Nearctic and Palaearctic ant species may soon be in need of conservation actions due to the emergence of a group of invasive ant species that mostly invades temperate climates (Chapter 13).

4.3.2 Conservation of ants in specific habitats

Some ant species are characterized as rare or endemic (see Section 4.4.1) because they are habitat specialists, restricted to a specific biome, habitat, or microhabitat (Underwood and Fisher 2006). These ant species may occur over a wide geographic area, but are restricted to a specific habitat type within this range and thus are vulnerable to local extinction if these habitats disappear. Some habitat specialists in the United States include *Camponotus mississippiensis*, known to nest only in the twigs and small branches of living white ash trees (MacGown *et al.* 2007), and *Temnothorax bradleyi* that live only in the bark of live pine trees and thus will be affected by threats to these tree species. Ant species that are restricted to nesting and foraging in the canopy of tropical forests, such as *Daceton armigerum* and many species of *Camponotus*, *Crematogaster*, *Polyrhachis*, and *Echinopla*, could face local extinction if forest canopy is opened or removed. Schulz and Wagner (2002) found that species of *Cataulacus*, *Leptothorax*, *Tetraponera*, and *Polyrhachis*, typical canopy-dwelling ant genera, had a significantly higher diversity and frequency in primary forest than secondary forest, where trees were often younger and lacking the dead branches and epiphytes important for ant colonization of trees.

In general, habitat specialist ant species tends to be found in more extreme biomes such as deserts, high-elevation forests, and grasslands, and unique habitats such as pine scrub. For example, Jourdan (1997) documented eight species of ants endemic to New Caledonia as specialists on low-productivity ultramafic vegetation areas of the island, and Johnson and Ward (2002) recorded specialist ant species in Baja California (México) inhabiting deserts and southern mountains. The UK biodiversity action plans (BAPs) target protection of heathland (a rare and threatened habitat type) and other sites where threatened ants are known to live (Mabelis 2007; UK BAP 2007).

Conserving places with an intact ant community is the best approach, since the conservation of individual species usually depends on the preservation of microhabitat and species interactions. The processes that determine species assemblage patterns can also have implications for conservation because species often depend on a micro-environment created by other ant species (Radnikova 2003).

Table 4.1 Ant species that may be considered rare in the United States

Subfamily	Genus	Species	Comments
Amblyponinae	*Amblyopone*	*orizabana*	One specimen known from the United States
	Amblyopone	*trigonignatha*	One specimen known
Myrmicinae	*Anergates*	*atratulus*	Rare parasite
	Aphaenogaster	*umphreyi*	Largest series had 16 workers
	Dolopomyrmex	*pilatus*	Known from fewer than a dozen collections, subterranean
	Pheidole	*aurea*	Known only from type series
	Pheidole	*bureni*	Known only from type series
	Pheidole	*elecebra*	Parasite
	Pheidole	*humeralis*	
	Pheidole	*marcidula*	
	Pheidole	*mera*	Known only from type series
	Pheidole	*nuculiceps*	Known only from type series
	Pheidole	*virago*	Known only from type series
	Pseudomyrmex	*leptosus*	Parasite
	Pyramica	*chiricahua*	One specimen known from the United States
	Pyramica	*inopina*	
	Pyramica	*reliquia*	
	Rogeria	*foreli*	
	Solenopsis	*phoretica*	Known only from one specimen
Ponerinae	*Cryptopone*	Arizona sp.	One specimen known from US

Source: Lloyd R. Davis, personal communication.

4.4 Which ants to conserve? Identifying priority species

In contrast to a biogeographic strategy, conservation efforts may also focus on individual species. The single-species approach is particularly common in the conservation of vertebrates, especially large mammals. Ant species may be identified as needing specific conservation action based on their global or local rarity, or the ecological roles that they fill.

4.4.1 Rare, threatened, and endemic species

Rare species have low abundances, and may either have a widespread or localized distribution (for discussions on species rarity, see Brandão *et al.* 2008; Kunin and Gaston 1997; Murray and Lepschi 2004; Rabinowitz 1981). Examples of naturally rare genera include *Kyidris*, *Mystrium*, and *Perissomyrmex* (Ward 2000). Table 4.1 lists ant species from the United States that are considered rare because they have been found and collected few times (L. Davis, personal communication). Other ant species and genera may become rare because their habitat or populations are threatened by human activities. For example, most of the endemic ants of Mauritius are now restricted to a small patch of upland native forest due to the loss of lowland forest and the presence of invasive ant species at lower elevations (Fisher 2005; see Figure. 4.1).

Rare and threatened species that are found in few sites, such as restricted range specialists, are often the target of conservation efforts since they are in danger of extinction if those sites are disturbed or destroyed. This is particularly true for endemic species, those found only in a specific locale, such as a mountaintop, island, country, or ecoregion. Loss of habitat could lead to the extinction of the species since they typically cannot be conserved elsewhere. If ants follow the patterns of other taxa, most endemic species will likely be identified from isolated islands and mountaintops, where they have speciated due to their isolation (see Section

Figure 4.1 Examples of threatened endemic ants on Mauritius: (*a*) *Discothyrea berlita*, (*b*) *Proceratium avium*, and (*c*) *Pyramica simoni* (Photos: www.AntWeb.org). Habitat transformation on Mauritius has reduced native habitat to small isolated remnant patches: (*d*) at the top of Le Pouce mountain and, (*e*) below Le Pouce. (Photos: Lori Lach)

4.3.1.). For example, Ward (1990) and Fisher (2005) identified nine ant species endemic to the island of Mauritius that are seriously threatened by introduced invasive ant species and habitat loss. In Fiji, 91 of the 138 ant species documented are endemic to the island (Ward and Wetterer 2006), and 11 of the 37 established species of ants in New Zealand are considered endemic to the island complex (Don 2008). The Malagasy region has a high level of endemism even at the generic level; of the 52 ant genera recorded, four are endemic to Madagascar (Fisher 1998). Peninsula areas also typically have many endemic ant species; Johnson and Ward (2002) documented 47 species (27.6% of the total fauna) endemic to Baja California.

For many ant species, however, it is difficult to know if they are really rare or appear rare because sampling has been limited in extent or intensity. Intensive surveys of the ant fauna of the La Selva Biological Station in Costa Rica by Longino *et al.* (2002) illustrate the problems defining rare ant species. Of the 437 ant species recorded, 51 species (12% of the total) were known from only one sample at the end of the inventory. Twenty of these species were 'methodological edge species' (species possibly abundant at the site but difficult to sample because of their microhabitat), and 'geographic edge species' (known to be common in habitats or regions outside of La Selva). Rarity of 31 species is unexplained. Most of the 51 rare species were known from additional collections outside of La Selva, either in other parts of Costa Rica or in other countries. Only six species were globally unique, known to date from only one sample on earth.

4.4.2 Ant species dependent on other species

Ant species dependent on direct interactions with other organisms for their survival, either as parasites, predators, or mutualists, may be vulnerable if their hosts, prey, or partners are threatened. Socially parasitic ants do not have colonies of their own – the queen infiltrates a nest of another ant species and lays eggs that are raised by workers of the host ant species. Many of these species exist as queens and have no worker caste; they are considered especially vulnerable to extinction because they are dependent on the survival of their host species. Many social parasites are listed on the IUCN Red List (see Section 4.4.5) because of this dependency, as well as their restricted range. Slave-making ants, such as *Polyergus* spp., *Rossomyrmex* spp., and many *Formica* species, that raid colonies of other ant species to steal their larvae and then raise the workers as slaves are thus also dependent on other ant species for their survival and are affected by population changes in these other species. Specialist predators, such as Cerapachyinae, Aenictinae, some Ponerinae, and several myrmicine tribes (e.g. Dacetini) may also be at risk if their prey organisms decline or disappear (e.g. Brandão *et al.* 1991).

Many ant species have mutualistic relationships with other insects, microbes, or plants. Those species that have obligate associations will be most vulnerable if the partner population declines. For example, some *Pseudomyrmex* species can only survive when inhabiting specific *Acacia* species that offer domatia, extrafloral nectar, and food bodies (Rico-Gray and Oliveira 2007). Obligate dependency for the ant has yet to be demonstrated for many other ant–plant relationships, but are especially likely among ant species showing extreme adaptations to their host plants (e.g. *Petalomyrmex phylax* on *Leonardoxa* plants, see Chapter 6). Much more research on this is needed, as the web of interactions among species is not always evident. For example, Palmer *et al.* (2008) found the exclusion of large herbivores from an African savanna disrupted an ant–plant mutualism.

4.4.3 Species with major impacts on the ecosystem

A keystone species has traditionally been defined as a species whose impact on its community or ecosystem is disproportionately large relative to its abundance (Paine 1995). Davic (2003) proposes several categories of species with major impacts on their ecosystems: '"keystone species" regulate local species diversity in lower trophic levels, "key species" regulate energy/nutrient dynamics, "intraguild competitors/predators" structure niche partitioning among closely related species, and "ecosystem engineers" modulate physical habitat. Some focal species may have multiple process roles in ecosystems. Many ant species fall into several of these categories. Red wood ants (*Formica* spp.) are targets of conservation action in Europe mainly to preserve the essential services they provide to their ecosystem: heavy predation on a wide range of insects including pest species, dispersing seeds of many myrmechorous plants such as *Viola* spp., providing habitat in their nests for a multitude of organisms (e.g. over 30 beetle species), and as an important food source for birds (especially woodpeckers) and other animals (Mabelis 2007). Some colonies of wood ants have been recorded to bring in over 456 g of insect prey per day (Holt 1955). Similarly, leaf-cutting ants (e.g. *Atta* spp.) are instrumental in structuring their environment as they move tons of soil, integrate nutrients, and aerate the soil while building their large nests (Costa *et al.* 2008).

Other important groups of ants include those that disperse seeds of plant species; where the association is disrupted, plant and ant communities can disassemble (Christian 2001; Sanders *et al.* 2003a). Other important species may be dominant members of the ant assemblage that structure the community through their interactions, such as army ants, driver ants (Boswell *et al.* 1998), and ants that tend hemipterans (Styrsky and Eubanks 2007).

4.4.4 Conserving phylogenetic diversity and ant phenomena

Recently, evolutionary history has been recognized as a key component of biodiversity conservation. Phylogenetic diversity is a measure of biodiversity based on the length of evolutionary pathways that connect a given set of taxa (sum of the length of branches of a phylogeny) (Forest *et al.* 2007). Species represent different evolutionary histories, so the extinction of a species in an older, monotypic, or species-poor clade would result in a greater loss of phylogenetic diversity than that of a young species with many close relatives (Isaac *et al.* 2007; Redding *et al.* 2007). Although species richness can be a good surrogate for phylogenetic diversity (e.g. Rodrigues and Gaston 2002), recent studies indicate that this is not the case in all areas (e.g. Cape Floral Kingdom, Forest *et al.* 2007). Ant species and groups that are in ancestral, species-poor clades, include several of the poneroids: Agroecomyrmecinae (*Tatuidris*), Paraponerinae (although the sole species, *Paraponera clavata*, is not currently highly threatened), Amblyoponinae (esp. *Apomyrma*), an Proceratiinae. The Leptanillinae are especially species-poor. Among the formicoids, species-poor or ancestral taxa include *Nothomyrmecia*, *Leptanilloides*, *Cerapachys sexspinus*, *Aneuretus simoni*, several new species of Dolichoderinae from Madagascar, *Aulacopone*, *Oecophylla*, *Notostigma*, *Gesomyrmex*, *Myrmoteras*, *Ankylomyrmex*, *Hylomyrma*, and *Lenomyrmex* (P. Ward, personal communication; Chapter 1).

Another group of ants of potential conservation concern may be considered as ant phenomena. A huge supercolony of the ant *Formica yessensis*, consisting of 45,000 nests, lives along the coast of Ishikari Bay, Hokkaido, Japan (Higashi and Yamauchi 1979). The colony may be 1,000 years old and is currently threatened by development plans along the coast (http://www.globalforestscience.org/research/projects.html?projectName=The_World%27s_Biggest_Supercolony_of_Ants_Under_Threat).

4.4.5 Ant species officially listed as threatened

Globally, the IUCN Red List of Threatened Species (IUCN 2008) is used as the guideline for determining priority species for conservation focus and action. Species on this list have been evaluated by expert scientists according to a strict set of criteria and are placed into categories based on the level of perceived or inferred threat for the species. The current IUCN Red List categories include Extinct (EX), Extinct in the Wild (EXW), Critically Endangered (CR), Endangered (EN), Vulnerable (VU), Near Threatened (NT), Least Concern (LC), and Data Deficient (DD, status not determined) (IUCN 2008). Having species listed by IUCN puts them on the global conservation map and provides important data to conservation decision makers (Rodrigues *et al.* 2006). Listing on the IUCN Red List is often a requirement for inclusion in broader conservation efforts and policy determinations and is now a Millennium Development Goals indicator for achieving global environmental sustainability (http://www.undp.org/mdg/basics.shtml), which gives this distinction considerable weight (J. Fellowes, personal communication).

The process of identifying and listing threatened species is a dynamic and iterative process, such that revisions, additions, and updates to the list help to determine which species may warrant listing or delisting. The 1983 IUCN Invertebrate Red Data Book (Wells *et al.* 1983) included six ant species of conservation concern: *Aneuretus simoni*, *Aulacopone relicta*, *Epimyrmy ravouxi*, *Formica yessensis* (one Japanese wood ant supercolony), *Leptothorax goesswaldi*, *Nothomyrmecia macrops*, and several species of European wood ants (*Formica aquilonia*, *F. lugubris*, *F. polyctena*, *F. pratensis*, and *F. rufa*) (Wells *et al.* 1983). A total of 149 ant species are listed in the 2008 Red List (IUCN 2008), which at first glance seems to be a good indication that ants are being taken into consideration in conservation. However, all these species were listed before 2001 and are now considered as 'needs updating', primarily because they were evaluated under previous Red List criteria (ver. 2.3, 1994), which were less rigorous than those in the current version (ver. 3.1). Few ant species have been evaluated for the current IUCN Red List. There are likely many more ant species under threat of extinction that we do not know about.

Three ant species are listed as CR: *Adetomyrma venatrix*, *Aneuretus simoni*, and *N. macrops* (IUCN 2008). *A. venatrix* (Dracula ant) from Madagascar was first described in 1994 from a single location and then an entire colony was found in 2001 (see the similar *Adetomyrma* sp. Mad01 in Plate 1); further sampling may expand its range. *A. simoni* is the only living species of the subfamily Aneuretinae (otherwise exclusively known as amber fossils). Prior to Wilson's rediscovery of the species in 1955 in Sri Lanka, *A. simoni* was known only from five or six specimens in museum collections. *N. macrops* (dinosaur ant) was first discovered near Esperance, Western Australia in 1931, and then not recorded again until it was rediscovered in South Australia in 1977 (see Box 2.1 and Figure 4.2). The Dinosaur ant is now known to occur at 18 locations along the Eyre Peninsula, South Australia and a study by Sanetra and Crozier (2003) found no evidence for a decline in populations, which would have resulted in low genetic diversity. This species is also listed as Protected Fauna under the Western Australian *Wildlife Act*. Six species of *Formica* wood ants potentially threatened by alteration of their habitat throughout Europe (see Mabelis 2007) are also listed as Lower Risk or NT.

Most of the ant species listed by IUCN are social parasites, thus their conservation depends on the protection of their host ant species and their habitats (Mabelis 2007). These species are currently categorized as VU D2 (ver. 2.3, 1994; IUCN 2008). However, this criterion has been changed (ver. 3.1, 2001; IUCN 2008) with the new criterion requiring populations to be restricted to an area of 20 km^2 or less, rather than 100 km^2. Previously listed species need to be re-evaluated based on the new criteria to determine whether they still require listing.

In addition to the IUCN list of globally threatened species, many countries have their own lists of threatened species, but few have included ants. The national governments of many countries have compiled lists of endangered species including the U.S. Fish and Wildlife Service, Australia's *Environment Protection and Biodiversity Conservation Act* 1999 (EPBC Act), and Brazil's list of endangered animal and plant species, but most of these lists do not include ants. The state of Victoria, Australia, however, does list an undescribed species of *Myrmecia* in its state conservation act (Victorian Flora and Fauna Guarantee Act 1988).

European countries are on the forefront of identifying and listing threatened ant species in their countries. Seven species of ants are listed as priority

Figure 4.2 This ancient extant ant (*Nothomyrmecia macrops*) from mallee habitats in South and Western Australia is the only living representative of an ancient lineage (the Prionomyrmecini). This so-called living fossil was famously rediscovered in 1977 (see Box 2.1). (Photo: Alex Wild)

Table 4.2 Ant species listed in the United Kingdom as priorities for conservation under the UK Biodiversity Action Plan (2007).

Ant species	Authority	Common name
Anergates atratulus	(Schenck, 1852)	dark guest ant
Formica exsecta	(Nylander, 1846)	narrow-headed ant
Formica pratensis	(Retzius, 1783)	black-backed meadow ant
Formica rufibarbis	(Fabricius, 1793)	red barbed ant
Formicoxenus nitidulus[a]	(Nylander, 1846)	shining guest ant
Tapinoma erraticum	(Latreille, 1798)	erratic ant
Temnothorax interruptus	(Schenck, 1852)	long-spined ant

[a] Species also listed as globally threatened by IUCN (2008, but in need of review).

species in the U.K. (Table 4.2). Scientists have developed conservation action plans for the first four species on this list and have issued a statement of conservation action for *Formicoxenus nitidulus* (UK BAP 2007). Most of these species are not globally threatened but are of conservation concern in the United Kingdom due to their localized distribution within the country and disturbance to or loss of their specific habitat (often heathland). *Anergates atratulus* is a social parasite of *Tetramorium caespitum*, and *Formicoxenus nitidulus* lives in nests of several *Formica* species; thus both are dependent on the conservation of their host ant species. These species and others are also protected in the Netherlands and several other European countries (Mabelis 2007).

4.5 Future directions

There are several areas in which important contributions to ant conservation and management can and must be made immediately. The future of ant conservation lies in research, management, communication networks, and action.

4.5.1 Compile current data on ant species richness and diversity

Collating existing data on ant species richness and diversity by bioregion is the first key step in the process of understanding ant conservation needs. Fisher (2005) outlines a practical and achievable approach for inventorying and cataloguing global ant species. Compilation of data will help answer many research questions that will guide conservation efforts, including: (*a*) identifying geographic and biome gaps in ant sampling, (*b*) identifying hotspots of ant species diversity and endemism (rather than just generic patterns), and (*c*) modelling future impacts of habitat conversion and climate change (Dunn *et al.* 2007c). Efforts currently underway that should be expanded and emulated include a global ant collaboration, which has compiled a database of ant species diversity data from more than 2,700 sample sites and 225 studies, as well as from all published literature since 1987 (Box 3.2; Dunn *et al.* 2007c). Similarly, Longino's (2008) Leaf Litter Arthropods of Mesoamerica (LLAMA) project, which investigates the species richness, degrees of endemism, and patterns of faunal turnover in ants across Mesoamerica is an example of the type of information that conservation planners urgently need. Additional information is becoming available for areas like Asia (through the Asian Ant Network, ANeT 2008), Madagascar, and the Mascarene Islands (AntWeb 2008), and globally, through global ant databases. (e.g. www.antbase.org, Agosti and Johnson 2005). It is however vital that they are continually updated to be of real value.

4.5.2 Incorporate ants into broader conservation efforts using existing data

Perhaps because ants are not as conspicuous, not as charismatic, not as large as most other organisms typically included in biodiversity monitoring projects (whether these issues are real or perceived), or because they are time consuming to collect and identify, ants are not incorporated into broader conservation efforts as often as they should be given their ubiquity. Some actions that can be taken immediately to facilitate their addition include:

1. Make existing ant distribution data available to global and local conservation planners to highlight key areas for ants that they may not be considering (see Box 3.2 for a list of databases already available). In a very positive step, ants have been included, along with several vertebrate taxa, in biodiversity maps that will guide conservation planning and reserve placement in Madagascar (Kremen *et al.* 2008).
2. Describe ant species new to science so that they have a name for comparisons.
3. Conduct studies of the effectiveness of other taxa as surrogates for ant diversity to assess whether other taxa can represent ants.
4. Conduct further studies of how the response of ant species or ant functional groups (Box 8.2; Andersen and Majer 2004) can be used as an indicator of habitat quality or disturbance. For example, for a particular region, are there ant species, functional groups, or levels of ant diversity that indicate an undisturbed forest or recovery?
5. Model ant species distributions to predict areas of highest ant diversity and endemism (e.g. Brooks *et al.* 2004; Dunn 2008a; Ferrier *et al.* 2004).
6. Build on current networks of myrmecologists to extend existing and construct new ant databases (see Section 4.4.3) to contribute to broader conservation strategies. This has already begun through the development of several networks, including the Global Ant Community Database (see Dunn 2008a), the Asian Ant Network (ANeT), the Global Ant Project (B. Fisher, personal communication) affiliated with the Encyclopedia of Life (EoL) project that aims to create a web page for every species on earth (EoL 2008), and a proposed new IUCN Ant Specialist Group (J. Fellowes, personal communication). Interested myrmecologists should get in touch with these networks to get involved.
7. Model future scenarios for ant species distributions with global climate change. Preliminary work is possible using currently available datasets (Dunn 2008a) and is especially important for extreme habitats that harbour many potentially threatened socially parasitic ant species, habitat specialists, and endemic ant species (particularly at higher elevations) that are most likely to be affected by global climate change.
8. Continue to develop new identification tools for ants such as electronic and pictorial keys, digital image libraries, DNA barcoding (Miller 2007a; Smith *et al.* 2006), and online type collections, which are all needed to facilitate taxonomy and systematics of ants as well as their inclusion in conservation agendas.

4.5.3 Monitor and address current threats to ants

Identifying, monitoring, and acting to alleviate threats are the mainstay of conservation. For example, habitat alteration and loss are major threats to ants, but for most ant species, we know little of their specific abiotic and biotic requirements. Research on the habitat, distribution, and natural history of most ant species is needed to understand how they will respond to future threats and what types of habitat (both micro and macro) are required to conserve them. Another principal activity should be mapping and monitoring of invasive ant species, which are one of the greatest threats to native ant species (Section 4.2.2; Chapter 15). Current efforts in using DNA barcoding for quick identification of invasive ant species at inspection stations could be pursued and applied at borders of countries and islands at most risk from invasive ant species (B. Fisher, personal communication).

With the widespread conversion of much native habitat to agriculture, pasture, and urban development, conservation of biodiversity within the land-use 'matrix' has become a key conservation strategy. Ants are prime candidates for this type of conservation due to their small size and the ability of many ant species to persist in the agricultural landscape (Philpott and Armbrecht 2006; see Chapter 8). While the habitat of many larger animals is highly modified by agricultural practices, many ant species can find their preferred microhabitat in small patches of forest or undisturbed open areas. Research and conservation efforts should continue in these matrix habitats and address questions such as: (*a*) how are different ant species affected by each land-use? (*b*) which ant species are most sensitive to land-use change, their presence thus serving as an indicator of change? And, (*c*) which land-use practices allow for the presence of an intact ant fauna? Gove *et al.* (2005) found that even small forest patches in the seasonally dry tropics of México harbour a significant ant fauna and thus contribute to biodiversity conservation where reserves are limited. Similarly, Monteiro *et al.* (2008) documented *Labidus praedator* army ants using field margins around cultivated areas as refuges, mini-corridors, and rich food patches.

The extinction of insects, especially ants, is of grave concern because insects play critical roles in most ecological processes. The loss of ecosystem function and the subsequent impacts on human livelihoods and health will be great. We need more information on current rates and mechanisms of insect extinctions to be able to predict the magnitude and locations of future extinctions and to prevent them. Dunn (2005) suggests actions that will contribute to our understanding of insect extinctions including:

1. Conduct repeated inventories of sites and regions to provide historical data on ant species distributions as well as the mechanisms and rates of local extinction;
2. Monitor species interactions over time to determine the degree of co-dependence and the rate of local extinction of partners; and
3. Focus site conservation measures on places with narrow habitat specialist ant species since they can be conserved only in those few sites.

4.5.4 Promote education and awareness of the ecological importance of ants

Ants are not a high profile, popular taxonomic group among the conservation community or the general public (Box 4.1). Increasing awareness of the ecological and economic importance of ants should be of high priority. Pressure to conserve particular species can often be generated by the public, and ants could become part of their scope with an increased understanding of their importance. Educational programmes that are aimed at both children and adults are best. Focusing on ants, the *Iimbovane Project*, run through the University of Stellenbosch in South Africa, is an excellent example of how scientific research on ant diversity and distributions can be combined with an outreach project that promotes environmental education and biodiversity awareness of school learners (see Braschler 2009).

Most people do not take note or interest in something until they develop a direct relationship with it. Sadly, few see the direct relationship we have with biodiversity and healthy ecosystems, even though the well-being of human populations is directly dependent on ecosystem services provided by ants and other organisms. Put simply, species matter, and they matter now more than ever. Healthy ecosystems are the prerequisite for thriving communities, and as habitats degrade and species disappear, these losses are having a profound effect on humans worldwide. Ants play a key role in this cycle, providing essential services such as nutrient cycling, seed dispersal, and natural pest control that benefit not only other species in nature, but also all of humankind. Getting the message of the direct links between our own welfare and the conservation of ants is essential not only for the preservation of ants and their habitats, but also for the health and future of humankind.

While most ant species are not considered very attractive or charismatic compared to cute and cuddly mammals, there are several ant species that could be used as flagship species to bring attention to the need for conserving ants. These species include leaf-cutting ants (e.g. *Atta*) and army ants (e.g. *Eciton* and *Neivamyrmex*) in the neotropics, and weaver ants (*Oecophylla*), driver ants (*Dorylus*), jumping ants (*Harpagnathus* spp.), and trap-jaw ants (*Odontomachus*) (see Box 12.1) in the African and Asian tropics, and bulldog ants (*Myrmecia*) in Australia. These ant taxa are common and conspicuous, and have fascinating life histories, morphological features or 'personalities' that the general public find extraordinary. They can serve to demonstrate the variety of life histories, cycles, and behaviours of ants, and to highlight the key roles they play in ecosystem functioning and trophic webs. Ant species that have common names, such as those listed in Table 4.2, are especially useful in reaching out to the general public and non-specialists since descriptive names provide a stronger connection and promote more interest in the ant species than do Latin names (Andersen 2002).

Box 4.1 Ants in museums
Ted R. Schultz

Ant collections support research in systematic biology

Museum collections of ant specimens represent a vast archive of largely unrealized research potential. Brandão (2000) lists 174 such collections in universities and museums around the world containing, in total, many millions of specimens. These collections have been built by, and are traditionally the domain of, ant systematists, who use them to (*a*) permanently document newly described species through the deposition of type specimens, (*b*) determine unidentified specimens, (*c*) understand the boundaries of species and higher taxa (e.g. genera and subfamilies), (*d*) understand the geographic distributions of species, and (*e*) reconstruct phylogenetic relationships (see Chapter 2).

Ant collections have unrealized potential for other disciplines

With increasing frequency, however, biologists working outside the realms of systematics and taxonomy are discovering the importance of museum specimens as a source of primary data for their disciplines. Every properly labelled specimen documents the occurrence of an organism at a particular place and time. Museums typically contain multiple such specimens collected at multiple sites and at multiple points in time. The oldest non-fossil ant museum specimens were collected over 250 years ago (e.g. in the Linnaean Collections in London), but most were collected during the past 150 years. Because the study of evolution is the study of organic change over time, and because many ecological, biodiversity, and conservation studies are concerned with evolutionary time scales well within this 150-year window, museum collections represent a rich source of data for these disciplines (Wandeler *et al.* 2007), e.g. for studies of habitat loss and biological invasions (Suarez and Tsutsui 2004). Historical distribution records are also increasingly employed for predicting changes in species ranges due to global climate change (Dunn *et al.* 2007d).

Two recent examples illustrate the potential of ant collections as sources of primary data for studies of invasive species. One study used museum specimens to reconstruct the historical spread of the invasive Argentine ant *Linepithema humile* across North America from 1891 to the present, documenting the mechanism of long-distance jump dispersal (Suarez *et al.* 2001). A second study used ant specimens intercepted by US Department of Agriculture (USDA) inspectors at US ports of entry between 1927 and 1985, archived in the Smithsonian Institution, to determine the correlation of opportunity (i.e. the number of times a species was intercepted) with the successful establishment of invasive ant species. Interestingly, those species with the most opportunity were not necessarily the most likely to become established (Suarez *et al.* 2005a). As techniques for extracting DNA sequences from museum specimens continue to improve, ant collections will become increasingly useful as sources of genetic data for studies such as these (Wandeler *et al.* 2007).

Ant collections are repositories for voucher specimens

Because the scientific method requires repeatability and accountability, specimens of organisms involved in population genetic, behavioural, ecological, conservation, and other biological studies should be held as voucher specimens in museum collections to ensure both the accuracy of the species identifications as well as the proper interpretation of the data. Unfortunately, the majority of such studies fail to lodge voucher specimens, so there is obviously room for significant improvement (Suarez and Tsutsui 2004). The rules for depositing voucher specimens vary with institutions, so the first step in the process is to contact the curator of an appropriate entomological collection for instructions. In recent years, ants have been increasingly employed as bio-indicators in biodiversity and conservation studies employing mass-collecting methods and generating thousands to tens of thousands of specimens per study (Agosti *et al.* 2000). This tidal wave of new specimens has rapidly expanded the ant collections of some museums, including the California Academy of Sciences, the Museu de Zoologia da Universidade de São Paulo (Brazil), the Instituto Nacional de Biodiversidad (Costa Rica), and the Smithsonian Institution (United States).

continues

Box 4.1 continued

Ant collections can be visited or accessed via the Internet

Most museum collections are designed to be visited and used by researchers. Because collections are understaffed, visiting researchers should be prepared to work autonomously, respecting and, where possible, improving the curation of the specimens they study. Increasingly, museums are making collections data available via the Internet. For example, major efforts are underway at a number of museums to make ant and other type specimens available online in the form of digital images (e.g. http://ripley.si.edu/ent/nmnhtypedb/public/). Such images can be extremely useful for identifying species.

Ant collections can engage the public

In contrast to their enthusiasm for fossilized dinosaur skeletons and stuffed polar bears, the general public is far less enthralled by museum exhibits of dead ant specimens. No doubt this is because ants are small and hard to see. Fortunately, a significant segment of the non-scientific public enjoys viewing the fantastic and baroque morphologies revealed when ant specimens are enlarged by scanning electron microscopy and by the relatively new technique of multiple-layered-focus microphotography. Museums now make such images available on the Web (e.g. at www.AntWeb.org). The lay public also enjoys watching colonies of live ants in museum exhibits, for example, at the Smithsonian National Museum of Natural History's Insect Zoo (the United States), which displays leaf-cutting ants (*Atta cephalotes*), acacia ants (*Pseudomyrmex ferruginea*), carpenter ants (*Camponotus pennsylvanicus*), and honeypot ants (*Myrmecocystus mendax*). Through such images and exhibits, museums and zoological parks increase public awareness of the roles of ants in natural ecosystems and thereby promote ant conservation.

Policy makers are perhaps the least likely to consider ants in their conservation plans, but they are most often persuaded by sound science. If robust data and sound arguments are presented to them, they often will consider ants in their policies and planning. Communication of key research results in an appropriate and accessible manner is crucial. Targeted communication via policy briefings and conservation reports, rather than dissemination of information through scientific publications only, provides a novel niche for science communicators and public servants.

4.5.5 Assess the status of listed and potentially threatened ant species

To ensure that ant data feed directly into the IUCN Red List and are used to guide global conservation programmes, the process and output of compiling ant distribution and biological information should be part of a targeted Global Ant Assessment done in collaboration with the IUCN and the myrmecological networks proposed in Section 4.5.1. A global assessment involves evaluating the distribution, biology, and threats of described ant species, and new species as they are described, based on specific criteria developed by IUCN (2008). Much groundwork has been laid by the previous Social Insects Specialist Group (SISG) of IUCN (Agosti and Johnson 2005). We can begin with an evaluation of ants using the new Sampled Red List Index (SRLI) process, in which a random set of ant species is assessed as a representative of the group. While a complete assessment is ultimately desired, this is a good first step for speciose groups such as ants.

4.5.6 Diversity surveys, taxonomy, and collecting ethics

Additional survey effort is needed for areas that have been under-sampled for ants, particularly those areas identified by Fisher (Chapter 2), and very hot and very cold biomes identified by the Global Ant Community Database: temperate rainforests, tundra, hottest subtropical deserts, and taiga (Dunn *et al.* 2007d). Although some of these areas would be expected to yield low species diversity, data are needed to fill in gaps in our knowledge of species' distributions.

Additional distribution data will help us to identify which species are truly endemic, rare, or range restricted, and then to address more fundamental ecological questions such as whether montane species are derived from nearby lowland species (evolutionary colonization of uplands) and which are derived from other montane species. Similarly, genetic studies, which inform us of a species' viability and population dynamics, will help us identify key regions of endemism for ants. In addition to species' locality data, additional natural history information is required so that appropriate habitats, microhabitats, food sources, or partner organisms can be conserved.

Fortunately, standardized methodologies for collecting ants have been developed and are widely used by field researchers (Agosti and Alonso 2000; see also Box 3.1) and there is a standard global catalogue of ant taxonomy (Bolton 1995a). However, one of the key challenges to ant conservation is overcoming the taxonomic impediment for ants (see, e.g., Figure 4.3). Species inventories turn up many species new to science, which ultimately have to be described and catalogued by only a few ant taxonomists. In addition, the phylogeny and taxonomy of many ant species groups are unresolved and are not being worked on – large genera such as *Camponotus*, *Pheidole*, and *Solenopsis* need taxonomic revision. Beta diversity comparisons (comparisons of species between sites and regions) cannot be done when the species are identified only to morphospecies, or if researchers are not all using the same global catalogue of ant species taxonomy. Parataxonomists, individuals who collect, mount, and identify specimens to morphospecies, can make a valuable contribution to this process. Importantly, parataxonomists are not an alternative to professional taxonomists, but by working alongside primary taxonomists (Janzen *et al.* 1993), parataxonomists can make the process of sample processing much more efficient (Basset *et al.* 2000; Basset *et al.* 2004). Thus while the training of new primary ant taxonomists ought to be a top priority, training of parataxonomists is also important in order to catalogue the extraordinary number of specimens waiting for a name. Already this approach has yielded substantial benefits in biodiversity projects in Costa Rica, Papua New Guinea (www.entu.cas.cz/png/index.html), and Guyana.

Due to the rapid destruction of habitats for ants and the disappearance of species, it is important that future ant collecting be done so that myrmecologists are not the cause of the demise of the last population of any ant species. Ant collectors need

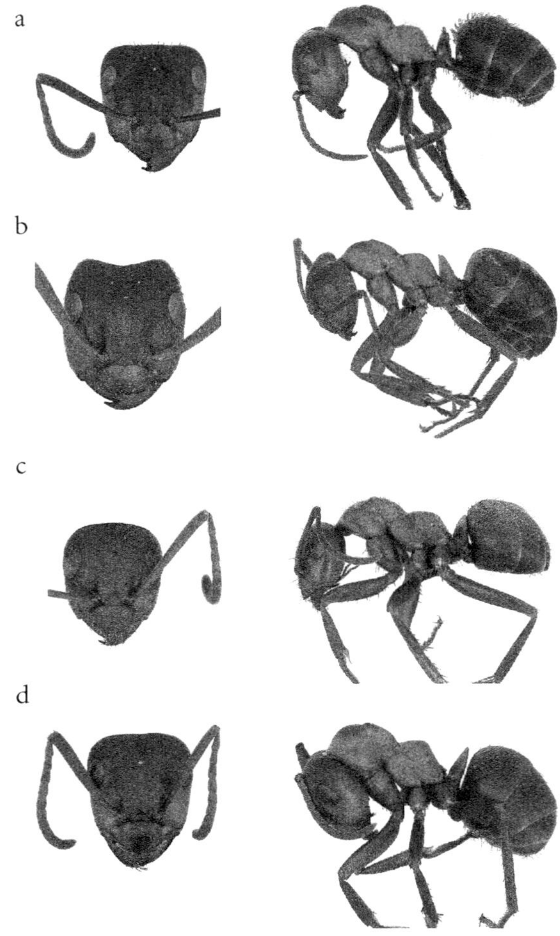

Figure 4.3 Identification to species level using morphological characters can be difficult, especially to those with little formal training. These species, all very similar in appearance, belong to the genus *Formica*: (*a*) *Formica pratensis*, (*b*) *Formica exsecta*, (*c*) *Formica rufibarbis*, (*d*) *Formica polyctena*. However, only *F. pratensis, F. exsecta,* and *F. rufibarbis* are listed as threatened or vulnerable. (Photos: www.AntWeb.org)

to be aware of the presence of priority ant species in an area when they set out to collect, and need to take care not to collect the last individuals or nest of that species. Methods targeted at the objectives of the study will be more effective. If a research study only requires workers to be collected, entire colonies should not be collected since this is the reproductive unit of ants. Sampling techniques such as pitfall traps or malaise traps should be used cautiously since they collect ants and other insects indiscriminately and often result in massive numbers of individuals of the same species. Since ants are difficult to identify in the field, it is necessary to collect, but this should be done ethically. This code of collecting is followed for other taxonomic groups and should also be followed by myrmecologists. Many ant collectors worry that drawing attention to threatened ant species, such as by listing species on the IUCN Red List, will bring restrictions on their collecting. This is a reasonable concern, but the species listed are in need of protection.

4.5.7 Develop targeted ant conservation action plans

For ant species that are highly threatened with extinction, it may become necessary to develop a conservation action plan for their survival. Such plans may take many forms and be developed by a variety of organizations, such as local or national governments (e.g. UK BAP 2007), local conservation groups, and international NGOs such as IUCN. An action plan addresses threats and recommends action to protect the species, taking into account the biology of the species. Action plans can contain a wide variety of conservation recommendations; one of the most effective ways to protect an ant species is to focus on conservation of its habitat by influencing land-use policies. There are several excellent examples of conservation efforts for threatened ant species in Europe, particularly in the United Kingdom. Some focus on protecting unique ant lifestyles, such as the social parasitism of *Anergates atratulus* (UK BAP 2007). Action plans for several species of red wood ant (*Formica* spp.) in Scotland focus on protecting the species in order to maintain essential ecological services provided by the ants (e.g. Forestry Commission 2007). Another good example to follow is the IUCN-The World Conservation Union Species Survival Commission (SSC) Action Plans that describe what it will take to ensure the survival of a particular species throughout its range (IUCN 2008).

Where an ant species of conservation concern has been under threat and had its population size reduced, reintroduction of the species to its native habitat may be possible; as has been done for red wood ants (*Formica* spp.), which control pest insects in many European forests (Mabelis 2007). For ant species officially designated as threatened, captive breeding is another approach to maintaining populations in danger of extinction in the wild. The Species Action Plan for *Formica rufibarbis* includes the development of a methodology for captive rearing (UK BAP 2007) at the Zoological Society of London with the long-term aim of reintroducing at least 40 captive-reared nests into the wild each year (Zoological Society of London 2007). While this approach may unfortunately become more necessary as more ant species become threatened, it will be costly (£50,000 was invested in *F. rufibarbis*).

4.6 Summary

The current data reveal that the Neotropical, Indomalayan, Afrotropical, and Australian bioregions contain the highest ant generic diversity and endemism and are thus key regions for ant conservation. However, these regions are large and finer-scale conservation priorities need to be identified for ants. Much more data are needed at the species level to accomplish this, but the recently initiated Global Ant Community Database project (Dunn 2008a) is a great start. Considering that preliminarily identified hotspots of ant diversity and endemism overlap with global biodiversity hotspots and priority areas for other taxa suggests that efforts to protect other taxa and habitats in these regions will also benefit ants. Other sites of particular importance for ants include critical habitats for key species as well as hotspots of ant endemism and threatened species. Endemic island ant faunas may face the greatest threats of all, mostly from invasive ant species.

At the species level, conservation efforts are needed to protect ants of conservation concern, including species that are endemic, threatened, habitat specialists, dependent on associations with other organisms, phylogenetically important, charismatic, or have major ecological impacts. Conservation action should start with compiling current data, incorporating ants into broader conservation efforts, identifying and monitoring current threats to ants, and promoting education and awareness of ant conservation. Longer-term actions must include collecting new data on ant species, assessing the status and biology of ant species, developing targeted ant conservation plans, and modelling future scenarios for ant conservation.

To really put ants at the centre of conservation efforts, the mymecological community needs to get involved. Large mammals and birds are the focus of conservation because primatologists are in the thick of conservation action and because there are thousands of ornithologists – both professional and amateur – who are active in inventories, monitoring, and conservation efforts. It is critically important, for both ants and ourselves, that we conserve the immeasurable ecological services provided by ants to ecosystems and to humans. These benefits need to be made known and appreciated. So speak up, make data available, and get involved with conservation efforts so that ants can get the attention they deserve. This is your call to action!

PART II:

COMMUNITY DYNAMICS

Species interact both in time and space. These interactions, which fall along the continuum from positive to negative, form the basis of community ecology. Key processes influencing community structure include mutualisms, competition, parasitism, and predation. Ants are involved in, and influenced by, all of these processes. Part II examines some of these processes in detail and explores how interactions among and between species and their abiotic environment influence the organization of ant assemblages.

Competition, a major theme within community ecology and one of the most important interactions within ant assemblages, is discussed by Parr and Gibb in Chapter 5. Described as the 'hallmark' of ant ecology, competition can influence assemblage structure in several ways; the outcomes of competition determine access to resources (e.g. food and nesting sites), which species can coexist, and whether they are organized into a dominance hierarchy.

Ants are involved in an astounding array of mutualistic interactions with other taxa and these are the focus of Chapter 6. Ness, Mooney, and Lach describe the currencies on which interactions between ants and trophobionts, ants and plant propagules, and ants, fungi, and bacteria are founded. Also discussed are the adaptations for participation in, and the context-dependency of these interactions, their consequences for the wider biotic community, and why ant assemblages are model systems for studying ecology.

Central to understanding the dominance, diversity, evolutionary, and ecological success of ants is an understanding of their resource requirements. What do ants eat, where do they nest, and why? Chapter 7 builds on understanding of variation in competitive ability and foraging behaviour to reveal the main requirements of, and adaptations and constraints to, food and nest-site resources. Blüthgen and Feldhaar include within their chapter sections on the specialized habits of leaf-cutting ants, army ants, and seed harvesters, and the importance of digestive enzymes and gut micro-organisms.

Habitat disturbance and transformation, whether natural or anthropogenically-induced, can substantially affect community structure and functioning. Philpott and colleagues reason that the degree to which they affect ant communities depends on the frequency and intensity of disturbance, the permanence with which habitats are transformed, and the distance propagules must travel to recolonize affected habitats. Chapter 8 describes a range of disturbances from fire to logging to agricultural intensification, and examines the mechanisms by which these disturbances affect ant assemblage structure and diversity and their effects on the ecosystem services provided by ants.

Chapter 5

Competition and the Role of Dominant Ants

Catherine L. Parr and Heloise Gibb

5.1 Introduction

Competition has had a long history in ecology, and is one of the more contentious issues, with much debate as to its importance in structuring assemblages and indeed even its detectability (Schaffer *et al.* 1979; Schoener 1982). It can be broadly defined as a negative interaction between individuals depending on a shared, limited resource and may be either asymmetrical or symmetrical. The outcome of competition between two species is an increase in fitness of one species over another due to superior ability to access a resource.

Ants with their often large, long-lived, sessile colonies possess many of the traits expected to generate competition both within and between species. As such, competition has been described as the 'hallmark of ant ecology' (Hölldobler and Wilson 1990), and is considered to play a key role in structuring local ant assemblages. The generally sessile nature of ant colonies means that a colony may have a significant local footprint. Because different species of ants often require similar resources, such as nest sites and food, they may be commonly observed to interact aggressively with each other (Andersen *et al.* 1991; Fellers 1987; Savolainen and Vepsäläinen 1988; Savolainen *et al.* 1989).

This chapter begins by setting out important definitions and concepts, including describing different types of competition. We then review the evidence for intraspecific competition, focusing on findings from mechanistic studies and on the distribution of nests. The remainder of the chapter emphasizes the evidence for interspecific competition and its possible role in structuring ant assemblages through dominance hierarchies. We consider the evidence that competition is the 'hallmark' of ant ecology, review factors mediating it, and finally examine some major theories based on competition: the discovery–dominance trade-off and the dominance–impoverishment rule. Because invasive species are considered in Part IV, we concentrate on competition in native ant assemblages. Included in this chapter are boxes that explore how dominance is defined for ant assemblages (Box 5.1) and that examine the use of co-occurrence matrices to test whether communities are competitively structured (Box 5.2).

5.2 Defining competition

Competition among individuals of the same species is referred to as intraspecific competition, while competition between individuals of different species is interspecific competition. For ants, both an individual ant worker or reproductive, and an ant colony, can be regarded as 'the individual' when considering competition. This is because ant colonies can be considered to be superorganisms (Hölldobler and Wilson 1990) and the reproductive success of the colony, which is a function of the outcomes for individual workers and reproductives, determines the evolutionary outcome for the species. Important distinctions between interspecific and intraspecific competitions are that individuals of different species do not usually require exactly the same resources and do not use resources in exactly the same way, such that interspecific competition is more likely to be asymmetrical and may be less intense.

There are three main mechanisms through which competition may occur, acting either separately or in conjunction: interference, exploitation, and apparent competition. These mechanisms operate either directly (in the case of interference competition) or indirectly (as with exploitation and apparent competition), and apply equally to intraspecific and interspecific competition. Interference competition (sometimes referred to as contest competition) occurs directly when individuals from one colony interfere with foraging, reproduction, or survival of another, and involves direct, aggressive encounters between individuals. Among ants, aggression includes biting (often legs or antennae), charging, spraying formic acid or other chemicals from the gaster, and stinging. Some of the more colourful examples of direct interference competition include nest plugging (used by *Aphaenogaster cockerelli* to limit the foraging period of its competitor *Pogonomyrmex barbatus*; Gordon 1988) and food robbing (e.g. observed in *Myrmecocystus mimicus*, which targets returning *Pogonomyrmex* foragers; Hölldobler 1986). In interference competition, the physical effects of resource stress are usually confined to a fraction of the population, although on occasion large-scale intense interference competition may result in colony extirpation.

Exploitation competition (sometimes referred to as scramble competition) occurs indirectly through a shared limiting resource, which acts as an intermediate. Here, the active use of a resource, including food or nest sites, depletes the amount available to other ants. Superior recruitment to a resource, whether recruiting many nestmates to a food resource or usurping space (e.g. establishment of a colony to the exclusion of other colonies) are examples of exploitation competition. Although interference competition is generally more easily detected than exploitation because of the more direct nature of the interaction (Roughgarden 1983; Schoener 1977, 1983), sometimes interference and exploitation competition act together; for example, interference behaviours may prevent another species accessing a resource which enables the first species to exploit the resource.

To conclusively demonstrate interspecific exploitation competition for food resources, a number of key factors must be established: (*a*) two (or more) species must share a resource; (*b*) survivorship and/or reproduction of the competing species must be limited by the availability of this resource; (*c*) the presence of one species must negatively affect the acquisition of this resource by the other (this can be symmetrical); (*d*) this reduced resource acquisition must result in negative demographic parameters (e.g. survivorship or fecundity) of the inferior competing species, causing a change in its distribution or abundance; and finally (*e*) interference mechanisms and other processes must be ruled out (Petren and Case 1996). Studies showing all of these factors, and thus providing definitive proof for exploitation competition, are rare.

A third kind of competition, apparent competition, is a negative effect of one species on the abundance or population growth rate of another species, mediated by shared predators or other natural enemies. It can occur whether or not the two species compete directly for resources. For example, this process can result in a decrease in the population growth of two prey species that do not compete for the same resource but do share the same natural enemy. The presence of two populations of prey species allows the predator to increase its population to levels where it may exclude of one of the prey species, for example, the one with a more vulnerable life history. Evidence for the importance of this form of competition in terrestrial invertebrate communities is scarce, and it is harder to identify for most ant species.

5.3 Intraspecific competition

Intraspecific competition occurs when two or more individuals of the same species compete for the same resources (e.g. for food, space, or access to mates); for ants, this means conspecific individuals belonging to different colonies. Such competitive interactions are important because they can strongly influence population development, fertility and survival, and population densities in an area. The most common approaches to studying intraspecific competition have involved examining and manipulating nest distributions and behavioural interactions.

5.3.1 Evidence for intraspecific competition

Broadly speaking, intraspecific competition can be manifested through death or emigration of whole colonies. If colonies are of relatively equal strength, a fluctuating stalemate may be observed, with aggressive protection of intra-colonial borders or territory. Although intraspecific competition can occur at any stage of the colony life cycle, it is likely to have its greatest effects on survival at the founding stage before the colony is fully established and can be defended (Chapter 9). Many foundress queens, for instance, are attacked and killed by conspecific workers from nearby colonies before they can establish a new colony (Pontin 1960). Intraspecific competition between colonies can also reduce the performance of established colonies resulting in altered foraging areas and territories, and reduced colony size and production of alate reproductives (e.g. Gordon and Wagner 1997; Wiernasz and Cole 1995). However, once colonies are established, intraspecific competition rarely results in colony mortality (Andersen 2008; Gordon and Kulig 1996). Section 5.3.1.1, on intraspecific competition, discusses the mensurative and experimental evidence for intraspecific competition with a focus on nest distributions and behavioural interactions.

5.3.1.1 Overdispersion: patterns and mechanisms

For ants, the most frequently cited example of intraspecific competition is overdispersion of nest sites (or foraging trails; e.g. Bernstein and Gobbel 1979; Wiernasz and Cole 1995). Overdispersion (the uniform spacing of mature colonies) was first documented by early ecologists (e.g. Brian 1956; Elton 1932), and has since been found to occur consistently across subfamilies and a range of vegetation types (see Figure 5.1). The presence of a uniform spatial distribution pattern is, however, not in itself proof of competition, and there is relatively little direct experimental demonstration of intraspecific competition. It should also be noted that, although nests of polydomous colonies (single colonies occupying several, spatially separated nests) are often overdispersed (Traniello and Levings 1986), the pattern does not reflect intraspecific competition, but strategic distribution of nests to reduce costs of foraging because the nests all belong to the same colony. Later, we discuss several studies that explore how competition may result in overdispersion.

Intraspecific competition is generally thought to be strongly density-dependent; the strength and frequency of competitive interactions should increase as population density (a function of colony size and number) increases (but see Folgarait *et al.* 2007). Ants can have high population densities and are frequently territorial and aggressive, so they often have been considered a group whose density is limited by space or resources (Cole and Wiernasz 2002; Chapter 7). Observations of the territorial ant, *Lasius flavus*, suggested that nest distribution patterns change with density; at low population densities, nests are randomly spaced, whereas at high densities, nests become overdispersed (Waloff and Blackith 1962).

Cushman *et al.* (1988) also found that the degree of nest uniformity of *Formica altipetens* (as measured by the nearest neighbour analysis) increased with nest density. They suggested that, as intraspecific nest density increases, resources become more limiting, and competition increases, thus increased uniformity of nests results as a mechanism to minimize negative interactions.

In an effort to better understand the mechanisms behind overdispersion, Ryti and Case (1986) examined four competing hypotheses (intraspecific competition, microhabitat selection by foundress queens, predation of foundress queens, and predation of established colonies) for two intraspecifically overdispersed harvester ant species (*Messor pergandei* and *Pogonomyrmex californicus*). Although primarily an observational study, their results suggested that resource competition and possibly predation on foundress queens by established colonies accounted for the intraspecific patterning of these species. A subsequent three-year study on the same species involving experimental removal of colonies as well as food addition found both territory use and alate production patterns were consistent with predictions of intraspecific resource competition: colonies with intact neighbours tended to forage away from neighbouring colonies at times of low resource availability (i.e. when competition would be expected to be higher), and they also produced

Figure 5.1 Overdispersed nest mounds of *Camponotus termitarius* in low-lying pasture in northern Argentina. (Photo: Alex Wild)

fewer alates per unit worker than colonies with neighbours removed (Ryti and Case 1988). Although territory use was affected, forager activity was not. Elsewhere, removal of the Colombian harvester, *Pogonomyrmex mayri*, resulted in a near doubling of the foraging area of the nearest *P. mayri* colony within a period of only four days (Kugler 1984).

In another removal experiment, Billick *et al.* (2001) tested the hypothesis that the recruitment of new *Pogonomyrmex occidentalis* colonies is influenced by the presence of conspecific established colonies. However, they found that colony removal significantly increased recruitment in some areas, but not others, suggesting that patterns of new colony establishment may be site-specific. The death of foundress queens or limited resource availability was offered as potential mechanisms underlying the effect of established colonies on new colony recruitment. Importantly, a study using simulation modelling demonstrated that, because there can be multiple causes of regular nest spacing in ants, the observation of spatial patterns alone cannot be used as a test for competitive interactions (Ryti and Case 1992). Indeed, colony spacing was shown to be influenced not only by the relative strength of intraspecific and interspecific competitions, but also by colony birth and death rates. The model was also useful in revealing that overdispersion and regular spacing can only result from neighbourhood competition where competition between established colonies is weak, and established colonies interfere with the establishment of conspecific foundress queens. This interference of foundress queens by established colonies thus offers a plausible mechanism for Billick *et al.*'s findings (2001).

While interference competition is often considered the main mechanism promoting overdispersion of nests, a long-term study of the seed harvester ant, *Pogonomyrmex barbatus*, revealed that founding colony survival and the spatial distribution of nests are more likely to be products of exploitative rather than interference competition (Gordon and Kulig 1996). Active nest choice by emigrating colonies is another mechanism for overdispersion; emigrating laboratory colonies of *Temnothorax albipennis* actively seek to distance themselves from conspecific colonies when choosing a new nest site (Franks *et al.* 2007a).

5.3.1.2 Intraspecific colonial conflict

Where colonies do not have well-defined territories, or lack territories entirely, foraging areas can overlap substantially; local interactions can therefore be frequent. In such instances, species such as *Messor aciculatus* may engage in ritualized combat and food robbing (kleptobiosis) of neighbouring colonies (Yamaguchi 1995). This type of combat (where aggression is ritualized with threat displays rather than actual combat) and food-robbing behaviour directly impedes foraging of the opponent colony. *Iridomyrmex purpureus*, the meat ant, also engages in ritualized combat with conspecifics at territory borders, however, where these ants encounter other species in the *I. purpureus* group, fiercer fighting can sometimes result, with a much graver outcome: the death of one of the protagonists. Ants defending territories are more likely than intruders to initiate fierce fighting (van Wilgenburg *et al.* 2005). Finally, caste ratios have also been demonstrated to change in response to levels of perceived intraspecific competition. In a laboratory experiment, *Pheidole pallidula* increased its investment in soldier production after perceiving the presence of foreign conspecific colonies across fine wire mesh (Passera *et al.* 1996).

Table 5.1 Behavioural dominance hierarchy matrices for a North American desert ant assemblage at fixed baits. Phorid flies parasitize two species of the nine most common ant species in the assemblage: *Pheidole diversipilosa* and *Pheidole bicarinata*. In the absence of parasitoids, species competing for resources are organized in a significantly linear dominance hierarchy, however, this linearity is lost in the presence of parasitoids.

	Pp	Pd	Pb	Me	Cs	My	Fg	Ds	Tn	Total	Proportion won[a]
Phorids absent											
P. perpilosa	-	1		4	1	*1*	3	1		11	0.92
P. diversipilosa	0	-	3	11	3	6	12	4	5	44	0.90
P. bicarinata	0	0	-	6		1	3	4	1	15	0.79
M. emersoni	0	2	0	-	1	2	11	1	7	24	0.50
C. sansabeanus	0	1		0	-	*	4	1	1	7	0.58
Myrmica sp.	1	1	0	0	*	-	6	*1*	2	11	0.41
F. gnava	0	0	1	3	0	5	-	*1**	1	11	0.22
D. smithi	0	0	0	0	0	*1*	*1**	-	1	3	0.19
T. neomexicanus		0	0	0	0	0	0	0	-	0	0.00
Total	1	5	4	24	5	16	40	13	18	126	*P* <0.006
Phorids Present	Pp	Me	Cs	*Pb*	My	Fg	Di	*Pd*	Tn		
P. perpilosa	-	4	1		*1*	3	1			10	0.91
M. emersoni	0	-	1	*1*	2	11	1	11	7	34	0.79
C. sansabeanus	0	0	-			4	1	1	1	7	0.78
P. bicarinata		*1*		-		*1*	2			4	0.67
Myrmica sp.	1	0			-	6	*1*	2	2	12	0.50
F. gnava	0	3	0	*1*	5	-	*1*	11	1	22	0.42
D. smithi	0	0	0	0	*1*	*1*	-	*1*	1	4	0.33
P. diversipilosa		1	0		3	4	*1*	-	1	10	0.28
T. neomexicanus		0	0		0	0	0	0	-	0	0.00
Total	1	9	2	2	12	30	8	26	13	103	*P* <0.18

Notes: Cells present the outcome of competitive interactions at baits. Rows list the species that won the interaction and columns list the loser. Interactions in **bold** type indicate the cells containing the majority of wins for that pair. Empty cells represent pairs for which no interactions were observed. Dyads where each species won an equal number of interactions are *italicized*. The species affected by phorid parastoids are underlined. * indicates pairs for which the optimized matrix did not resolve their relative dominance. Pp: *Pheidole perpilosa*, Pd: *P. diversipilosa*, Pb: *P. bicarinata*, Me: *Monomorium emersoni*, Cs: *Camponotus sansabeanus*, My: *Myrmica* sp., Fg: *Formic gnava*, Ds: *Dorymyrmex smithi*, and Tn: *Temnothorax neomexicanus*. The proportion of interactions won was used to rank these species, and all species in non-significantly linear hierarchies.

[a]The proportion of all interactions a species was observed in that it won. *P* values test whether the observed matrix is more linear than would be expected by chance.

Source: LeBrun (2005).

5.4 Interspecific competition

Although detection of competition is sometimes difficult, and its importance in structuring communities has been questioned, interspecific competition is still regarded as one of the most fundamental processes in ecology, affecting not only the current distribution and success of species, but also their evolution. Competition was first described as the 'hallmark of ant ecology' by Hölldobler and Wilson (1990), but its importance as a key process structuring ant assemblages was widely accepted decades before this. Despite this acceptance, critiques of the competition literature for a wide range of taxa suggest that it is wise to be cautious when attributing outcomes to interspecific competition because it is a process that is very difficult to conclusively demonstrate (e.g. Connell 1983; Schoener 1983); this wisdom is also likely to apply to studies of ant communities.

5.4.1 Evidence for competition as the 'hallmark of ant ecology'

Support for an important role for interspecific competition in structuring ant assemblages includes agonistic behaviour such as physical and chemical aggression at resources and territorial borders and resource partitioning, in terms of both food and space. Competition is also thought to result in the structuring of ant assemblages into dominance hierarchies (see Box 5.1). Previous authors have defined species types within these hierarchies based on the abilities of ant species to defend nests, resources, and territories. However, evidence for competition suggests that it might not be as clear-cut as previously thought, and it is especially important to distinguish between individual behaviour and the effects of one population on another. Thus, while agonistic behaviours, such as biting and charging, are often cited as evidence of interference competition, such observations do not imply this type of competition is having an important effect at the population level because the process of competition acts on the unit of selection: in the case of ants, the colony (Pontin 1961; Ribas and Schoereder 2002).

5.4.1.1 Dominance hierarchies

Looking first at competition for food resources, interference competition is most easily demonstrated with the use of baits. Here, inferior competitors are displaced by more dominant species (see also Box 5.1). This scenario is found globally and forms the basis for the construction of dominance hierarchies. Many animals including hummingbirds, fish, and ants are organized in dominance hierarchies based on competitive ability (Des Granges 1979; Fellers 1987; Savolainen and Vepsäläinen 1988; Werner 1976), with the most dominant, competitive species at the top and submissive, subordinate species at the bottom of the hierarchy. These hierarchies have been found to be dynamic, with factors such as resource type, temperature, or even the presence of parasitoids affecting the position of ants within the hierarchy (see Section 5.4.3 and Table 5.1).

On the basis of interference competition at baits, three types of ants with distinct behaviours have been recognized (Hölldobler and Wilson 1990; Wilson 1971): opportunists, extirpators, and insinuators. Opportunists are species that are able to locate food resources very rapidly; they are often the first species to arrive at baits, but they are very timid in the presence of more aggressive, dominant ant species. Extirpators may take longer to find the resource, but they recruit quickly and fight to gain control. These species often have well-developed soldier castes (e.g. *Pheidole* spp.) and are usually able to capture and dominate the resource. Finally, insinuators are species that are able to gain access to a resource but usually go unnoticed by the extirpators. This they do through their small-body size, and small numbers recruiting to baits. These behaviours also link directly to the trade-off described by the discovery-dominance hypothesis (see Section 5.4.4).

Based on work in Finnish boreal forests, a similar approach to the description of ants was put forward by Savolainen and Vepsäläinen (1988) based on competitive interactions in relation to spatial behaviours. Territorial ant species are described as those that maintain and defend absolute territories (an area is continuously defended whether it contains food, nest sites, or other resources) (e.g. *Formica rufa* group

Box 5.1 Defining dominance
Catherine L. Parr and Heloise Gibb

In broad ecological terms, dominance occurs when a single species makes up a large proportion of community biomass or numbers. In ant ecology and in the context of competition, this simple term has several interpretations, being defined variously as behavioural, numerical, or ecological.

Behavioural dominance is commonly determined using observations of interspecific interactions at food baits. Species that exhibit aggressive behaviour that causes other species to retreat or avoid them are considered behaviourally dominant (Bestelmeyer 2000; Cerdá *et al.* 1997; Davidson 1998; Fellers 1987). Aggressive behaviours of ants include charging, biting (most often legs or antennae), and spraying noxious chemical compounds on a competitor. The relative behavioural dominance of different species can be compared by calculating an overall dominance score or index. This involves calculating the number of times each species is behaviourally dominant in interactions with other taxa (i.e. it forcibly expels another species from a bait, or it prevents another species gaining access to a bait) or submissive (i.e. it is expelled from or withdraws from a bait, prevented from accessing a bait, or avoids interaction) (Bestelmeyer 2000; Fellers 1987). Dominance scores are calculated as the proportion of encounters in which a species is dominant in all its interspecific encounters. The dominance scores for each species can then be used to construct a dominance hierarchy (see Table 5.1). Dominance is also sometimes calculated as turnover rates and persistence (a particular species being observed over consecutive observations). The number of shared baits at the time of observation, or coexistence interactions can also be calculated and provides additional information on interactions.
Territoriality in ants is also associated with behavioural dominance because territorial ants aggressively defend not only food resources and nests, but also mutually exclusive territories. Well known examples of such species include the epigaiec wood ants of the boreal region (*Formica rufa* group), and meat ants in Australia (*ridomyrmex purpureus* group), and arboreal territorial dominant species include the weaver ant (*Oecophylla* spp.) in tropical forests.

Numerical dominance refers broadly to dominance due to greater numbers, biomass, and/or frequency of occurrence, and is often, but not exclusively, used with reference to baits. Although when used broadly in ecology it refers to abundance, here in the context of competition it has a wide range of definitions. Usually several measures of numerical dominance are considered simultaneously in order to provide an overall indication of dominance. Numerical dominance measures include (*a*) abundance (number of individuals of a species) at baits or in traps, (*b*) frequency of occurrence of species at baits or in traps, (*c*) number of baits controlled (a minimum number of individuals are required to control a resource, e.g., >10 individuals (Bestelmeyer 2000)), and (*d*) number of baits monopolized (only one species is present). Combinations of abundance, control, and monopolization have also been used; for example, a species is deemed to control a resource bait only when it has recruited a minimum number of individuals to the bait and is the only species at the bait (Parr *et al.* 2005; Santini *et al.* 2007). The threshold abundance for bait control is debatable: It may be dependent on body size (fewer large-bodied ants can physically fit around a bait) (Bestelmeyer 2000), and foraging behaviour (e.g. mass versus individual recruitment) (Santini *et al.* 2007).

Ecological dominance was first defined by Davidson (1998), who suggested that ecologically dominant ants were those that combined superior interference behaviours (behavioural dominance) with superior exploitative competitive ability, such that they were first to discover a resource and best at defending it (see Section 5.4.4). More commonly, however, ecologically dominant ants are considered to be those that have a higher abundance at baits, relative to that in pitfall traps (sometimes called the Andersen index, Andersen 1992) (see also Cerdá *et al.* 1997). Ecological dominance is thus a ratio of foraging success to general abundance in the

continues

Box 5.1 continued

environment (LeBrun 2005). Variants include frequency-based approaches, for example, the ratio of the number of baits a species has captured to the number of pitfall traps in which the species was present, or baits at which the species was observed (see LeBrun 2005). Using frequency instead of abundance allows species to score highly regardless of whether they recruit workers to the baits (or fall into pitfall traps in high numbers), and gives a relatively high ecological dominance rating to species that do not mass recruit or are large-bodied. Behavioural and numerical dominance are commonly correlated (Blüthgen and Fiedler 2004a; Savolainen and Vepsäläinen 1988), although there are some exceptions. For example, *O. smaragdina* is behaviourally dominant at baits, but not numerically dominant (Basu 1997).

When referring to competition, the term 'dominance' is used with the aim of describing which species has control of a resource (whether food, nest site, or territory), thus whether the dominant species has captured the resource through numerical or behavioural dominance is of less importance. Using a combination of measures to assess dominance at the scale of interest would therefore ensure that all dominant species can be identified.

in northern Europe, or *Azteca* species in Panamanian mangroves [Adams 1994]; and *Oecophylla longinoda* in Africa forests), while encounter species are those that defend food and nests, but not territories (e.g. *Lasius niger*), and submissive species only defend their nest, but do not have territories, nor do they defend food resources (e.g. *Formica fusca*). Of course, not all ants fit neatly into these categories; many species defend territories, but only against conspecifics or closely related species (e.g. Greenslade 1987).

5.4.1.2 Agonistic behaviours and territoriality

In addition to aggressive contact behaviour, such as biting and charging (Plate 2), some species are known to employ chemical defence to win control of resources (most typically small food resources such as dead insects). Small, slow-moving, and non-aggressive ant species are known to use chemical defence to repel other ants and gain access to resources that would otherwise be controlled by more dominant ant species. For example, *Monomorium 'rothsteini'* group in northern Australia uses venom alkaloids to deter highly aggressive species of *Iridomyrmex* (Andersen *et al.* 1991). However, although (and perhaps because) chemical defence is widespread among taxa (e.g. *Solenopsis, Tetraponera, Wasmannia*, and *Forelius*; Braekman *et al.* 1987; Howard *et al.* 1982), it does not guarantee behavioural dominance or necessarily infer superior competitive powers (Davidson 1998).

At a larger scale, some dominant ant species with distinct territories may engage in large aggressive interspecific wars to defend and retain control of their territory (see also Chapter 12). Examples include battles between the tropical ant, *Oecophylla smaragdina* and large-bodied meat ants, *Iridomyrmex* spp. (including *Iridomyrmex sanguineus* and *reburrus*), which engage in fierce battles often characterized by many injuries and deaths (C.L. Parr, personal observation). Elsewhere, mound-building *Formica* species are known to exhibit spatial territorial competition, and defend territory boundaries (Hölldobler and Wilson 1990). Further evidence of spatial competition includes expansion of one species following removal of competitor species (Haering and Fox 1987; see Section 5.4.2).

5.4.1.3 Resource partitioning

The partitioning of resources used by different ant species is thought to represent the 'ghost of competition past' (Connell 1980), where species have switched from shared to unshared resources as a result of strong competitive forces. In reality, it is impossible to determine if resource partitioning results from competition or some other factor, but partitioning of resources in ant assemblages is common. There are many dimensions in which the resources in any natural system can be divided up. For example, food resources can differ through their nutritional composition, quality, size, spatial distribution, and temporal persistence. Success rates of competing species differ with resource type (e.g. protein and carbohydrate baits; Sanders and Gordon 2003) and resource size (e.g. LeBrun 2005). Other examples include partitioning of diurnal forager activity due to differences in temperature tolerance (Cerdá *et al.* 1997) and partitioning of nest sites (see Chapter 7). The partitioning of resources and microhabitat space and the creation of a plethora of niches may thus be key to the occurrence of such high ant species richness in ecosystems worldwide. However, given that partitioning results partially from past events that may or may not have involved competition, the role of competition in promoting resource partitioning is difficult to prove.

5.4.1.4 Spatial ant mosaics

Ant mosaics, a type of spatial partitioning, are spatial patchworks of two or more dominant ant species that have non-overlapping territories. Subordinate ants and other arthropods commonly show positive or negative associations with particular dominant ants (Blüthgen and Stork 2007; Davidson *et al.* 2007; Majer 1972; Room 1971). Such mosaic patterns in the arboreal ant fauna have received considerable interest since they were first described in plantation crops (e.g. cocoa, oil palm, and citrus) in the tropics (see Greenslade 1971; Leston 1973; Majer 1972; Room 1971; Samways 1983). Evidence of mosaics in natural forests has been divided, with several studies questioning the existence of mosaics in undisturbed habitats (e.g. Floren *et al.* 2001; Floren and Linsenmair 2000). The existence of mosaics in ground-active ant assemblages has not received as much attention although microhabitat mosaics around different food resources may exist.

Recently, studies using a null model approach (Gotelli 2000; Gotelli and Graves 1996) to re-examine patterns found in plantation studies have challenged the statistical validity and the conclusions of previous studies (Ribas and Schoereder 2002; Sanders *et al.* 2007a; see Box 5.2). Theory predicts that if interspecific competition is important in structuring assemblages, species should co-occur less often than expected by chance (Diamond 1975). Although null model co-occurrence analyses provide a method for quantifying and detecting co-occurrence patterns, it should be remembered that a significant checker board pattern (non-random co-occurrence) does not necessarily mean that competition is the structuring mechanism (see Box 5.2). Other factors such as habitat variation, food, or nest site availability may also be important. Ideally, this technique should therefore be used in conjunction with others to detect and provide conclusive evidence of competition. A recent review by Blüthgen and Stork (2007) provides an excellent critical examination of the ant mosaic concept, and highlights some important issues that must be addressed if we are to distinguish whether mosaics are absent from, or just more difficult to detect, in complex natural systems. Suggestions include tests with improved statistical power, and analyses at appropriate spatial scales.

5.4.2 Field tests of interspecific competition

Although there has been considerable debate on the importance of competition in ecological communities (Connell 1983; Schoener 1983), its importance in structuring ant assemblages has often been uncritically accepted. Experimental manipulations of ant assemblages are difficult to perform, particularly in the field. As a consequence, much of the evidence suggesting an important role for interspecific competition, both in studies of ant communities and of other organisms, is non-experimental and attributes variations in morphologies, behaviours, habitat use, and distributions to resource partitioning as a result of competition (Connor and Simberloff 1986). The problem with this 'natural experiment' approach is that the existence of non-random patterns does not provide evidence that the process responsible for the pattern was competition (Connell 1983; Schoener 1983; Underwood 1986). Even when populations are experimentally manipulated, problems such as a lack of replication, lack of appropriate controls, and confounding in the experimental design can result in incorrect

Box 5.2 Co-occurrence analyses: what are they and how do they work?

Heloise Gibb and Catherine L. Parr

Co-occurrence analyses test for non-random patterns of species co-occurrence using a presence—absence matrix. Co-occurrence analyses have been used to explore patterns in ant mosaics (e.g. Sanders *et al.* 2007a), between different habitat types and at different scales (e.g. Gotelli and Ellison 2002), as well as to investigate the effect of invasive species on community structure (Sanders *et al.* 2003a).

Working on assemblage rules and bird occurrence patterns on islands, Diamond (1975) first coined the term 'checkerboard distribution' to describe a pattern where a species pair never co-occurs. As such, Diamond suggested that a high prevalence of checkerboard pairs in a community is indicative of a competitively structured community. At the community level, the checkerboard pattern was first tested against a null hypothesis of random community assembly by Connor and Simberloff (1979). Their finding that these assemblage patterns can arise as a result of random colonization and in the absence of interspecific competition resulted in considerable debate about the assumptions of the null models used to test co-occurrence and the appropriateness of particular null models (see, e.g., Gotelli 2000; Gotelli and Entsminger 2007; Manly 1995; Schmid-Hempel 1998).

However, Stone and Roberts (1990) 'C-score', which compares favourably with a range of different techniques for testing co-occurrence (Gotelli 2000), has become the favoured technique for examining species co-occurrence patterns. The C-score test quantifies the number of checkerboard units found for each species pair (see Figure 5.2.1). If the C-score is large, compared with the null distribution, it suggests that avoidance between species predominates in the assemblage. If it is small, species tend to be aggregated. However, high C-scores may result from high levels of both aggregation and avoidance, simultaneously. This is because extreme 'checkerboardness' is produced by forces leading to both species aggregation and avoidance (Stone and Roberts 1992). While this leads to difficulties in identifying whether associations tend to be negative or positive, a mixture of such associations is precisely what one might expect in a strongly structured assemblage.

C-scores can yield more useful information when subsets of species, for example, behaviourally dominant ants, rather than an entire assemblage are used. However, selection of species of interest can often lead to bias if more objective measures are not used. To determine the importance of aggregation in an assemblage with a high C-score, Stone and Roberts (1992) also developed a *T*-statistic, which determines the 'togetherness' or aggregation of species among sites, and an *S*-statistic to describe the number of shared sites.

The C-score method has several limitations, including its potential to be overly conservative as a result of its failure to account for differences in abundances at shared sites (Blüthgen and Stork 2007). In addition, it is sensitive to statistical independence and spatial autocorrelation, so care must be taken when drawing conclusions about the patterns observed. However, problems related to statistical independence can be accounted for to some extent by taking care to design appropriate sampling regimes. One commonly ignored issue is that similar species are likely to have similar habitat preferences, so controlling for habitat affinities by testing within, rather than between, habitat types may be critical in detecting meaningful patterns (Schoener and Adler 1991). Methods are available that allow one to distinguish between shared habitat associations and competitive interactions by using data on both habitat and species abundances (see Schoener and Adler 1991; Sfenthourakis *et al.* 2005). Competing species pairs are expected to covary negatively in the co-occurrence matrix, but to share habitat preferences (Sfenthourakis *et al.* 2005). However, interactions with unrecorded abiotic factors, as well as mutualisms or predation may also be causative factors, so observation of significant

Box 5.2 continued

co-occurrence patterns should always be followed up with studies testing their causes (Schluter 1984).

Freeware to conduct co-occurrence analyses is available in the EcoSim package (Gotelli and Entsminger 2007; http://www.garyentsminger.com/ecosim/ecosim.htm). Sfenthourakis *et al.* (2005) have produced an add-on to this programme, 'COOC', which allows inclusion of environmental data.

a

	1	.	.	.	.	.	.	.	*m*
1	0	0	0	0	0	1	1	1	1
.	0	0	0	0	0	1	1	1	1
.	0	0	0	0	0	1	1	1	1
.	0	0	0	0	0	1	1	1	1
.	0	0	0	0	0	1	1	1	1
.	0	0	0	0	0	1	1	1	1
.	1	1	1	1	1	0	0	0	0
.	1	1	1	1	1	0	0	0	0
.	1	1	1	1	1	0	0	0	0
.	1	1	1	1	1	0	0	0	0
.	1	1	1	1	1	0	0	0	0
n	1	1	1	1	1	0	0	0	0

b

	1	.	.	.	.	.	.	.	*m*
1	1	0	0	1	1	0	0	0	1
.	0	0	1	0	1	1	0	0	0
.	1	1	0	1	0	0	1	1	1
.	1	0	0	0	0	1	1	0	1
.	1	1	0	0	0	1	0	1	1
.	1	0	1	0	1	0	1	0	0
.	0	1	1	1	0	1	0	1	0
.	1	1	0	1	0	0	1	1	1
.	1	0	1	0	1	1	0	0	0
.	0	0	1	0	1	1	0	1	0
.	1	0	0	1	0	0	0	1	1
n	0	1	1	0	0	1	1	0	1

Figure 5.1.1 The C-score test quantifies the number of checkerboard units found for each species pair. The number of checkerboard units for each species pair is $(R_i - S)(R_j - S)$, where R_i is the number of occurrences for species i, and R_j is the number of occurrences for species j, and S is the number of sample plots in which both species occur. A checkerboard unit is large if the species pair shares few sites, and small if they share many (i.e. they are aggregated). The C-score is the average of all checkerboard units for all species pairs. The matrices above show the arrangement of species among sites in the situation: (*a*) where there is a high level of co-occurrence of species pairs among sites, and (*b*) where co-occurrence of species pairs among sites is low.

inferences being drawn (Underwood 1986). Here, we briefly examine the experimental evidence for competition in ant assemblages in the field.

Experimental manipulations of ant assemblages have addressed the role of competition either by introducing or removing key species. In most cases, the species removed have been selected due to their behavioural and/or numerical dominance (e.g. Andersen and Patel 1994; Gibb and Hochuli 2004; Majer 1976; Rosengren 1986) or because they are invasive (King and Tschinkel 2006; LeBrun *et al.* 2007). Some of the earlier studies involved removals or additions of a series of species (e.g. Cole 1983; Kugler 1984; Majer 1976; Pontin 1969), and one involved removal of an entire assemblage of granivorous ants (Valone and Kaspari 2005). Most studies have thus targeted species expected to show the strongest competitive effects.

Colony removal methods have included applying hot water directly to or excavating nests (e.g. King and Tschinkel 2006; LeBrun *et al.* 2007), caging ants into their nests (Gibb and Hochuli 2004; see Figure 5.2), cutting down arboreal nests (e.g. Majer 1976), poisoning (e.g. Valone and Kaspari 2005), and fencing (e.g. Andersen and Patel 1994). Experiments have been maintained from between one day (Kugler 1984) and 15 years (Valone and Kaspari 2005), and populations have been monitored using a range of techniques, including counting nest entrance holes, pitfall trapping, visual surveys, and baiting. The success of removal experiments has also varied immensely, with a 40–100% reduction in the abundance of the excluded species compared with control sites. That many experiments do not achieve complete removal (e.g. King and Tschinkel 2006), combined with the range of habitats and diversity of ant assemblages in which these experiments have been performed, makes it unsurprising that the authors' conclusions have varied from an absence of competition (King and Tschinkel 2006), through functional group-specific competition (Gibb and Hochuli 2004) to the existence of a competitive network or hierarchy (Rosengren 1986).

Many of these studies have focused on the outcomes of ant resource use by recording interactions at baits. Studies using baits to measure responses have tended to observe stronger patterns than those using measures of activity independent of baits (e.g. pitfall traps). For example, northern meat ants, *Iridomyrmex sanguineus*, affected ant abundance and species richness at baits, but similar patterns could not be detected in pitfall traps (Andersen and Patel 1994). This may have been a result of the short-term nature of the study, but it is also likely that the apparent effect of *I. sanguineus* was artificially inflated at baits because other species were excluded. Resource type also mediates the effects of competition; the success of *Aphaenogaster cockerelli* was only affected by *Myrmecocystus* species when foraging at preferred protein baits (Sanders and Gordon 2003). Testing the role of meat ants, *I. purpureus*, Gibb (2005) also detected more pronounced effects of competition at protein baits when compared to carbohydrate baits. Responses of the invasive *Linepithema humile* (Argentine ant) and *Solenopsis invicta* (red imported fire ant) to site-level removals of each other in their native habitat were roughly symmetrical at baits, and other ant species also captured more bait stations when one or the other of these species was removed (LeBrun *et al.* 2007).

Responses to more passive sampling methods are often less clear. Using visual surveys, Gibb and Hochuli (2004) showed that *I. purpureus* most strongly affected ecologically similar species (other species of *Iridomyrmex*) on rock outcrops, with limited effects on other species (see Figure 5.3). Similarly, Haering and Fox (1987) showed that removal of one species of *Iridomyrmex* affected the area occupied by a competing congeneric and

Figure 5.2 Exclusion experiment showing a cage built around nest of *Iridomyrmex purpureus* in southeastern Australia. (Reproduced with permission, from Gibb and Hochuli 2004).

vice versa, and that this turnover was related to habitat succession. A partial removal of the invasive *S. invicta* from an assemblage dominated by other introduced species also had little effect on ant abundance and species richness in a disturbed habitat (King and Tschinkel 2006). However, it is possible that the relatively short-term nature of these experiments (up to two years) and incomplete removal did not allow sufficient time for an assemblage response. Work by Pontin (1969) suggests that increasing both inter- and intraspecific competition can have short-term effects on alate production in mainland populations. Introduction experiments on islands, where habitat area and heterogeneity are low, suggest a more important role for competition than exclusion experiments on mainland habitats. Cole (1983) showed that the colonization success of transplanted ants on mangrove islets often depended on which other ant species were present, while Rosengren (1986) suggested that effects of transplanted *Formica truncorum* cascaded across small island ant assemblages.

Clearly, competition plays some role in structuring ant assemblages, but, as with other taxa, the biased nature of species selection for competition studies, both mensurative and manipulative, may have enhanced its apparent prevalence. It is also likely that factors such as environmental stochasticity and inherent differences in food, temperature, and habitat preferences play a significant role in reducing the impact of competitive interactions on ant populations. We consider the impacts of these factors in Section 5.4.3.

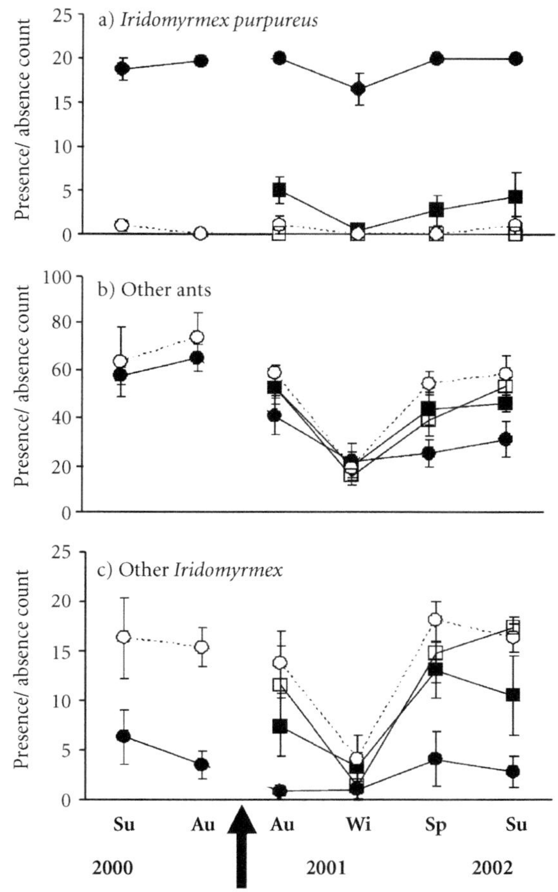

Figure 5.3 Exclusion of the dominant ant, *Iridomyrmex purpureus* resulted in increased activity of functionally similar species (Other *Iridomyrmex* that are also dominant), and a more limited response of all 'Other ants'. Data are mean (± SE) presence/ absence counts per treatment (N = 4 sites) of (*a*) *Iridomyrmex purpureus*, (*b*) Other ants, and (c) Other *Iridomyrmex*. Symbols: ● sites with *I. purpureus*, ■ exclusion sites, ○ sites without *I. purpureus*, □ procedural control sites. Arrow represents commencement of exclusion experiment, and sampling times are abbreviated as Su (summer), Au (autumn), Wi (winter) and Sp (spring). (Reproduced with permission, from Gibb and Hochuli 2004).

5.4.3 Factors modifying competitive interactions

Differences in the findings from different types of experiments highlight the point that the outcome of competitive interactions is likely to be highly conditional and dependent on abiotic factors including temperature, humidity, habitat structure, and habitat disturbance, and biotic factors, such as parasitism and resource size and composition.

Temperature and humidity are usually interdependent, so most studies examine the effects of changes in temperature and humidity simultaneously, or concentrate on temperature-related effects. Similar to plant communities (Grime 1979), there may be a trade-off between stress tolerance, disturbance, and competition in ant assemblages, with competitive interactions being most intense under the least stressful and disturbed conditions (Andersen 1995). For ants, lowered temperatures as a result of vegetation and climatic conditions may provide the most stressful conditions, and thus the least competitive environment. High temperatures (e.g. above 35°C) may also present a stressful environment to many species, with behaviourally dominant ants being active at times of the day when temperatures are only moderately high (Bestelmeyer 2000). In some habitats, temperature extremes result in partitioning of foraging times, thus reducing opportunities for competitive interactions (e.g. Mediterranean habitats; Cerdá *et al.* 1988). However, competition may occur between spatially co-occurring species that differ in activity rhythms (i.e. nocturnal and diurnal species) if they alter their activity patterns in the presence of a persistent food source (e.g. Mercier and Dejean 1996). Importantly, the interaction between temperature and competition may explain the failure of invasive ants, such as the Argentine ant, *L. humile*, to extensively invade otherwise suitable habitats in some areas to which they have been introduced. For example, in Australia, the native *Iridomyrmex* species, is able to tolerate higher temperatures than *L. humile*, and thus retain greater control of resources (Thomas and Holway 2005).

Parasitism, most commonly by flies of the family Phoridae (most prevalent in the Americas), has been shown to influence the outcome of competitive interactions (Feener 1981). Parasitoids commonly reduce the competitive success of ant species, such that the outcome of competitive interactions becomes less predictable, and a greater range and abundance of species gain access to resources (LeBrun 2005; LeBrun and Feener 2002; Philpott 2005b) (see Figure 5.4 and Table 5.1). Escape from parasitism has often been cited as an explanation for the incredible success of invasive species such as *S. invicta*, allowing these species to break the discovery–dominance trade-off (see Section 5.4.5 and Chapter 14) (Porter *et al.* 1997). Much of the research on the effect of parasitoids on competitive interactions has thus resulted from the search for biological control agents.

Parasitoid presence at food resources can reduce the occurrence of competitive encounters by inducing behavioural changes in the host, associated with

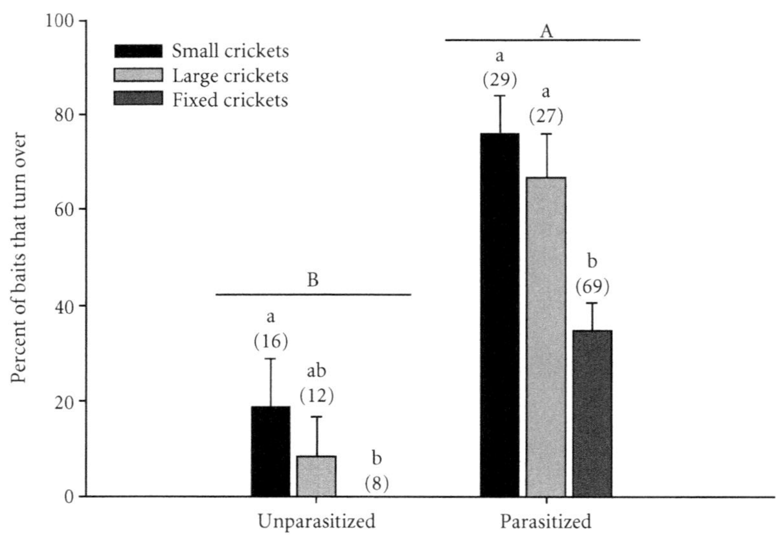

Figure 5.4 Presence of parasitoids at food resources influences the success of ants in capturing food resources when competing with other ant species. LeBrun and Feener (2002) found that unparasitised ants (*Pheidole diversipilosa*) were more successful at retaining control of food baits in the presence of competitor species than parasitised individuals. Furthermore, food resource size also affects the ability of the host ant to capture the resource; where baits cannot be carried off the host ant was best able to defend and retain control of the resource. Bars represent mean percent of baits (± SE). Significant differences are indicated with letters. With kind permission from Springer Science and Business Media: LeBrun and Feener (2002).

avoidance of parasitism; for example, *Solenopsis geminata* adopts a stationary, curled, defensive posture in response to its parasitoid (Morrison 1999). Parasitoids can also affect the outcome of competitive interactions because ants trade-off their need for food, success in competitive interactions, and avoidance of parasitism. For example, *Solenopsis richteri* responded to the presence of *Pseudacteon* flies by reducing feeding activities by the same factor, independent of the size of the resource (Folgarait and Gilbert 1999). Conflict between host ants and competitors may also improve the ability of parasitoids to locate their hosts, as a result of chemicals released during interspecific encounters (LeBrun and Feener 2002; Orr *et al.* 2003). However, behaviour may also affect attack rates; parasitoid flies have no effect on the outcome of interference competition, but a large effect on the outcome of exploitative competition between *S. invicta* and *S. geminata* (Morrison 1999, 2000). While most studies have detected significant parasitism–competition interactions, some have not, or have found that they are conditional on abiotic factors, such as humidity (Ramirez *et al.* 2006).

We were unable to find studies investigating the effects of non-ant predators on competitive interactions among ants. Army ants, *Neivamyrmex compressinodis*, may indirectly affect competition between *Wasmannia auropunctata* and other native species in French Guyana. This is because *W. auropunctata* defends its nest less effectively than other species and thus suffers a severe population reduction when attacked by army ants, making it a less-effective competitor (Le Breton *et al.* 2007). Intra-guild predation and slave-raiding can obscure the distinction between interspecific competition and predation or appropriation because both processes can have effects on potential competitors (Hölldobler 1983; Wilson 1976; Zee and Holway 2006). Intra-guild predators prey on species within the same functional guild, that is, species with which they would normally compete. Slave-raiding, on the other hand, involves elimination of the competition by incorporating part of the population into the aggressor's colony.

Habitat structure has also been shown to alter the outcomes of competitive interactions, although the body size of the competitively dominant ant can determine how changes in habitat complexity affect its success. When the competitively dominant species is small relative to gap sizes, simple habitats may impede its success in competition (Wilkinson and Feener 2007). In contrast, for relatively large species, complex habitats may eliminate a competitive advantage (Gibb 2005; Sarty *et al.* 2006). Highly heterogeneous microhabitats may impede the success of aggressive behaviours of large behaviourally dominant ants and provide more hiding places for other species, therefore reducing the impact of behavioural dominance. Natural or anthropogenic disturbances often simplify habitat structure, resulting in changes in competitive interactions (Chapter 8). Disturbance-favoured competitively dominant species may utilize the simple habitat of access roads to colonize national parks (Gibb and Hochuli 2003), may be favoured by fire (C. Parr, unpublished data) or by soil disturbance from agriculture (Folgarait *et al.* 2007) or promoted by habitat fragmentation (e.g. areas of edge habitat favour Argentine ants, possibly allowing them to outcompete native species, Suarez and Tsutsui 2008). Heterogeneity in resources may also contribute to species coexistence. In an ant-*Acacia* guild, competitively dominant ants were shown to be more likely to supplant subordinates on trees close to resource-rich termite mounds than those at greater distances (Palmer 2003). This was because they were better able to exploit protein resources that were abundant close to the mounds. Competitively subordinate species thus persisted by occupying the less desirable areas, with lower resource abundance.

The composition, quality, and size of natural food resources vary enormously, and the intensity of composition for a resource depends both on the nutritional value of the resource and its desirability for a particular ant species. Blüthgen and Fiedler (2004a) showed that nectars used by dominant weaver ants, *Oecophylla smaragdina*, were characterized by higher sugar and amino acid concentrations than those used by other nectarivorous ant species. They suggested that 'competition for composition' meant that other species were only successful competitors at suboptimal resources. Using the qualitatively greater contrast of protein and seed baits, Sanders and Gordon (2003) showed that resource type mediates the effects of competition, with the success of *Aphaenogaster cockerelli* only affected by

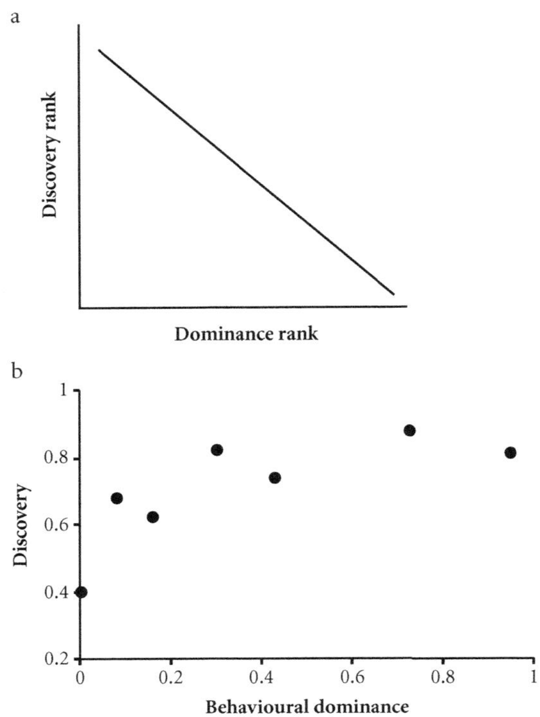

Figure 5.5 (*a*) The discovery–dominance relationship describes the trade-off between the ability to discover food resources and capture food resources through dominance (Davidson 1998; Fellers 1987). Dominant species (low dominance rank) are considered poor discovers of food (high discovery rank), while species that discover food quickly (low discovery rank) have low dominance (high dominance rank).(*b*) Santini *et al.* (2007) found the discovery–dominance relationship to be reversed in Spain, and suggest it is due to habitat type and thermal regimes. (Reproduced with permission, from Santini *et al.* 2007).

Myrmecocystus species when foraging at preferred protein baits.

Resource size has also been shown to affect the outcome of competitive interactions, with small, dispersed resources favouring quick discovery and removal times, while large, clumped resources favour recruitment over a longer harvesting period. Aggressive behaviours are more likely to be more successful and rewarding at larger resources that can be more easily defended (e.g. Kaspari 1993; Gibb 2005; LeBrun 2005), than at small resources that can be carried away by individual workers (see Figure 5.4). The relationship between success at resources and behavioural dominance may therefore depend on resource size, with species that rapidly discover resources avoiding interference competition at small, but not large baits (Adler *et al.* 2007; LeBrun 2005). Kaspari (1993) showed that bird droppings with large numbers of seeds were used by a subset of the ant assemblage that exploited droppings with fewer seeds. This suggests that resource defence by aggressive species prevented access of other species to the more desirable larger food sources. Many of the studies of ant assemblage dynamics discussed in Section 5.4.2 used large baits for ease of observation (i.e. baits that were too large to be harvested by single individuals and thus required recruitment, usually at least 1 cm^3). They were thus very likely to detect interference competition, but how strongly this reflects the true prevalence of competitive interactions is unclear.

The outcome of competitive interactions is thus clearly highly conditional, suggesting that focusing on effects at just one bait type, bait size, or microhabitat might limit our understanding of the importance of competition in structuring ant assemblages.

5.4.4 The discovery–dominance trade-off

Interspecific trade-offs are thought to be a requirement for species coexistence, particularly at small spatial scales (Kneitel and Chase 2004, MacArthur 1972, Tilman 1982; see also Hubbell 2001). A trade-off between a species' ability to dominate resources and to discover them is analogous to a competition–colonization trade-off (Hastings 1980; Levins and Culver 1971), and may occur if a species' ability to excel at interference competition results in specialized morphological, behavioural, and physiological characteristics that somehow reduce that species' ability to discover resources in the first place (Case and Gilpin 1974). Such a trade-off in an ant assemblage was first described by Fellers (1987), who suggested that the discovery–dominance trade-off (see Figure 5.5*a*) explains how so many ant species apparently depending on similar resources can coexist. Subsequent authors have suggested that their findings are consistent with the presence of a trade-off (e.g. Holway 1999; Le Breton *et al.* 2007; Morrison 1996; Perfecto 1994; Sarty *et al.* 2006). However, several studies have suggested that such a trade-off may not be a consistent characteristic of ant assemblages. Davidson (1998) reignited the issue, suggesting that an ant garden assemblage failed to show such a trade-off because some species were able to break the trade-off through enhanced abilities to access carbohydrate resources. Davidson suggested that higher energy levels could then lead to greater activity, resulting in faster ants, a high density of active workers, rapid discovery of resources, and an enhanced ability to capture prey. Greater energy levels could also increase the ability of a species to physically defend spatial territories or resources. This feedback mechanism would mean that ants defending a stable carbohydrate resource, such as hemipteran honeydew, could therefore be both faster at discovering resources and better at defending them. Species that break the trade-off could be considered 'ecological dominants' (Davidson 1998) or 'superspecies' (Tilman 1982).

Invasive species provide another example of species breaking the discovery–dominance trade-off (Chapter 14). Holway (1999) showed that, while the assemblage of native ants at his study sites displayed a clear discovery–dominance trade-off, the introduced Argentine ant, *L. humile*, excelled at both dominance and discovery. Its enhanced abilities may have resulted from its evolution in a different competitive environment. Another possibility is that species that usually experience other pressures (e.g. parasitism) in their home ranges may appear to have superior abilities to dominate resources if they invade enemy-free space (LeBrun and Feener 2002). Such species may therefore lie outside the usual relationship (i.e. they may be both the quickest to discover a resource and the most competitively dominant at it), thus breaking the trade-off. In their native ranges, however, it is not always clear that there is a significant discovery–dominance trade-off; for example, *S. invicta* showed a significant trade-off in only one of three study locations (Feener *et al.* 2008). LeBrun and Feener (2007), also showed clearly that resource size influenced the discovery–dominance relationship because, for smaller resources, the discoverer usually also succeeded in rapidly removing the resource, thus resulting in a positive relationship, rather than a trade-off.

Habitat complexity may also play an important role in determining rates of discovery and dominance. In an ant assemblage composed mainly of introduced species, Sarty *et al.* (2006) show that a trade-off between dominance and discovery is least apparent in simple habitats. In complex habitats, large, aggressive species were almost as quick to discover baits, but their ability to aggressively defend them may have been compromised by the greater complexity, which provided more hiding spaces for other species and less room for them to manoeuvre. However, the outcome may have been different if smaller ant species had been behaviourally and numerically dominant.

A clear positive relationship between dominance and discovery abilities, rather than just the existence of an outlying 'ecological dominant', has been observed in Mediterranean ant assemblages (Santini *et al.* 2007) (see Figure 5.5*b*). They described their findings as a 'complete reversal of perspective', suggesting that habitat and thermal factors may be responsible for the apparent anomaly. One aspect to consider in testing the discovery–dominance trade-off is that more abundant species are more likely to be first to discover food resources, just by chance. LeBrun and Feener (2007) offered a relative measure of resource discovery that adjusts for differences in species abundances and might help account for this bias. Clearly, the discovery–dominance trade-off and factors that modify it require further investigation, with too many exceptions and conditions to establish a general assumption of such a relationship in all systems.

5.4.5 Dominance–richness relationships across scales

Another key idea that has dominated competition theory in ant ecology has been the 'dominance-impoverishment rule' (Hölldobler and Wilson 1990), and associated dominance-richness relationships (see also Box 5.1). The dominance-impoverishment rule describes the relationship between ant species richness in a community and dominant species: The fewer ant species in a local community, the more likely it is to be behaviourally dominated by one or two species with large aggressive colonies. The rule is thought to hold across a variety of environments (Greenslade 1971; Vepsäläinen and Pisarski 1982), although at a regional scale exceptions include the tropics of Australia, where both local-species richness and dominance are high (Andersen 1995).

Most competition studies have investigated the effect dominants have on species richness (the dominance-richness relationship), not vice versa (e.g. Andersen 1992; Andersen and Patel 1994; Farji-Brener *et al.* 2002; Morrison 1996; Retana and Cerdá 2000). Although it is difficult to disentangle cause and effect, succession or colonization studies can shed light on the causality as the assemblage develops. Dominance-richness relationships have been documented most commonly at the level of baits, where dominant species regulate 'momentary' diversity (Andersen 1992), or the numbers of species in attendance at a bait, at a given time. Such 'small-scale' regulation of diversity has been found in several ant baiting studies, including those in tropical savannas (Andersen 1992; Andersen and Patel 1994; Parr *et al.* 2005), French Polynesia (Morrison 1996), the boreal taiga biome (Savolainen and Vepsäläinen 1988), and tropical agroecosystems (Perfecto and Vandermeer 1996).

The effect of dominant ants on local-species richness was examined by Andersen (1992), who suggested that for baits, the full relationship between species richness and dominance is unimodal: Species richness is low at very low levels of dominance, and as dominance increases species richness also increases until a point is reached after which species richness declines as dominance increases (see Figure 5.6). The mechanisms behind the unimodal dominance-richness pattern are varied. The ascending portion of the curve has been shown to be largely constrained to take this form as a result of the shape of abundance frequency distributions (Parr *et al.* 2005), although environmental stress (increasing habitat favourability) also plays a role. The descending part of the curve is due to interspecific competition, where high abundances of dominant ants reduce species richness through competitive exclusion (Andersen 1992; Parr *et al.* 2005).

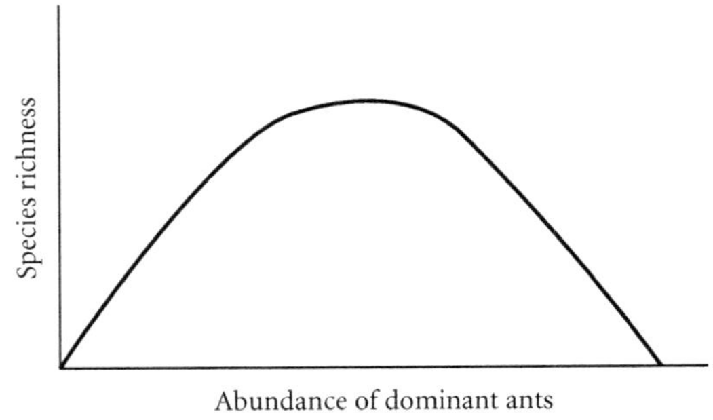

Figure 5.6 The dominance–richness relationship (Andersen 1992) where species richness is low at very low levels of dominance, and as dominance increases species richness also increases until a point is reached where they abundance of dominant ants becomes so great that they exclude other ant species and reduce species richness.

The extent to which this unimodal relationship applies at the broader scale of the assemblage has had few explicit tests (see Morrison 1996). This is problematic because exclusion from baits does not necessarily imply competitive exclusion by dominant ants is taking place at the assemblage level (see Andersen and Patel 1994; Gibb and Hochuli 2004; Ribas and Schoereder 2002; Sanders and Gordon 2002). Dominant ants may control bait resources but, because at a broader scale resources may either be not limited or species in the assemblage use a variety of only partially overlapping resources, it may not be possible to predict with confidence whether competition is also the main structuring force for the local assemblage. Similarly, it is not clear from pitfall trapping alone, which, if any, species are behaviourally dominant and thus exert a competitive influence on the assemblage, or whether some other factor is responsible for the patterns found. Several methods may be needed to detect competition and provide evidence of assemblage structuring. For example, data from a combination of baiting and pitfall trapping in South African savanna habitats demonstrated that where there were high abundances of dominant ants, species richness was reduced, strongly suggesting that dominant ant species can control assemblage richness patterns (Parr 2008). While competition has been shown to be important in regulating 'momentary' diversity, at the assemblage level, the form of the dominance-richness relationship and its causal mechanisms still require further investigation across a range of systems.

5.5 Future directions

Competition is certainly one of the most accessible and fascinating aspects of ant ecology for anyone who has sat and watched behavioural interactions of ants enticed by a tasty food item. Potential avenues for research relating to intraspecific competition include investigating nest spatial distributions in relation to habitat heterogeneity and resource availability, and how patterns alter with scale. Given the potential importance of competition in structuring local assemblages, the impact of disturbances (natural and anthropogenic) on competition and dominant ants deserves more attention.

Although competition is well studied in ant communities relative to other taxa, studies show a wide variety of conclusions on the role of interspecific competition, ranging from little or no competition, to communities entirely structured by competitive interactions. Results clearly indicate that the strength of competitive interactions is context-dependent and, as such, a better understanding of the role of factors such as habitat complexity and biotic interactions in regulating competition will allow us a better predictive ability. Specifically, does the degree of competitive structuring decrease with habitat complexity or heterogeneity in resources? Such questions could be addressed with straightforward experiments manipulating habitat complexity or resource heterogeneity. Much of the research on competition now focuses on invasive species, and invaded or disturbed communities. Such research is of critical importance in understanding how to prevent homogenization of ant faunas, but it is equally important that we understand the role of competition in more natural communities and how it differs from invaded communities.

In determining the degree to which communities are structured, studies that more thoroughly dissect ant assemblages by removing a series of species and that also use suitable controls may help to disentangle its importance across the assemblage, and enable the concept of competitive hierarchies to be more thoroughly tested. Such manipulative experiments allow a much stronger mechanistic understanding of how communities are structured and, in the long term, they provide more conclusive results than mensurative studies, even if they are more laborious and costly to conduct.

Exciting developments in our understanding of ant nutritional ecology suggest that ecologically dominant ants may be reliant on plant-derived sugars (such as nectar and hemipteran honeydew) to maintain their active and aggressive lifestyles (Davidson *et al.* 2003; Grover *et al.* 2007). Key competition-related questions in this area centre on control of and access to these sugar resources. Previous studies show that ecologically dominant ants exclude other ants from higher quality sugar resources (e.g. Blüthgen and Fiedler 2004a), but it is not clear how important the spatial and temporal availabilities of resources are for the effectiveness of aggressive resource protection strategies by dominant ants.

More specifically, of particular relevance for future work on the discovery–dominance relationship is investigation of the biotic and abiotic conditions under which these trade-offs are found; presumably, such trade-offs are only possible in systems where no native or invasive 'ecological dominance' occurs. Finally, if the ability to excel at interference competition results in specialized morphological, behavioural, or physiological characteristics that somehow reduce that species' ability to discover resources in the first place, then what characteristics might species have traded-off?

5.6 Summary

Competition occurs when different species or individuals require the same limiting resources and can only be shown definitively using experimental methods. It can occur within colonies for reproductive rights, between colonies of the same species (intraspecific competition) and between populations of different species (interspecific competition). For ants, both an individual ant and an ant colony can be regarded as 'the individual' when considering competition, although ultimately the process of competition acts at the population level (the colony). Intraspecific competition can be especially intense given that often colonies must search for the same resources within a limited shared locality. Outcomes of intraspecific competition include prevention of colony foundation, reduced colony size and production of reproductive alates, nest overdispersion, and reallocation of castes in response to new conspecific neighbours. Both intraspecific and interspecific competition can be characterized by territoriality and physical and chemical aggression at resources. Additional evidence for interspecific competition includes spatial ant mosaics and dominance hierarchies. However, experiments show that interspecific competition is highly conditional on biotic and abiotic conditions, and the functional similarities of species. A trade-off in the abilities of ants to dominate or discover resources suggests a possible mechanism for coexistence for so many apparently similar species, although it is not yet clear how widespread this discovery–dominance trade-off is. The abundance of competitively dominant ants can drive species richness of assemblages at a local scale, with studies showing a unimodal relationship between the abundance of dominant ants and species richness at baits. Recent work indicates that in some systems this relationship at baits can scale up to the assemblage level. Although often hard to demonstrate conclusively, competition remains an important factor in ant ecology making a key contribution to the structuring of ant assemblages across a range of levels.

Chapter 6

Ants as Mutualists

Joshua Ness, Kailen Mooney, and Lori Lach

6.1 Introduction

The historical emphasis on the ecological and evolutionary importance of antagonistic interactions such as competition, predation, and parasitism is increasingly informed by a recognition of facilitative and mutualistic interactions where one or both participants receive a net benefit (Bertness and Callaway 1994; Bruno *et al.* 2003; Grosholz 2005; Stachowicz 2001). Interactions between ants and their partners provide some of the best examples of the reciprocally beneficial interactions (Bronstein 1998) and, in particular, the mutualisms that play critical roles in structuring community composition and functioning (e.g. Christian 2001; Kaplan and Eubanks 2005; Mooney 2007; O'Dowd *et al.* 2003; Poulsen and Currie 2006; Wimp and Whitham 2001). Interactions between ants and their partners date to 45–60 Mya (Poulsen and Currie 2006; Stadler and Dixon 2005) and are critical to understanding the evolution and ecological success of ants as a taxon. The rewards provided by mutualists can increase the survival and reproduction of ants and colonies, provide the fuel that allows ants to collect new resources and engage in aggressive behaviours (Davidson 1998), and encourage colonies to reallocate resources towards particular responsibilities and/or locations. Here, we describe the currencies and dynamics of these mutualistic interactions, and highlight recent developments in our understanding of ants' participation in mutualisms.

The complexity and breadth of this topic warrant two caveats. First, the dynamics of particular ant mutualisms have been the focus of substantive reviews (e.g. refer to plant protection in Bronstein 1998, Heil and McKey 2003; insect tending in Pierce *et al.* 2002, Stadler and Dixon 2005, Way 1963; seed dispersal in Giladi 2006; ant–plant symbioses in Davidson and McKey 1993; Heil and McKey 2003; and ant–fungi–bacteria in Poulsen and Currie 2006) and are featured in several books (e.g. Beattie 1985; Huxley 1991; Rico-Gray and Oliveira 2007; Stadler and Dixon 2008). We encourage readers to seek out these more in-depth works. Second, the mutualisms we describe often include currencies based on antagonistic interactions and/or access to food. Competition, predation, and parasitism of (and by) ants are treated in other chapters (see Chapters 5, 12, 9, 10, and 11, respectively, and Box 6.1), and aspects of ant diet and shelter are the focus of Chapter 7. In many cases, dissecting mutualistic interactions requires an understanding of those currencies.

We begin by describing mutualisms on the basis of the resources and services being traded. We focus on trophobiotic interactions (Section 6.2), wherein ants receive access to food resources in exchange for services provided to the reward producer (whether plant or insect; bacterial endosymbionts are discussed in Chapter 7), interactions where ants receive nutritive profit while dispersing plant propagules (seeds and pollen) (Section 6.3), and the tripartite mutualism among ants, fungal cultivars, and bacteria, in which food, protection, and dispersal are the currencies (Section 6.4). In each case, we identify instances in which these interactions can have consequences for the larger biotic communities and identify characteristics of ants that make them particularly well suited for participation in the interaction. We then take a synthetic approach to explore elements of context

Box 6.1 'Berry' ants: an eye-popping symbiosis from the rainforest canopy
Stephen P. Yanoviak

Successful transmission to a terminal host is one of the biggest challenges in a parasite's life cycle. Consequently, parasites have evolved a variety of mechanisms to change the behaviour and appearance of intermediate hosts to facilitate their consumption by the subsequent hosts. Several remarkable examples of this phenomenon involve ants as intermediate hosts, including the grass-climbing behaviour of *Formica* spp. infected with the fluke *Dicrocoelium dendriticum*, the yellow colour of cestode-infected *Leptothorax* spp., and the distended gasters of fluke-infected *Camponotus* spp. These and many other examples are summarized in reviews by Schmid-Hempel (1998) and Moore (2002). Evolutionarily, these changes in host appearance or behaviour are often interpreted as extended phenotypes of the parasites (Dawkins 1982; Hughes *et al.* 2008).

Recently, a striking case of ant manipulation by a parasite was uncovered in the rainforest canopies of Panama and Peru. Workers of the arboreal ant *Cephalotes atratus* infected by the nematode *Myrmeconema neotropicum* have red gasters containing several hundred worm eggs (Poinar and Yanoviak 2008; Yanoviak *et al.* 2008b). The life cycle of the nematode is closely linked to the life cycle and temporal polyethism of the ant; peak redness occurs when the ant is spending large amounts of time outside the nest, and coincides with the presence of infective nematode larvae within the eggs. The colour change is not caused by the deposition of red pigments. Rather, it results from localized exoskeletal thinning or leaching of pigments by the developing worms. This dramatic change in appearance is accompanied by continuous gaster-flagging and a substantially weakened postpetiole, characteristics not found in healthy ants. During the latter stages of infection, the parasitized ant becomes sluggish and assumes an erect posture (Plate 4).

In combination, these changes likely facilitate the consumption of ant gasters by frugivorous or omnivorous birds, which presumably mistake the red gasters for ripe fruit (Yanoviak *et al.* 2008b). Unlike the examples mentioned earlier, ants are the final hosts for this parasite, and birds function as paratenic hosts (i.e. animals that transmit parasites to new hosts without becoming infected themselves; Moore 2002). Unfortunately, direct evidence for bird predation on infected gasters is lacking. However, given what is known of the natural history of *C. atratus* (especially their frequent foraging on bird faeces; reviewed by de Andrade and Baroni-Urbani 1999) and circumstantial evidence from field experiments (Yanoviak *et al.* 2008b), fruit mimicry remains the most parsimonious explanation. Many Neotropical angiosperms have small red fruits available at different times of year, and it is logical that a bird foraging on such fruits would sample any similar red object in its vicinity.

A plausible alternative hypothesis to fruit mimicry is that the red gasters make *C. atratus* workers more conspicuous to predators. Such 'increased conspicuousness' strategies are common among parasites, although few have been studied experimentally (Moore 2002). Increased conspicuousness is not supported in this case for at least two reasons. First, *C. atratus* is already one of the most conspicuous arboreal ant species in Neotropical lowland rainforests. Aside from non-selective foraging by tropidurid lizards, the workers are generally ignored by insectivorous vertebrates (de Andrade and Baroni-Urbani 1999; S. Yanoviak, personal observation). Thus, although infected workers stand out from healthy workers, this difference is unlikely to greatly increase predation on a common but unpalatable ant that is already an easy prey.

Second, the colour red is generally aposematic in insects. To overcome this strong negative signal, infected ants should resemble non-insects, or red gasters should provide a tasty reward. At the peak of infection (Plate 4), parasitized workers are practically immobile. They resemble ants morphologically, but not behaviourally. Given that nematode eggs pass through birds undigested (Yanoviak *et al.*

continues

Box 6.1 continued

2008), there is no obvious reward (nor penalty, excluding effort) associated with consuming an infected ant. Thus, a fundamental assumption of the increased conspicuousness hypothesis – that an attractive signal is associated with valuable resources – is not supported. Likewise, if there is no negative consequence of gaster consumption (a sting or noxious chemical), this mistake should persist in the bird's behavioural repertoire.

Symbioses between ants and other organisms are common and well documented; ant–plant and ant–fungal mutualisms support entire research programs, books, and conference symposia. In contrast, ant symbioses with nematodes (Poinar *et al.* 2006) are under-investigated. Such parasitism is frequently overlooked or mistaken as a taxonomic variety, as occurred with red-gastered *C. atratus* over a century ago (Poinar and Yanoviak 2008). 'Berry' ants exemplify the remarkable interconnectedness of species in tropical forests, and hopefully will stimulate additional research on ant–parasite interactions.

dependency in these interactions (Section 6.5), and put this variation in the context of macroevolutionary variation (Section 6.6). Finally, we highlight the utility of these interactions for addressing questions fundamental to the field of ecology (Section 6.7) and conclude (Section 6.8) by identifying promising areas of future research.

6.2 Ants providing protection for food

Trophobiotic interactions involve the consumption of a food reward, often in return for protection from natural enemies. For ant-loving hemipterans, caterpillars, and most plants, these rewards almost invariably involve a sugary and/or nutrient-rich liquid, one that is collected by the foragers that patrol the area surrounding the resource (Plate 3). Highly specialized ant-plants (myrmecophytes) offer additional food rewards and provide ants with a domicile.

6.2.1 Sap-feeding hemipterans

Many ant species engage in mutualisms with herbivorous hemipterans (Stadler and Dixon 2005; Way 1963), and 41% of ant genera include trophobiotic species (Oliver *et al.* 2008). 'Myrmecophily' (ant-loving) occurs within most families of the Sternorrhyncha and Auchenorrhyncha (Hemiptera), including aphids (Aphididae), coccids and scales (Coccoidea), and membracids (Membracidae) (Stadler and Dixon 2005). Within aphids, 40% of species are ant-tended, and many aphid genera include both tended and untended species (e.g. Mooney *et al.* 2008). These hemipterans tap into host plant phloem sap, which is rich in carbohydrates but relatively poor in nutrients and amino acids. As a consequence, sap-feeding hemipterans must dispose of large quantities of processed, but nonetheless sugar-rich, fluid. Many ants collect this sugary liquid waste, commonly referred to as honeydew. Ant attendance often results in larger hemipteran colonies (Way 1963) and greater fecundity (Bristow 1983; Del-Claro and Oliveira 2000). Ants that otherwise prey upon arthropods do not attack the sap-feeding herbivores, or at least do so more rarely. However, the incentives to view some proportion of an aphid colony as prospective prey rather than mutualistic partners may increase as honeydew-supplied carbohydrates become less limiting with colony growth (Cushman 1991; see Figure 6.1). In addition to this occasional predation, hemipterans may also bear yet unrevealed ecological or physiological costs from their mutualisms with ants. For example, in the absence of predators, ants can reduce aphid reproduction (Stadler and Dixon 1998, Yao *et al.* 2000).

The mechanism of ant benefits to tended hemipterans is most often presumed to occur via protection from natural enemies (Buckley and Gullan 1991; Stadler and Dixon 2005; Way 1963). Ants provide protection against enemies that are frequently

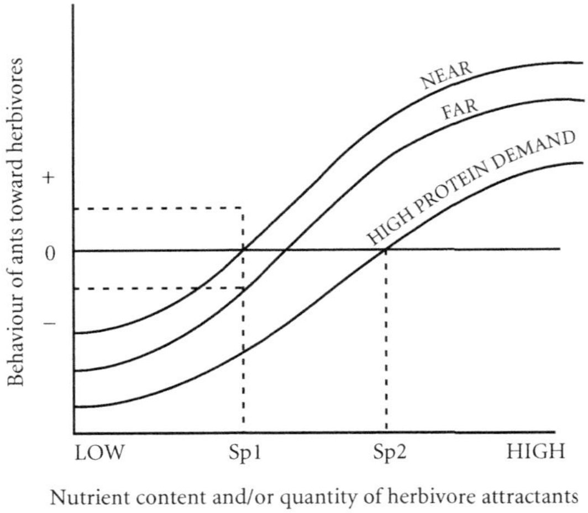

Figure 6.1 An ant's response to prospective mutualists may vary with the prospective benefits and costs of interacting with that partner. Cushman (1991) proposed that incentives for antagonistic interactions towards ant-tended herbivores (e.g. predation) will increase with travelling costs, the ant colonies' demand for protein, and where the quality or quantity of rewards offered by the prospective mutualists is low. When reward quality is high (e.g. Sp2 relative to Sp1), protein demand is low, and prospective partners are close to nests, ants are more likely to act as mutualists. (Reproduced with permission, from Cushman 1991).

somewhat specialized as hemipteran predators, including ladybird beetle larvae and adults (Coleoptera: Coccinellidae), syrphid fly larvae (Diptera: Syrphidae), lacewing larvae (Neuroptera: Chrysopidae), and parasitoid wasps (Hymenoptera: Braconidae), although ants can also provide protection against more generalist predators such as spiders (Cushman and Whitham 1989; Del-Claro and Oliveira 2000). Other benefits can include reduced fouling from honeydew accumulation (Bach 1991), reduced competition from other herbivorous insects (Smith *et al.* 2008), and allowing aphids to divert resources away from predator avoidance or parental care, and towards feeding, growth, and reproduction (Abbot *et al.* 2008; Bristow 1983; Flatt and Weisser 2000). Such non-protective benefits are rarely studied, and their frequency or importance compared to protection from predators is not well understood.

It would seem that the entire benefit of tending hemipterans for ants is nutritional. Aside from water, carbohydrates are the dominant constituents of hemipteran honeydew. However, sugar type, nutrients, and plant secondary compounds all influence the attractiveness and presumably nutritive value of honeydew for ants (Blüthgen *et al.* 2004b; see Figure 6.1 and Chapter 7). Because direct measures of ant fitness are rare (but see Helms and Vinson 2008), we can only infer that honeydew attractiveness is indicative of nutritive value and value to the colony.

Because ants are effective and abundant predators of many arthropods, ant–hemipteran mutualisms have been defined as a 'keystone interaction' (Styrsky and Eubanks 2007) where variation in the strength or occurrence of the interaction has far-reaching consequences for the community in which it is embedded (see Bishop and Bristow 2001; Kaplan and Eubanks 2005; O'Dowd *et al.* 2003; Wimp and Whitham 2001). With respect to herbivores, hemipteran-tending ants increase mutualist abundance while often preying upon untended herbivores (e.g. Bishop and Bristow 2001; Mooney 2007). Key questions for the ecology of ant–hemipteran mutualisms have been whether the net effect of tending ants is to increase or decrease total herbivore abundance, and to what indirect effect on plant growth and fitness (e.g. Horvitz and Schemske 1984). Thus far, the literature suggests that hemipteran-tending ants typically depress the local abundance and species richness of several guilds of chewing herbivores, often to the host plant's benefit (Styrsky and Eubanks 2007). There are, however, spectacular exceptions to this generalization (e.g. Box 15.1; O'Dowd *et al.* 2003).

6.2.2 Lepidopterans

Approximately 70% of Lepidoptera in the family Lycaenidae (an estimated 6,000 species) whose life histories are known engage in associations with ants (Eastwood *et al.* 2006; Fiedler 2006; Pierce *et al.* 2002). Although some species of lycaenid may act as parasites of ants, we focus on the mutualistic species (Travassos and Pierce 2000). The lycaenid larvae and/or pupae attract the attention of ants by producing nitrogen-rich secretions (e.g. Agrawal and Fordyce 2000; Devries 1991), in some cases complemented by chemical and acoustic signalling

(Devries 1991; Travassos and Pierce 2000). Perhaps as a result of the nitrogen investment in reward production, species that feed on nitrogen-fixing plants or on nitrogen-rich plant parts such as flowers and seed pods are more likely to be ant-tended than are species that feed on other plant types or parts (Pierce 1986 but see Billick *et al.* 2005; Fiedler 1995). As with ant-tended hemipterans, ants benefit lycaenids by reducing attacks by parasitoids and predators (e.g. Devries 1991; Pierce and Mead 1981; Wagner and Kurina 2003). Lycaenids may add mass and reach maturity more quickly when protected by ants (Cushman *et al.* 1994), and some species preferentially oviposit in sites where ant densities are high (Wagner and Kurina 2003).

6.2.3 Extrafloral nectary-bearing plants

Plant species in over 90 families attract ants to nectaries not associated with flowers (hereafter, extrafloral nectaries or EFNs; Koptur 1992). These structures typically produce carbohydrate-rich nectar that can also include trace amounts of nitrogen or amino acids (Koptur 1992; Chapter 7). The nectar-attracted foragers may subsequently patrol the plant in search of further nectar, to deter competitors, and/or to consume alternative food items. This combination of foraging, deterrence, and consumption can decrease or alter the distribution of herbivory (Chamberlain and Holland 2009; Heil *et al.* 2001; Koptur 1992; Ness 2003a; Oliveira *et al.* 1999; Rudgers and Strauss 2004; and see appendices in Rico-Gray and Oliveira 2007), typically to the net benefit of most ant-tended plant partners (Chamberlain and Holland 2009).

The interaction between EFN-bearing plants and ants often includes multiple ant species sympatrically foraging on the same plants or plant species in the same population (e.g. Cuatle *et al.* 2005; Oliveira *et al.* 1999; Schemske 1980). These foraging bouts may be segregated in space or time (i.e. within particular branches, diurnal versus noctural, within particular temperature ranges or seasons; Díaz-Castelazo *et al.* 2004; Oliveira *et al.* 1999; Rico-Gray 1993; Schemske 1980). As a result, the EFNs on a particular plant may provide an important resource for an ant community even if it is not particularly important for any one population or colony (Blüthgen *et al.* 2004b; Díaz-Castelazo *et al.* 2004; Oliveira *et al.* 1999).

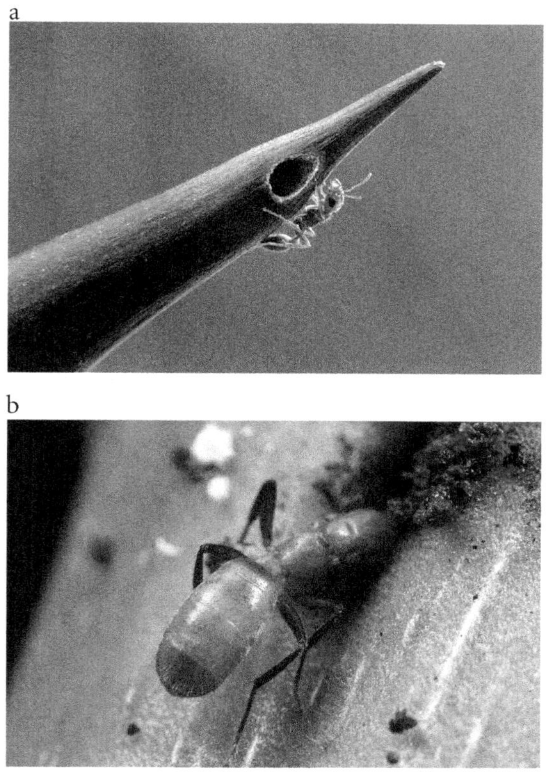

Figure 6.2 Myrmecophytes offer domatia in a variety of forms, many requiring that the ants gain access to a hollow structure by cutting through plant tissue. (a) A *Pseudomyrmex spinicola* worker at the entrance of its nest on a swollen thorn *Acacia*. The thorn is hollow but the ants must cut a hole to gain initial entrance. (b) An *Azteca isthmica* queen cutting into the soft tissue of a *Cecropia* tree to start a nest in the hollow center. (Photos: Alex Wild)

6.2.4 Myrmecophytes (ant-plants)

Plants in over 100 tropical genera host ants in specialized structures such as swollen thorns, hollow stems, and leaf pouches, collectively termed domatia (Bronstein *et al.* 2006; see Figure 6.2). These myrmecophytes or 'ant-housing plants' can also provision ants with lipid and protein-rich food bodies and/or nectar (Heil and McKey 2003; O'Dowd 1982), or host honeydew-producing

Figure 6.3 Myrmecophytes differ in the types of food they offer to their resident ants. (a) Lipid- and glycogen-rich Mullerian food bodies on a *Cecropia* tree. (b) A *Pseudomyrmex spinicola* worker on a swollen thorn *Acacia* harvesting a protein-rich food body to feed to the colony's larvae. (Photos: Alex Wild)

hemipterans (Gaume *et al.* 1998; Palmer *et al.* 2008) (see Figure 6.3). Some of the best-studied myrmecophytic relationships include those of *Acacia* with *Pseudomyrmex* ants (e.g. Janzen 1966, 1967a), *Cecropia* with *Azteca* spp. and other ants (e.g. Folgarait and Davidson 1994, 1995; Longino 1989), and *Piper* with *Pheidole* ants (Letourneau 1983; Letourneau *et al.* 2004) in the Neotropics, and *Macaranga* with *Crematogaster* ants in southeast Asia (e.g. Feldhaar *et al.* 2003; Fiala *et al.* 1989; Itino *et al.* 2001). Myrmecophytic *Acacia, Macaranga, Cecropia*, and *Piper* provide their ant partners with domatia and food bodies (see reviews in Davidson and McKey 1993; Rico-Gray and Oliveira 2007), whereas myrmecophytic *Hirtella* spp. and *Tachigali myrmecophila* do not produce food rewards but instead support hemipterans that nourish the resident ant colonies (Rico-Gray and Oliveira 2007). Myrmecophytic interactions are much less widespread than facultative ant–plant associations involving EFNs (Heil and McKey 2003) and exhibit much greater specialization by both partners. For example, some myrmecophytic *Acacia* offer extrafloral nectar high in invertase (sucrose-cleaving enzyme) and low in sucrose, which corresponds to the preference of resident *Pseudomyrmex* ants for sucrose-free nectar that is unpalatable to other ants (Heil *et al.* 2005). Morphological adaptations include the prostoma (unlignified organ at the tip of the domatia) of *Leonardoxa* plants, the shape, and size of which corresponds strongly to the head of mutualistic ants (Brouat *et al.* 2001) and the wax crystals on the stems of some *Macaranga* that exclude ants not adapted to the slippery surface (Federle *et al.* 1997).

Plant-dwelling ants may provide nutrients, and/or protect their hosts from invertebrate and vertebrate herbivores, plant pathogens, and encroachment by competing plants (Bronstein *et al.* 2006; Davidson and McKey 1993; Heil and McKey 2003). These benefits can be pronounced. For example, a successful ant–plant symbiosis can create large monospecific 'devil's gardens' within otherwise diverse tropical rainforests (Frederickson *et al.* 2005). Plants that house ant residents may also (or instead) benefit from greater access to nitrogen and CO_2 as a result of the activities of its plant-dwelling ants (Sagers *et al.* 2000; Treseder *et al.* 1995). The benefits of its resource transfers may exceed the value of any protection provided by the ants in some systems and/or ecological settings.

6.2.5 The best ant partners

From the trophobiont's perspective, the ideal protectors are competitively dominant ants capable of aggressive behaviours (biting and stinging) and mass recruitment (e.g. Buckley and Gullan 1991) that might deter the partner's natural enemies. Ant characteristics that lessen the costs of foraging or patrolling can also increase the likelihood of mutualistic interactions by allowing a trophobiont that produces modest rewards to, nonetheless, engage ants in a beneficial manner (see Figure 6.1). For example, foraging costs are reduced if the prospective ant partners are capable of establishing satellite nests at the base of plants with EFNs or hemipteran aggregations, or if the plant itself is the domicile of the colony. Because trophobiosis invariably involves the collection of sugary and/or nutrient-rich liquid, key adaptations include the capacity for trophallactic exchange of liquid food among members of the colony (Fiedler 2006) and morphological changes to the ant's proventriculus and gaster, which enable them to carry large amounts of sugary fluids (honeydew and nectar) and regulate the digestion of these fluids (Davidson *et al.* 2004). This is one explanation why subfamilies such as Formicinae and Dolichoderinae (and, more rarely, Myrmicinae), whose members have some subset of these adaptations, are the most common tenders, and why these foragers may tend to more than one of these partners within a particular habitat (e.g. Blüthgen *et al.* 2000; Devries 1991; DeVries and Baker 1989; Rico-Gray 1993).

From the perspective of the myrmecophyte, the ideal ant partner is quick to detect and deter would-be plant antagonists at a minimal cost to the plant. However, because the ecologies of myrmecophyte and plant-ant are largely inseparable, among-system transplants that could allow scientists to contrast the benefits of particular pairings are impossible. That is, we cannot test whether *Acacia* plants might do better hosting the *Azteca* ant associates of *Cecropia*. However, the diverging natural histories of particular systems provide clues to explain why the favoured (or at least realized) characteristics may differ among systems. Fast-growing pioneer trees with rapid rates of resource supply, such as *Acacia*, *Macaranga*, and *Cecropia*, often host an active, aggressive workforce of large ants (Davidson and McKey 1993). Smaller trees and shrubs, such as *Leonardoxa* and *Piper*, often host smaller, more timid or sluggish workers that can nonetheless be effective against very small herbivores, eggs, and microbes (Gaume *et al.* 1997; Letourneau 1983). Food (Itino *et al.* 2001) and nesting site (Fonseca 1993) resources impose limits to the hosting capacity of plants and imply a trade-off between the number and size of ants that can be hosted. If the

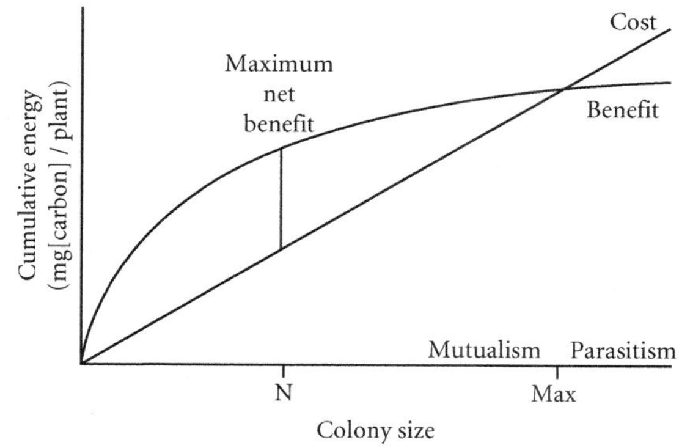

Figure 6.4 Graphic model of the cost and benefit to a myrmecochore of hosting an ant colony of varying size. The model describes a system wherein the cost of maintaining and housing ants increases linearly with colony size, while the benefits that ants provide saturate. N is the colony size that provides the greatest difference between benefit and cost (i.e., greatest net benefit) to the plant. Beyond a threshold ant colony size (Max), plant costs exceed plant benefit. Ant colonies at this stage may experience disproportional net benefit and/or be subject to plant reprisals. (Reproduced with permission, from Fonseca 1993).

incremental benefit to a plant of hosting additional ants progressively lessens as ant density increases and the structural or metabolic costs of hosting those ants increases linearly (i.e. each ant costs the same), the net benefit of hosting ants could lessen as colonies increase in size (Figure 6.4). This ant–plant conflict can influence the density of plant-ants that can occupy a given plant, population or community (Fonseca 1999), and may well dictate the best partner for particular settings.

The worst partner ants decrease the fitness of their partners. For example, in some settings, ants may consume more aphids than they protect from natural enemies (see Figure 6.1). Other costs might be more subtle. Highly aggressive ants that visit EFNs may also deter pollinators as effectively as they do to natural enemies (e.g. Ness 2006), leading to conflict between the defensive and reproductive mutualisms. Some ant residents also prune the flowers of their myrmecophytic hosts (Stanton *et al.* 1999; Yu and Pierce 1998) in an apparent effort to reallocate host resources towards ant rewards, enabling increased colony size at the plant's expense (see Figure 6.4). Ant-tended plants may

Figure 6.5 Ants from the genus *Rhytidoponera* are important seed dispersers. Here, *Rhytidoponera metallica* carries a seed with elaiosome attached. (Photo: Benoit Guénard)

limit these indirect costs by including ant-deterring compounds in their flowers (e.g. Ness 2006; Willmer and Stone 1997a).

6.3 Ants provide dispersal for food

6.3.1 Seeds

Myrmecochory is the dispersal of ant-adapted seeds by ants. Over 90% of the >3,000 ant-dispersed plant species are found in the South African fynbos and in areas of Australia dominated by sclerophyllous plants (Berg 1975; Bond and Slingsby 1983). Most of the remaining identified ant-dispersed species are spring ephemerals in the temperate deciduous forests of northern Europe, Japan, and North America; myrmecochores account for 40% of the herbaceous species and 60% of emergent stems in portions of temperate deciduous forests of the eastern United States (Beattie and Culver 1981; Handel 1981).

Myrmecochorous seeds have an attached, lipid-rich food reward, called an elaiosome, which attracts ant foragers (Figure 6.5). Because the elaiosome's fatty acid composition is similar to that of insect prey (Hughes *et al.* 1994), the diaspore (seed + reward) is attractive to omnivorous foragers. Ants may preferentially collect seeds with larger elaiosomes or more favorable elaiosome-to-seed ratios (Mark and Oleson 1996), and some elaiosomes also include compounds that elicit collection behaviors by workers (e.g. 1,2-diolein in Hughes *et al.* 1994). As a result of the nutritive value and chemical signalling component of the elaiosome and a durable seed coat, ants that might otherwise act as *seed* consumers are perhaps converted into *elaiosome* consumers, and hence, seed dispersers (Rico-Gray and Oliveira 2007). Interestingly, plants may also co-opt the attention of carnivorous ants; the most avid collectors of elaiosome-bearing seeds rarely include plant material in other aspects of their diet (Hughes *et al.* 1994). Myrmecochorous species are found in >80 plant families, and the morphological features associated with myrmecochory have evolved at least 20 times in the monocots (Dunn *et al.* 2007a). This estimate may be conservative, as some seeds that rely on ants for dispersal lack food rewards. For example, the diaspores of some 'ant garden' plants use odorants, rather than food rewards, as ant attractants (e.g. Youngsteadt *et al.* 2008). Whether that collection provides sufficient benefit to the ants to qualify as a mutualism is unclear.

The conventional forms of myrmecochory benefit the ant colony by providing a food resource that can enhance the colony's reproductive output (Gammans *et al.* 2005; Morales and Heithaus 1998). Myrmecochores produce their seeds in seasons where seed collection by ants is most likely to occur. This is the early summer in temperate deciduous forests, when ant forgers are both highly active and have dietary preferences that make elaiosomes attractive (Oberrath and Bohning-Gaese 2002), and plants have few opportunities for interactions with avian frugivores (Thompson 1981). The benefits to the seed include protection from granivores (e.g. Bond and Slingsby 1984; Christian 2001; Turnbull and Culver 1983) and/or fire (Christian 2001), directed dispersal to atypical microsites (such as nutrient-rich ant middens; Davidson and Morton 1981, but see Rice and Westoby 1986), dispersal away from parent plants and siblings (e.g. Bond and Slingsby 1984; Horvitz and Schemske 1986; Kalisz *et al.* 1999; Ness *et al.* 2004), and enhanced germination rates (e.g. Cuatle *et al.* 2005). The cumulative effect can be greater fitness for those seeds that are collected by ants (Hanzawa *et al.* 1988). Although these consequences are often studied in isolation, multiple benefits may be derived from any one ant–seed interaction (Giladi 2006).

6.3.2 The best ant partners

Changes in the seed-dispersing ant communities can alter seed survival rates, mean and maximum seed dispersal distances, and the distribution and composition of mature plant communities (Andersen and Morrison 1998; Bond and Slingsby 1984; Christian 2001; Ness *et al.* 2004; Ness and Morin 2008; Parr *et al.* 2007; Chapters 8 and 15). From the perspective of a myrmecochorous plant, ideal ant partners share several characteristics. High quality dispersers are typically solitary, omnivorous foragers that range far from their nest, disperse diaspores at substantial distances to those nests (the criteria for 'substantial' may be defined by the size of plants and the scale of soil heterogeneity within the site), feed on the elaiosome while leaving the seed intact, and bury the seeds shallowly in microsites where they can respond to germination cues (Giladi 2006). Low quality dispersers may be granivorous, disperse seeds at insufficient distances to avoid competition with maternal and sibling plants, dissect diaspores in situ (i.e. 'elaiosome robbing') rather than carry them to the nest, and cache high densities of seeds deep underground where germination is unlikely (Giladi 2006). Well-studied 'high quality' ants that collect a disproportionate amount of myrmecochorous seeds are *Rhytidoponera* spp. in Australia (Figure 6.5; Andersen and Morrison 1998; Gove *et al.* 2007; Hughes *et al.* 1994) and the *Aphaenogaster rudis* complex in North America (Beattie and Culver 1981; Ness and Morin 2008).

6.3.3 Pollen

The ubiquity of ants and their diverse interactions with plants begs the question of why ants so rarely act as pollinators. Several characteristics make ants poor candidates: maximum foraging distances are short relative to winged visitors, ant territoriality may decrease the likelihood of outcrossing among plants, and exposure to ants can reduce the viability of pollen (likely due to ant-borne antibiotics; Beattie *et al.* 1984). In so far as these shortcomings decrease the success of both male and female plant function (e.g. Galen and Butchart 2003), flower-visiting ants may be unwelcome 'parasites' of the interaction between plants and legitimate pollinators. These consequences are the likely selection pressures for the chemical and physical impediments that can deter ants from entering flowers (e.g. Galen and Butchart 2003; Ness 2006).

Plant characteristics that can favour pollination by ants (or increase the incentives for ant pollination) include living in sites where ant activity is high (and/or other pollinators are rare), few synchronously blooming flowers per plant (to minimize intra-plant pollination or stigma-clogging for self-incompatible plants), pollen volumes insufficient to elicit grooming behaviours by the ants, and nectar rewards sufficiently unrewarding to discourage visitation by alternative, more expensive, pollinators (Hickman 1974). Although rare, such systems do exist. There are also a few plant species that receive pollination services by mimicking opportunities for ant copulation (e.g. *Leporella fimbriata* orchids are pollinated by male *Myrmecia urens*, Peakall 1989).

6.4 Ants, fungi, and bacteria

Originating 50 Mya (Schultz and Brady 2008), the tripartite association among ants, fungal cultivars, and actinomycete bacteria is perhaps the most highly evolved and complex set of mutualisms in ant ecology. More than 210 species in 13 genera of Myrmicine ants in the New World Attini tribe cultivate basidiomycete fungi as their main food source by collecting and preparing an appropriate fungiculture substrate (Currie 2001; Poulsen and Currie 2006). The ant genera vary in their choice of fungiculture substrate, colony size, and polymorphism. The more basal or 'lower' attines utilize insect corpses, faeces, or plant detritus as fungal-growing substrates, and tend towards smaller, monomorphic colonies. In contrast, the more derived or 'higher' attines utilize plant detritus or fresh plant material (e.g. leaf-cutting ants; Figure 6.6), can display extreme polymorphism, and may achieve colony sizes of several million individuals (Currie 2001; Poulsen and Currie 2006). Queens and larvae of attine ant colonies feed exclusively on the fungus, while workers may supplement their fungal diet with plant sap (Quinlan and Cherrett 1979). In the case of leaf-cutting ants, the fungi convert

a

b

Figure 6.6 *Atta* are among the most conspicuous ants in the Neotropics and their colonies can number millions of workers in multiple subcastes. (a) An *Atta cephalotes* worker carries its harvest back to the nest to feed the colony's fungal cultivar. (b) *Atta cephalotes* workers tend the colony's fungal garden. (Photos: Alex Wild)

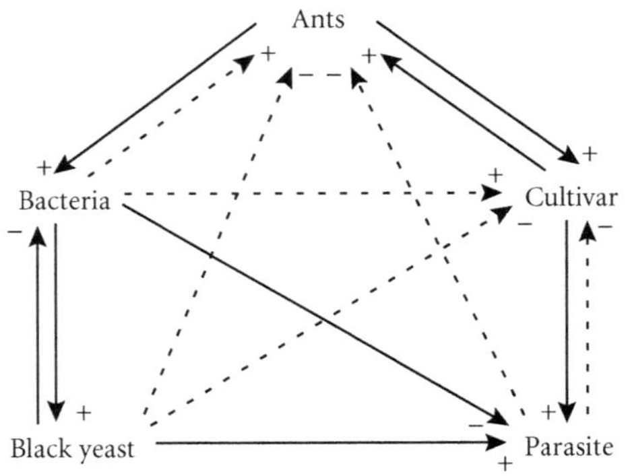

Figure 6.7 A diagram of the direct and indirect interactions of the attine ant-microbe symbiosis. Solid lines represent direct effects, dashed lines represent indirect effects, requiring the presence of an intermediary species. Cost (-) or benefit (+) deriving from the interaction is indicated at the tip of the arrowhead. Cultivar = fungal cultivar; parasite = specialised fungal parasite, *Escovopsis*; bacteria= actinomycete bacteria, *Pseudonocardia*, hosted on the ants; and black yeast= parasite of the bacteria. (Modified with permission, from Little and Currie 2008).

inedible plant material into lipid and carbohydrate-rich gonglydia, making the monophagous ants 'ecologically polyphagous' (Rico-Gray and Oliveira 2007). As a result, the ant–fungal composite feeds on a great diversity of widely distributed plants that would otherwise be inaccessible to the fungi and/or inedible to the ants.

Fungus-cultivating ants have an elaborate set of behaviours and traits that facilitate fungal cultivation. The selection of an appropriate substrate is key to fungal growth. Leaf-cutting ants avoid harvesting from plants with incompatible chemistry, possibly via feedback from the fungus (North *et al.* 1997). Attines further promote the growth of their fungal cultivars by pruning, redistributing fungus-produced proteolytic enzymes around the fungal garden, and maintaining the garden chamber at the appropriate temperature and humidity (Poulsen and Currie 2006). The ants employ behavioural and chemical means to protect their fungal gardens from other microbes. The use of a platform by founding *Atta* queens reduces the risk of infection by microbes in the soil (Fernández-Marín *et al.* 2007). Weeding and grooming by workers also reduce contamination by non-mutualist microbes (Currie and Stuart 1991). Metapleural gland secretions provide effective general antibiotics and defend the fungal cultivars from an array of microbes (Poulsen *et al.* 2002; see Box 9.1).

The weeding and grooming behaviours and metapleural gland secretions are not effective against specialized fungal parasites in the genus *Escovopsis*. Thus the ants and their fungi depend on another mutualist, actinomycete bacteria (Figure 6.7). These actinomycetes, in the genus *Pseudonocardia*, are reared in specialized, elaborate crypts present in genus-specific locations on the cuticles of attines (Currie *et al.* 1999, 2006). The bacteria produce antibiotics that selectively inhibit the growth of *Escovopsis* (Currie *et al.* 1999) and are associated with all attine ants that have been examined (Currie *et al.*

2006). In *Acromyrmex*, the bacteria are most abundant on major workers that are most active at the bottom of the fungal garden where *Escovopsis* is most likely to be encountered (Poulsen *et al.* 2002). The actinomycetes further benefit the fungal cultivars by providing growth-promoting compounds (Currie *et al.* 1999) and may also directly benefit the ants by protecting them from pathogens (Currie 2001). Benefits conferred on *Pseudonocardia* by the ants include dispersal (by virgin queens during the nuptial flight), provision of a unique habitat in the cuticular crypts, and nourishment from specialized glands (Currie 2001; Currie *et al.* 2006). However, symbiotic black yeasts can parasitize *Pseudonocardia*, reducing its growth and decreasing the ability of the ants to suppress *Escovopsis* infection of their fungal gardens (Figure 6.7; Little and Currie 2008).

The synergism between leaf-cutting ants and leaf-digesting fungi provides perhaps the most dramatic example of the community-wide consequence of ant mutualisms. This ant-fungal collaboration exploits up to 50% of the plant species (Cherrett 1968, 1972) and can remove >10% of total leaf production in some Neotropical forests (Hölldobler and Wilson 1990). Few other herbivorous animals approach this breadth or magnitude of impact. Not surprisingly, this consumption can profoundly constrain plant recruitment and direct the nature and pace of plant succession (Vasconcelos and Cherrett 1997; Wirth *et al.* 2003), and the consolidation of resources in and around leafcutter nests can alter soil properties and the distribution of in-soil resources (Farji-Brener and Illes 2000; Wirth *et al.* 2003).

Fungiculture in non-attine ants is much less studied. Ants in the Old World *Lasius* genus, in the subgenera *Dendrolasius* and *Chthonolasius*, utilize ascomycete fungi to bind shredded wood or soil to reinforce nest walls. The ants nourish their fungi with honeydew and protect them from competing fungi, possibly through grazing (Schlick-Steiner *et al.* 2008).

6.5 Context dependency and stability

Because mutualisms are, by definition, reciprocally beneficial interactions, it is reasonable to ask what forces stabilize these interactions and regulate the populations of the participants. That is, might the positive feedbacks derived from these interactions encourage these populations to grow progressively larger ad infinitum? At least three explanations clarify why this 'orgy of mutual benefaction' is so rarely observed (but see 'invasional meltdown' as in O'Dowd *et al.* 2003 and Box 15.1). First, other forces external to the mutualism, including natural enemies attracted by the success of one participant, intraspecific competition, or abiotic factors, may eventually limit the populations of at least one partner. For example, the black yeast symbionts that exploit the mutualism between fungus-growing ants and their actinomycete bacteria decrease the ability of the ants to protect their fungal gardens from the *Escovopsis* parasite (Little and Currie 2008), to the detriment of fungal garden health (Currie 2001). Second, the resources on which the interactions are based may collapse, as could be the case if a synergistic ant–aphid interaction was overly detrimental to the host plant. Third, the benefits conferred by participating in the interaction may saturate. For example, from the perspective of a plant or honeydew-producing aphid aggregation, the distinction between being tended by 5 ants versus 10 may be of negligible importance if 3 ants are sufficient to provide services required to increase their population (Ness *et al.* 2006). Additional ants may even be worse, if the costs of reward production increase linearly with partner abundance (Fonseca 1993; see Figure 6.4). Likewise, the carbohydrate rewards provided to these ants may become progressively less important to the well-being of the colony once access to protein, rather than carbohydrates, becomes limiting (see Figure 6.1).

Case studies of ants and their 'trophobionts' have shown remarkably disparate costs and benefits when studied in different settings. The benefits provided by ants to aphids have been shown to decline with aphid colony size (Breton and Addicott 1992). Because aphids can bear costs for provisioning mutualist ants (Stadler and Dixon 1998, Yao *et al.* 2000), the net benefit of ants can be predicted to depend on the risks from natural enemies, honeydew fouling, and competition from other herbivores. Moreover, aphids can compete intra- and interspecifically for ant attendance, and whether ants are mutualists for a particular aphid clone depends on the availability of other sources of honeydew and nectar (Cushman and Addicott 1989). There is also evidence that host plants can mediate not only the strength of ant benefits to aphids, but also the direction of ant effects (Mooney and Agrawal 2008). Variation in phloem sap quantity or quality may be responsible for mediating these ant–aphid interactions (see also Figure 6.1).

Context-dependency might be particularly prevalent in interactions where ants primarily provide benefit by conferring protection. Most ant visitors (or at least visits) may not benefit the myrmecophile because: (*a*) the workers do not protect the partner (i.e. when ants are timid or ineffectual against ant-adapted herbivores) or (*b*) the partner's need for protection does not coincide with interactions with particular ant species or populations (e.g. EFNs: Schemske 1980; hemipterans: Cushman and Whitham 1989; myrmecochores: Fedriani *et al.* 2004). Perhaps as a result of selection pressures to increase the likelihood that ants can provide appropriate services when that service is required, the spatio-temporal distribution of reward production is correlated with the plant's vulnerability to natural enemies. For example, EFNs are often located in areas where the consequences of herbivory could be severe (e.g. young leaves, at the base of reproductive units; Horvitz and Schemske 1990; Oliveira *et al.* 1999; Schemske 1980). Likewise, myrmecochores drop seeds during the day, when foraging by granivorous rodents is lessened and the likelihood of seeds being collected by ants is greatest (Cuatle *et al.* 2005; Turnbull and Culver 1983).

If mutualist ants, or the subset of ants that are particularly effective, are a limited resource, prospective partners will compete for their services and a subset may suffer from decreased service. For example, experimental augmentations of membracid aggregations decreased overall tending rates by ants due to the decrease in the ratio of ants to membracid. The consequence of this decrease in service was a >90% decrease in the production of membracid adults (Cushman and Whitham 1991). A shortage of mutualists, or the disincentives of supporting partners when they are unnecessary, may explain why some myrmecophillic partners have adaptations to help them attract additional ant partners. Some of the methods employed to promote additional ant attendance include greater production of extrafloral nectar (Heil *et al.* 2001; Ness 2003a), ant domiciles (Stanton *et al.* 1999), lycaenid food rewards (Agrawal and Fordyce 2000), or acoustical (Morales *et al.* 2008a) and chemical (Del-Claro and Oliveira 1996) signalling to ants during times of need.

Some level of constancy in partner identity, quality, and benefit is essential if local evolutionary specialization for particular mutualisms is to occur at the population level. Some myrmecophiles will occur in settings where interactions with ants are more necessary, and hence, more beneficial (Rudgers and Strauss 2004). Among-site variation that is stable through time could result in local evolution if the effective population size of partners is small relative to the area occupied by a particular partner taxon (Horvitz and Schemske 1990; Rudgers and Strauss 2004), whereas temporal variation can only lead to diffuse selection by the assemblage of mutualists ('the interaction') rather than particular partner species. There is some evidence that the identity of ants that act as prospective mutualists varies greatly among sites but can be more consistent over time within sites than are other mutualisms (e.g. see EFN tending ants versus pollinators in Horvitz and Schemske 1990). This may be due to the longevity (and immobility) of individual colonies, relative to some of their partners.

6.6 Macroevolutionary patterns in the face of variation

Although over half the ant subfamilies do not include species known to engage in mutualisms with trophobiotic insects, mature plants (EFN-bearing or myrmecophytic) or fungi, the incidence of all three interactions are positively correlated with one another among the remaining, vigorously mutualistic, subfamilies (Oliver *et al.* 2008). Within those subfamilies, however, mutualisms with trophobiotic insect or plants are negatively correlated with those with fungi at the genus level. One explanation is that ant lineages need to specialize in one type of mutualism when the adaptations for service or receiving benefit in one mutualism diverge from another. Specifically, the characters that favour collecting fungal substrates and maintaining suitable fungal growing conditions in return for edible mycelia differ from those of aggressive defense and ingesting sugary secretions (Oliver *et al.* 2008).

Among facultative associations, there is evidence that the adaptations favouring myrmecophily are evolutionarily labile (i.e. can be acquired and/or lost at the species level of resolution). Hemipteran traits associated with ant tending include modification of honeydew chemical composition, aggregated feeding, longer proboscis length (Bristow 1991, Shingleton *et al.* 2005), and loss of defensive structures and predator avoidance behaviours (Stadler and Dixon 2005). A complete understanding of hemipteran adaptations to ant-tending is still forthcoming; some of the observed associations between traits and ant-tending are known from single hemipteran lineages, while associations among taxonomically disparate species do not distinguish between evolutionary convergence (as is presumed) and common ancestry. Nevertheless, that myrmecophily is not constrained to any single lineage suggests multiple origins and high lability for mutualism with ants. Consequently, many untended hemipteran species may be only a few evolutionary or ecological steps away from such mutualisms, and adaptations may be subtle. Among ants, adaptations that are correlated with, and perhaps favour, trophobiosis include a modified proventriculus, polygyny, and polydomy (Oliver *et al.* 2008).

Ant–myrmecochore interactions were not included in the aforementioned phylogenetic analyses. In so far as elaiosomes are dead insect analogues, their collection and utilization by ants may require little specialization or trade-offs with other mutualisms. Further, the repeated independent origins of myrmecochory (Dunn *et al.* 2007a) and diversity of elaiosome shapes, weights, histological origins, caloric and nutritional content, and manner of diaspore presentation in that guild imply great generalization by the plants. Nonetheless, the existing field observations demonstrate that two ant genera collect a majority of the myrmecochorous seeds in sclerophyllous Australia (*Rhytidoponera* spp., Figure 6.5; Gove *et al.* 2007) and temperate North America (*Aphaenogaster rudis* complex). Whether this constitutes 'specialization' by the myrmecochorous guild (to say nothing of coevolution) is unclear.

The strongest evidence of coevolution and 'partner-filtering' occurs in the interactions among myrmecophytes and their partners (see 6.2.4, Bronstein *et al.* 2006; Brouat *et al.* 2001; Federle *et al.* 1997; Heil *et al.* 2005; Janzen 1966) and between fungus-farming ants and their symbionts. All of the studied fungus-growing ants have phylogenetically specific modified exoskeletons for housing and feeding, for example, *Pseudonocardia* bacteria; closely related ant species lack these modifications (Currie *et al.* 2006 but see Kost *et al.* 2007). That all fungus-growing ants host a strain of *Pseudonocardia* (Currie *et al.* 2006), suggests that there is a yet-to-be-discovered mechanism for preventing establishment by other bacteria that may not act as a mutualist to the ant or the fungal cultivar (Kost *et al.* 2007). Similarly, although the ant–fungal cultivar relationship is now thought to be indicative of more diffuse, rather than pairwise, coevolution (Mikheyev *et al.* 2006), the incompatibility of alien fungal strains and hostile ant behaviour towards alien fungal fragments can prevent the introduction of competing fungal clones (Poulsen and Boomsma 2005).

6.7 Model interactions for ecology

Ant mutualisms have several characteristics that make them 'model systems' for addressing questions regarding mutualism and plant defense. We highlight these advantages later, and propose promising research questions in Section 6.8.

First, ant attendance and behaviour can be monitored in real time. Ants that forage on the surface of plants, leaf litter, and soil can be counted. As a result, variation in the number or behaviours of foragers allocated to a particular task and turnover in the species performing a task can be accurately described. These measures can provide information about the costs and benefits received by each participant in the prospective mutualism, and how these vary over time, space, or in response to experimental treatments.

Second, ants can be excluded from particular microsites. Many studies of ant protection mutualisms use sticky substances (e.g. Tanglefoot™) to experimentally exclude ants from some subset of their partners, or from portions of particular partners (e.g. control versus treatment branches). Remarkably, these exclusion treatments can even be performed at the scale of hectares (e.g. poison baits in Abbott and Green 2007). Ant densities can also be depressed by adding ant predators (e.g. Letourneau *et al.* 2004).

Third, many individual partners interact with one ant colony for sustained periods of time. This largely occurs as a result of a combination of the central-place foraging requirements of a (largely) immobile ant colony and territorial interactions between colonies and/or species. Although there are many exceptions to this gross generalization and the very definition of 'sustained' will depend on the lifespan of the partner, some ant mutualisms are believed to have continuously functioned for centuries (Frederickson *et al.* 2005). Irrespective of duration, this dynamic of sustained interactions between two individuals, or at least one individual and one superorganism, is more common in ant mutualisms than in those involving more mobile partners, such as pollination.

These three attributes of ant mutualisms have facilitated much of our understanding of ant and non-ant mutualisms (Bronstein 1998; Heil and McKey 2003), as well as plant defense, food web structure, and the dynamics of symbiotic interactions. For example, an ecologist's ability to reliably count and exclude ants (or ant-occupied thorns) from particular branches makes it possible to quantify and manipulate plant defenses to an extent that is nearly impossible (or at least terribly expensive) for chemical plant defenses such as tannins or alkaloids.

6.8 Future directions

In the following text, we highlight several ecological topics that we perceive as particularly critical and promising for better understanding the role of ants as mutualists.

6.8.1 Diverse partners

What are the consequences of interacting with a variety of prospective ant partners? There are many examples of great variation in partner quality (e.g. Buckley and Gullan 1991; Horvitz and Schemske 1986; Miller 2007b; Ness *et al.* 2004; Ness *et al.* 2006), and striking examples where a greater frequency of interactions with suboptimal partners lessens the benefits to the ant's partner (Christian 2001; Palmer *et al.* 2008). Nonetheless, how often interactions with suboptimal partners constrain selection for the mutualism is unclear.

Can the inclusion of suboptimal partners be beneficial, and are there settings in which the diversity of a partner assemblage itself confers benefits? For myrmecochores, a more diverse disperser assemblage could increase the variety of sites where seeds are deposited and, perhaps as a result, decrease the influence of detrimental density-dependent processes. The synergistic effect of multiple predators is well documented in other systems (Cardinale *et al.* 2003; Sih *et al.* 1998); that it occurs in ant protection mutualisms is a reasonable (Beattie 1985; Rico-Gray and Oliveira 2007), albeit largely untested, hypothesis. Further, repeated interactions with suboptimal partners, or interactions with many of those partners, can remedy the mediocrity that may be so pronounced on a per capita or per interaction basis (Ness *et al.* 2006). If partner diversity does confer benefits, the costs of participating in a mutualism that typically includes a diverse assemblage may only become apparent when a prospective mutualist is simultaneously deprived of those diverse partner assemblages and limited to interacting with one or a few partners that are mediocre (or outright parasitic) in all settings. That combination of homogeneity and inadequacy may be a historically rare phenomenon in the natural world. Or, put differently, myrmecophillic organisms may be rare in habitats where such conditions are the norm in the natural world. We predict such pairings may occur increasingly frequently due to anthropogenically induced disruptions in ant faunas in the face of landscape conversion, global climate change, and exotic ant invasions.

Partner diversity and specificity of the ant-fungi-bacteria mutualism are also ongoing subjects of investigation. Recent discoveries of filamentous actinomycete bacteria on non-attine ants that also inhibit *Escovopsis* growth have called into question the specificity of the attine ant–actinomycete mutualism and whether ants have any mechanism to control bacteria on their exoskeletons (Kost *et al.* 2007). In addition, much more is to be learned about the non-attine ants that cultivate fungi for architectural purposes. Is there a similar complex interplay of mutualists and parasites as has become evident in the attine ant fungal gardens?

6.8.2 Benefits to ants that engage in mutualisms

The vast majority of studies that explore the interactions between prospective mutualists have focused almost exclusively on the consequences of the interaction for the non-ant partner. This discrepancy may be partly attributable to (*a*) the logistical difficulties of measuring ant fitness (but see Cushman *et al.* 1994; Lach *et al.* 2009; Morales and Heithaus 1998); (*b*) the assumption that the participation of the more mobile participant (ant) is evidence of choice, and thus, benefit to that participant; and (*c*) the ease of quantifying other variables relevant to the ant's partner (e.g. defoliation, aphid mortality). Ant-fungal mutualisms are an exception; microbe partners or substrates can be manipulated, and outcomes for colony growth or survival can be measured relatively easily (e.g. Fernández-Marín *et al.* 2007; Seal and Tschinkel 2007a). One solution to this widespread shortcoming is to utilize the modular organization of ant colonies, and to more fully describe the effects of these interactions on individual modules (i.e. individual ants, see Cushman *et al.* 1994; Lach *et al.* 2009). Stable isotope techniques are one new promising technique to quantify the benefits that ants receive (e.g. Sagers *et al.* 2000; Box 7.1). Davidson *et al.* (2003) used stable isotopes to infer that access to extrafloral nectar and hemipteran exudates in the rainforest canopy fuel the spectacular diversity and abundance of ants in those habitats. Critically, stable isotope techniques highlight the integration of rewards into ant tissues or particular castes rather than measuring fitness, and their correct interpretation requires a comprehensive knowledge of the natural history of the system.

Barring obligate ant–myrmecophyte or ant–fungal interactions, we know of no studies that have sought evidence (much less demonstrated) that among-site variation in ant communities is attributable to variation in the availability of their mutualists (but see Díaz-Castelazo *et al.* 2004; O'Dowd *et al.* 2003). The inherent asymmetry in many ant–mutualist interactions offers one explanation for that absence. Ants often provide their plant and trophobiont partners with protection at critical demographic stages (e.g. protection of seeds and aphid nymphs from predators), and they often literally transport those partners into their 'sphere of influence' (e.g. carrying seeds and aphids closer to the nest). In contrast, the benefits to the ants often are limited to the augmentation of resources that are already found in the regular diet of the ant.

6.8.3 Costs and cheating

To answer whether the net effect of an interaction is beneficial, our interpretation of the spatio-temporal heterogeneity in benefits should be balanced by an appreciation for the magnitude and variability of the costs of participating in the interactions. In so far as the benefit accrued by one partner translates into the cost experienced by the other, conflicts of interest between ants and their prospective mutualists may seem unavoidable (e.g. Section 6.2.5 and Figure 6.4, see also Palmer *et al.* 2008; Stanton *et al.* 1999; Yu and Pierce 1998). However, three largely untested hypotheses explain why this need not be the case:

The resources or strategies that benefit one participant may come at negligible cost to its partner. In such situations, the benefit to the recipient may not come at a commensurate cost to the provider, and one can get 'something for nothing'. The costs of plant-produced rewards can be quite minor (O'Dowd 1979, 1980), and are lessened when plants decrease or curtail extrafloral nectar production in the absence of perceived threats (Lach *et al.* 2009; Ness 2003a). Indeed, the multiple prospective benefits of ant attendance to hemipterans beg the question, why do not all hemipteran species exchange their waste products for ant attendance? Similarly, some of the benefits ants provide likely incur negligible cost to the colony. For example, some plants benefit from access to the debris accumulated by foraging ants and the CO_2 they exhale (e.g. Sagers *et al.* 2000; Treseder *et al.* 1995; Wagner 1997). Last, although participation in particular mutualisms may incur costs, those solutions are often cheaper than the alternatives (e.g. myrmecochory versus frugivory in nutrient-poor habitats; Westoby *et al.* 1991b).

The resources traded in these interactions may be less important than are other components of the interaction. For example, many ant-collected seeds include odorants that elicit collection behaviours by workers (e.g. Hughes *et al.* 1994), and in some cases an elaiosome reward is entirely absent (e.g. Youngsteadt *et al.* 2008). In some respects, these interactions may function more as 'behavioural usurpation' than a reciprocal exchange of resources. Ecologists have yet to ask whether diaspores will be collected when these compounds are experimentally disassociated with the seed, although we know that other non-rewarding substances impregnated with the volatiles will be collected by workers (e.g. Hughes *et al.* 1994).

The outcome that benefits the myrmecophile may be a product of strategies that best suit the ant. In such a case, the concept of 'cheating' becomes meaningless. For example, if an ant colony is capable of the vigorous defence of a resource against real or perceived competitors, be they rival colonies, herbivores, or carnivores, it will do so. If it cannot, the opportunity to harvest that resource may well be usurped by a more aggressive colony that provides even greater protection to the reward (plant or insect). On a different vein, Ness *et al.* (2009) demonstrated that sustained collection of carbohydrate-rich resources changes ant dietary preference, and inferred that an abundance of one resource highlights the relative absence of complementary resources (here, provision of abundant carbohydrates elicit attacks on relatively nitrogen-rich prey). Last, from a myrmecochore's perspective, the most important characteristic of an ant is that it does not 'cheat' by removing the elaiosome and abandoning the denuded seed (to predators, competitors, etc). For subordinate ants that specialize in discovering but not dominating resources, the most advantageous behaviour may be to immediately collect the elaiosome (with the seed attached) rather than engaging in the time-consuming task of separating reward from seed, and hence risk losing the resource to a competitor.

6.8.4 Inter-mutualism conflict

Relatively little is known of the interactions between mutualisms, including those in which ants are involved. Such inter-mutualism dynamics may play a central role in shaping the ecology and evolution of ant–mutualist interactions. For example, it has been proposed that EFNs have evolved as a means of distracting ants from tending hemipterans (Becerra and Venable 1989) and collecting floral nectar (Wagner and Kay 2002). However, ant-tended insects are disproportionately common on EFN-bearing plants (Offenberg 2000), and some even ingest extrafloral nectar (DeVries and Baker 1989). The rewards provided by hemipterans can also supplement the ant rewards provided by myrmecophytes (Fonseca 1993; Palmer *et al.* 2008). In so far as ants have greater control over hemipteran densities than they do direct plant rewards, the involvement of these third parties can affect the functioning of the symbiosis (Gaume *et al.* 1998).

Some plants also utilize ants within the context of multiple mutualisms involving protection and seed dispersal (*Turnera ulmifolia*: Cuatle *et al.* 2005; *Urera baccifera*: Dutra *et al.* 2006). The most thoroughly studied of these systems, and perhaps the most reticulate, is *Calathea ovandensis*; this tropical herb has EFNs, is attacked by ant-tended Lepidoptera, and relies on myrmecochory for seed dispersal (Horvitz and Schemske 1984, 1986). How often the coterie that participates in one interaction is well suited for the other, or interacts with that counterpart, is unknown (but see Cuatle *et al.* 2005). Explorations of these inter-mutualism dynamics may provide important insight into the evolution of ant mutualisms generally.

The multiple mutualisms and complex interactions occurring within the nests of fungal gardening ants are a rich area for exploring potentially competing mutualisms and the effects of parasites. The mutualism between actinomycete bacteria and ants and the parasitic black yeast-actinomycete bacteria and *Escovopsis*–fungal cultivar relationships have only recently been discovered to substantially shape the dynamics of the ant–fungal cultivar mutualism. Questions remain about the mechanisms through which some effects are seen. For example, given that actinomycete bacteria are stimulated by the presence of *Escovopsis* (Currie *et al.* 2003), are black yeasts as well? And if so, are there feedback mechanisms by which black yeasts facilitate *Escovopsis* infection? It is likely that new microbes that may further affect the costs and benefits of the multiguild relationships await discovery.

6.8.5 Biotic interactions on an abiotic stage

Experimental studies have begun to explore the importance of variation in abiotic resources on the evolution and functioning of ant mutualisms. In so far as these resources are limiting, they can alter the incentives for particular interactions. For example, carbon-rich resources such as extrafloral nectar and ant domatia should be less costly for plants to produce where carbon is in excess (Folgarait and Davidson 1994). Perhaps as a result, EFN-bearing plants are common in sunlight-rich habitats such as rainforest canopies (Blüthgen *et al.* 2000), forest edges (Bentley 1976), and deserts (Pemberton 1988). The influence of resource limitation (or surplus) is also detectable at smaller spatio-temporal scales. Nitrogen fertilization of host plants can increase tending rates of some trophobionts (e.g. lycaenids: Billick *et al.* 2005, but see Morales and Beal 2006 re. membracids), and alter plant investment in indirect defences (Folgarait and Davidson 1995). Ant mutualisms can also reorganize abiotic resources. The construction, maintenance, and feeding of ant colonies and nests often concentrate resources, expose buried nutrients, and alter moisture retention rates (Moutinho *et al.* 2003), perhaps to the benefit of their partners (Davidson and Morton 1981; Giladi 2006; Wagner 1997). Remarkably, there is also evidence that tending by ants can alter the nitrogen content of tended hemipterans and their host plants (Kay *et al.* 2004, but see Abbot *et al.* 2008). The generality of these documented patterns, and how such modifications will influence the incentives for ant mutualisms, is largely unknown. To make matters more complex (and worthy of attention), the availability of nitrogen and CO_2 continues to increase at scales ranging from individual plants to the biosphere as a result of anthropogenic influences.

6.8.6 Putting ant mutualisms in their place

Ant mutualisms are unevenly distributed across habitats. Perhaps problematically, the settings for the research that underpins our understanding of these interactions rarely occur in proportion to the density or diversity of these interactions. How might our understanding of these interactions change if we studied them in the settings where they most often occur? Most studies of myrmecochory (and all that quantify benefit to the ants) focus on temperate deciduous myrmecochores; what do the costs and benefits described in this nutrient-, moisture-, and granivore-rich biome tell us about the >90% of myrmecochores that reside in dissimilar biomes in Australia and South Africa? Might our sense of the costs, benefits, and selection pressures on ant-tended insects and plants differ if we studied them in communities such as tropical rainforest canopies or some deserts where tending by ants is the modal interaction? The characteristics of ants, prospective natural enemies, and competition for services may be sufficiently different in these settings to profoundly alter those interactions. We recognize that particular systems offer advantages for studying particular ecological and/or evolutionary phenomena. However, we propose that the wealth of studies that comprise the current literature and inform reviews, meta-analyses, and our gestalt sense of how interactions function may describe the range and modal version of the interactions only in so far as those studies occur in comparable environments. Our understanding of those interactions will change as we better place our questions in the context of the larger environment.

6.9 Summary

Ants are perhaps the most common and dominant animal mutualists in terrestrial environments. As a result, better understanding the dynamics of these interactions should be a priority for those who hope to understand the taxon, their role in communities, and mutualism as a widespread interspecific interaction. These mutualisms include interactions with ant-loving plants, insects, fungi, and bacteria, with the ants typically receiving food and/or shelter, and their partners receiving food, protection, and/or propagule dispersal. Context dependency, wherein the magnitude of costs and benefits incurred as a result of participation in the interactions varies with the ecological setting may be particularly prevalent in interactions involving protection by ants. Adaptations that enable effective participation in one type of mutualistic interaction may preclude a species from participating in others, but may also make it more difficult for non-beneficial interactors to intrude. Because ant interactions with their mutualists are relatively easily monitored, manipulated, and are sustained over time, ant mutualisms are model systems for understanding mutualisms and plant defence. We encourage future work that explores the influence of partner diversity, better quantifies costs and benefits to participants, and addresses how contemporary interactions and abiotic resources alter these interactions.

Acknowledgments

This work was greatly improved by the comments of Aaron Gove, Paulo Oliveira and Kirsti Abbott. Joshua Ness was supported by sabbatical funds from Skidmore College during the composition of this chapter.

Chapter 7

Food and Shelter: How Resources Influence Ant Ecology

Nico Blüthgen and Heike Feldhaar

Over 2,500 years ago in ancient Greece, Aesop wrote a fable in which he characterized ants as prudent animals. In contrast to grasshoppers that leisurely enjoy their days without wasting a thought on the hard days to come, ants invest most of their time on building a shelter and storing food. This view, indeed identified the key features of ant ecology and other social insects. The ants' nest functions not only as a shelter for the queen and the brood, but also enables the storage of food as an insurance against variable conditions. Its location also determines where ant workers can forage. Ant workers are wingless and slender, both allowing them to use even the smallest crevices as nests. Being wingless facilitates slender body forms since no thoracic wing muscles are required, but on the other hand it strongly limits the ants' foraging range. The ability to actively construct nests enables many ants to move their home into the proximity of their preferred food resources. The interplay of nesting habit and food preference contributes directly to niche differentiation in ant communities, and is important for our understanding of how various ant species can coexist in a habitat. Even closely related species may differ in food niches or nest types. For example, carnivorous ponerine ants of the genus *Leptogenys* show only minor differences in nesting habits, but pronounced differences in dietary preferences (Steghaus-Kovac and Maschwitz 1993). In contrast, other ant genera such as *Polyrhachis* differ strongly in nesting behaviour which, in concert with factors such as colony size and foraging strategy, determines species coexistence, whereas nutrition habits play a relatively minor role (Liefke *et al.* 1998).

This chapter focuses on *adaptations* that permit ants to effectively consume certain kinds of food and overcome limitations in nesting sites and on *constraints* that compromise the utilization of other potential sources. We first outline more general aspects of ant diets, nutrient requirements and digestive enzymes, and micro-organisms, before we focus on specific food sources and their peculiarities. For nests, we contrast the more static ground and tree nests with highly dynamic carton nests and bivouacs that allow ants to follow their food sources. Where possible, we relate the species-specific differentiation in nutrition and nesting habits to processes at the ant community level and species coexistence, but such aspects have received relatively little attention so far, and require more community-wide studies before general conclusions can be drawn. Apart from niche partitioning of food and nest sites, additional factors that maintain diverse ant assemblages include macroecological drivers (see Chapter 3), competitive abilities (see Chapter 5) and variation in foraging behaviour (see Chapter 12) – these are only treated briefly in the present chapter. Moreover, ant nutrition and shelter provide the basis of mutualistic associations between ants and plants or ants and other insects (see Chapter 6).

7.1 Ant food

7.1.1 Trophic position

Most ants are considered to be omnivores, combining predation, scavenging dead animals and faeces, and consumption of plant-based resources to a variable extent (Stradling 1978). More specialized hunters, primary consumers of various plant diets and even fungivores are also found among ant species. However, when we consider ant nutrition, we should keep in mind that ants are holometabolous insects with different food requirements of larval stages and adults. In addition to foraging for food for their own metabolic needs, workers mainly forage for food items to feed larvae. The larval diet is crucial for growth and biomass composition of pupae, while adult diets are only required for maintenance of body functions.

Apart from qualitative and quantitative observational data on ant diets, natural-abundance stable isotope analyses have recently contributed to our knowledge of ant diets (Box 7.1). The ratio of heavy-to-light nitrogen isotopes (^{15}N to ^{14}N, displayed as $\delta^{15}N$) in the structural body mass of an ant is used as an indicator of its trophic position, or more precisely, its nitrogen source (Blüthgen *et al.* 2003; Davidson *et al.* 2003). Carbon isotopes ($\delta^{13}C$) may additionally reveal insights into carbon sources, particularly if they are highly variable as between C3 and C4 plants (Box 7.1). Usually, adult ant workers are measured and compared because these are most easily sampled. However, it should be considered that measurements of the stable isotope or molecular body mass composition of adults mainly reflect their larval diets, because their structural biomass is derived from their larval, not from their adult nutrition. This is particularly true if not only honey crop and gut contents, but also fat bodies and gland contents of the workers are removed (e.g. by cutting of the ant's gaster prior to isotope analysis) that may otherwise interfere with measures of the structural body mass composition (Blüthgen *et al.* 2003; Fiedler *et al.* 2007; Tillberg *et al.* 2006). The analysis of pupae instead of adults would be best suited to circumvent gland and crop contents. Ants often carry about half of their (wet) body mass in their crops; including this content in analyses of entire ant bodies severely overestimates carbon/nitrogen ratios and underestimates $\delta^{15}N$ (Blüthgen *et al.* 2003).

Stable isotope analyses generally confirm that larval diets of ant species vary considerably in their trophic position and cover the whole continuum from chiefly primary consumers to first- or second-order predators. Not only many arboreal ants in tropical rainforests (Blüthgen *et al.* 2003; Davidson *et al.* 2003), but also a number of ground-dwelling temperate species (Fiedler *et al.* 2007) derive their nitrogen to a large extent from plant food such as nectar and honeydew. Honeydew can be considered as a largely plant-based liquid as well, although obtained indirectly as excretion from plant-sucking insects. Across these three studies covering four continents, the highest dietary contribution of plant-based nitrogen (lowest $\delta^{15}N$) is found for Formicinae (e.g. *Camponotus* and *Polyrhachis*, more moderate values in *Formica*, *Lasius*, and *Oecophylla*) and Pseudomyrmecinae (*Tetraponera*, *Pseudomyrmex*), followed by Dolichoderinae (e.g. *Azteca*, *Dolichoderus*, *Tapinoma*, *Technomyrmex*). In contrast, most Myrmecinae (e.g. *Pheidole*, *Myrmica*, but not *Cataulacus*, *Cephalotes*, and some *Crematogaster*) and particularly Ponerinae, Ecitoninae, and Aenictinae have the highest $\delta^{15}N$, as expected for predominantly or entirely carnivorous taxa.

The capacity to access plant-derived nitrogen may explain the high abundance of formicines and other ants that forage intensively on honeydew, nectar, seeds, or other plant diets. In the trophic pyramid, the total biomass of primary consumers strongly exceeds that of higher trophic levels. Ants in tropical forest represent a large proportion of the animal biomass, and this corresponds to their basal trophic position as largely primary consumers (Davidson 1997; Tobin 1994). However, if developing ant larvae are largely sustained with plant-derived nitrogen – much like real 'herbivores' that feed on foliage or suck on phloem – how do they cope with the poor nitrogen availability? Nectar and honeydew contain very low concentrations of amino acids and proteins compared to arthropod prey. Therefore, worker ants must harvest large volumes of these liquids in order to sustain protein demands of growing larvae. A number of morphological, behavioural, and

Box 7.1 Stable isotopes in the study of ant diets

Kailen Mooney and Chadwick V. Tillberg

Many ants feed in multiple trophic roles as herbivores (direct consumers of plant resources), indirect herbivores (collection of hemipteran honeydew), primary predators (feeding upon herbivores), secondary predators (feeding on predatory arthropods), and omnivores (feeding on more than one trophic level). Ant diet can vary not only among species, but also within species based upon ecological setting. Disentangling such a reticulate network of trophic connections is a challenging task, especially because the liquid component of an ant's diet, including plant nectar, hemipteran honeydew, and prey haemolymph, can be difficult to directly observe and quantify. Ant exclusion studies reveal the net effect of ants on their food resources, but these effects include both direct and indirect effects. Additionally, ants can have non-consumptive effects, so knowing their impacts on resource abundance does not accurately reveal what they have consumed.

Stable isotope analysis provides a means of efficiently tracing the pathways of organic matter among consumers (Peterson and Fry 1987). This technique measures the ratio of heavy to light isotopes of biologically relevant elements such as nitrogen and carbon. To calculate 'δX' (or 'delta' X) for an element (e.g. N or C), the heavy/light isotopic ratios of the samples are compared to an element-specific standard as follows: $\delta X = ((R_{sample}/R_{standard}) - 1) \times 1000$. R_{sample} and $R_{standard}$ refer to the ratio of heavy to light isotopes of the sample and standard, respectively. The equation yields a 'per mil' (‰) value, where $\delta X = 0$ means no difference in isotopes in relation to the standard, where a $\delta X = 1$ means a 1 part per mil increase in the heavy isotope in relation to the standard.

Knowing the isotopic composition of an ant in relation to its resource base reveals the relative contribution of these different resources to the ant's diet. For example, nitrogen stable isotopes (15N and 14N) participate in physiological reactions at different rates, and are thus incorporated into the biomass of the consumer at different rates, a process called 'fractionation.' This results in the isotopic ratio of consumers becoming increasingly enriched in the heavier isotope relative to their prey with each trophic transfer. Carbon stable isotopes (13C and 12C) differ substantively among primary producers, but show relatively little or no trophic fractionation in consumers. Arthropod δ 15N typically is enriched by ca. +1‰ to +2‰ above that of its food resources, whereas δ 13C differs substantively between food webs based on C3 and C4 plants, but shows less trophic fractionation (ca. + 0.5‰ to + 1.0) (McCutchan et al. 2003). The degree of stable isotope enrichment can differ based upon trophic levels, dietary differences of consumers and consumer, and resource physiologies.

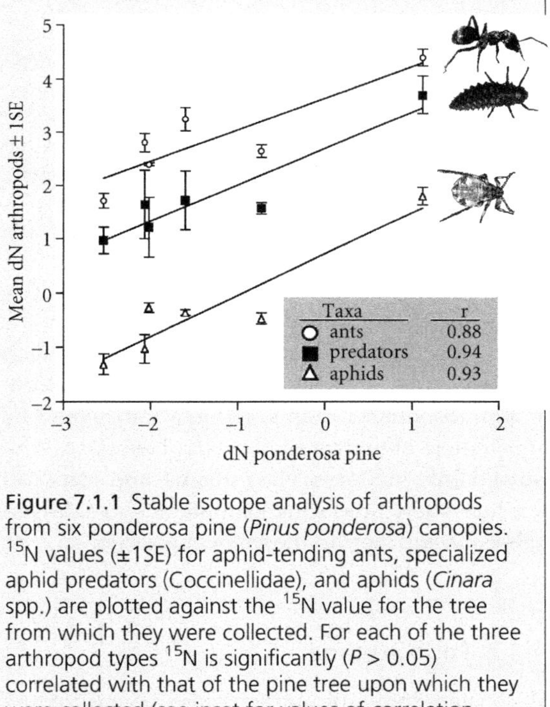

Figure 7.1.1 Stable isotope analysis of arthropods from six ponderosa pine (*Pinus ponderosa*) canopies. ^{15}N values (±1SE) for aphid-tending ants, specialized aphid predators (Coccinellidae), and aphids (*Cinara* spp.) are plotted against the ^{15}N value for the tree from which they were collected. For each of the three arthropod types ^{15}N is significantly ($P > 0.05$) correlated with that of the pine tree upon which they were collected (see inset for values of correlation coefficients). The trophic position of ants, estimated by comparing ant enrichment to that of specialized aphid predators, ranged from 3.2 to 3.6 among pine canopies, with 3.0 being that of primary predators and 4.0 being that of secondary predators. Analysis based on data from Mooney and Tillberg (2005).

continues

Box 7.1 continued

Isotope mixing models allow one to infer the relative contributions of an animal's dietary resources that differ in $\delta\ ^{13}C$ and $\delta\ ^{15}N$ (or other stable isotopes), but the details of the sampling methods used can strongly influence results. Estimating feeding relationships with this technique depends upon accurate values for trophic enrichment either from published accounts or, ideally, from system-specific feeding trials using the organisms and resources to be studied. Colonies of a single species can vary in their dietary inputs, and a sampling regime that replicates at the level of the colony is necessary to capture this trophic range. Finally, individual ants or ant prey items may be too small to analyse individually, requiring bulk analysis of multiple individuals (Tillberg et al. 2006).

Mooney and Tillberg (2005) estimated the trophic position of the ant *Formica podzolica* by comparing $\delta\ ^{15}N$ for ants with that of pine, herbivores, and primary predators from six ponderosa pine canopies (Figure 7.1.1). Taking the enrichment from herbivores to predators as indicative of a single trophic level, it was estimated that ants fed at a trophic position mid-way between that of primary and secondary predators. In addition, there was variation in $\delta\ ^{15}N$ of the pine trees at the base of each replicate food web. Consequently, $\delta\ ^{15}N$ for arthropods depended upon the individual pine tree upon which the food web was based, as indicated by significant correlations between $\delta\ ^{15}N$ for pine and each arthropod group (see inset in Figure 7.1.1 for values of correlation coefficient r). The level of arthropod trophic enrichment above that of pine was similar for arthropods collected from each pine tree, although there was significant variation in ant trophic position; on some replicates, ants fed at close to the level of primary predator, while in others they fed at or near the level of secondary predators. These results demonstrate the utility of stable isotope analysis for inferring the diet of omnivorous ants, as well as how variation in isotopic signatures at the base of a food web can be traced across multiple trophic levels.

physiological adaptations in ants thus facilitate the effective collection, monopolization, and digestion of plant diets. Chewing leaves or sucking phloem is not an option for ants, though both would be much more abundant than nectar. Leaves are consumed only indirectly via a fungus, and plant sap via honeydew-producing hemipterans. We discuss these specific diets in the following sections.

7.1.2 Food storage

A constant influx of nutrients may be ensured by storing food within the nest. In addition to the production of trophic eggs and cannibalism of larvae in times of food shortage (Wheeler 1994), workers can function as a buffer by storing nutrients in their fat bodies (Hasegawa 1993a; Stradling 1978; Yang 2006). Nutrients may also be stored as liquids in the crop of replete workers, whose gasters may be conspicuously enlarged, for example, in honey pot ants of the genus *Myrmecocystus* and others (Borgesen 2000; Rissing 1984; Plate 9). In addition to internal storage (e.g. in the fat body) ants can store food reserves in the nest. In seed harvesting ants like *Pheidole* or *Pogonomyrmex*, the lipid- and sometimes protein-rich seeds are utilized for long-term storage of food or as a buffer against fluctuating food intake (Judd 2006; Mackay and Mackay 1984).

Recently another sophisticated storage technique has been discovered. *Solenopsis invicta* dries and stockpiles dead insect prey in warm and dry parts of its mounds just beneath the surface. In the absence of fresh insect prey, this 'insect jerky' is fed to fourth-instar larvae that are able to liquefy the dried food item with their salivary secretions, before the nutritious liquid is distributed within the colony

(Gayahan and Tschinkel 2008). Apart from storage within the nest, monopolization of food sources, for example, by building shelters over trophobionts or extrafloral nectaries, can ensure a steady influx of food over longer periods of time (Anderson and McShea 2001b).

7.1.3 Nutrient requirements and balance

For insects in general, proteins and lipids are only required for larval growth, ovary development, and egg production, but neither for metabolism in adults, nor for sperm production, which require only carbohydrates (Nation 2002). Correspondingly, adult ant workers mainly use carbohydrates as an energy source (Markin 1970a). Egg-producing queens do require a more continuous nitrogen intake, just as other 'income breeders' among holometabolous insects where females are long-lived and utilize their adult diets for ovary development. Queens thus receive a particularly nutritious diet.

To date, little is known about minimum dietary requirements of ants. The ten essential amino acids are principally the same across most animals, despite some minor variation (Nation 2002) and are thus assumed to be the same for ants. However, the amounts of micronutrients required as well as their distribution in the colony are largely unknown. Sterols are required as hormone precursors and as components of cell membranes, but cannot be synthesized by insects and thus need to be obtained from food or symbionts (Nation 2002). Some vitamins have been shown to stimulate feeding of ants if offered in water solutions, e.g., folic acid, B12, and inositol (Ricks and Vinson 1970). As known for macronutrients, adult workers have a lower need for micronutrients than growing larvae (Judd and Fasnacht 2007). On the other hand, some gland products in adult ant workers may require specific diets, either the glandular substances themselves, their precursors, or at least as donors of specific molecules. For instance, only specific amino acids provide carbon for formic acid production (Hefetz and Blum 1978). Thus, adult ants may have specific dietary requirements in addition to their energetic needs.

Apart from viewing each of the nutrients or diets separately, nutrient balance is an important key to understanding ant ecology and evolution. Ants with a largely monotonous diet are expected to have developed mechanisms to cope with particular nutrient deficiencies. On the other hand, omnivorous ants may select their food in response to nutrient imbalances and forage in a way that optimizes a complementary nutrition. Most notably, the carbohydrate: protein balance (sometimes referred as CHO: N balance) has been examined in a number of studies (Dussutour and Simpson 2008; Kay 2002; Markin 1970a; Nonacs 1991). Davidson (1997) suggested that a high carbohydrate: protein ratio in ant diets had selected for a low level or lack of nitrogen-based defences, analogous to other animals or plants where nitrogen limitation may shape defences or biomass composition (Fagan *et al.* 2002; White 1993). Correspondingly, formic acid, acyclic ketones, terpenes, iridoids, or other nitrogen-free toxic compounds are common in ants that largely live on N-poor plant diets (Davidson 1997). Additionally, protein-starved ants may save nitrogen by a lower investment in N-based sclerotized cuticles and often lack a strongly developed sclerotized sting. In turn, stings and protein-based toxins are mainly found in largely predatory subfamilies such as most ponerines, myrmicines, and all army ants (Davidson 1997). Apart from this variation among species, elemental composition may have pronounced plasticity within an omnivorous ant species. The carbohydrate: protein ratio of the larval ant diet strongly affects the body composition of carbon, nitrogen, and phosphorous in larvae until the pupation, and this variation is retained in adult workers (Kay *et al.* 2006).

A more dynamic aspect of nutrient balance is a change in behaviour: ants with high dietary carbohydrate: protein ratios maintain a higher level of energy-demanding foraging activities (Davidson 1997). Correspondingly, *Linepithema humile* showed a reduced activity level and lower aggressiveness when deprived of carbohydrates, but not proteins (Grover *et al.* 2007). In contrast, when fed a diet with surplus proteins, colony growth has been shown to decrease in *Camponotus floridanus*, although queens produced more eggs. This result implies a colony-level effect of nutrient balance, as maintaining more brood may be a costly activity for workers, thereby decreasing their survival (Nonacs 1991).

Accordingly, colony growth of *Solenopsis invicta* was substantially reduced when honeydew sources were unavailable, even when insect prey was provided *ad libitum* (Helms and Vinson 2008). Scarcity of dietary sugar also led to higher worker mortality and production of smaller workers in *Tetramorium caespitum* colonies (Kay *et al.* 2006).

Omnivorous ants may even actively control their colony's nutrient balance and alter their preferences for either carbohydrate- or protein-based diets, showing a higher preference for items that are currently limited. This plasticity has been demonstrated experimentally not only for colonies with different feeding regimes, but it also occurs across different species with variable natural diets (Kay 2002). Dynamic responses of omnivorous ants to carbohydrate: protein balance thus range from changes in worker size and mortality, colony growth, and body mass composition, to altered foraging activity and aggressiveness.

7.1.4 Digestive enzymes and gut micro-organisms

Digestive capabilities differ among ant species and constrain the spectrum of food sources that are available to a colony. This variation provides an important mechanism to explain niche differentiation among ants on broader (e.g. trophic level) and finer scales (e.g. which types of honeydew are harvested by different species). Among the carbohydrates commonly available, sucrose is most preferred by most ant species (Blüthgen and Fiedler 2004b; Cornelius *et al.* 1996; Lanza *et al.* 1993), and accordingly, the enzyme required to hydrolyze sucrose, invertase, seems to be present among most ants (Ayre 1967; Boevé and Wäckers 2003; Ricks and Vinson 1972). However, some specialized plant-ants lack invertase and consequently do not feed on sucrose (Heil *et al.* 2005). In a comparison across five ant species, Ayre (1967) showed that the activity of specific enzymes corresponds to feeding habits; non-predacious ant species showed a lower activity of proteases and lipases than predacious ants. Moreover, amylase is required for breakdown of starch as well as glycogen, and allows digestion of starch-containing seeds or glycogen stores in insect prey among other sources. Correspondingly, amylase was highly active in predatory black imported fire ants, *Solenopsis richteri* (Ricks and Vinson 1972), although this activity was not confirmed for some other predatory ant species (Ayre 1967).

Most enzymes may be specific to certain glands, body parts, and life stages. Notably, in adult ants proteinase seems to be absent in the salivary and maxillary glands and the crop, but active in the midgut (Ayre 1967; Ricks and Vinson 1972). The midgut is only reached by minute particles that pass through the narrow proventriculus, unlike larger chunks of prey. However, adult insects may not require nitrogenous compounds at all or only in very low concentrations (Nation 2002). Larvae are much less constrained in their protein consumption (Petralia *et al.* 1980; Ricks and Vinson 1972), and thus it has been suggested that the larvae function as a 'digestive' caste (Erthal *et al.* 2007; Hölldobler and Wilson 1990) by providing liquid secretions for the nutrition of adults (Ricks and Vinson 1972). Larvae of red imported fire ants (*S. invicta*) show protease and amylase activity both in labial glands and the midgut, and may also perform extraintestinal food digestion (Petralia *et al.* 1980). Fungus-growing ants (Section 7.1.8), particularly their larval stages, harbour a diverse spectrum of enzymes associated with the degradation of fungal substrate (d'Ettorre *et al.* 2002b; Erthal *et al.* 2007). Ayre (1967) even suggested that the activity of amylase may help *Camponotus* species to digest fungal mycelia that are rich in glycogen. Fungal diets, however, have received little attention apart from leaf-cutting ants.

Gut microsymbionts supply vitamins, essential amino acids, and sterols for various insects such as blood-feeders, plant sap feeders, or species that feed on cellulose (Dadd 1985; Nation 2002). The few studies on gut microflora of ants have concentrated on ant genera that mainly feed on sources poor in nitrogen and essential amino acids, and which presumably gain most from a beneficial microbial community in their gut. In the ants studied to date, the gut microflora seem to comprise a low bacterial diversity and only few bacteria that are specific to ants (Feldhaar *et al.* 2007; Li *et al.* 2005; Sameshima *et al.* 1999; van Borm *et al.* 2002). Several species of the genus *Tetraponera* have a gut pouch opening out into the midgut that is filled with specific bacteria (Billen and Buschinger 2000) that are closely related to the Rhizobiales *Bartonella* and *Rhizobium* (Stoll *et al.* 2007). These bacteria have been found in brood as well as in workers, and seem to form a stable gut microflora. Interestingly, a closely related bacterium has also been found in several species of *Acromyrmex* (van Borm *et al.* 2002), *Atta* (Feldhaar unpublished data) and *Dolichoderus* (Stoll *et al.* 2007). *Tetraponera* species that do not possess a gut pouch also harbour a specific gut microflora of gamma-Proteobacteria (Stoll *et al.* 2007). Functional studies are still lacking for these bacteria, although the presence of *nifH*, a structural subunit of the dinitrogenase, in both the Rhizobiales and the Proteobacteria suggests a possible role in the fixation of aerial nitrogen (Stoll *et al.* 2007). In other insects, it has been shown that fixation of aerial nitrogen by gut bacteria contributes to the insect's nitrogen intake (Behar *et al.* 2005).

Specific intracellular endosymbionts residing in specialized cells, the bacteriocytes, are only known from the closely related genera *Camponotus*, *Colobopsis*, *Polyrhachis*, and *Echinopla* (Sameshima *et al.* 1999; Sauer *et al.* 2000; Stoll *et al.* 2007). Their endosymbiont *Blochmannia* has been shown to upgrade the nutrition of its host by recycling nitrogen and providing essential amino acids (Feldhaar *et al.* 2007). Apart from supplying endogenous nitrogenous compounds, the bacterium should enable the ants to utilize urea and possibly also uric acid from faeces since it possesses a functional urease (Feldhaar *et al.* 2007). Thus, *Blochmannia* should confer a competitive advantage to these ants, rendering their larval growth less dependent on a steady influx of prey items into the colony.

7.1.5 Honeydew

Honeydew is the liquid excretion from the alimentary tract of plant-sucking hemipterans (Delabie 2001). Many ant species collect honeydew directly from the anus of the sucking insect (Figure 7.1a; Plate 3), maintaining a close association (trophobiosis) that is often a mutualism of ants and honeydew producers (Chapter 6). This holds true also for lycaenid caterpillars and some other ant-tended insects. Lycaenids secrete a sugary solution from their glands, which is also rich in amino acids (Daniels *et al.* 2005). Moreover, the ants' protective services may even extend to the host plant, where untended herbivores suffer from the high activity of aggressive and partly predatory ants (Box 7.2).

The dominant compounds in the dry mass of honeydew are carbohydrates. Moreover, honeydew contains amino acids, amides, minerals, lipids, and sterols and therefore, has been thought to provide a complete diet for growing ant larvae (Stradling 1978), although this idea has rarely been examined in greater detail. The array of amino acids in some frequently consumed honeydews may be wider than what is found in typical floral or extrafloral nectars that are consumed by ants (Blüthgen and Fiedler 2004a). Indeed, migrating herdsmen ants (*Dolichoderus* spp. in southeast Asian forests) seem to rely solely on honeydew as a nutritional resource (Dill *et al.* 2002). Although nutrient flows have rarely been quantified in detail, honeydew seems to be the main nutrient source for several other ant species (e.g. Abbott and Green 2007; Horstmann 1974).

In plant nectar, carbohydrates are usually confined to sucrose, and the monosaccharide components glucose and fructose. Honeydew is more diverse and may contain a number of mono-, di- and trisaccharides. The trisaccharide melezitose is most widespread and prominent in honeydews regularly consumed by ants (Blüthgen and Fiedler 2004a; Völkl *et al.* 1999), although it has proven unpalatable, if not toxic, for various insects (Zoebelein 1956). However, some ants (Boevé and Wäckers 2003) as well as some other hymenopterans (Wäckers 2001) are able to digest such sugars – a potential key innovation for successful utilization of many honeydew sources. Certain enzymes (glucosidase and galactosidase) that hydrolyse oligosaccharides may thus be important for honeydew-feeding ants (Boevé and Wäckers 2003), but their activity has not been compared across ants of different lifestyles to date.

Digestive capabilities for oligosaccharides may influence which hemipterans are tended or not. By and large, each ant species may be associated with a variety of hemipteran partners, and vice versa. Hence, the level of specialization in ant–hemipteran associations is relatively low (Blüthgen *et al.* 2006).

Figure 7.1 Ants use a range of resources to fulfil their food requirements: (a) Honeydew-feeding *Anoplolepis gracilipes* (Photo: Nico Blüthgen); (b) *Myrmecia pyriformis* attacks a wasp (Photo: Alex Wild); (c) A worker of *Polyrhachis abdominalis* collecting bird droppings from a leaf surface. (Photo: Heike Feldhaar)

However, several cases can be found where ant species substantially differ in their partners, and this resource partitioning may be explained by their acceptance of oligosaccharides. For example, weaver-ants (*Oecophylla smaragdina*) readily accept melezitose unlike their sympatric competitors *Anonychomyrma gilberti*; this differentiation corresponds to the composition of their main honeydew sources (Blüthgen and Fiedler 2004a). Melezitose has even been suggested to represent a specific ant attractant, but this does not seem to be a general rule, since most ants do not prefer melezitose over other common sugars such as sucrose (Blüthgen and Fiedler 2004b; Boevé and Wäckers 2003; Cornelius *et al*. 1996).

The specific role of enzymes and micro-organisms in the digestion of honeydew compounds may present an important key to understanding the success of honeydew acquisition in ants. Variation in honeydew composition can be pronounced among hemipterans, probably corresponding to their feeding mode, i.e., whether they suck on xylem, phloem, parenchyma, or a mixture thereof (see Carver *et al*. 1991), or on plant tissues of variable age (Douglas 1993). Such ideas have been poorly explored to date, and clearly more research is needed to understand how honeydew composition regulates, and is regulated by, tritrophic interactions among host plants, hemipterans, and tending ants.

Apart from digestive traits, key adaptations that facilitate an efficient use of large amounts of honeydew or nectar include a modified proventriculus, a muscular valve that controls the flux of liquids between the midgut and 'honey' crop where large

Box 7.2 Ants as biological-control agents in the horticultural industry
Renkang Peng and Keith Christian

As predators of pests, some species of ants are useful in integrated pest management (IPM) programmes. A total of 24 species of ants from 10 genera are recognized by farmers and researchers in tropical locations as beneficial for about 16 agricultural crops and some timber tree species (CAB Abstracts 1910–2007). Among these, about 12 species in 8 genera are beneficial to horticultural crops (Table 7.2.1). *Oecophylla* spp. have been used more extensively than other genera of ants and are the most effective biocontrol ants known (Table 7.2.1). *Oecophylla smaragdina* and *O. longinoda* control more than 50 and 15 pest species, respectively, belonging to 18 families, on 8 crops; whereas fewer pest species on a narrow range of crops are controlled by other ant species (Table 7.2.1). There are four factors that commonly constrain the use of ants as biocontrol agents: (*a*) suppression of only a few pest species on each crop, (*b*) disturbance to people, (*c*) unstable populations due to biological factors, and (*d*) protection of some honeydew producing pests.

These four constraints have been extensively studied and solved in *Oecophylla* ants. An assessment of the effectiveness of *Oecophylla* ants in controlling insect pest assemblages showed that, except for some honeydew producing pests, the ants were able to control all the main pests of cashew and mango in Australia (Peng and Christian 2005; Peng *et al.* 2004), of citrus in Vietnam (Van Mele and Cuc 2003), and of cocoa and coconuts in Papua New Guinea, the Solomon Islands, Tanzania, and Ghana (Stapley 1980; Way and Khoo 1992). However, these ants can annoy farmers with their aggressive nature, particularly during harvest. Spraying clean water on trees prior to harvest can reduce *Oecophylla* ant activity on mango trees by 88% for the first 20 min, and by 61% after a further 30 min (Peng and Christian 2005). Some fruit growers rub wood ash on tree branches to reduce ant activity on part of the tree when harvesting, or they rub their hands and arms with ash to prevent *Oecophylla* ants from biting (Van Mele and Cuc 2003).

The mechanism by which the ants in Table 7.2.1 protect tree crops is mainly through catching and deterring pests while foraging for extrafloral nectar secreted by flushing terminals (see Chapters 6 and 7) because these ants are both sugar and meat eaters. To get enough food, the ants frequently patrol flushing terminals and prey on the pests (or scare them off), giving them little chance to feed or lay eggs on or in the vulnerable parts of the plants. This foraging activity further results in clean and shiny nuts on cashew trees because the extrafloral nectar secreted by nuts is continuously removed by ants; otherwise, the nectar accumulates on the nuts, resulting in the growth of sooty mould and dull nuts. To achieve continuous control of pests, it is important to keep ant populations high and stable (Peng *et al.* 1999; Stapley 1980; Way and Khoo 1992). The unstable populations in *Oecophylla* are caused by boundary fights between *Oecophylla* colonies due to their antagonistic behaviour (Peng *et al.* 1999) and by competition between *Oecophylla* ants and other ant species (e.g. Stapley 1980; Way and Khoo 1992). This can be solved by reducing the fights by separating and monitoring the ant colonies, managing queen ants, creating a mixed-cropping system, and by reducing populations of other competitive ant species with ant baits (Peng and Christian 2005).

Because some insect pests, such as mealybugs, scales, and aphids produce honeydew, they are farmed by ants as a part of their food source. As a result, these pests may damage flowers and fruits. Several organically approved chemicals are effective in reducing these pest populations in mango orchards without being detrimental to *Oecophylla* ants (Peng and Christian 2005).

Three IPM programmes for cashew growers and for conventional and organic mango growers in northern Australia have been developed. These IPM programmes use *Oecophylla* ants as a key element, together with farming tactics and organically approved chemicals (Peng and Christian 2005; Peng *et al.* 2004), allow growers to produce high quality fruits and nuts, and provide them

continues

Box 7.2 continued

Table 7.2.1 The number of ant species used in the horticultural industry in the world according to CAB Abstracts (1910–2007) and Huang and Yang (1987).

Ant species	Number of pest species controlled	Horticultural crops	Country	Number of articles that report the use of the ant and the pests controlled by the ant during different periods of time						
				304-877	958-1401	1600-1900	1910-1940	1941-1970	1971-1999	2000-2007
Oecophylla smaragdina	>50 species of hemipteran bugs, beetles, fruit flies, caterpillars, thrips, pest ants, and leafhoppers	Cashew, citrus, cocoa, coconut, mango, and oil palm	China, Australia, Papua New Guinea, Solomon Islands, Vietnam, Philippines, India, Thailand, Sri Lanka, Indonesia, Malaysia, Fiji, and Cambodia	3	3	5	13	12	29	13
Oecophylla longinoda	>15 species of hemipteran bugs, beetles, pest ants, fruit flies, and caterpillars	Cocoa, coconut, citrus, coffee, mango, and sour fig	Tanzania, Ivory Coast, Ghana, Benin, Congo, Gold Coast, Kenya, Nigeria, and Cameroon	N/A	N/A	N/A	1	13	20	2
*Dolichoderus thoracicus**	4 spp of hemipteran bugs and 1 sp of caterpillar	Cocoa and sapdilla	Malaysia, Indonesia, and Vietnam	N/A	N/A	N/A	3	0	11	1
Azteca spp* (including *chartifex*)	1 sp of pest ant, 1 sp of thrips, and various pests	Cocoa and citrus	Brazil and Trinidad	N/A	N/A	N/A	1	0	2	0
*Wasmannia auropunctata**	A few spp of hemipteran bugs, 1 sp of weevil, and 2 spp of pest ant	Cocoa, coconut, and banana	Cameroon, Solomon Islands, and Venezuela	N/A	N/A	N/A	0	1	2	0
*Anoplolepis gracilipes**	2 spp of hemipteran bugs and 3 spp of beetle	Coconut and cocoa	Solomon Islands, Papua New Guinea, and Seychelles	N/A	N/A	N/A	2	3	6	0
*Anoplolepis custodiens**	1 sp of hemipteran bug	Coconut	Tanzania	N/A	N/A	N/A	0	1	1	0
Crematogaster spp*	A few spp of hemipteran bugs and date palm pests	Date palm, cocoa, and oil palm	Yemen Arab Republic, Congo, and Colombia	N/A	N/A	N/A	0	1	2	0
*Pheidole megacephala**# & *Tetramorium guineese*#	2 spp of weevil beetle and 3 spp of pest ant	Banana	Cuba, Nicaragua, and Dominican Republic	N/A	N/A	N/A	N/A	N/A	2	1

Species with '*' are harmful in some other places; Species with '#' have been extensively used in Cuba and Nicaragua as a part of IPM programme to control the main pests in agricultural crops of sweet potato and sugarcane (Levins 2008).

Box 7.2 continued

with the opportunity to produce organic products. The use of these ants allows farmers to avoid environmental and health problems associated with insecticides.

The other ant species in Table 7.2.1 are either reported as being beneficial in some places or harmful in other places because of the damage they can cause to homes, public health, domestic animals, plants, indigenous insect communities, and/or competition with beneficial ants. For these species, the aforementioned constraints need to be extensively studied before the use of these ants can be promoted.

amounts of liquids can be stored. Proventriculi in formicines and some dolichoderines are more sclerotized and passively dam the efflux of crop contents, and in some myrmicines they are particularly narrow (Davidson 1997). These modifications are associated with a more rapid uptake and larger storage of liquid volumes (Davidson *et al.* 2004). Some other intensively honeydew-feeding ants such as *Dolichoderus* do not have such modified proventriculi, but at least partly compensate for this by storing fluids in the hindgut (Cook and Davidson 2006; Davidson 1997). Instead of sucking liquid food as most other ants, ponerines have been shown to lick fluids and transport liquid food as droplets between their mandibles. The latter technique is more efficient when sugar concentrations are high, but less efficient than sucking at low sugar concentrations (Paul and Roces 2003). Corresponding to the improved storage of liquids, trophallaxis seems to be particularly pronounced in excessively honeydew- or nectar-feeding ants. Other modifications in the digestive system of mainly liquid-feeding ants are discussed in Cook and Davidson's review (2006).

Nectar secretion rates of plants may be accelerated by more frequent consumption by ants and herbivore attack to some extent (Heil *et al.* 2004b), but the overall nectar productivity of a plant is limited to a relatively low level. In contrast, the amount of honeydew available on a plant can reach very high levels. Honeydew quality and quantity can be considerably controlled by ants by the intensity of tending, protecting, relocating, or preying on their hemipteran partners (Stadler and Dixon 2005). It thus seems likely that honeydew, as opposed to nectar, fuels the high abundance and colony size of many ant species, including numerous tropical arboreal ants (e.g. from the genera *Camponotus*, *Polyrhachis*, or *Dolichoderus*) (Davidson *et al.* 2003) as well ground-nesting ants (*Formica polyctena*) from temperate zones (Horstmann 1974). Given the considerable overlap of potential hemipteran partners, interspecific competition between ants for honeydew can be severe. Profitable honeydew sources on a tree or shrub are usually monopolized by a single ant colony and defended against competitors (Blüthgen and Fiedler 2004b; Blüthgen *et al.* 2006). Such spatio-temporally stable honeydew sources often provide the basis of ant territories and lead to a mosaic-like distribution of dominant ants (known as 'ant-mosaics', see Dejean and Corbara 2003 and Chapter 5). Since larger ant colonies are often not only competitively superior, but also more dependent upon honeydew intake, there may be a positive feedback between number of hemipterans, the size of ant colonies, and their control over this resource (see Grover *et al.* 2007; Helms and Vinson 2008; Chapter 6).

Honeydew that has been flicked off by or dropped down from the hemipterans onto the foliage or the ground may function as a cue for ants scouting for new trophobiotic partners (Del-Claro and Oliveira 1996). Such fallen honeydew may potentially contribute to an important part of the ants' nutrition, but this has not been quantified. Honeydew accumulation can be pronounced under some trees and contribute considerably to nutrient fluxes and microbial activity (Stadler *et al.* 1998).

7.1.6 Nectar

Apart from honeydew, one of the most conspicuous liquid food sources for ants, at least in tropical forests, is provided by extrafloral nectaries (EFNs). They are located on leaves, stems, or around flowers (circumfloral nectaries) and occur across a wide variety of plant families and species, which may constitute a significant proportion of local tropical floras (e.g. Oliveira and Freitas 2004). In turn, ants are the most frequent consumers of such extrafloral nectars, where they often contribute to anti-herbivore defences (Heil and McKey 2003; Rico-Gray and Oliveira 2007; Chapter 6). The often mutualistic association is usually highly opportunistic, as illustrated by the diversity of ants that may be observed collecting extrafloral nectar from a plant species (Oliveira and Freitas 2004). In a single Amazonian forest site, 52 ant species from five subfamilies have been recorded on four EFN-bearing epiphytic *Philodendron* species alone, representing the majority of ant species collected in canopy traps (Blüthgen *et al*. 2000). Morphological or physiological constraints that might prevent utilization of nectar seem to be virtually absent among most ants. Even specialized predators such as *Strumigenys* or *Odontomachus* feed on extrafloral nectar (Blüthgen and Fiedler 2004b), and most ants examined so far readily accept sucrose, fructose, and glucose (but see Heil *et al*. 2005). Most EFNs are open structures and readily accessible to ants, in contrast to many myrmecophytes, where trichomes, waxes, or domatia architecture prevent access to a number of species (Davidson *et al*. 1989; Heil and McKey 2003; Chapter 6). The difference in accessibility may explain why both ant-plant 'networks' differ in their degree of specialization: ant–EFN associations are much more generalized than myrmecophytic ones (Blüthgen et al. 2007). Several ant species may simultaneously forage for nectar on the same individual plant, which is unusual among other ant–plant associations such as myrmecophytes or trophobioses (Blüthgen and Fiedler 2004b). Moreover, the relatively low and discontinuous nectar flow of EFNs temporally prevents more stable associations, as found for ants at honeydew sources.

Extrafloral nectars mainly contain carbohydrates, usually confined to sucrose, glucose and fructose, amino acids, and traces of other compounds, and are highly variable in quality and quantity. Apart from pronounced preferences for certain sugars, ants generally prefer solutions containing multiple amino acids (Lanza *et al*. 1993). Some nectars resemble nutritious honeydews in their amino acid composition and are indeed more often monopolized by competitively superior ants than nectar sources of poorer quality. Consequently, competitively inferior ants may switch to less nutritious EFNs (Blüthgen and Fiedler 2004a), although their fundamental preferences in the absence of competition are similar to those of the superior species (Blüthgen and Fiedler 2004b). Asymmetric competition thus plays a role in the distribution of ants on EFN-bearing plants.

Apart from EFNs, several tropical plant species possess more solid lipid-rich food bodies that are readily consumed by ants (Buckley 1982; Fischer *et al*. 2002; O'Dowd 1982). Food bodies (FBs) are highly variable in nutrient composition (Buckley 1982; Heil *et al*. 2004a). On myrmecophytic plants (Chapter 6), FBs are common and particularly rich in lipids and proteins (Fischer *et al*. 2002; Heil *et al*. 1998), but FBs also occur on many non-myrmecophytic plants. Myrmecophytic *Macaranga* species provide more protein-rich FBs for the resident ants, embedded within stem-clasping stipules, whereas FBs on the foliage of some non-myrmecophytic plants have higher concentrations of (less costly) carbohydrates (Heil *et al*. 1998). Compared to EFNs, composition and consumption of FBs have received much less attention, despite their importance in tropical communities.

The regular consumption of extrafloral nectars contrasts with the ants' relatively infrequent use of floral nectar. The conspicuous lack of ants on floral resources stimulated the search for mechanisms that function as constraints. Janzen (1977) proposed that floral nectar may be toxic to ants. His hypothesis provoked several subsequent tests on the palatability of floral nectars to ants. In essence, some cases of repellent nectar were confirmed, but these were exceptional, since the vast majority of nectars were attractive to ants and immediately consumed when offered outside the flowers (e.g. Junker and Blüthgen 2008; Koptur and Truong 1998). In some flowers, nectar is protected morphologically by narrow tubes or other concealing structures (Beattie 2006), or a wax layer on the plant stem or inflorescence may prevent the access of ants and other crawling insects (Harley 1991). However, a more general explanation for the conspicuous absence of ants on accessible flowers may be provided by a repellent effect of floral odours. For instance, *Acacia zanzibarica* flowers were observed to repel ants only during anthesis, but not during development or maturation of buds and seeds when protective services by ants are beneficial (Willmer and Stone 1997b). Recently, olfactometer experiments confirmed that naturally emitted floral scents provide an effective mechanism that determines the visitation and avoidance of flowers by ants (Junker and Blüthgen 2008). Visited flowers, e.g., umbellifers, produced floral scent bouquets that did not affect the ants' foraging decisions, while unvisited flowers emitted strongly repellent odours and may effectively protect their nectar from ants. Ants were effectively repelled by different terpenoids that are common in flower scents and by floral scent bouquets from a wide spectrum of plants (Junker and Blüthgen 2008). Analogously, some terpenoids are also known to be produced by bees or other insects as defensive compounds to deter ants and other predators (Cane 1986).

While virtually all ants consume nectar, most species are unable to digest pollen, except some species of the tribe Cephalotini, which seem to collect mainly wind-dispersed pollen from the foliage, rather than directly from flowers (Baroni Urbani and de Andrade 1997). These ants are equipped with a diverse gut microflora (Jaffé *et al.* 2001) that may help to overcome digestive barriers of some diets such as pollen or bird droppings.

7.1.7 Granivory and seed collecting

Seeds from a large number of plant species bear elaiosomes: appendages that mainly contain not only lipids, but also proteins, carbohydrates, and essential sterols (Fischer *et al.* 2008; Gammans *et al.* 2005). They usually attract a variety of ant genera, and in most cases the resulting myrmecochory is a relatively generalized and facultative mutualism rather than an obligate and species-specific one (Gorb and Gorb 2003; Chapter 6). Seeds are usually harvested from the ground rather than from the plant (Buckley 1982). Ants often (but not exclusively) remove the elaiosomes from the undamaged seeds inside their nest. While elaiosomes are then completely consumed, the seeds may be disposed in garbage piles where they often find a suitable substrate to germinate and grow.

Nutritional benefits of elaiosomes for ants can be substantial and translate into a higher production of female reproductives (Morales and Heithaus 1998) or total brood production in experimental colonies (Fokuhl *et al.* 2007; Gammans *et al.* 2005). However, the importance of nutrient flows of elaiosome-bearing floras to ant communities are largely unknown. It can only be assumed that elaiosomes from geophytes provide important resources for temperate woodland ants particularly in spring, given their nutritious value (Fischer *et al.* 2008; Gammans *et al.* 2005). Apart from nutrition, other substances in elaiosomes, such as oleic acid, trigger a stereotypical carrying response of ants, similar to corpse transport (Brew *et al.* 1989). Chemical cues of seeds may also include volatiles. In Neotropical ant-garden associations, *Camponotus femoratus* ants are attracted to the scent emitted by epiyphte seeds that they then collect and incorporate in their nests as 'hanging gardens' (Youngsteadt *et al.* 2008). However, it seems unlikely that olfactory cues are important in other, more generalized cases of myrmecochory.

In contrast to elaiosomes that function as rewards for seed-dispersing ants, the seeds themselves are only consumed by a limited number of ant species (myrmicines such as *Messor*, *Monomorium*, *Pheidole*, and *Pogonymyrmex* among other ant taxa; Plate 5). These include omnivorous as well as granivorous ant species that largely depend on seeds (Andersen 1991; Buckley 1982). Ant granivory is particularly common in arid and semi-arid habitats, where dry seeds can be stored underground to bridge seasonal bottlenecks in seed availability (Buckley 1982; Rico-Gray and Oliveira 2007). Seed size, morphology, and availability are important features to explain which seeds are harvested by ants (Andersen 1991). Preferred seed size often corresponds to the ant's body size (Traniello 1989). In the presence of superior competitors, however, niche shifts in terms of seed choice or spatio-temporal activities (Mehlhop and Scott 2008) or between seed and prey collection (Sanders and Gordon 2003) can be observed.

Tough seed coats prevent the use of seeds by most ants, as strong mandibles are required to open them. For breakdown of starch stored in seeds, amylase is necessary. Amylase has been rarely tested and so far only recorded from maxillary and salivary glands of *Camponotus* (Ayre 1967). It remains unclear as to how ants overcome defensive secondary metabolites that are common in seeds. A pronounced preference for grass seeds may suggest an avoidance of defensive compounds in several other plants. However, endophytic fungi in grass seeds may lower their palatability due to alkaloid production (Knoch *et al.* 1993). Fleshy pulps or arils of diaspores are frequently consumed by ants as well, particularly on forests floors in tropical rainforests (Rico-Gray and Oliveira 2007), and seeds consumed from bird or mammal droppings are the basis of secondary dispersal services by ants. Eggs of several phasmids are similarly dispersed and buried by ants, motivated by appendages analogous to elaiosomes (Hughes and Westoby 1992), but the nutritional or semiochemical basis of this interaction is currently unknown.

7.1.8 Growing and eating fungi

Although they occasionally lick plant sap from wounds, ants – unlike classical folivores – do not chew plant foliage. Reasons for this limitation may be manifold and include putative constraints by digestive, morphological, or detoxification capabilities. However, fungus-growing ants of the tribe Attini utilize plant foliage to nourish their associated fungus; ant larvae are fed with specific fungal staphylae that may provide a full diet (Martin 1970). Each ant species may be associated with a variety of fungal cultivars from different clades within the family Lepiotaceae (Mueller *et al.* 1998). Worker ants, their larvae, and the associated fungus are equipped with a diverse range of enzymes required for the degradation of leaf material as substrate for the fungus (d'Ettorre *et al.* 2002b; Erthal *et al.* 2007), with complementary digestive abilities among adults, larvae, and the fungus (d'Ettorre *et al.* 2002b). Chitinase is particularly important in this regard, and is probably obtained from the fungus in addition to several other digestive enzymes (Martin 1970). A number of studies have shown that the selection of leaves harvested by these ants is strongly influenced by their secondary metabolites (e.g. Howard 1988). Decisions on which resources are collected largely reflect their suitability to the fungus (Herz *et al.* 2008; Seal and Tschinkel 2007b). Apart from this selectivity, about half of the available woody plant species are actually harvested in a territory of *Atta cephalotes* (Cherrett 1968; Rockwood and Hubbell 1987; Vasconcelos and Fowler 1990), classifying *Atta* as one of the most generalized 'herbivores' known. Hundreds of tree species, shrubs, climbing plants, herbs, and hemiepiphytes have been recorded as being harvested by *Atta* and *Acromyrmex* ants (Farji-Brener 2001; Vasconcelos and Fowler 1990).

Fungus gardening is one option, harvesting naturally grown mushrooms another. Witte and Maschwitz (2008) recently discovered in the Malaysian rainforest that *Euprenolepis procera* ants are specialized fungal feeders. They harvest numerous species of fungi on the forest floor at night. As mushrooms are their almost exclusive diet and occur irregularly in space and time, these ants have adapted a fully nomadic lifestyle (Witte and Maschwitz 2008). To date, nothing is known on how these *Euprenolepis* digest the fungus.

7.1.9 Predation and scavenging

Whereas the subfamilies Dolichoderinae, Formicinae, and Myrmicinae comprise species that predominantly feed on liquids, army ants and most species of the subfamilies Ponerinae, Leptanillinae, and numerous other Myrmicinae have a largely or solely predatory lifestyle. Predatory ants typically feed on other invertebrates, with some ants being specialized on certain dominant groups such as termites or collembolans. Moreover, ants often feed on other ants, and the level of cannibalism or intraguild predation is pronounced. The frequency of predatory taxa varies among ant communities. In tropical rainforests in Borneo, predators were found to decrease faster in abundance than omnivorous ants with increasing altitude, which corresponds to a decrease of prey items such as termites or other ground-dwelling ants (Brühl *et al.* 1999).

Foraging strategies of predatory ants mainly fall into two categories (also see Chapter 12). Small prey items are captured by either single workers utilizing their mandibles or sting, or groups of ants forage cooperatively, forming large raiding groups or swarms that enable them to overwhelm large prey items or other social insects like termites or ants (Hölldobler and Wilson 1990). Individual ant workers are constrained in their choice of prey size, generating a correlation between body size and prey size across different ant species and polymorphic castes for individually retrieved prey items. Cooperative prey handling, however, enables ants to overcome this limitation and contributes to niche overlap between ants of different size (Traniello 1987, 1989). Such cooperative food recruitment does not occur among all ant species, and thus contributes to niche partitioning in ant communities. Solitary hunting ant species often show highly specialized mandibles like pitchforked mandibles or trap-jaw mechanisms (e.g. *Anochetus* spp, *Plectroctena mandibularis, Myrmecia* spp., Figure 7.1b). The movements of trap-jaws are among the fastest movements in the animal kingdom and the striking force may directly lead to the death of prey items and may also be used in defence (e.g. Patek *et al.* 2006) (see Box 12.1). The mandibles' closer muscle in such trap-jaw ants is adapted to maximize velocity and carries a higher proportion of fast muscle fibres in comparison to granivorous ants that require powerful mandibles to crack seeds (Gronenberg *et al.* 1997). A second adaptation of predatory ants is the paralysis or storage of living prey items. Workers of the ponerine *Harpegnathos saltator* have been shown to preserve insect prey for a period of two weeks. Moreover, the ponerine *Cerapachys turneri* stored paralysed larvae of *Pheidole* that were captured by raiding nests over a period of two months without visible signs of growth or death of the larvae (Hölldobler 1982; Hölldobler and Wilson 1990).

In order to successfully paralyze, but not kill their prey, ants need to limit the force of the sting. Hence, the sting's penetration in *Cerapachys* is controlled by mechanoreceptors on the pygidium (Hölldobler and Wilson 1990). Apart from prey items, predatory ants can still utilize liquid food sources and sometimes carry haemolymph in their crop. Food niches of ants are most likely much broader than currently known, and probably include diets other than the more 'simple' resources described earlier. Bird droppings, and mammal faeces and urine are used by several ants (Figure 7.1c), and are rich in nitrogen. Many species appear to be regularly grazing over leaf surfaces in tropical forests, but it is largely unknown which diets are gathered during this 'leaf foraging'. Davidson *et al.* (2003) suggested that they may obtain epiphyllic fungi and other primary 'vegetarian' diets apart from the known exudates such as nectar, honeydew, and wound sap.

7.2 Ant nests

Ants are often regarded as central place foragers for whom the nest represents a highly persistent central location for sheltering the queen, rearing brood, storing food, cultivating fungi, and exchanging food among workers. Selection of nest sites and nest structure are influenced by various factors, the two most important being shelter and ensuring optimal conditions for the brood. In addition, the location and structure of the nest influences foraging range and foraging strategy when workers can be recruited from decentralized nest structures. Suitable nest sites and characteristics (e.g. single defendable entrances) increase the protection of a colony against interference competition with other ants and other antagonists. Environmental conditions (temperature, soil, vegetation) and species-specific traits, such as foraging behaviour, also contribute to nest site preferences. In addition, the size and structure of colonies (monodomous or polydomous, Chapter 10), as well as the capability to utilize different substrates for nest building, strongly influence both the form and the spatial distribution of ant nests (Figure 7.2).

Nest sites are generally a limited resource for ants (Philpott 2005b), and requirements for nests change throughout a colony life-cycle. First, foundresses need to localize a suitable nest site for rearing the first brood. Whereas founding chambers may be quite small, more space is required with increasing colony size. This can be achieved by actively excavating the nest in the case of ground- or wood-nesting ants. Some arboreal ants are able to actively enlarge their nest by adding carton or leaf structures. Ants that do not build structures themselves may need to relocate their nest to larger crevices or may move into larger preformed cavities such as another adjoining hollow branch.

In temperate regions, the majority of ant species nest underground, since temperatures and humidity are less variable in the soil than in more exposed sites. Cold air temperatures in winter and hot temperatures during summer may prevent the survival of brood and workers above ground (Hölldobler and Wilson 1990). In tropical forests, arboreal nesting habits are just as common as ground nesting ones. The ability to actively build carton nests made some arboreal ants independent of natural occurrences of nest sites in this habitat and allowed the establishment of large colonies in the canopy – in proximity to food sources such as trophobionts or extrafloral nectar. Arboreal ants use very diverse building materials such as mud particles, spider silk, plant material, or detritus (subsumed under 'carton'). Some nests are formed mainly of silk produced by larvae (e.g. *Polyrhachis*) or of living leaves that have been connected with such silk (*Oecophylla*) (Figure 7.2). In certain cases, such carton nests may contain nutrient-rich substrates and soil and function as ant gardens into which seeds from certain epiphytes are planted, which then grow and additionally stabilize the nest.

Whole colonies frequently move towards new nest sites following disturbance, when food supply is lacking around the nest, or when pressure by pathogens or enemies becomes too high (e.g. McGlynn *et al.* 2004; Smallwood 1982) or when moving implies a benefit, i.e. when a more suitable nest site is found and emigration to this site is not too costly (Dornhaus *et al.* 2004). However, the energy and time that a colony requires for nest-building activity may be substantial. For example, Mikheyev and Tschinkel (2004) have estimated that up to 20% of a *Formica pallidefulva* colony's energy intake, and at least 6% of worker time per year, is spent on excavating nests. In contrast, army ants or migrating herdsman of the genus *Dolichoderus*, minimize their costs for nest building, since the main nest sheltering the queen and brood is formed by the workers' bodies. Nonetheless, even in bivouacs of army ants, the core temperature within the nest is regulated accurately to ensure optimal conditions for the brood (Franks 1989). The lifestyles of these ants represent extremes in terms of their food niche. Both army ants and migrating herdsmen ants rely on food sources that are patchily distributed in space and time, requiring frequent nest relocations. Army ants need to move when insect prey becomes rare in the vicinity, and migrating herdsmen species need to find young, freshly growing leaf parts to which their trophobionts, the *Allomyrmococcini* mealybugs, are adapted (Dill *et al.* 2002). Such bivouacs represent the upper extreme of nest dynamics, while large terrestrial nest mounts of some territorial ground-nesting ants are much more static, as they may persist over several years.

Building behaviour frequently extends from nests to extensive other structures that shelter valuable food sources or provide protection during foraging against desiccation, predators, or competitors. Such sheltering structures include galleries, arcades, or pavilions and are usually built from the same substrate as the nest substrate of the respective ants (Anderson and McShea 2001b). Building such structures may be a time-consuming task. However, ants are able to monopolize a food source by covering (and hiding) it at a time when it is not yet attractive to competitors, e.g. small but growing aggregations of honeydew-producing hemipterans. In addition, *Crematogaster* ants cover the flowers of *Ipomoea* before they start producing extrafloral nectar and keep them sheltered until nectar production ceases (Beckmann and Stucky 1981). Thus, ants that are inferior in defending food sources against competitors by force may be able to exploit such food sources over long time periods if they build shelters before their behaviourally dominant competitors arrive.

7.2.1 Ground nests

Ground nesting was most probably the ancestral nesting habit in ants and facilitated by the key innovation of the metapleural gland, the secretions of which often inhibit growth of pathogenic microorganisms (Hölldobler and Wilson 1990) that are abundant in soils. Ground nests can be excavated by the ants themselves, often in a species-specific shape (Tschinkel 2003). However, preformed

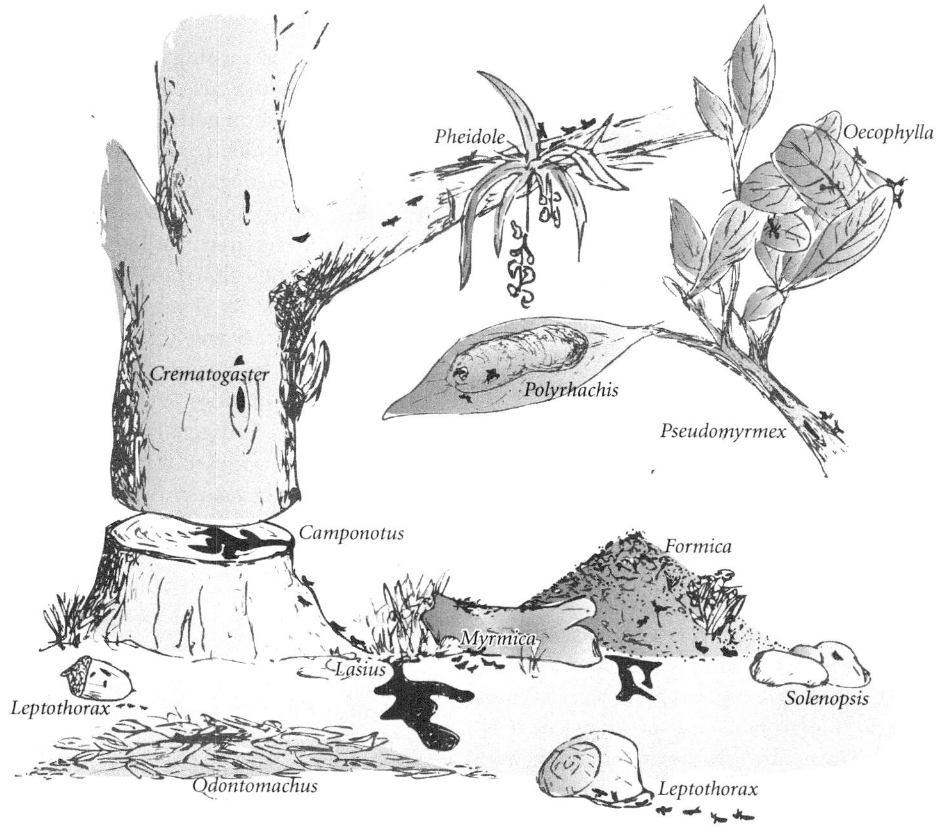

Figure 7.2 Overview of ant nests with a typical representative ant genus. Arboreal nests occur in trunks, under epiphytes, inside twigs, inside myrmecodomatia, and as silk and carton nests or woven leaf nests. Terrestrial nests are found as mounds, in the soil or leaf litter, under rocks, underneath or inside dead wood, snail shells, or acorns. (Drawing: Nico Blüthgen)

cavities like abandoned termite mounds or burrows created by other animals may also be used opportunistically. Ground nests predominate when environmental conditions prohibit nesting above ground, but also frequently occur in tropical regions. Regulation of temperature in brood chambers is facilitated by building chambers at an adequate soil-depth. In addition, structures such as nest mounds or rocks under which colonies nest, buffer extreme temperatures and may serve as supplementary heat sources in spring (e.g. Thomas 2002) allowing an earlier onset of brood production.

Ants can also actively regulate microclimatic conditions by altering nest architecture (Kleineidam and Roces 2000) or by including decaying plant material to warm the nest (Coenen-Staβ *et al.* 1980). In addition, *Solenopsis invicta* and several species of the genera *Formica* and *Lasius* have been shown to build their oval mounds in a particular orientation such that the long sides of the mound are exposed to the sun early in the morning and before sunset. Like in the famous 'magnetic-termites' of Australia, the ants' mounds warm up more rapidly in the morning and cool down more slowly in the afternoon (Hubbard and Cunningham 1977). Relocation of the brood within the nest to regions with adequate temperature (e.g. actively moving brood towards the soil surface or deeper into the ground) is a common strategy to ensure optimal conditions for the brood (Anderson and Munger 2003; Bollazzi and Roces 2002; Penick and Tschinkel 2008).

Ants have developed behavioural adaptations in addition to glandular secretions to keep pathogens in nest chambers at bay. Leaf-cutting ants dispose of waste from the fungus garden and nest chambers and collect it in special waste chambers. Waste is handled mostly by older workers that are less valuable for the colony (Bot *et al.* 2001a). *Formica* wood ants in temperate regions collect plant resin with antimicrobial properties and incorporate this into the nest mound (Chapuisat *et al.* 2007). Avoidance of empty nest sites that contain dead workers when choosing a new home may also contribute to reduce the risk of contamination with pathogens (Franks *et al.* 2005). If ground nesting poses a higher threat of pathogens to ant colonies due to higher humidity and close contact with micro-organisms in the soil, then ground-dwelling ants should invest more into their immune functions with increasing persistence of the nest in comparison to closely related ants with arboreal nesting habits (Boomsma *et al.* 2005b). That more humid conditions of the nest substrate lead to a higher pathogen pressure has been shown in termites, where a dampwood species has a larger pathogen load in comparison to drywood termites (Rosengaus *et al.* 2003). However, comparative studies on the strength of the immune functions (immune defences and gland secretions) are still lacking.

Ground-nesting ants frequently build aboveground trenches or arcades that may extend over large areas (Anderson and McShea 2001b; Kenne and Dejean 1999). These structures can help to protect ants from sunlight and dessication when they need to leave their nest for foraging activities. Such structures can also be built in order to cover large food items that cannot be taken back into the colony by foraging workers to provide exclusive access to the food source. For instance, numerous *Pheidologeton* workers often build a dirt covering over sugar baits (Blüthgen and Fiedler 2004b), fruits, or even the cadaver of a 25 cm long snake within two days (H. Feldhaar, personal observation). In all cases, the colony manages to wholly exploit its resource without interference by competing ant species.

7.2.2 Tree trunks and plant cavities

While most wood-nesting ants live in softer degraded wood where they use or excavate cavities, some species from several genera also utilize solid wood in living tree trunks. For instance, many species in the genus *Crematogaster* not only inhabit dead wood, but some can also excavate their own large cavities in living tree trunks (e.g. Longino 2003) or utilize cavities that have been pre-formed by wood-boring insects (Tschinkel 2002). Apart from *Crematogaster*, other ants that actively cut out living wood include large genera that inhabit myrmecophytes (e.g. *Camponotus* and *Pseudomyrmex*). Ants are unable to digest wood, however. Cellulolytic capacities are confined to relatively few specialized insect taxa and often accomplished by symbiotic micro-organisms (Martin 1991). For trunk-nesting ants, fungi or other micro-organisms might, however, play a role in wood decay and wood detoxification, and thereby facilitate excavation of solid wood, possibly in conjunction with an increased surface due to channelization by ants (like in termites, Delaplane, and LaFage 1990). Moreover, facilitated entry into the trunk by wood-boring insects seems to be a relatively common phenomenon, which has been poorly explored to date.

Competition for nesting space in tree trunks or branches that are accessible via preformed openings is intense; trees with natural openings to hollow stems and myrmecophytes that have been abandoned by their specific partner ants have been found to be colonised by opportunists with high frequency (Moog *et al.* 2002; Philpott 2005b). Myrmecophytic plants (Chapter 6) usually possess morphological barriers such as epicuticular wax covers, trichomes, or specific sizes and shapes of domatia entrances, which help to exclude opportunistic ants from utilizing the preformed nesting space that they offer (Davidson *et al.* 1989). However, in spite of the coevolution between ants and plants, the partners may have conflicting interests regarding resource allocation. Thus, colonies of plant-ants have been found to be restricted in colony size by limited nesting space (Fonseca 1993) or limited production of food bodies, which appear as the more important investment from the plant's point of view (Heil *et al.* 2002). By castrating the flowers of hosts, the ants can manipulate their host to allocate more resources to the ants (Yu and Pierce 1998).

7.2.3 Carton and silk nests

Carton nests are here defined as nest structures that have been actively constructed by ants (often formicines, dolichoderines, and myrmecines) in vegetation. Besides nests that shelter the brood or the queen, the term is usually also commonly used for pavilions that give shelter to trophobionts. Protection of trophobionts within the nest or in specific shelters may reduce the effects of extreme weather conditions, and thereby contribute to a more constant food supply for the colonies. In addition, the decentralized nest structure enables the ants to place carton structures in proximity to food sources, thus minimizing the costs for foraging and facilitating an easier monopolization of trophobionts. This suggests a close connection between honeydew-feeding and the decentralized nesting style typical for carton building and weaving ants (Anderson and McShea 2001b; Weissflog 2001). In addition, the ability to weave silk nests may have released arboreal ants from the pathogen pressure ground nesting species face. Johnson *et al.* (2003) found a strong phylogenetic correlation between weaving and the absence of the metapleural gland that has been shown to produce antimicrobial substances in numerous ant species.

While wasps produce sticky proteinous secretions in their labial glands that are mixed with plant material to improve nest stability, ants have not yet been found to do this. Instead, ants utilize a broad variety of materials for building such as plant fibres, mud, silk, fungi, or detritus (Liefke *et al.* 1998; Weissflog 2001). Depending on the main substances that are used for nest construction, four groups of carton nests have been identified by Weissflog (2001) in an extensive study on carton nest-building ants in southeast Asia: (*a*) silk nests, (*b*) nests stabilized by fungal hyphae, (*c*) nests built mostly of dead plant material and (*d*) ant gardens where the nest substrate is stabilized by the roots of epiphytes (see Corbara *et al.* 1999; Davidson 1988; Kaufmann and Maschwitz 2006). The main substrate component was often stabilized by the incorporation of plant trichomes, larval silk, spider silk collected by workers, or fungal hyphae. Fungal growth, as well as growth of epiphytes was supported by workers fertilizing the substrate with faeces and other nutrient-rich substances (see also Davidson 1988; Kaufmann and Maschwitz 2006). Except for nests formed with living leaves that had been connected with larval silk by *Oecophylla*, the height of the brood chambers in carton nests rarely exceeded the height of a worker. Limited mechanical stability of the carton and the sizes of the plant structures to which nests are attached may pose an upper limit to the size of each carton nest, thus forcing the ants to divide their brood among several nests. Most carton-nest-building species were found to have polydomous nests, with nests belonging to the same colony being distributed over a single or few trees (Weissflog 2001). Interestingly, most of the carton-building arboreal ants were found to have a lower number of actual brood nests in relation to pavilions sheltering trophobionts (Weissflog 2001).

Weissflog (2001) did not detect a specialization of carton-building ants towards a particular plant species, and the variety of trophobionts tended exceeded that of the specialized migrating herdsman species (Dill *et al.* 2002). Thus, competition among ant species with carton nests that often represent dominant arboreal ant species rather than specialization towards particular plants may be the most important factor determining the spatial distribution of nests on plants. It will be interesting to unravel whether the suitability of different plants, or plant parts for trophobioses, shapes the distribution of such carton nests. To date, extensive comparative studies on carton-nest building with respect to ant ecology (excluding ant-gardens where ants use the network of epiphyte roots mixed with substrate) are lacking from other regions like Africa or South America.

7.3 Future directions

In order to understand ecological success, dominance, and diversity of ants, we need to understand what they eat, what they do not eat, and why. Many ants utilize plant-based resources such as nectar, honeydew, or seeds to a variable extent, but a number of other, less visible resources may require more attention in future studies, e.g. faeces, fungi, and flicked-off honeydew. Several adaptations have been proposed that characterize ants that live on nitrogen-poor plant diets. With an increasing contribution of less 'simple' resources, gut micro-organisms may play a crucial role in facilitating nitrogen uptake, nutrient balance, or food detoxification. The role of digestive enzymes and micro-organisms in niche differentiation has been poorly explored. How do granivorous ants cope with defensive secondary compounds in seeds? Even for the oligosaccharides typically found in honeydew, the variable responses of ants cannot be sufficiently explained to date. Does the characteristic enzymatic or microbial activity of an ant explain which ants attend which hemipterans?

Apart from the wide array of diets used, it is also important to understand why certain other resources such as leaves or pollen are never or only poorly exploited by omnivorous, and virtually omnipresent, ant species. What are the constraints? How do different morphological or behavioural constraints and digestive capabilities of ants translate into food-niche partitioning in ant communities? Niche partitioning has been frequently inferred from studies that were based on artificial food baits such as tuna, honey, or jam. Spatial and temporal niche partitioning of ants on baits have been shown to be pronounced, and differences in competitive abilities, recruitment, or variation in diurnal activities have been suggested to promote species coexistence (Albrecht and Gotelli 2001; Fellers 1987). However, since baits poorly match the properties of ants' natural diets in terms of density and composition, more studies on natural resources are required to examine the importance of niche partitioning in ant community organization.

Fundamental questions of ant nutrition remain unanswered to date, partly because nutritional biology has generally become less fashionable. Nutritional requirements have not been characterized enough with respect to macro- and also micronutrients, where larval and adult nutrition needs to be distinguished. Micronutrients have been almost completely ignored in the nutritional ecology of ants. Quantitative information on actual ant diets is only available for a very few species. Most quantitative estimates of food intake in ant colonies date back to early work on wood ants (e.g. Horstmann 1974). While stable isotope techniques have contributed to a recent renaissance of basic studies on ant nutrition, such analyses do not provide a complete and entirely conclusive picture of the actual diets, unless accompanied by classical observations. Furthermore, the influence of gut micro-organisms on stable isotope signatures is also unknown. Digestion is another topic that deserves more attention. For instance, most enzymatic studies were performed some decades ago (e.g. Ayre 1967; Martin 1970; Ricks and Vinson 1972), and few recent studies have focused on digestive enzymes in ants (d'Ettorre *et al.* 2002b; Erthal *et al.* 2007; Heil *et al.* 2005). However, basic studies on digestion and nutrient flows may provide important contributions to our understanding of food niches in ant communities, and may describe and explain subtle niche differences among ant species.

Individual foragers from a single colony often exploit different resources. At the colony level, an imbalance in food resources collected by individual foragers may thus be buffered by complementary contributions of different individuals. To date, it is largely unknown, however, how ants maintain an influx of nutrients that is in accordance with the individuals' and the colonies' requirements. Nutrient balance in omnivorous ants is an emerging field that warrants further investigations. A variety of responses have been found for carbohydrate/protein ratios. These responses need to be integrated into a more holistic, dynamic view of colony performance, ranging from egg production, larval growth, sex and caste determination, sex ratios, biomass composition, immunity responses, glandular products, and foraging and defensive behaviour. Do omnivorous ants optimize complementary nutrition in the field? The carbohydrate: protein ratio is important, but more subtle balances have not been investigated so far. Studies on other animals or humans have demonstrated that dietary balance is more fine-tuned than suggested by carbohydrate: protein ratios alone, and also involves single components or even non-essential components such as an optimal ratio of essential to non-essential amino acids (Nation 2002). While it is evident that amino acids play a key role for determining ant food preferences, it is unclear how single amino acids are detected and differentiated.

Further studies on resource limitation, regulation, and nutrient balance in ants are needed to understand the consequences of nutrition to colony performance. Such studies may, among other methods, utilize artificial diets for ants that allow variation of ingredients (Dussutour and Simpson 2008; Straka and Feldhaar 2007). Appropriate experimental studies should be conducted over a long time frame since the brood and worker-stored resources function as a buffer, and some nutrients are required in extremely low concentration only, and may be transmitted via the eggs into subsequent generations (Dadd 1985).

The pronounced omnivory, extreme diversity of resources used, and variable dietary specialization in ants provide a unique opportunity for comparative analyses of evolutionary pathways using molecular phylogenetics. This huge variation occurs within a single insect family – across species of a comparable body organization (*bauplan*). The fact that different ant species cover an enormous body size range over three orders of magnitude in terms of body mass (Kaspari and Weiser 1999) may stimulate promising studies that test predictions based on body size, e.g. allometric relations. For example, metabolic theory has usually been examined across a broad taxonomic spectrum of highly variable morphology and life histories, but not in ants.

With respect to nests, not only advantages of realized nest types, but also limitations of potential nest sites need to be considered. To date, comparative work on building costs of different types of nest are still lacking. In contrast to the few data where energy or time expenditure on building ground nests has been measured, nothing is known about costs of excavating wood or building carton structures as nests. Nest longevity and maintenance costs may differ strongly among nest types. Moreover, potential conflicts between nest sites and locations of food sources or avoidance of competition, predation, and parasitism can be crucial. Are there trade-offs in the security provided by a nest and the costs of building it? Arboreal and terrestrial nests may differ especially, not only in costs of building, but also with respect to microclimate and defendability against parasites and competitors. The capability to actively construct nests may lower the extent of nest-site limitation for an ant species and increase the importance of food limitation in structuring ant communities.

Similar to their important role in nutrition, microorganisms may also be crucial in the suitability of nests. They may either function as pathogens, forcing ants to defend or to relocate their nests, or as mutualists that, for example, stabilize carton structures (Schlick-Steiner *et al.* 2008), or that are grown as a biotic defence against pathogens (Currie 2001a). These interactions warrant more attention in future studies. Thus, actively constructed carton nests may be built by ants either to overcome nest site limitation, or to escape predators and pathogens. It is currently unknown whether micro-organisms help ants to excavate living tree trunks and branches, and a comparative analysis of wood excavation is needed.

7.4 Summary

Food and nesting space are the most important resources in ant ecology, and contribute strongly to the structure of ant communities. Some ant communities may be influenced more strongly by the limited availability of suitable nesting sites, whereas competition for food may be more important in others. Because ants are central place foragers, food and shelter are linked: the size, location, and distribution of nest sites, and whether nests are static or dynamic, affects the diets available to ants, given their limited foraging range.

Most ants can be considered omnivores that can utilize a large variety of nutritional resources, and only few species are specialized on a particular food. However, differences in morphology and digestive capabilities of ants constrain the availability of food sources and contribute to fundamental niche differentiation. For instance, the ability or inability to digest trisaccharides may contribute to species partitioning of honeydew sources. The size of food items collected by individual ant workers may be constrained by their body size, but for large prey, cooperative handling in several ant species serves to overcome this limitation. Realized niches of submissive ants differ under the influence of competition by dominant ants.

The exploitation of 'new' resources, for example, direct consumption of fungi by ants, or the acquisition of endosymbionts that enables ants to survive on nitrogen-poor food resources may have driven the evolution of new lifestyles in ants and promoted niche differentiation. Moreover, differences of macronutrients in ant diets (particularly the carbohydrate: protein ratio) have been demonstrated to affect other aspects of ant ecology (e.g. competition and territorial behaviour). In turn, distance between nests and food sources are closely linked to the ability to discover and dominate resources against competitors. When food is available continuously, territoriality and permanent nests, may be favoured, while short-lived food sources require more frequent nest relocation. Consequently, nest types are highly variable, ranging from relatively persistent nests in the ground or wood cavities to dynamic, flexible bivouacs formed only by the worker's bodies. Decentralized nests and extended nest structures incorporating food sources, such as pavilions sheltering trophobionts, facilitate monopolization of valuable resources over longer periods of time. Investments into permanent shelter such as ground or wood nests may pay off as defence against enemies and competitors, but at the cost of flexibility in resource use and higher risk of pathogen infection.

Fundamental knowledge about the nutritional and nesting ecology of ants is crucial to understand the organization of ant communities, not only with respect to niche partitioning and species coexistence, but also foraging strategies or territorial behaviour – a challenge for future studies and syntheses.

Chapter 8

Ant Diversity and Function in Disturbed and Changing Habitats

Stacy M. Philpott, Ivette Perfecto, Inge Armbrecht, and Catherine L. Parr

8.1 Introduction

Habitat transformation and disturbance are significant threats to biodiversity conservation and ecosystem function. Disturbance is generally defined as any event that removes biomass (Townsend and Hildrew 1994), and is distinguished from habitat transformation or stress, which reduces available resources or changes the microclimate or structure of the habitat (Andersen 2000; Pickett and White 1985). Habitat disturbance and transformation affect communities in many ways either by altering the balance of competitive interactions, often in effect resetting the process of competitive exclusion, or by clearing space for colonization of new organisms. The degree to which habitat disturbance and transformation affect animal communities in general, and ants in particular, depends largely on the frequency and intensity of disturbance, the permanence with which habitats are transformed, and the distance from which propagules travel to recolonize affected habitats.

Ant habitats of all kinds are modified by natural disturbances, such as fire, forest gap formation, hurricanes, and flooding, which vary in their extent, magnitude, and frequency. Furthermore, many terrestrial ecosystems, especially in tropical regions, have been altered by human activities including deforestation, urbanization, agriculture, agricultural intensification, grazing, and mining. At the same time, ants themselves are also instigators of habitat modification via their roles as mound builders and ecosystem engineers (Decaëns *et al.* 2002; Folgarait 1998). The impacts of habitat disturbance and transformation for ants are widespread, yet they vary with region and ecosystem. Ants can be very sensitive to habitat transformation and disturbance, and for this reason have been extensively used as indicator species (Hoffmann and Andersen 2003; see Box 8.1). Because ants are colonial organisms, removal of individuals (mortality) caused by habitat disturbance or transformation may not translate to extirpation of the colony from the habitat (Andersen 2000). This may mean the responses of ants to disturbance may differ from other terrestrial animals and plants that may become locally extinct after disturbances. Disturbance effects on ant communities include loss of diversity, changes in species composition, alteration of interspecific interactions, changes in trophic interactions with ant-plants and honeydew-producing hemipterans, and modification of ant-provided ecosystem services such as seed dispersal, predation, and soil modification. Virtually all habitats are subject to some sort of disturbance, although the disturbance will obviously vary in origin (natural or human-induced), in scale, and in magnitude. Many insights to basic ecology are thus gained by investigating ecology in disturbed habitats. For example, seminal work examining predator-caused disturbance in intertidal zones has formed the basis for the field of disturbance ecology (e.g. Paine 1996). Similarly, research in tropical forests affected by hurricanes and tree-fall gaps has shaped our knowledge

Box 8.1 Using ants as indicators of ecosystem change
Alan N. Andersen

The sensitivity of ant communities to environmental disturbance, combined with their great functional importance and ease of sampling, makes them powerful monitoring and assessment tools in land management. The use of ants as indicators of ecosystem change is particularly widespread in Australia, especially in the context of mine site rehabilitation, but also for a variety of other land-use situations such as off-site mining impacts, forest management, and pastoralism (Andersen and Majer 2004). Ants have been strongly championed as indicators in Australia, but could equally be used as such in most other parts of the world, wherever they are diverse and abundant.

What to measure

Ant monitoring programmes typically focus on changes in species composition rather than diversity, as the latter can remain relatively constant in the face of major compositional change and therefore be uninformative (Kaspari and Majer 2000). Moreover, species diversity can respond in unpredictable ways to disturbance, or in ways that are highly situation-specific, such that changes can be difficult to interpret. If species-level responses to disturbance are well understood, then monitoring can focus on changes in the abundance of individual species. For example, several Australian ant species have been shown to increase or decrease consistently in abundance in relation to disturbance (Andersen *et al.* 2004a). However, in most cases reliable species-level information is unavailable so broad species compositional change is monitored.

Majer and Beeston (1996) have developed a protocol for scaling-up local information on the effects of disturbance on ant species composition to address regional scale effects of different land uses. The effects on ant species composition is multiplied by the proportional area affected for each land use, and the sum of these scores becomes a 'biodiversity integrity' index for the region. On this basis, Majer and Beeston (1996) concluded that the land use causing most biodiversity loss in Western Australia was intensive agriculture, followed by rangeland grazing, with mining having a negligible impact despite its dominant economic contribution.

Reliability

The use of ants as bioindicators is founded on the assumption that the extent of ant community change reflects broader ecosystem change. How valid is this assumption? The few relevant studies all suggest that ants do indeed reflect broader ecological change, rather than providing idiosyncratic responses that are as uninformative as they are unrepresentative. For example, a range of mine site rehabilitation studies show that patterns of ant recolonization reflect those of other invertebrate groups and of key ecosystem processes such as nutrient cycling (Andersen *et al.* 2004a). However, this is a ripe area for further research, as it is important to understand what ecosystem components and processes ants are representing, and what they are not. For example, many vertebrate groups are likely to respond to different habitat variables, and at different spatio-temporal scales, than those driving ant (and other invertebrate) community dynamics.

Feasibility

Concerns are often expressed by land managers that invertebrate monitoring is too difficult and too time-consuming to be cost effective. However, a critical evaluation of different monitoring options suggests that this is not the case. Majer *et al.* (2007) have compared the performance of various invertebrate groups with plants and vertebrates as indicators of restoration success at Western Australian mine sites. Assemblage composition of a range of invertebrate groups, including ants, all reflected trends in the composition of other groups to a greater extent than did either plants or vertebrates. Invertebrates were much more efficient than vertebrates in terms of information yield per unit time in the field and laboratory.

continues

Box 8.2 continued

Moreover, several studies have shown that ant sampling and processing can be greatly simplified – by recording species presence or absence rather than abundance, or considering only a subset of species, without losing indicator effectiveness (Andersen and Majer 2004). For example, presence–absence data for large species only, gave comparable results to comprehensive ant surveys in terms of detecting off-site mining impacts at Mount Isa in northwestern Queensland. Indeed, selected species can actually improve indicator performance, as has been shown for small subsets of genera in terms of discriminating land condition in relation to livestock grazing in western New South Wales (Andersen and Majer 2004).

The future

There is an ongoing need for further research on ant responses to different disturbances in different places, and on how broadly these responses represent general ecological change. However, there is already a strong body of knowledge, and the use of ants as bio-indicators in land management is limited more by a land management tradition of ignoring invertebrates altogether. Land managers could profitably learn from their aquatic colleagues, who for decades have been effectively using invertebrates as bio-indicators of river health (Hawkins *et al.* 2000). Ants are acting as environmental monitors in most terrestrial habitats – we just need to be asking them what is going on!

about community assembly (Hubbell *et al.* 1999; Vandermeer *et al.* 2000). More recently, knowledge about relationships between diversity and ecosystem function has advanced by studying predatory effects of birds across a range of agricultural management systems (Van Bael *et al.* 2008). Specifically for ants, research in agroforests set the groundwork for studying the spatial ecology of ant communities and mechanisms underlying pattern formation (e.g. ant mosaics) (Leston 1973; Majer 1976). Studies in a range of agricultural and forest habitats have revealed the relative importance of competition and environmental characteristics of habitats in assembling ant communities. Furthermore, changes in resource availability in disturbed ecosystems have provided important insight into factors that are essential in maintaining the diversity of tropical ants. Especially because of the prevalence of habitat disturbance and transformation in nearly all biomes and ecosystems, understanding whether and how disturbance alters ant behaviour, diversity, composition, and subsequent changes in ecosystem services is critical.

In this chapter, we summarize the effects of natural and anthropogenic disturbance on ant species and community structure, mechanisms causing biodiversity loss with habitat transformation, and subsequent implications for trophic interactions and ecosystem services provided by ants in altered habitats (see Table 8.1). We also examine the role of ants as ecosystem engineers. In the concluding remarks, we generalize what is known about the impacts of disturbance on ant communities. Throughout the chapter, we provide information about management or conservation recommendations (also see Chapter 4) useful or necessary to restore ant communities to states present before drastic human-induced habitat disturbance and transformation. Finally, we will present an agenda for future research that will advance our understanding of this important field.

8.2 Agents of habitat disturbance and transformation and effects on ant communities

8.2.1 Fire

Fire is a frequent and widespread disturbance in many of the world's major biomes, including savannas, grasslands, boreal forests and

Table 8.1 Key findings relating habitat disturbance and effects on ant communities.

Type of disturbance/ transformation	Effects on ant community	Reference
Fire	Resilience and resistance to fire	Parr *et al*. (2004); Parr and Andersen (2008)
	Shifts in composition	Andersen *et al*. (2006)
	Negative impact on arboreal and cryptic litter-dwelling species	Arnan *et al*. (2006)
	Enhanced seed dispersal by ants	Parr *et al*. (2007)
	Habitat type determines the extent of (burning) effect on ants	Barrow *et al*. (2007); Farji-Brener *et al*. (2002); Ratchford *et al*. (2005)
	Increased abundance for particular functional groups (Opportunists)	Hoffmann and Andersen (2003)
Flooding and inundation	Reduction in species richness	Ballinger *et al*. (2007); Milford (1999)
	Development of survival and behavioural mechanisms	Ballinger *et al*. (2007); Klein *et al*. (1993); Lude *et al*. (1999); Maschwitz and Moog (2000); Nielsen (1997); Nielsen *et al*. (2006)
	Swimming or surfing ants	Adis (1982); Jaffé (1993)
Forest tree-fall and gap creation	No detectable response	Feener and Schupp (1998)
	Context of gap (primary or secondary forest) affecting impact of herbivory by leaf-cutting ants	Peñaloza and Farji-Brener (2003)
Deforestation and logging	Decreases in ant species richness	Dunn (2004); Majer *et al*. (1997, and references therein)
	Changes in ant composition	Nakamura *et al*. (2007); Palladini *et al*. (2007); Vasconcelos (1999a)
	Transition from stochastic to deterministic processes of community assembly in disturbed habitats	Floren *et al*. (2001)
	Increased vulnerability to ant invasions	Suarez *et al*. (1998); Vasconcelos *et al*. (2000)
	Selective logging favours species richness	Azevedo-Ramos *et al*. (2006)
Agricultural intensification	Reduction of species richness	Perfecto *et al*. (2007)
	Favours herbivory by leaf-cutting ants	Blanton and Ewel (1985)
	Increased foraging activity by fire ants (*Solenopsis geminata*)	Nestel and Dickschen (1990)
	Possible functional disruption on soil food web interactions for maintenance of soil fertility and structure	Amador and Gorres (2007)
Grazing	Relative proportions of functional groups change according to grazing practices	Bestelmeyer and Wiens (1996); Hoffmann (2000)
	No response to intensive pulses of cattle grazing	Read and Andersen (2000)
	Increased arboreal ant species richness with trees	Majer and Beeston (1996)
	Fodder banks increase predatory ant richness	Ramírez *et al*. (2007)
Mining	Increased richness with rehabilitation time	Andersen *et al*. (2003); Majer *et al*. (1984)
	Reduced richness with sulphur deposits	Hoffmann (2000)
Urbanization	Richness decline in natural habitats inside urban areas with respect to rural areas	Lessard and Buddle (2005); Pacheco and Vasconcelos (2007); Yamaguchi (2004)
	No change	Gibb and Hochuli (2003)
	Opportunistic or non-native species persisting in urban sites, compositional changes	Carpintero *et al*. (2003); Gibb and Hochuli (2003); Holway and Suarez (2006)

sclerophyllous vegetation (Pyne 1997). Studies on the effect of burning on biota have been conducted in a correspondingly broad range of habitats. While there are several aspects of these fires that can be studied (e.g. season, frequency, intensity, size, type), most fire and ant studies have focused on either a comparison of burnt versus unburnt areas, or on the effect of applying repeated fires. Here we focus on areas that burn naturally.

Overall, ant assemblages exhibit striking resilience and resistance to burning (Barrow *et al.* 2007; Parr *et al.* 2004) with fires generally having little effect on ant abundance, species richness, assemblage composition, or structure. Even where ant abundance was found to decrease immediately post-fire (ten days post-fire, Andersen and Yen 1985), in the longer term there was little obvious negative effect.

The degree of response of ant assemblages to burning has been linked to habitat type (Barrow *et al.* 2007; Farji-Brener *et al.* 2002; Ratchford *et al.* 2005), with variation in resilience linked to the aridity of a site (Arnan *et al.* 2006), and the degree to which the habitat is modified post-fire (Barrow *et al.* 2007; Parr *et al.* 2004; see Figure 8.1). Exceptions to this extreme resilience occur either in systems that burn extremely infrequently (e.g. low flammability vegetation such as rainforest), or those that undergo a major shift in vegetation composition and structure post-fire (e.g. in the Mediterranean, *Pinus nigra* forest is converted to shrubland, Rodrigo and Retana 2006). Composition of ant assemblages may also sometimes differ in burn/no burn comparisons but these instances are usually where repeated, rather than once-off fires have been applied (Parr *et al.* 2004), or at the other extreme, where there has been long-term fire exclusion in a highly flammable environment. For example, a shift in composition toward more rainforest-associated ant species was reported where fire was excluded from tropical savanna in northern Australia (Andersen *et al.* 2006). In terms of habitat modification, clearly the consumption of dead wood or litter by fire is likely to negatively affect arboreal or cryptic litter-dwelling species (Arnan *et al.* 2006). Although changes in ant assemblages due to fires have primarily been attributed to changes in vegetation structure, there are few studies that directly test *how* a change in

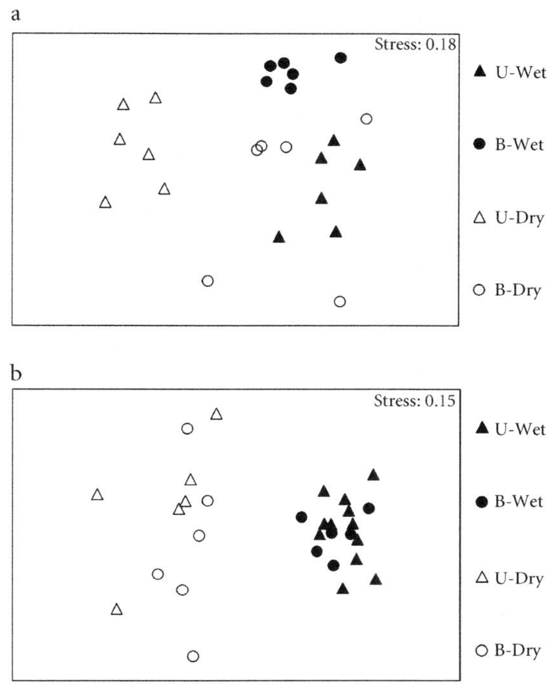

Figure 8.1 Multi-dimensional scaling ordination for burnt and unburnt sites for (*a*) spinifex and (*b*) sandplain habitat at Purnululu National Park in Western Australia in both the wet and dry season sampling periods (U = unburnt, B = burnt, wet = wet season, dry = dry season). Each point of the ordination represents a sampling grid in the respective habitats. In the spinifex habitat, which undergoes pronounced structural change with burning, there is a significant difference in ant assemblage composition between burnt and unburnt sites. A more limited difference in vegetation structure in the sandplain habitat results in only a slight difference in ant assemblages. (Reproduced with permission, from Barrow *et al.* 2007).

vegetation structure or insolation level affects ant assemblages.

Although ants in some habitats exhibit remarkable resilience to fire, there can be striking changes in the abundance of different functional groups. For example, burning increases the abundance of some particular functional groups (Hoffmann and Andersen 2003) (see Box 8.2 for an introduction to functional groups). Elsewhere, processes such as myrmecochory (seed dispersal by ants, see Chapter 6) are affected by burning; savanna fires in northern Australia can enhance rates of seed removal and significantly

Box 8.2 Functional groups in ant community ecology

Alan N. Andersen

Ecologists often classify species into functional groups as a way of reducing ecological complexity and allowing for comparative analyses of ecological systems with little or no species overlap. No particular functional group scheme can serve all purposes, and groups based on different 'functions' will have different applications. One approach to functional groups in ant community ecology is to classify species according to niche dimensions such as diet, nest location, and time of foraging. This is particularly useful for detailed analyses of particular communities. However, such schemes tend to be purely descriptive, and often reveal little insight into fundamental ecological, biogeographical, and evolutionary processes driving community structure. A commonly used scheme that aims to redress this is based on global-scale responses of ants to environmental stress (factors affecting productivity) and disturbance (factors removing biomass), operating at the genus or species-group level (Table 8.2.1). These groups originated from P. J. M. Greenslade's pioneering studies in arid Australia (Greenslade 1978), and have since been modified and extended for continental and intercontinental analyses of biogeographical patterns of ant community structure and their responses to disturbance (Andersen 1995, 1997, 2000, 2003; Hoffmann and Andersen 2003).

The most important functional groups in this global scheme are Dominant Dolichoderinae, Generalized Myrmicinae, and Opportunists, because they respectively represent the three primary ecological types in relation to stress and disturbance from a global perspective – dominant, subdominant, and ruderal (Grime 1979). They strongly parallel the three primary plant life-forms used in vegetation analysis and classification: trees, shrubs, and grasses (Andersen 1995). See Figure 8.2.1 for representatives of some functional groups.

From a global perspective, dominant species are those at the top of dominance hierarchies in the most productive environments. For ants, maximum productivity occurs where the sun's thermal energy in hot, open and structurally simple environments combine with the metabolic energy of carbohydrates from plant exudates, especially honeydew (Andersen 1995;

Table 8.2.1 Ant functional groups based on global-scale responses to environmental stress and disturbance, and their major constituent taxa.

Functional group	Major taxa
Dominant Dolichoderinae	*Anonychomyrma*, *Azteca*, *Dolichoderus*, *Dorymyrmex* (*bicolor* group), *Forelius*, *Iridomyrmex*, *Liometopum*, *Linepithema*, *Papyrius*, *Tapinoma* (*nigerrimum* group)
Generalized Myrmicinae	*Crematogaster*, *Monomorium* (part), *Pheidole*
Opportunists	*Aphaenogaster*, *Dorymyrmex* (*insanus* group), *Ectatomma*, *Formica* (*fusca* group), *Lepisiota*, *Myrmica*, *Paratrechina*, *Rhytidoponera*, *Tapinoma*, *Technomyrmex*, *Tetramorium*
Subordinate Camponotini	*Camponotus*, *Opisthopsis*, *Polyrhachis*
Hot-Climate Specialists	*Cataglyphis*, *Melophorus*, *Meranoplus*, *Messor*, *Monomorium* (part), *Myrmecocystus*, *Ocymyrmex*, *Pogonomyrmex*
Cold-Climate Specialists	*Anoplolepis* (part), *Formica* (*rufa* and *exsecta* groups), *Lasius*, *Lasiophanes*, *Temnothorax*, *Monomorium* (part), *Notoncus*, *Prolasius*, *Stenamma*, *Stigmacros*
Tropical-Climate Specialists	Many taxa characteristic of tropical rain forest, including Dorylinae, Ecitoninae, and Attini; it also includes the fire ants (*Solenopsis* subgenus *Solenopsis*), and the behaviourally dominant genus *Oecophylla*
Cryptic Species	Many genera of small-sized and small-eyed myrmicines and ponerines
Specialist Predators	*Anochetus*, *Cerapachys*, *Leptogenys*, *Myrmecia*, *Odontomachus* (part), *Pachycondyla*

continues

Figure 8.2.1 (*a*) *Iridomyrmex* sp. from Australia is behaviourally dominant and typifies the Dominant Dolichoderinae group; (*b*) Species of *Camponotus* belong to the Subordinate Camponotini; (*c*) Specialist Predators include the genus *Pachycondyla*; (*d*) Generalized Myrmicinae, such as species of *Monomorium* are classified as subdominant according to the Functional Group scheme devised by Andersen (1995). (Photos: Alex Wild)

Davidson 1997; Kaspari 2000; Kaspari and Weiser 1999). This coupling of thermal and metabolic energy powers the large colony sizes and high rates of activity that are characteristic of behaviourally dominant species, and is best developed in the canopies of lowland tropical rainforest (Blüthgen *et al.* 2000; Davidson *et al.* 2003; Tobin 1994), and on the ground in arid Australia (Andersen 2003). The behaviourally dominant ants in these habitats are typically dolichoderines, notably species of *Iridomyrmex* in the Australian arid zone (Greenslade 1976), species of *Anonychomyrma, Philidris*, and *Dolichoderus* in the Old World tropics (Huxley 1982; where *Philidris* is referred to as the *cordatus* group of *Iridomyrmex*), and species of *Azteca, Dolichoderus, Linepithema, Liometopum, Dorymyrmex*, and *Forelius* in the New World (Andersen 1995; Davidson 1997; Tobin 1994). It should, however, be noted that not all dolichoderines are behaviourally dominant, with many being behaviourally submissive Opportunists (see Table 8.2.1).

It is noteworthy that although behaviourally dominant dolichoderines occur in all climatic zones of Australia, they are absent in cool-temperate regions elsewhere in the world. In these regions, behavioural dominance has evolved in Cold-Climate Specialist formicines, such as *Formica* (*rufa* and *exsecta* groups) throughout the Holarctic, and *Anoplolepis* (*custodiens* group) in southern Africa. No such behaviourally dominant, cold-adapted formicine occurs in Australia. Behavioural dominance also occurs in the arboreal Tropical-Climate Specialist *Oecophylla*, which occurs throughout the Old World Tropics; it has no parallel in the New World Tropics.

Subdominant Generalized Myrmicinae have a much broader distribution in relation to environmental stress and disturbance than do Dominant Dolichoderinae, and tend to predominate in moderately, rather than highly, productive environments for ants (Andersen 1995). They are often extremely successful at recruiting to and defending food resources, but compared with Dominant Dolichoderinae have lower rates of activity, smaller colony sizes, and smaller foraging territories, and tend to be less aggressive. Generalized Myrmicinae are often the most abundant ants in warm environments where Dominant Dolichoderinae

continues

Box 8.2 continued

are absent, such as the tropical savannas of southern Africa (Parr *et al.* 2004) and Brazil (Campos *et al.*, unpublished data), and in leaf litter of lowland tropical rainforest throughout the world (Ward 2000). Opportunists are unspecialized, behaviourally submissive species, often with wide habitat distributions. They predominate only at sites where stress or disturbance severely limits ant productivity and diversity, and consequently where behavioural dominance is low.

The functional group scheme described here is designed for biogeographical-scale analyses of ant community structure and dynamics. It is also useful for other large-scale studies involving ants, such as analysing ecological attributes of pest ant species (McGlynn 1999b) and the ant partners of lycaenid butterflies (Eastwood and Fraser 1999). However, such generality inevitably comes at the cost of precision. The functional group scheme can provide a useful framework for analysing the broad structure of particular communities, but it is not designed for studies of community dynamics at local scales that require a detailed understanding of the ecology of individual species. The scheme is particularly limited for local-scale analysis in regions where relatively few functional groups are represented, such as in cool-temperate regions of the northern hemisphere, or where one functional group is particularly diverse both taxonomically and ecologically, such as Tropical-Climate Specialists in lowland tropical rainforest. A functional approach is still highly useful in such situations, but requires functional groups that are designed for the specific purpose (e.g. Delabie *et al.* 2000).

increase the distance of seed dispersal, more than doubling it for some ant species (Parr *et al.* 2007).

8.2.2 Flooding and inundation

Floods cause major habitat changes by scouring flood plains and removing vegetation, soil, and litter. These catastrophic, large-scale, stochastic flood events 'reset' areas of flood plain, and dramatically alter habitat availability and quality. For example, immediately after floodwaters recede, ant species richness and abundance can be reduced (Ballinger *et al.* 2007; Milford 1999). Over longer timescales (several years), duration of inundation is an important factor. Richness and abundance are lower in areas where inundation duration has been longer (Ballinger *et al.* 2007). Floodplain species tend to be opportunists that can recolonize disturbed areas quickly, with some species specifically colonizing young gravel bars devoid of vegetation (e.g. *Formica selysi* in braided rivers in the Alps, Lude *et al.* 1999). Survival mechanisms include evacuating nests to higher ground or in trees (Adis *et al.* 2001; Ballinger *et al.* 2007; Lude *et al.* 1999), and forming rafts (comprising the queen, several dozen workers, and brood) that are carried by the current to dry ground. Where more predictable seasonal flooding occurs, species richness tends to be lower (Majer and Delabie 1994), and soil-, litter-, and shrub-associated species are most negatively affected.

Ant species occurring in frequently wet areas have extraordinary ways of dealing with flooding. For example, mangroves are regularly inundated with sea water. Although most ants in the wettest mangrove areas in northern Australia are arboreal (e.g. *Crematogaster*), *Polyrhachis sokolova* nests in the mud and relies on trapping air in nest galleries to survive inundation periods longer than 3 h (Nielsen 1997). Extreme physiological adaptation enables *Camponotus anderseni*, which nests in the twigs of mangrove trees, to survive hours of inundation. Because the head of soldiers plugs the nest entrance when the tide comes in, gas exchange is prevented, and in response, the ants partly switch to anaerobic respiration (Nielsen *et al.* 2006). Important to point out is that this type of flooding is a consistent disturbance that differs from unexpected or stochastic events and may elicit physiological adaptations on evolutionary, rather than ecological timescales.

At a smaller scale, heavy rain can cause flooding of nests. Ants exhibit a range of responses to this threat including plugging nest entrances with their heads, and removing water that has leaked into the nest by ingesting it, and either regurgitating (Klein *et al.* 1993) or excreting a droplet outside the nest (Maschwitz and Moog 2000). Some ant species are able to continue foraging when their habitat is flooded. For example, leaf-cutting ants reportedly 'walk' on water when foraging during flooded periods (Adis 1982). In the intertidal zone, *P. sokolova* has been described as 'swimming' as the tide comes in (see Box 10.1) and even a surfing-like behaviour has been observed in ants that forage in the intertidal zone whereby the ants adopt a 'nymphal' position enabling them to ride the wave until the sea water is absorbed into the sand, and then they walk off (Jaffé 1993).

8.2.3 Forest tree-fall gap creation and hurricanes

In tropical forests, gap creation from falling trees is an important disturbance shaping ecological and evolutionary dynamics. Although gap creation has been shown to influence plant and bird dynamics, Feener and Schupp (1998), working in Panama, found little to suggest that ant assemblages respond significantly to tree-fall gap formation. They found no differences in species richness, abundance, composition, or rates of resource discovery between gaps and the surrounding forest. Likewise, Shure and Phillips (1991) found no differences in ant abundance in recently created forest gaps differing in size from 0.016–10 ha. Instead, larger-scale processes may be of greater importance, and seasonality and habitat difference may reduce the impact of gaps on ants. Herbivory caused by leaf-cutting ants (mainly *Atta cephalotes*) varies with the age of the surrounding forest matrix, with foliar damage in gaps adjacent to old-growth forest a magnitude higher than that in gaps adjacent to secondary forest (Peñaloza and Farji-Brener 2003). Consequently, disturbance from tree-fall and the formation of gaps may play a vital role in providing islands of palatable resources in an ocean of less palatable forest.

Hurricanes may also affect ant assemblages. For example, Morrison (2002a) examined ant communities on 17 Bahamian islands before and after a hurricane that caused significant damage to vegetation and soils. He found no ant species that went locally extinct as a result of the hurricane, but observed substantial decreases in overall ant abundance and changes in the composition of species visiting baits (Morrison 2002a). Hurricanes may also alter ant–plant mutualisms over landscape scales. During the five years following a hurricane in the Nicaragua, proportions of *Cecropia* spp. trees occupied by *Azteca* spp. were greatly reduced, likely leading to high mortality of this ant associate (Ferguson *et al.* 1995).

8.2.4 Logging

Logging is a globally important threat to biodiversity. However, the degree of tree removal varies from complete extraction (deforestation) to selective logging, where only certain species of trees are targeted for removal. Ant species richness may decrease in logged areas (King *et al.* 1998), increase in recently logged stands (Palladini *et al.* 2007), or experience no change with selective logging (e.g. Kemel *et al.* 2001; Vasconcelos *et al.* 2000). Where deforestation occurs, resulting in varying levels of habitat fragmentation, there are usually changes in ant communities, namely, changes in species composition (e.g. Nakamura *et al.* 2007; Palladini *et al.* 2007; Vasconcelos 1999a,b) and demography (e.g. Kemel *et al.* 2001) of the resident species. In boreal forests, colony abundance of red wood ants (*Formica rufa* group) declined drastically in deforested areas that were also ploughed, but in areas where some trees were left standing, colony abundance was similar to mature forests (Domisch *et al.* 2005). Ant assemblage composition in logged forests may become similar to that of primary forest but it can require several decades, or even centuries, of natural regeneration (Floren *et al.* 2001; Palladini *et al.* 2007). Changes in forest composition affect ant dynamics, possibly causing a transition from stochastic to deterministic – driven processes of community assembly in disturbed habitats (Floren *et al.* 2001). Such demographic disruptions caused by deforestation may increase vulnerability of native ant communities to ant invasions (Suarez *et al.* 1998; Vasconcelos *et al.* 2000). In addition, deforestation may be accompanied by fire, which further severely negatively affects rainforest ant species richness within the burned area (MacKay *et al.* 1991) since most species are not adapted to fire. While deforestation (and concomitant conversion to agriculture) usually leads to decreases in species richness, selective logging has a less drastic effect on ant species richness (Dunn 2004). Further, practices such as reduced-impact logging may have less of an adverse effect on ant species richness and composition (Azevedo-Ramos *et al.* 2006), likely because the reduced impact practices maintain a forest structure similar to an unlogged forest.

8.2.5 Fragmentation and edge effects

It is difficult to distinguish between effects due to habitat loss and effects of fragmentation because they often go hand in hand (Debuse *et al.* 2007). Perhaps the largest and longest-running forest fragmentation experiment is the Biological Dynamics Forest Fragment Project (BDFFP) initiated in 1979 in the Amazon basin. Results of 20 years of studies in this project reveal that fragmentation effects are diverse and responses of different species and taxonomic groups are highly individualistic (Laurence *et al.* 2002). Studies of fragmentation have shown variable effects on ants with regard to species diversity and composition. However, most studies report a decline in species richness and nest density within fragments (Brühl *et al.* 2003; Carvalho and Vasconcelos 1999; Vasconcelos 1999b), as well as a higher number of non-native, invasive, or tramp species in fragments as compared to continuous forests (Suarez *et al.* 1998). At the landscape level, total abundance of ants tends to increase due to the availability of young successional areas (Vasconcelos *et al.* 2001).

Studies show no consistent effects of fragment size on species richness but edges tend to have higher species richness than forest interiors (Dejean and Gibernau 2000; Majer *et al.* 1997; Vasconcelos *et al.* 2001, but see Golden and Crist 2000). Likewise, many studies across a range of ecosystems including tropical, temperate, and boreal forests have reported changes in species composition with fragmentation, especially due to edge effects (Carvalho and Vasconcelos 1999; Debuse *et al.* 2007; Suarez *et al.* 1998; Vasconcelos *et al.* 2001). For example, in central Amazonia, Carvalho and Vasconcelos (1999) reported large changes in ant species composition in edges (up to 200 m) and forest fragment interiors, but saw no differences in species richness. In rainforest areas this edge effect has been attributed primarily to increases in leaf litter in the forest edge as compared to forest interior, but microclimatic and vegetation changes could also be implicated (Carvalho and Vasconcelos 1999; Perfecto and Vandermeer 1996). The diversity of ant and hemipteran mutualists also increases in edge habitats, presumably due to higher plant productivity towards forest edges (Dejean and Gibernau 2000).

One of the most consistently reported effects of fragmentation is the increase of non-native, invasive, or aggressive large colony weedy or invasive species (Brühl *et al.* 2003; Dejean and Gibernau 2000; Lessard and Buddle 2005; Ness 2004; Suarez *et al.* 1998). This has detrimental effects on the native ant fauna, decreases fragment habitability for other ground-dwelling arthropods, and may directly or indirectly affect plants and their associated arthropods (see Chapter 15).

One of the most interesting emerging results on the effects of fragmentation on ant communities is that the quality of the matrix surrounding forest fragments is important, and that matrices that are more similar to the forest structure will better promote inter-fragment connectivity (Belshaw and Bolton 1993; Byrne 1994; Vasconcelos 1999a). For example, immigrants from external source populations may help maintain local populations of twig-nesting ants in forest fragments. Thus, ant populations in isolated fragments (i.e. surrounded by pastures) may be more extinction-prone (Byrne 1994). Carvalho and Vasconcelos (1999) proposed that fragmentation effects for ants likely diminish with forest re-growth in pastures, because many ant species use these habitats (Belshaw and Bolton 1993). This is also true for coffee agroforestry systems as forest ant diversity is better maintained in high-quality matrices resembling natural vegetation

(Perfecto and Vandermeer 2002). These results link well to disturbance ecology theory in general in that distance to sources of propagules to re-colonize disturbed or transformed habitats is critical (e.g. MacArthur and Wilson 1967). Given that there is local species extinction, even in large fragments, the results from the BDFFP and other studies on forest fragmentation strongly suggest a need to transform highly degraded agricultural matrices to those of a style that favour migration among patches.

8.2.6 Agricultural intensification

Although some primary forest leaf litter ant species may survive in agricultural landscapes, such as cocoa plantations (e.g. Belshaw and Bolton 1993), permanent transformation of forests into agriculture dramatically reduces ant species richness (e.g. Majer *et al.* 1997), as does agricultural intensification. Intensification is generally associated with crop specialization, increasing mechanization, and generalized use of agrochemicals and other external inputs in the crop field. Yet because agricultural landscapes form the matrix surrounding forest fragments, understanding the impacts of agricultural intensification on biodiversity is valuable for conservation purposes. There is a growing awareness that agroecosystems should be a priority in the biological conservation agenda because some agroecosystems are repositories of high levels of biodiversity including ants (Perfecto *et al.* 2007). Ants are a robust group as ecological indicators, and constitute a rare example of the adoption of invertebrates as indicators of land management (Andersen and Majer 2004; see Box 8.1).

In temperate regions, ant richness and abundance are strongly affected by agriculture, and effects may vary depending on common agricultural practices and landscape components. For example, in a study in Virginia and North Carolina, Peck *et al.* (1998) found that ant species richness and colony density for most species were lower in more disturbed crop fields than in field margins, in areas practicing conservation tillage, and in areas where fewer insecticides were applied. In Germany, active agricultural lands offer habitat for a relatively low number of ant species (7), but where the landscape includes some meadows, fallow lands, and edge habitats species richness increases to 19 (70% of the species known from the area) (Dauber and Wolters 2004). Thus, maintaining habitat heterogeneity in the landscape may be very important for maintaining ant diversity in disturbed agricultural landscapes.

In the tropics, differences in ant diversity have been used to assess the consequences of agricultural intensification in coffee and cacao crops, specifically – mainly involving different levels and varieties of shade trees (Perfecto *et al.* 2007). Thousands of hectares of traditionally shaded agroecosystems in the tropics have been transformed into plantations with little or no shade (e.g. sun coffee). Intensification of coffee plantations significantly reduces the associated biodiversity; for instance, 18 of the 22 studies on ants examined by Perfecto *et al.* (2007), showed that ant diversity declined with agricultural intensification. This trend also applies for most studies including those on plants, arthropods, and vertebrates. Some of the lost ant fauna may be important biological control agents (Perfecto *et al.* 2007) and might positively affect soil fertility and quality (Amador and Gorres 2007). Because not only the number of ant species decreases with the removal of shade trees, but also the abundance, case-specific analyses are needed in order to evaluate the impacts of such changes in ant assemblages. For example, *Solenopsis geminata*, a voracious predator of other insects, is extremely common in sun coffee plantations of Mexico, but the same species is a seed predator in some open agroecosystems (Nestel and Dickschen 1990). Further, open agroecosystems may also favour economically detrimental ants such as *Atta cephalotes*, which cut 3.5 times more leaf tissue in a cassava monoculture and in plots of non-indigenous plant species than in diverse successional plots (Blanton and Ewel 1985).

Even though ants have been increasingly used as the focus group in many studies regarding changing agroecosystems, natural systems, rehabilitation, and other land management systems, there is a real

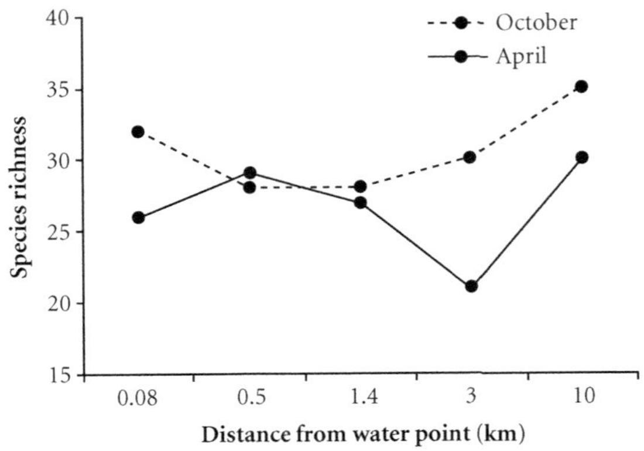

Figure 8.2 Species richness at varying distances from a waterpoint at Kidman Springs cattle station in northern Australia. The increasing distance from water represents decreasing grazing pressure. Pitfall sampling was conducted in October and April. (Reproduced with permission, from Hoffmann 2000).

need to establish reliable sampling protocols for using ants in conservation monitoring (see Chapter 4). Ants provide invaluable information about constantly disturbed habitats such as agroecosystems in intensification or rehabilitation processes, in a relatively short time and for low cost (Underwood and Fisher 2006).

8.2.7 Grazing

A large fraction of anthropogenically modified landscapes is designated for cattle pasture. There is an increasing concern about intense and constant disturbance associated with unsustainable management of pasture lands. Ant richness is dramatically higher in tropical (or subtropical) forests compared with intensively grazed neighbouring grasslands (Quiroz-Robledo and Valenzuela-Gonzalez 1995). Increases in grazing intensity may also result in declines of ant species richness, especially of litter-inhabiting cryptic species and specialized predators (Bestelmeyer and Wiens 1996), and strong changes in species composition, although the relative proportions of different functional groups appear somewhat resilient to grazing pressure (Hoffmann 2000; Rivera and Armbrecht, unpublished data). However, in arid areas of Australia, ant species richness, and particularly the richness of soil dwelling ants does not change with intensive pulses of cattle grazing (Read and Andersen 2000). Thus, there are no clear trends about how grazing affects ant richness (see Figure 8.2). Ranching and grazing practices may be improved, from a biodiversity standpoint, using several techniques. In Brazil, arboreal ant species richness in isolated trees embedded in tree pastures (and especially in large trees with epiphytes) increased with proximity to forest patches (Majer and Delabie 1999). Converting intensive pasture lands to silvopastoral systems (pastures with trees) by planting a diverse selection of trees and shrubs and thereby increasing canopy cover might increase predatory ant richness (Ramírez *et al.* 2007). As practiced in Colombia, silvopastoral systems include frequent pruning to generate fodder banks for cattle. Such fodder banks are extremely labour intensive, but involve very high plant biomass production in short periods of time (a few months). The drastic changes in vegetation associated with cyclic foliage pruning and regrowth alter the physical–physiological conditions of the habitat and may hasten colonization or displacement processes in ant communities (Ramírez *et al.* 2007). Most of what is known about the impacts of heavy grazing on ant communities is from tropical systems, although some temperate studies have been done. For example, Dauber *et al.* (2006a) investigated ant richness and composition in regenerating grasslands of varying size and condition used for grazing for centuries. They found that remnant grassland size did not affect ant richness, and that smaller patches did not contain subsets of ant species, but that habitat condition, namely tree and vegetation cover, had strong impacts on differences in species composition.

8.2.8 Mining

Mining represents perhaps the most extreme form of habitat disturbance and transformation, resulting essentially in complete habitat loss. Ants have been widely used as an indicator group for mine rehabilitation work (see Box 8.1). Studies on mine site rehabilitation aim to determine how natural undisturbed habitat and its complete associated ant fauna can be restored. Typically these studies

Figure 8.3 Mine site rehabilitation is a process that takes many years with little guarantee that the habitat will return to its exact former state. These photos from Nhulunbuy, northern Australia, illustrate how sites of different ages since rehabilitation differ in vegetation; this change in habitat strongly influences the ants: (*a*) un-mined reference site, (*b*) 2 year old rehabilitated site, (c) 5 year old rehabilitated site, (*d*) 24 year old rehabilitated site. (Photos: Benjamin D. Hoffmann)

compare a range of rehabilitation sites varying in age with undisturbed reference sites. An important principle to take into account is that proximity to a source of colonizing species influences recovery from severe disturbances. Species richness is generally positively associated with time since rehabilitation (Andersen *et al*. 2003; Majer *et al*. 1984) and may also increase with increases in habitat heterogeneity and cover of litter and vegetation (Majer *et al*. 1984). In this regard, rehabilitation sites with a monoculture of either indigenous or introduced plant species tend to be less successful than those with mixed vegetation. Importantly, although species richness increases with rehabilitation age, with some rehabilitation sites having the same species richness as undisturbed reference sites, the composition of these species can differ significantly. In some systems the recreation of an intact pre-mining assemblage can take a substantial period of time (e.g. >20 years in Mediterranean woodlands;

Ottonetti *et al.* 2006; see Figure 8.3), and in cases where the ecological community tends along a different trajectory (e.g. due to stochastic events), complete re-creation may be impossible. Mining can therefore leave a lasting impression on local biota including ant assemblages. Finally, pollution caused by mining can also negatively affect ant communities; dry sulphur deposits from mining emissions significantly reduce ant richness and abundance, and dramatically alter assemblage composition (Hoffmann *et al.* 2000).

8.2.9 Urbanization

Urbanization is a driving force behind habitat destruction, and has dramatic impacts on ant richness and composition. Ecological studies of urban ants generally focus on investigating changes in species richness and species composition in different urban habitat types, urban habitat fragments of different size or age, or along urban to rural gradients (Gibb and Hochuli 2003; Lessard and Buddle 2005; Pacheco and Vasconcelos 2007; Yamaguchi 2004). Ant species richness sometimes declines with reduced size and increased age of habitat fragments embedded in urban areas (Yamaguchi 2004), along rural to urban forest gradients (Lessard and Buddle 2005), or from parks at urban edges to inner city squares (Pacheco and Vasconcelos 2007). In contrast, others have found that ant richness does not decline with increases in urban sprawl or with decreasing size of natural habitat fragments in urban areas (Gibb and Hochuli 2003). Nearly all studies, however, do find clear changes in ant species composition in urban habitats compared with nearby natural areas.

There is support for two main groups of factors that influence species richness and composition of ants in urban areas: habitat and landscape factors and competitive interactions. Disappearance of necessary nesting resources or food items in urban habitats may affect specialist ant species. For example, due to a decline in the abundance of rotting wood resources, generalist ants tend to dominate these nesting resources excluding dead wood specialists from urban areas of Helsinki (Vepsäläinen *et al.* 2008). Likewise, ants more frequently colonize artificial nesting resources most similar to the most limiting cavity nest resources in urban habitats (Friedrich and Philpott 2009). Increased soil temperature and decreased soil moisture may enhance the establishment potential of some invasive species and reduce the abilities of some native species to persist (Yamaguchi 2004). In contrast, other invasive species such as the Argentine ant (*Linepithema humile*) thrive with higher soil moisture, allowing them to displace native species in irrigated or watered urban areas (Holway and Suarez 2006). Dispersal limitation may be important in community assembly in urban areas as founding queens may not arrive in urban centres from source populations (Pacheco and Vasconcelos 2007). Those ants that do persist in urban habitats tend to be generalist and opportunistic species, competitive dominants, and ants with large, aggressive colonies (Carpintero *et al.* 2003). Additionally, factors that cause losses of some native species may facilitate invasion of non-native tramp species in urban areas (Holway and Suarez 2006; see Chapter 14). Finally, some urban areas are dominated by smaller-bodied ants (Holway and Suarez 2006) and by phytophagous rather than predatory species (Gibb and Hochuli 2003) indicating that urbanization may alter ant species composition thereby affecting the ecological function of the ant community in urban habitats.

8.3 Mechanisms causing change with habitat disturbance

The effects of habitat disturbance and transformation are brought about through changes in one or several local-scale factors; these include behaviours (e.g. competitive interactions, predator avoidance, parasitism, and colonization ability), soil type, and resource availability (Kaspari *et al.* 2003). Many of the factors influencing ant assemblages are indirect, rather than direct, and are linked to habitat alteration. For example, disturbances alter habitat structure, which then influences microclimate.

Removal of vegetation or growth of weedy plants following disturbance can have a significant effect on ant assemblages through changes to the microclimate. Although little work has been done on mechanisms behind disturbance impacts, differences in microclimatic conditions including moisture gradients, temperature regimes, and exposure to wind affect ants physiologically and may potentially reduce foraging times and their success. Furthermore, seasonality and vertical stratification may also contribute to ant species assembly in Neotropical forests (Feener and Schupp 1998). Some types of disturbance are inherent to ecosystem dynamics and constitute mechanisms of biodiversity generation and maintenance, while other long-term, large-scale disturbances may challenge species survival. Further, anthropogenic disturbance may change the relative balance of bottom-up versus top-down population regulation.

Resources, such as nest sites, food, and refuges, among others, are important for ant community assembly (see Chapter 7) and may become increasingly limited as natural habitats are disturbed or converted to productive agroecosystems. At the soil level, local patches of litter naturally suffer continuous disturbance, and evidence shows that plant succession occurring in these differently disturbed patches may lead to changes in species composition and may constitute a mechanism of diversity maintenance in tropical forests (Campos *et al.* 2007). For instance, leaf litter ant assemblages of several Neotropical forests may be partially limited by local resources, such as patchy food availability (McGlynn 2006; but see Torres 1984). Using twig augmentation experiments, Kaspari (1996) found some evidence of nest site limitation on the forest floor of four forests in Panama and Costa Rica (but see Carvalho and Vasconcelos 1999). As natural forests are converted into agroecosystems and consequently simplified, nest site limitation of both litter-dwelling and arboreal ants increases. In agroecosystems, ant diversity may be lost due to a shortage of animal (e.g. shells) or plant derived (e.g. twigs and seeds) nesting resources (Armbrecht and Perfecto 2003; Philpott and Foster 2005; but see Torres 1984). Furthermore, ant nest-site limitation may be stronger in more intensively managed sites (Philpott and Foster 2005) or in sites with a lower diversity of twig resources (Armbrecht *et al.* 2004).

Disturbance can alter competitive interactions and colony dynamics. For example, in forests, local disturbances such as caused by army ant (*Eciton*) raids may prevent leaf litter or soil ants from reaching densities high enough to saturate nesting and food resources (Kaspari 1996). The loss of habitat complexity associated with agricultural transformation of land or other disturbances (e.g. fire) causes changes in ecological relationships among ant assemblages (via parasitism by phorids, competitive interactions, refuge presence) and determines the coexistence of the various ant populations (Wilkinson and Feener 2007). When soil-nesting ant assemblages, especially those in temperate regions, reach high densities, they may self-thin via competition, driving trends towards fewer and larger adult colonies. Likewise, top-down processes (predation in this case) may partially control the litter ant community while bottom-up organization (competition, resources) may be controlling soil-nesting species in tropical forests (Kaspari 1996).

Dispersal limitation or slow recolonization may also contribute to changes in ant assemblages. At large spatial scales, forest patches embedded in disturbed landscapes (e.g. urban areas) may not receive necessary colonists to maintain the communities found in forest patches (Pacheco and Vasconcelos 2007). But even at small spatial scales, colonization and colony presence may highly influence ant assemblages. Patterns of diversity are sensitive to spatial scale, for which careful quantification of diversity gradients at different grains is necessary (Kaspari *et al.* 2003). At the local level, heterogeneous habitats may provide patchy distributed resources, which may derive from small-scale disturbance processes. For instance, artificial disturbances at the ~0.1 m^2 scale in a Brazilian forest showed that the recolonization of leaf litter ant species was limited by the colonizing abilities of ant species and not by the limitation of litter resources (Campos *et al.* 2007). Because ant richness is positively correlated with ant colony abundance at small scales (1 m^2), factors limiting colony abundance may ultimately determine ant species presence, abundance, and richness (Kaspari *et al.* 2003).

8.4 Consequences of habitat disturbance and transformation for trophic interactions and ecosystem services provided by ants

8.4.1 Ant–hemipteran–plant interactions

Disturbance may alter trophic interactions involving ants, such as predation, symbioses, scavenging, leaf-cutting activity by attine fungus-growing ants, and foraging on extrafloral nectaries or hemipteran-secretions. These changes consequently alter ecosystem services that ants provide, both in natural and disturbed ecosystems. Humans often perceive ants either as dangerous pests that form associations with sap-sucking insects or as beneficial predators of pests (Philpott and Armbrecht 2006). Studies evaluating how habitat transformation may affect ant–hemipteran interactions are scarce but suggest that interactions are highly disrupted by human disturbance, even creating or exacerbating potential pest problems. In Indonesia, Ozaki *et al.* (2000) described sharp declines in scale insect populations (*Aulacaspis marina*) due to predation by *Monomorium floricola* and *Paratrechina* sp. in mangroves (*Rhizophora mucronata*). Remarkably, planted mangroves were severely infested while in neighbouring natural forest ants controlled the scale insect. Habitat disturbance may also facilitate invasion of exotic species that alter ant–hemipteran associations (see Chapters 14 and 15). For example, *Paratrechina fulva* is a pest in Colombia, because it associates with hemipterans and because it depletes native invertebrate fauna (Gómez *et al.* 2002). Spread of this species tends to be highly favoured by anthropogenic disturbances around lagoons and sugar cane intensive monocultures (Chacón *et al.* 2000).

Ant–hemipteran–plant interactions may be extremely rich and non-specific, indicating that a high variety of ant responses are possible with disturbance. Comparing ant–scale interactions in two types of agroecosystem provides some evidence for this. In intensive coffee plantations in Venezuela, *Crematogaster* and *Camponotus* species are considered pests because they tend scale insects (*Coccus viridis*) (Hanks and Sadof 1990). In contrast, in an organic, shaded coffee plantation in Mexico, species of these same two genera tend *C. viridis*, but the scales are not considered pests in the plantation. Instead, a mutualistic relationship between *Azteca instabilis* and *C. viridis* effectively protects coffee plants from attacks by coffee's most severe pest, the coffee berry borer (*Hypothenemus hampei*) (Perfecto and Vandermeer 2006). However, *A. instabilis* is negatively affected by shade tree pruning (Philpott 2005a), suggesting that even minor habitat disturbances may influence ant–hemipteran–pest interactions.

8.4.2 Ants as biological control agents

Ants have been used for biological control of insect pests and fungal pathogens in agricultural, agroforestry, and forestry systems for centuries (Perfecto and Castiñeiras 1998; Philpott and Armbrecht 2006; Way and Khoo 1992; see Box 7.2). However, the impact of disturbance on the ecosystem function of ants as biological control agents has not received as much attention (Philpott and Armbrecht 2006). The most obvious disturbances that can alter the biological control activity of ants are those associated with agricultural intensification, such as pesticide application, tillage (de Bruyn 1999), and reduction of plant diversity (Armbrecht and Gallego 2007; Armbrecht and Perfecto 2003). All of these factors reduce predatory activity of ants.

Coffee agroecosystems have received detailed attention in recent years, especially from the point of view of understanding how agricultural intensification affects biological control provided by ants. Several studies demonstrate reductions in ant diversity with intensification of coffee systems, but only a few of these examine how the predatory activity of ants is affected (see Philpott and Armbrecht 2006). In a few studies, ant removal of pests diminishes with coffee intensification (Armbrecht and Gallego 2007; Armbrecht and Perfecto 2003), but one study did not find any change in ant effects along a coffee intensification gradient (Philpott *et al.* 2008a). Armbrecht and Gallego (2007) demonstrated that ants have stronger predatory effects on the coffee berry borer in shaded coffee farms than in sun coffee. Whereas one species, *Gnamptogenys sulcata*, a shade-loving ant, is an effective predator of this pest, but is rare in sun coffee systems. Finally, a number of arboreal ant species are important biological control agents in coffee and cacao farms and will be entirely lost if the shade is eliminated (Perfecto and Castiñeiras 1998). In particular, *A. instabilis* is a keystone species, associated with the regulation of three main coffee pests (green coffee scale, coffee berry borer, and coffee leaf rust) in Mexican plantations. The elimination of shade trees will most likely eliminate this species along with its biological control function (Perfecto and Vandermeer 2008b). Aside from the effects of individual ant species as predators, some evidence demonstrates that behavioural diversity of ants may be important in the predatory role of ant assemblages (Philpott *et al.* 2008b), thus any disturbance affecting ant diversity may also affect ecosystem function.

8.4.3 Ants as seed dispersers

Generally, seed dispersal is affected by biotic factors (vegetation structure, ant composition, ant size, nest density, and competition for resources) and abiotic factors (temperature and seed desiccation rates) (Guitian *et al.* 2003; Ness 2004). As such, both natural and human-caused habitat disturbances can have strong impacts on seed dispersal by ants. Because myrmecochory is a mutualism involving non-specific sets of partners, ant abilities to disperse seeds, as well as rates and distances of seed dispersal, will depend strongly on the ant species present (see Chapter 6). Several studies have examined impacts of habitat modification and fire on seed dispersal by ants. In highly disturbed sites devoid of vegetation, dispersal rates decrease drastically (Andersen and Morrison 1998; Guitian *et al.* 2003; although see Parr *et al.* 2007). Similarly, Guitian *et al.* (2003) found lower seed dispersal rates in open woodlots and hypothesized that this was due to lower ant activity, and quicker seed desiccation. Yet in highly disturbed crop areas seed dispersal rates may increase (Heithaus and Humes 2003), perhaps due to lower quantities of available seed resources compared with native habitats. In disturbed sites where species composition shifts towards small-bodied ants, dispersal distances generally decline (Heithaus and Humes 2003; Ness 2004), but where composition shifts towards large-bodied ants with larger foraging ranges, dispersal distances can increase (Andersen and Morrison 1998; Parr *et al.* 2007). Finally, seed dispersal in disturbed sites may be affected by high abundance of exotic ant species (Ness 2004; see Chapter 15). Because seed predation also increases in disturbed and invaded habitats (Andersen and Morrison 1998; Ness 2004) care should be taken in equating seed removal with seed dispersal – some seeds removed by ants may not be dispersed far enough to convey an advantage.

8.4.4 Ant effects on soil processes and nutrient cycling as ecosystem engineers

Ants also act as agents of disturbance and modification to soils due to their role as ecosystem engineers. Ecosystem engineers are organisms that directly or indirectly modify the availability of resources for other species by causing physical changes in biotic or abiotic material. Ants, by their construction of nests, perturbation of soils and interactions with many other organisms are important ecosystem engineers and keystone species (Decaëns *et al.* 2002; Folgarait 1998). Although the impacts of ants on soil physical and chemical structure is not as well known as for termites and earthworms, their effect is likely to be substantial (Vandermeer and Perfecto 2007).

The most visibly obvious effect of ants as ecosystem engineers is bioturbation of soils through the formation of mounds, subterranean galleries and chambers, and the movement of soil particles along the soil profile (Folgarait 1998). These soil modifications directly and indirectly affect the energy flow, habitats, and resources for other organisms, especially plants and soil micro-organisms. Through formation of underground galleries, ants increase the drainage and aeration and reduce the bulk density of the soil. Through transformation of organic matter by storing food and accumulating faeces and corpses, ants provide habitat for soil micro-organisms and enhance soil nutrient conditions (Brian 1978; Folgarait 1998). Ants can move up to ten tons of soil per hectare per year in moist subtropical and temperate systems (Paton *et al.* 1995). Leaf-cutting ants are among the most important agents of soil modification in the tropics, moving biomass, altering chemical composition, and altering soil structure with complex galleries (Folgarait 1998). In Brazil, a single colony of *Atta sexdens* deposited 40 tons of soil on the surface (Autori 1947). Perfecto and Vandermeer (1993) estimated that *Atta cephalotes* could cause complete soil turnover in as little as 200 years in a lowland rain forest in Costa Rica. In terms of global rates of animal perturbation ants are considered second only to earthworms (Folgarait 1998), probably due to their high biomass.

Many studies have investigated the roles of ants in creating nutrient and soil heterogeneity and modifying soil structure (reviewed in Folgarait 1998). Some studies in disturbed habitats provide information about how the interaction of habitat disturbance and ant activity affect soils. Both grazing and tillage of agricultural fields affect nest densities in agricultural landscapes with subsequent effects on soil processes (Beever and Herrick 2005; de Bruyn 1999). *Atta* (a genus with large impacts on tropical soils) are strongly affected by edge effects, deforestation, and presence of roads (Vasconcelos *et al.* 2006), but resulting impacts on soil processes have yet to be quantified. Decaëns and colleagues (2002) examined how conversion of native savanna to crops and pasture affects ants and subsequent influences on soil structure. They found a significant decline in diversity of bio-structures (e.g. tunnels and other structures created by invertebrates) in crop areas compared with grasslands and that the presence of a high density of ants and other ecosystem engineers maintained a diverse assemblage of soil particle sizes. In restored grasslands, ants and ant mounds generate soil heterogeneity (in variables such as soil texture, bulk density, soil temperature, and soil moisture) and create microsites for plant colonization in restored grasslands, but these effects may change with age since recovery (Lane and Bassir-iRad 2005). Thus, there are isolated examples of ant effects of soils in some disturbed (and recovering) habitats, but there is still a general lack of understanding of how changes in ant richness, composition, and nest densities will affect soil processes in disturbed habitats. Many research objectives posed by Folgarait (1998), including comparing effects of ants in areas with impoverished diversity due to the entrance of exotic species and consequent loss of native ant fauna, or the impact of habitat disturbance are still valid topics for further research.

Ants also alter soil chemistry and affect nutrient immobilization, indirectly affecting plant and microbial communities (Dauber *et al.* 2006b; Lugo *et al.* 1973). Most studies show an increase of organic matter and N, P, and K in ant mounds, as compared to adjacent soil samples (Decaëns *et al.* 2002; Folgarait 1998). Decomposition processes by fungi and ammonifying bacteria increase, while humification is delayed due to declines in Actinobacteria abundance in mounds as compared to adjacent soils (Folgarait 1998). In Panama, *Atta columbica* increases the flux of 13 chemical elements by 38 times compared to surrounding areas (Haines 1978), and in Puerto Rico leaf-cutting ant activity is associated with higher plant productivity, presumably because of an increase in phosphorous availability (Lugo *et al.* 1973).

All the disturbances that affect ant communities can also alter the impact of ants as ecosystem engineers. Tillage, pesticide use, and decreases in organic inputs associated with agricultural intensification affect macrofaunal activity and diminish the contribution of ants as soil modifiers (Decaëns *et al.* 2002). Likewise, the increase of invasive species associated with some disturbance events can also alter ecosystem engineering by ants. The effect of disturbance on the ecosystem engineering activities of ants has seldom been studied directly.

8.5 Future directions

There are several topics covered here that deserve attention in future studies. Given phylogenetic differences among global regions, ant ecologists should make it a priority to establish classifications for different regions to assess which ant species, species groups, and genera will be affected most by different types of disturbance. As mentioned in Box 8.2, functional groupings have been developed for predicting community level responses to disturbance and stress at a global level. Yet, there have been some criticisms that these functional classifications are somewhat specific to Australasia. For most mutualisms and ecosystem services described (fungus cultivation, hemipteran-tending, seed dispersal, and biological control of pests), the way in which habitat disturbance affects the interaction relies very heavily on the composition of the ant species present in the disturbed habitat. But as for other taxa, it is now becoming increasingly recognized that functional classifications can provide important tools for determining how species groups with certain traits respond to disturbance or other ecological factors.

A much more detailed understanding of how disturbed ant communities perform ecosystem functions would be beneficial for both maintaining or even increasing the ecosystem services provided by ants and for restoration. Outside of coffee agroecosystems, few studies have examined how biological control services provided by ants are affected by agricultural intensification. The studies have also not examined whether habitat changes cause behavioural changes in particular ant species thus altering their relative impact as hemipteran-tending 'pests' versus biological control agents. There is little work examining how habitat disturbance per se influences ant effects on soils. Other than understanding that tillage has extreme impacts on soil ants, what other more minor disturbances affect ants? How do the roles of ants differ in conventional versus sustainable agricultural systems? Additionally, it would be very useful to know whether ants could eventually be used as functional agents in the recovery of degraded landscapes via seed dispersal. A more detailed understanding of the importance of increased seed dispersal distance in disturbed habitats should be investigated. Further, despite a relatively large number of studies examining seed dispersal in disturbed habitats, few mention how dispersal is affected with relative changes in seed densities across disturbance gradients or how the relative proportion of myrmecochorous seeds changes with disturbance. How might humans manipulate ants to benefit restored grasslands or mines?

Ecological studies relating to the assembly and maintenance of ant communities could be especially fruitful in disturbed and transformed habitats. For example, there are several mechanisms driving the observed changes in ant communities with disturbances, but a more detailed understanding of those factors (e.g. resource availability, microclimate, ecophysiological conditions, changes in predator or parasite populations) is needed. There is ample evidence that both fragmentation and habitat disturbance influence ant assemblages. Working in disturbed landscapes may help elucidate questions such as, how do habitat configuration or other landscape factors affect ants? What is the relative importance of local vs. landscape factors in determining ant diversity and composition? It is long known that dispersal of colonists is important for the recovery of original communities (e.g. MacArthur and Wilson 1967), and that the size of the disturbance and distance to source populations affects colonization success. With suitable distributions along disturbance gradients, or at increasing distances from source populations, studies of dispersal and recruitment limitation of ants in disturbed areas may enlighten investigations into the assembly rules of ant communities generally.

Finally, we lack research examining the synergistic effects of multiple disturbances on ant communities and ecosystem services. Most of the work examining ants is limited to particular regions, habitat types, or commonly studied assemblages. Although challenging, research that replicates experimental designs in multiple habitat types or in multiple regions may reveal those processes important for the maintenance of ant assemblages and the important services that they provide.

8.6 Summary

There are some generalizations that can be made about effects of habitat disturbance and transformation on ant communities, but several areas deserve much more attention. Fire, flooding and inundations, forest tree-fall gap creation, hurricanes, logging, fragmentation, agricultural intensification, grazing, mining, and urbanization can have very drastic effects on ant assemblages. Generally, disturbances that directly cause colony mortality will have different effects on ant communities than disturbances that have indirect effects through alteration of plant biomass (Hoffmann and Andersen 2003). Some natural disturbances, such as fire, short floods, and tree-fall gaps, although drastic in their immediate effects, may have few long-term impacts on assemblages if entire colonies are not lost, or if foundresses colonize disturbed sites quickly. Yet natural disturbances that occur more frequently than colonization and establishment may significantly exclude ant species not adapted to disturbed conditions. Generally, human disturbance result in greater changes in ant species composition than natural disturbances (e.g. mining, urbanization, and repeated agriculture have especially severe effects). As with other animal communities, disturbance impact will depend on frequency and intensity of disturbance and the time over which habitats recover from perturbation.

The specific effects of disturbance on species composition and functional groups based on the studies presented here may seem specific to the type of disturbance and the study region. Yet, in most cases, disturbed sites are dominated by opportunistic or generalist species presumably because these species can take advantage of changing resource bases, especially when disturbance puts them at a competitive advantage (Hoffmann and Andersen 2003). In most habitats affected by human disturbance, invasion by exotic or tramp ants is often reported and prevalent, but in ecosystems with frequent natural disturbance (fires and floods), only certain native ant species seem adapted to local conditions, sometimes precluding invasion.

Several mechanisms are implicated in changes in ant species richness, abundance, and composition with habitat disturbance and transformation. Ants may be highly sensitive to changes in microclimate brought about by changes to the dominant vegetation structure of a particular habitat and may be affected by changes in availability of food or nesting resources. Change in competitive interactions or in colonization processes may also affect ant assembly in disturbed areas. Thus, habitat disturbance and transformation have an impact on local ant assemblages both indirectly through changes to habitat structure, and directly, through reduced resource availability and removal of colonies. Although some mechanisms have been examined, much more work is needed to understand the details.

Finally, ants provide essential ecosystem functions such as biological pest control, seed dispersal, and soil modification, many of which are affected by habitat disturbance and transformation. Interactions of ants with hemipterans may significantly change in disturbed habitats. Some evidence shows that biological control capabilities of ants decline in disturbed habitats either due to a loss of diversity and change in vegetation structure, or because of shifts in abundance or composition of ants. In some disturbed habitats, seed dispersal capabilities increase, whereas in other habitats they decrease. More work is needed to examine the implications for plant regeneration in disturbed and transformed habitats. Finally, ants have very strong impacts on soils, adding to nutrient enrichment, nutrient cycling, and to the biophysical structure of soils, but much more research is needed to understand the intersection of ant effects as ecosystem engineers and habitat disturbance.

PART III

POPULATION ECOLOGY

The sterility of the worker caste initially presented problems for Darwin in formulating his theory on natural selection. And though he eventually saw 'no great difficulty in this being effected by natural selection' (Darwin 1859), the mechanisms remained to be elucidated. Today, our knowledge of the complexities of kin selection is testimony to how far science has progressed our understanding of population dynamics and social evolution in ants. Eusociality is no longer an obstacle to our enquiry into evolutionary processes. Division of labour, haplodiploidy, and kin conflict are indeed integral to our understanding of the life history strategies of ants. The following chapters review the current state of knowledge on fundamental aspects of life history in ants. In following the theme of the book, the authors leave aside theoretical population dynamics, and instead take an ecological brush to colony foundation, growth, and reproduction, colony structure, nestmate recognition, and foraging and defence strategies in ants.

Chapter 9 raises important issues about the colonial life cycle, which can be broken into three stages – foundation, growth, and reproduction. It is colony foundation, however, on which the authors focus their attention, considering all stages are significantly influenced by colony beginnings. The traditional view of colony foundation has been one of the independent queen making the journey to a new nest site and hunkering down to produce her first brood. However, Peeters and Molet show that this strategy has been replaced in many species, across all subfamilies, by dependent colony foundation, (also called 'budding' or 'fission' in the literature). Chapter 9 clarifies the confusion in terminology on this subject, and highlights the dearth of field studies that investigate the processes of either independent colony foundation or dependent colony foundation.

Chapter 10 tackles the complexities of colony structure – its caste, demographic, genealogical, and spatial constitution. Steiner, Crozier, and Schlick-Steiner consider the characters and character states of colony structure while emphasizing the interconnectedness of ecology and evolutionary pathways and processes. Underscored throughout the chapter is that explanations on the origins and interrelationships between characters and their states will most likely change dramatically with future research. In addition, one of the most recent and hotly debated topics in ant ecology is the evolution of supercoloniality (also see Chapter 14 and Box 14.1), which the authors establish as a paradox of colony structure. This chapter is a timely review of these concepts.

Chapter 11 broadly examines current knowledge of ant recognition systems. Beginning with clarifying key terminology, d'Ettorre and Lenoir then lead us to mechanistic explanations of the recognition systems they review. Importantly, they highlight that *kin* and *nestmate* recognition are different, and only coincide in ants when colonies are headed by one singly mated queen, without queen turnover, which we learn from Chapter 9 is not always the case. Moreover, this distinction is relevant to the current debate on the evolution of eusociality (not covered in this volume, but see Hölldobler and Wilson 2008 and Pennisi 2009 and references therein). Mechanisms of recognition including cuticular hydrocarbons (CHCs), the genetic basis of CHC profiles, and the possible role of volatile cues are placed in ecological contexts and discussed in detail.

Chapter 12 uncovers the diversity of individual and collective strategies in foraging and defence.

Considering that ants are usually foraging or defending something, this chapter explores the fundamentals of sustaining and protecting the colony. Dornhaus and Powell explain that foraging strategies are of use only if the defensive strategies that have evolved to safeguard harvested resources are robust and generally fail-safe. They illustrate the complexities of decision-making during the acquisition of resources outside the nest, and point out that because ants are eusocial, foragers do not operate alone, but are part of many colony-level foraging strategies, the diversity of which is astounding. The authors advocate the need for more explicit integration of ecological context and environmental variation into empirical studies of foraging and defence strategies.

Chapter 9

Colonial Reproduction and Life Histories

Christian Peeters and Mathieu Molet

9.1 Introduction

The spectacular ecological success of ants can be attributed to the benefits of both division of labour and morphological specialization among adults. Unlike solitary insects that sequentially perform the various tasks leading up to successful reproduction, in social species different tasks are performed in parallel by individuals with distinct behavioural profiles. Furthermore, in the ants, this division of labour is more efficient because functional differences between queens and workers are amplified by morphological differences. Queens (usually winged) start new colonies and produce offspring, while workers (never winged) raise the brood, build the nest, and forage for food. The degree of size dimorphism between queen and worker castes varies dramatically across ant species, and there is a clear phylogenetic pattern behind this. Molecular data provide very strong support for a 'formicoid' clade, not revealed by previous morphological studies (Ward 2007c). This clade comprises 14 of the 21 extant ant subfamilies (including Myrmicinae, Dolichoderinae, and Formicinae) and about 90% of all described ant species. The 'poneroid' group, comprising five subfamilies (including Amblyoponinae and Ponerinae), exhibits a larger proportion of ancestral traits (see Chapter 1). Moreover, winged queens and wingless workers in the poneroids are often similar in size and morphology, and this is associated with small colony sizes (e.g. *Harpegnathos saltator*; Peeters *et al.* 2000). The degree of queen–worker dimorphism generally increases sharply in the formicoids, but not always; in many species one large highly fertile queen produces numerous tiny workers, which leads to a dramatic increase in colony size.

The colonial life cycle can be conveniently broken into three stages – colony foundation, growth (production of workers), and reproduction (production of sexuals) (see Figure 9.1) – and we review existing knowledge about these. The relative importance of these three stages is deeply affected by the way colonies begin. Thus, it is impossible to understand life cycles without contrasting the two main strategies of colony founding. Independent colony foundation (ICF) – where queens found new colonies alone – is widespread in all taxonomic groups except the army ants. Yet, in many species across all subfamilies, this strategy has been repeatedly selected against, and replaced by dependent colony foundation (DCF), where queens cannot found a colony without the help of workers from the natal colony. In addition to a substantial decrease in dispersal distance, this shift led to crucial adaptations of individuals (queens become wingless or they are replaced by reproductive workers) and colonies (more workers are produced as part of the reproductive investment). We illustrate how each stage of the colony life cycle is influenced by the strategy of colony reproduction, and we analyse the causes and consequences of the frequent shifts from ICF to DCF.

9.2 Colony life histories, mating biology, and dispersal

Hölldobler and Wilson (1990) as well as Bourke and Franks (1995) reviewed the tremendous diversity in life histories and mating patterns in ants. Young winged queens ('gynes') and males are generally reared once a year. The gynes remain in the maternal nest for a variable period, during which they usually accumulate metabolic reserves. Following a change in weather conditions (e.g. temperature, humidity), gynes and males exit the nests synchronously. According to species, the sexuals disperse at various distances from the natal nests (Table 9.1). At one extreme, gynes stay close to their natal colony and wait on the ground or low vegetation for foreign males flying singly. The latter usually have larger eyes than the queens, suggesting the importance of vision to locate single gynes or nests. Males also detect pheromones released by the gynes, hence the term 'female calling'. Following copulation, the females can disperse farther by flight to avoid local competition and reach new habitats, or in other species, shed their wings and initiate new colonies in the vicinity. At the other extreme, gynes and males converge to fixed aggregation sites where mating occurs. Large-scale nuptial flights are of limited duration and involve sexuals from many colonies; thus, they result in maximum dispersal. They seem possible only in species with relatively dense colonies, and a large seasonal production of gynes. In contrast, because the number of sexuals produced is too low for mating aggregations to be possible (Peeters and Ito 2001), female calling is found in ants exhibiting small colonies, including many Ponerinae (references in Peeters 1991). Permanently wingless ('ergatoid') queens also remain near their natal nests and wait for flying males (see Section 9.6.4). The dichotomy between 'female calling' and 'male aggregations' focuses on the mechanism of locating sexual partners and is not a reliable indicator of dispersal distances. Gynes that mate close to their natal nest can either fly away to do ICF in some species, or re-enter the nest in other species (Table 9.1). Once back in their natal nest, they can either stay, or leave alone (ICF), or with a group of workers (DCF), depending on species. In *Carebara vidua*, gynes fly to a distant male aggregation, but then settle on vegetation and 'call' for males (Robertson and Villet 1989). Aerial dispersal that precedes mating is very unlikely to result in inseminated gynes returning to their natal nest (Bourke and Franks 1995); hence it is always

Table 9.1 Comparison of various mating strategies in ants. The category 'female calling' is heterogeneous and cannot predict dispersal distance or founding strategy. Only gynes that do not fly prior to mating can be involved in DCF.

	Historical terminology		
	Female calling		**Male aggregation**
	Location of mating (relative to gyne's origin)		
	Inside/outside of nest	**Outside of nest**	**Away from nest**
Behaviour of gynes	Wait for foreign males (if outside: can return inside after mating)	Fly, then wait for foreign males	Fly and search for aggregation
Behaviour of males	Fly individually, search for trails, nests or ants	Fly individually, search for ants	Fly and search for aggregation, select partner
Synchronized exit of males and females?	No	Yes	Yes
Dispersal distance	Low	Moderate?	High
Risk from predators	Low (exit not predictable, protection of nest)	Low (not predictable, no aggregation)	High (mass exit from nest, aggregation)
Colony foundation strategy	DCF (rarely ICF)	ICF	ICF
Gyne morphology[a]	EQ, G, BQ, AQ (can dealate before mating)	AQ	AQ

[a] EQ = ergatoid queen, G = gamergate, BQ = brachypterous queen, AQ = alate queen

associated with ICF. Mating in aggregations likely leads to greater mortality of gynes (e.g. predators, desiccation) compared to mating near natal nests. It appears that aggregations were selected against in various species and replaced by female calling, although in other species, female calling is an ancestral trait. To understand evolutionary diversification in reproductive strategies, mortality associated with dispersal and mating needs to be considered separately from mortality during colony founding because the selective pressures are distinct.

Unlike various social bees and wasps with annual colonies, all ants have perennial colonies, even in temperate and boreal environments. Colonies are started in one of two completely different manners: ICF or DCF. During ICF, queens work alone for several weeks up to a few months (depending on species), feeding and caring for the brood until these become adult workers. This founding stage is reminiscent of the life of solitary insects, because failure in any of the steps needed before the first workers become active outside the nest is fatal. In contrast is DCF (also called 'fission' or 'budding' as discussed in Section 9.4), where an established colony divides into two or more autonomous colonies. DCF is obviously possible only in insects that are social, and it should lead to much decreased mortality rates of queens (which are never without the workers). However, DCF in ants necessitates a loss of aerial (i.e. long-range) dispersal, because ant workers are strictly wingless. This founding strategy is unlike DCF in social wasps and bees, where the winged workers can disperse together with the queens (i.e. swarming) (Peeters and Ito 2001). Nevertheless, thousands of ant species exhibit DCF, and this strategy is obligate in species with permanently wingless ('ergatoid') queens (at least 60 genera out of 283; C. Peeters, unpublished data). DCF occurs across all the subfamilies, in species having both tiny and huge colonies and irrespective of monogyny or polygyny. A literature review indicates that it evolved in numerous unrelated taxa as a substitute for ICF (C. Peeters, unpublished data). Indeed, ICF is ancestral in ants, which is intuitive since they evolved from solitary wasps (the synapomorphy of ants is the metapleural gland that produces antimicrobial substances necessary for underground life; see Box 9.1).

9.3 Independent colony foundation (ICF): queens work without nestmates

We consider ICF to include all instances of queens that found colonies without the help of workers originating from their own colony. ICF foundresses disperse individually, and are often alone until the first brood become adult workers. According to the species, foundresses exhibit a variety of strategies to improve their survival: foraging outside the nest (Section 9.3.1), expanded wing muscles and other metabolic reserves (Section 9.3.2), cooperating with randomly encountered conspecific foundresses (Section 9.3.3), invading established colonies of other ants (Section 9.3.4), or exploiting other living organisms as a food source (Section 9.3.5). All these adaptations make use of available resources in the environment in order to increase founding success.

9.3.1 Non-claustral ICF

In hundreds of species, founding queens perform non-claustral ICF, i.e., they need to hunt outside their nests in order to feed the first generation of workers. This is the ancestral strategy in ants, and occurs widely in poneroid species (e.g. Ponerinae and Amblyoponinae; Peeters 1997). These ants typically exhibit a low queen–worker dimorphism; given that they must raise workers that are almost the same size as they are (Plate 7), the queens lack sufficient metabolic reserves (notably obtained from the histolysis of wing muscles; Roff 1990; Zera and Denno 1997) to feed their first brood without foraging outside the nest. Once the first workers have emerged, they start performing all non-reproductive tasks from brood care to hunting, and queens can shift to egg-laying only. Non-claustral ICF is intuitively associated with a low probability of foundress survival (Brown and Bonhoeffer 2003; Peeters 1997). Ant workers suffer high mortality during foraging (e.g. Schmid-Hempel and Schmid-Hempel 1984) and similarly, hunting queens are vulnerable. Survival is particularly poor in environments that provide low or unpredictable resources. Moreover, the brood is

Box 9.1 Antimicrobial defences in ants: pure and applied science
Andrew J. Beattie

Ants possess a variety of antimicrobial defences including two biochemical systems, one external and one internal. Antimicrobial compounds are secreted externally by the paired thoracic metapleural glands (Beattie *et al.* 1986). Protein and lipid products have been characterized, but little is known about how they vary or about the genes behind their production or regulation. A few genera (e.g. *Camponotus*), and male ants, do not possess these glands, possibly because their tenure in the nest is relatively brief. Male ants isolated from gland-bearing workers quickly develop large microbial loads, illustrating the hazards of not possessing them.

Internally, ants have a sophisticated immune system. Workers of *Myrmecia gulosa*, when challenged with strains of *Escherichia coli*, produced antibacterial peptides, rich in proline with *N*-acetylgalactosamine *O*-linked to a conserved threonine. The synthetic non-glycolsylated form was ineffective, showing that glycosylation was necessary for maximum antibacterial activity (Macintosh *et al.* 1998). These biochemical properties were of both evolutionary and commercial interest (see later). A defensin gene has been identified from *Formica aquilonia* and compared with the defensin genes of the Hymenoptera *Apis mellifera* and *Bombus ignitus* and with three other insect species. It differs from each of them with respect to the number and length of the introns and exons, providing insights into the role of selection in the evolution of defensin in ants.

The case of leaf-cutting ants

The regulation of subterranean fungal cultures by antibiotics is well known, but the complexity of the ant—microbial interactions is still being revealed. Currie *et al.* (1999) showed that *Acromyrmex octospinosus* harbours a mutualistic, antibiotic-producing bacterium, *Pseudonocardia* on the cuticle, and that its secretions are potent against the parasitic fungus *Escovopsis* (see Chapter 6). Mutualistic bacteria with similar functions have been discovered in other Hymenoptera, hence the question arises as to how much antibiotic production is ultimately microbial. *Acromyrmex octospinosus* also exhibits immune responses to fungal pathogens.

Behavioural defences

Ants are known to forcibly remove diseased and moribund individuals from the colony, and the colonies of many species have obvious trash heaps and/or graveyards with workers assigned to their care. Ants groom one another, removing fungal spores, in some cases storing them in an infrabuccal cavity. Incomplete disposal of the infrabuccal pellet can result in fungal contagion of the colony; a mechanism that has been used to spread control agents in colonies of pest ants. This cleaning activity may also extend to the nest infrastructure. Ants may avoid areas of high infection or even move the colony in response to microbial invasion (Roy *et al.* 2006). The leaf-cutting ant *Atta columbica* responds to unwanted fungi by physically removing the spores from their fungal gardens and, if the substrate is already infected by them, the ants remove it in a highly specialized set of behaviours known as 'weeding'. These activities can be very effective against generalist fungal invaders, but the specialist parasite, *Escovopsis*, appears to have evolved counter adaptations that result in its persistence in the gardens (Currie and Stuart 2001).

Antimicrobials and the evolution of sociality

The idea that the evolution of sociality required the synchronous evolution of increased defences against microbial attack has recently been addressed in detail. Pursuing the hypothesis that increased crowding and decreased genetic diversity within colonies provided ideal conditions for microbial pathogens, Stow *et al.* (2007) demonstrated that the strength of antimicrobial compounds increased along a gradient of solitary, semisocial, and eusocial native Australian bee

continues

Box 9.1 continued

species. Such increased defences appear critical to the evolution of eusociality as the most primitively semi-social bee species showed an increase in antimicrobial strength, an order of magnitude greater than solitary species. Such comparisons cannot be made among ant species, which are exclusively eusocial. However, a similar evolutionary pathway may be inferred by the presence of the two chemical antimicrobial defence systems in ants. The presence of 'social immunity', demonstrated in termites, where previously unchallenged individuals are more resistant to disease if they have been in contact with immunized nestmates, has not been demonstrated in ants.

Commercial aspects of research into ant antimicrobials

Ecologically driven bioprospecting provides many commercial possibilities, not least among ants (Beattie 1995). The field relies on a deceptively simple question: Where has the desired product already evolved? Among insects, as Stow *et al.* (2007) have shown, strong antimicrobials have evolved especially in social species, suggesting that commercial bioprospecting should be most profitably focused on social species. Ants have yielded two patents, one based on metapleural secretions, the other from immune peptides (Macintosh *et al.* 1998). Bioprospecting for biologically active molecules in ants and other social insects is still in its infancy, but holds the promise of chemical novelty, a prerequisite for the new generation of antibiotics. In an interesting role reversal, patents have been taken out for fungal entomopathogens to control invasive ant species such as the red imported fire ant (*Solenopsis invicta*).

exposed to predators and parasites whenever the nests are left unattended.

9.3.2 Claustral ICF

In many species belonging to formicoid subfamilies, founding queens never leave the incipient nests because they are able to feed the first brood using just their metabolic reserves. Such winged queens are often larger than workers (Plates 6 and 7); they have enlarged wing muscles and are also provisioned with large amounts of lipids (Passera and Keller 1990) and storage proteins (Wheeler and Martínez 1995). Claustral ICF relies on capital breeding (i.e. resources are acquired prior to the reproductive period), whereas non-claustral ICF uses income breeding (resources are acquired during the reproductive period) (Johnson 2006). Increased isolation from the outside world likely results in a higher success rate relative to non-claustral queens, although the vagaries of aerial dispersal remain a constant for both types. Yet, reversals from claustral ICF to non-claustral ICF have occurred in several formicoid ants that live in habitats where resources (mostly seeds) are massively available and predictable, and predation is low (e.g. *Messor andrei*; Brown 1999; *Pogonomyrmex californicus*; Johnson 2002). In these habitats, foundresses can raise more workers of higher quality than if they were confined inside their nest.

9.3.3 Unrelated queens unite together

Following aerial dispersal, foundresses meeting at random can increase their success by cooperating together. Such 'pleometrosis' is beneficial because division of labour allows for a quicker production of either more workers or higher-quality workers (Bernasconi and Strassmann 1999). Generally, once the first workers have emerged, only one queen survives while the others are killed by the workers. Only in a few species do foundress associations persist and lead to polygyny, i.e., several queens reproduce concurrently (e.g. *Pachycondyla* cf *inversa*; Heinze *et al.* 2001). Pleometrosis is relatively uncommon in ants, being confined mostly to formicoid subfamilies with claustral ICF (Choe and Perlman 1997). However, there are exceptions, as in the non-claustral species *Pachycondyla villosa* and *Pogonomyrmex californicus*, where pleometrosis is an optional strategy (Johnson 2004; Trunzer *et al.* 1998). Pleometrosis involves both cooperation and conflict among foundresses, and individual contributions to the production of workers are highly variable. Cooperation can evolve among foundresses that are unrelated because survival probability needs to go up only slightly faster than the number of queens in a group (see game theory; Dugatkin and Reeve 2000). Pleometrosis can even involve foundresses from two species, as found in *Azteca* nesting inside *Cecropia* trees (Choe and Perlman 1997).

9.3.4 Solitary queens that parasitize established ant colonies

Some species have queens that parasitize the nests of closely related ants. Once the parasite queen has succeeded in entering a host colony (sometimes killing the host queen), she is protected and fed by the host workers. She benefits from their labour to produce her own workers and sexuals. In the most derived species (e.g. *Pogonomyrmex anergismus*; Johnson 1994), the worker caste has disappeared and the parasite queen only produces queen-destined eggs, making her completely reliant on her host (inquilinism).

Other interspecific associations are less parasitic. They involve species that are phylogenetically distant and generally exhibit very different body sizes (e.g. Kaufmann *et al.* 2003). *Pyramica maynei* (Myrmicinae) nests in close vicinity to *Platythyrea conradti* (Ponerinae) and collects food from the host's chambers (Yéo *et al.* 2006). Founding queens thus forage in a protected environment (i.e. inside the *Platythyrea* nest), which is a big advantage during non-claustral ICF.

9.3.5 Solitary queens that are mutualistic with other organisms

In various formicoid genera (e.g. *Acropyga, Aphomomyrmex, Cladomyrma, Tetraponera*) that are highly dependent on mutualistic associations with hemipterans, founding queens need to carry one of the mutualists during the nuptial flights. Foundation is fully claustral. Plant-sucking hemipterans feed on roots or in galls; given their parthenogenetic reproduction, a new group is quickly generated and provides additional food (honeydew) to the incipient ant colonies (references in Passera and Aron 2005). Similarly, claustral foundresses of fungus-growing ants need to bring the spores of their symbiont; their faeces are used to nurture the fungus, and the latter will be food for the first generation of workers. In both cases, the added nutrition is likely to increase the success of claustral foundation. Such associations do not seem to exist in the case of non-claustral founding.

9.4 Dependent colony foundation (DCF): Queens and nestmate workers cooperate

In many ants, the young queen(s) start a new colony together with nestmate workers. There is no solitary phase, because the queen is continuously protected and her offspring cared for. Mating occurs close to or inside the maternal nest. The new nesting site may already be known by workers that have explored the surroundings of the maternal nest, so the dispersal phase can be quick and efficient. Because workers perform all non-reproductive tasks, this caste determines the success of incipient colonies and needs to be present in sufficient numbers. DCF is the only mode of colonial reproduction in many ants, among both poneroid and formicoid groups (Peeters and Ito 2001), while it exists as an alternative to ICF in a minority of species (see Section 9.6.1).

9.4.1 Does queen number affect DCF?

A persistent belief in the ant literature is that DCF is inherently linked to polygyny. This confusion stems from the fact that various congeneric species are either monogynous and exhibit ICF, or polygynous and exhibit DCF (e.g. Keller 1991; Rosengren *et al.* 1993). However DCF is also common in monogynous species spread across all subfamilies (Table 9.2). New gynes can be produced before or after colony division: either they mate in the presence of the old queen and coexist for a short time (e.g. *Cataglyphis cursor*; Lenoir *et al.* 1988; army ants;

Table 9.2 A selection of species that exhibit DCF (in few species, as an alternative to ICF) as observed in the field ('obs'), or otherwise inferred from population genetic data, colony demography, or morphological characteristics of reproductives (AQ: alate queens; EQ1: sole-purpose ergatoid queen; EQ2: multi-purpose ergatoid queen; BQ: brachypterous (i.e. short-winged) queen; G: gamergate; m: monogyny; p: polygyny).

	Morphol. of reprod.	Mono- or Polygyny	Colony size (mean ± s.d.)	Direct or indirect evidence for DCF	References
Poneroids					
Mystrium 'red'	EQ2	p	56 ± 42	Size EQ < workers	Molet *et al.* (2009)
Onychomyrmex hedleyi	EQ1	m	850 ± 341	Nomadic; ≤ 6 gynes/colony	Miyata *et al.* (2003)
Diacamma cyaneiventre	G	m	214 ± 80	DNA markers	Doums *et al.* (2002)
Leptogenys kraepelini	EQ1	m	21 ± 7	Frequent nest emigration	Ito (1997)
Odontomachus coquereli	EQ1	m	19 ± 11	?	Molet et al. (2007a)
Pachycondyla marginata	AQ	m/p	881 ± 332	Obs; mostly ICF	Leal and Oliveira (1995)
Pachycondyla (Megaponera) analis	EQ1	m	583 ± 174	Frequent nest emigration ≤ 50 m	Longhurst and Howse (1979)
Formicoids					
Dorylus wilverthi	EQ1	m	$>10^6$	Obs; nomadic	Gotwald (1995)
Rhytidoponera mayri	G	p	577 ± 281	DNA markers	Tay *et al.* (1997)
Gnamptogenys striatula	AQ + G	p	372 ± 298	Obs; DNA; fat reserves	Giraud *et al.* (2000)
Dolichoderus cuspidatus	EQ1	m	$>10^4$	Obs; nomadic	Maschwitz and Hänel (1985)
Cataglyphis cursor	AQ	m	654 ± 596	Obs; queens do not fly; limited dispersal (DNA)	Clémencet *et al.* (2005); Lenoir *et al.* (1988)
Proformica longiseta	EQ	p	480 ± 41	Obs 3.2 m	Fernández-Escudero *et al.* (2001)
Aphaenogaster senilis	BQ	m	1260 ± 69	≤3 gynes per colony	Boulay *et al.* (2007b)
Leptothorax sp. A	AQ + EQ	m	<100	Size EQ < AQ	Heinze and Buschinger (1987)
Monomorium nr *rothsteini*	AQ + BQ	?	50,000	Obs; 16 daughter colonies after one split; ICF also	Briese (1983)
Myrmecina sp. A	EQ2	p	130 ± 96	Size EQ = workers	Ito (1996)
Ocymyrmex picardi	EQ2	m	360, 108	Obs; 1 mated young EQ in daughter colony (150 m away)	Bolton and Marsh (1989)

Gotwald 1995), or they mate after the old queen has left with a group of workers (e.g. *Aphaenogaster senilis*; Ledoux 1973). A distinct mechanism of DCF is exhibited in some polygynous species that have polydomous colonies, i.e., distributed over several nests that are separated in space (e.g. *Formica* and *Myrmica*, Rosengren *et al.* 1993; Walin *et al.* 2001). It seems that polydomy evolves initially as a foraging strategy; by increasing the number of nests occupied by the colony, a larger territory can be explored and more resources gathered. Workers, brood, and food are readily exchanged among nests. Eventually, however, when exchanges stop between two or more constituent nests, these can become the nucleus of a new colony. This phenomenon, often termed as 'budding' (see Section 9.4.2) is facilitated by polygyny. However, most polygynous ant species are monodomous, i.e., they have a single nesting site, and colony division is not different from DCF in monogynous species. In this case, exchanges between daughter colonies probably stop quickly.

Field descriptions of DCF are scarce in ants (Table 9.2), unlike wasps and bees where open nest constructions make observations easier (see Peeters and Ito 2001). In the polygynous *Proformica longiseta*, 26 DCF events were observed over a period of 30 days, with brood and adults carried over a distance of 3.25 $\pm$ 0.33 m (mean $\pm$ SE). All DCF propagules originated from populous parental nests (1,185 $\pm$ 236 workers, $N = 4$) that were more than twice the average size for this species (480 $\pm$ 41 workers, $N = 50$) (Fernández-Escudero *et al.* 2001). Daughter colonies had 262 $\pm$ 42 workers ($N = 11$) with either 1–2 adult queens or 21 $\pm$ 6 gyne pupae. After 3–4 days, all exchanges of workers had stopped, but there was much excavation in the new nests. Chance observations of single DCF events have also been reported for *Cataglyphis, Monomorium, Ocymyrmex*, and *Pachycondyla* (Bolton and Marsh 1989; Briese 1983; Leal and Oliveira 1995; Lenoir *et al.* 1988). In army ants, where observations are possible because the nests are above ground (a huge number of workers simply cluster around the queen and brood), a few gynes are produced each year. However, only one is selected by the workers, and after she mates, the colony divides into two groups of equal size that migrate in opposite directions. The old queen remains in one group, while the new queen joins the other (Gotwald 1995).

DCF allows for faster colony growth, and accordingly earlier colony reproduction. In addition, it facilitates local spreading and competition with incipient ICF colonies. It is linked to the success of various invasive species that are initially transported by humans (Hee *et al.* 2000; see Chapters 13 and 14 and 'jump dispersal' in Suarez *et al.* 2001).

9.4.2 The terms 'fission' and 'budding'

The terms 'fission' and 'budding' are both commonly used in the literature to describe DCF. 'Fission' was originally intended for monogynous species, like army ants, where colonies split into two independent parts, while 'budding' was meant for polygynous species that produce new nests that initially remain interconnected (polydomy). The definitions of both these terms have evolved over the years (see Peeters and Ito 2001), and authors have variously emphasized dispersal distance, number of queens, or size of the propagules. Both terms are sometimes used interchangeably. Ultimately, the consequences of budding and fission are the same: limited dispersal and enhanced local propagation. Since DCF has replaced ICF in many unrelated taxa, the exact processes are likely to be highly diverse. Propagules can vary a lot in their composition because multiple individuals from two castes are involved. Thus, propagules can range from very small (a trait originally used to define budding) to large (a trait used in the definition of fission), with all possible intermediates. Accordingly it is difficult to categorize DCF with a dual terminology (i.e. fission or budding). Some species, e.g. *Proformica longiseta*, exhibit traits from both 'budding' (small propagules) and 'fission' (no proximity of the mother and daughter colonies and no durable exchanges) (Seppä *et al.* 2008). Thus, describing DCF on a case-by-case basis remains necessary before we can hope to define general patterns. Given our lack of empirical knowledge in almost all ants, we advocate the use of only one neutral term, dependent colony founding (DCF). Determining whether DCF is obligate or facultative is crucial, and morphological specialization of queens (especially permanent loss of wings), as well as investment in sexuals may give reliable information about this (see Section 9.6).

9.4.3 Parasitism is not DCF

We think it is not heuristic to consider parasitism (see Section 9.3.4) as a special type of DCF. Host colonies are nothing more than a resource of the environment to be exploited, similar to insect prey. Parasitism is another adaptation to improve success during ICF, just like pleometrosis (see Section 9.3.3). The crucial point is that the founding queen is not helped by relatives from her natal colony, hence the interests of protagonists diverge. Parasitism is a risky strategy, because a foundress needs to enter an alien colony by deceiving its workers and sometimes killing its queen. Workers from the parasitized colony have no interest in rearing unrelated brood, leading to the evolution of defensive mechanisms. An arms race is expected between the host and the parasite, the latter evolving ways to escape detection. Similarly, pleometrotic associations are based on selfish strategies to access direct reproduction, and only one queen generally survives once workers have emerged. In both pleometrosis and parasitism, queens disperse on their own without the protection of nestmates. Thus, foundresses have a high failure rate, and many individuals need to be produced annually. This matches the investment of species that rely on non-claustral or claustral ICF; large numbers of queens are produced to compensate for mortality, and nestmate workers are not a part of the reproductive investment. In contrast, with DCF, all participants have convergent interests because they are highly related (young queens and workers are generally sisters), and evolution favours utmost cooperation.

9.5 Colony growth and reproduction

Offspring number and quality (e.g. size, weight, and metabolic reserves) have crucial consequences on the capacity of living organisms to spread in their local environment and to colonize new habitats. Solitary insects can adjust the traits of their offspring to a limited extent, by trading-off quantity and quality, which are generally constrained by limited resources (e.g. Fox and Czesak 2000). Social insects with morphological castes have more complicated trade-offs, because quantity and quality of either the gynes or the workers can vary independently. In ants reproducing by ICF, the number and size of gynes produced by a colony follow the classic quantity or quality trade-off found in solitary organisms. Dramatic changes in individual size are impossible because functionality must be retained (Molet *et al.* 2008). However, by shifting to DCF, a much wider range of offspring phenotypes becomes possible, i.e. colonies can produce propagules with more or fewer workers and queens, and these can vary independently in quality. Hölldobler and Wilson (1990) described the colony life cycle as 'an orchestration of energy investments, in which workers are multiplied until [. . .] it is profitable to convert part of the net yield into new queens and males'. This maturation point differs markedly across species, and it is heavily influenced by the occurrence of either ICF or DCF.

9.5.1 Investment of resources in workers

Ants are holometabolous insects, meaning that the juvenile form (larva) hatching from the egg is very different from the adult (Plate 8). The larva is fed and grows during several weeks. Weight increase is continuous, but size increase is discrete and occurs only during regular moulting events, i.e. three to five larval instars depending on the species (e.g. Onoyama 1982). Once a critical size has been reached, the larva can undergo metamorphosis. By then, environmental factors have channelled female larvae towards either queen or worker caste. Nutrition determines castes in honeybees (quantity and quality of royal jelly), but in ants, abiotic parameters (e.g. cold) also seem to be involved. Once individuals are adults, there is no further growth (no more moulting). The size of the workers is thus fixed and cannot change as they age.

Only workers are reared during most stages of the colonial life cycle (Figure 9.1). Species with small colonies typically have monomorphic workers (i.e. exhibiting little variation in size and shape) because these must perform all tasks efficiently. In species with larger colonies, polymorphic workers can be produced because specialization for specific tasks becomes advantageous (Wilson 1983). Thus workers can vary markedly in size, either continuously or as discrete subcastes. In the latter, 'soldiers' are allometrically distinct from the rest of the workers, and these occur only in 42 (41 of which are formicoid) of the 283 genera of ants (Oster and Wilson 1978). Workers in poneroid ants exhibit much less variation in size, and division of labour is then based mainly on age (Peeters 1997). In many formicoid ants, founding queens produce a first brood of nanitic (i.e. undersized) workers

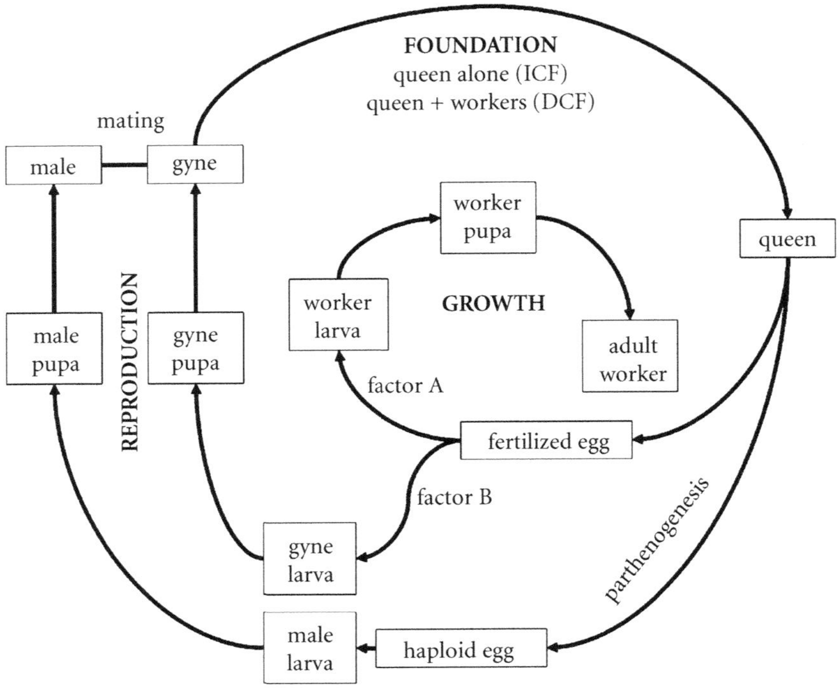

Figure 9.1 Typical life cycle of an ant colony. Newly founded colonies go through a growth phase during which all larvae are reared into workers. Once a critical size is reached, sexuals can be produced (reproductive phase): larvae hatching from the queen's fertilized eggs develop into gynes (exposure to biotic and abiotic environmental factors), while larvae emerging from unfertilized eggs develop into males. Males and gynes mate with foreign partners, and gynes then found new colonies either alone (ICF) or helped by nestmate workers (DCF).

termed 'minims' (Porter and Tschinkel 1986). Incipient colonies can thus reach a greater number of workers in a fixed time. Workers from subsequent broods progressively reach a normal size. In red imported fire ants (*Solenopsis invicta*), as colonies grow to about 200,000 workers, mean worker mass increases fourfold (Tschinkel 1993). In some species, colonies can alter the ratio between subcastes according to environmental influences. For instance in *Pheidole pallidula*, more soldiers are produced when high competition or predation makes nest defence a priority (Passera *et al.* 1996).

9.5.2 Investment of resources in sexuals of ICF species

During a certain period of the year, female larvae are reared into gynes instead of workers. In addition, unfertilized eggs are laid and develop into males by arrhenotokous parthenogenesis (Figure 9.1). This is due to the 'complementary sex determination' system, where individuals that are heterozygous for the sex determination gene develop into females, whereas those that are not (e.g. are haploid) develop into males (Heimpel and De Boer 2008). Colonial investment in winged gynes varies tremendously across species that exhibit ICF and is affected by colony size, degree of dimorphism of gynes relative to workers, and the amount of reserves accumulated by gynes prior to dispersal (e.g. *S. invicta* gynes gain 290% in mass, mostly fat, before dispersal; Tschinkel 1993). In poneroid species, colonies generally have several dozens or hundreds of workers, these are not much smaller than queens, and the latter have few reserves. In contrast, in formicoid species, colonies can have thousands to millions of workers, these are often much smaller than queens, and the latter are provisioned with more metabolic reserves. In both cases, gynes represent a large proportion of colonial resources, and this is necessary due to their poor individual success rate at dispersing and founding new colonies. For instance, among poneroids, *Harpegnathos saltator* colonies can produce as many gynes in one season as there are workers (e.g. 58 in a colony of 60 workers, Peeters *et al.* 2000).

The production of this large number of sexuals needs to be completed over a short time of the year (sexual activity is often limited to a few weeks or days), so it necessitates integrated colony operations. In some species, fat is stored in workers outside the reproductive season and is utilized to feed sexual larvae. This is the case in *Pogonomyrmex badius*, where a large number of sexuals are produced in early spring, at a time when foraging brings too few resources (Smith 2007; Tschinkel 1998). If stored resources are still insufficient, then investment in gynes is deferred to the following year, and more workers are produced instead. In *Solenopsis invicta*, colonial investment in sexuals does not rise gradually, but suddenly, once colonies reach about 50,000 workers, and then it remains constant at 33% (Tschinkel 1993). Workers keep on being produced once the colony has reached this maturation point to compensate for mortality or even allow for additional growth.

As we see in Section 9.6.2, the investment in gynes decreases considerably in DCF species.

9.5.3 Short-lived workers but long-lived colonies

In many poneroid ants, queens have a relatively short lifespan (e.g. 1.8 years in *H. saltator*; Liebig and Poethke 2004), whereas several formicoid taxa have evolved queens that can live much longer (up to 20 years or more) than conspecific workers (Keller 1998). These variations in the queen's lifespan lead to major differences in colony life expectancy, because they are often directly connected. In various monogynous ants, death of the original foundress is commonly thought to be followed by rapid extinction of the colony (after a brief period of male production from worker-laid eggs). This is surprising, considering the high value of the nest and worker force (Myles and Nutting 1988), as well as the generally poor success of new queens at founding colonies. In *Nothomyrmecia macrops*, orphaned workers accept that one daughter queen (sister to them) takes over reproduction following the queen's death; this is likely the best reproductive option for newly mated queens, considering that ICF is so risky (Sanetra and Crozier 2001).

Such colony inheritance by a related secondary reproductive may be more common than currently perceived. The short lifespan of workers relative to queens may be responsible for the failure to detect successive queen replacements (serial polygyny), because the workers produced by a previous queen quickly disappear, making the colony appear monogynous in genetic studies (e.g. André *et al.* 2001). Colony lifespan can also be extended by adoption of unrelated queens in species where the dead queen's sexual brood could not otherwise be raised to adulthood. In *Camponotus ligniperdus*, sexuals overwinter twice in the colony before leaving on their mating flight, and workers benefit from accepting unrelated queens that will produce additional workers and ensure that the previous queen's sexuals survive (Gadau *et al.* 1998). More studies are required on this topic. One outstanding question is whether queen replacement can influence the occurrence of DCF events. Indeed, as long as the mother queen is alive, workers who leave their natal colony give up raising sisters because they start raising nieces produced by a sister queen that mates with a foreign male. However, once the mother queen has died and been replaced by a sister queen, workers face no additional cost if they leave the colony with another sister queen.

Polygynous species probably have colonies with higher longevity because multiple queens succeed each other in time, and there are no gaps in brood production (Frederickson *et al.* 2005). For instance, the lifespan of *Formica montana* colonies can exceed 33 years (Henderson *et al.* 1989). Polygyny can result either from pleometrotic associations that persist beyond the founding phase (primary polygyny, rare), or from adoption of daughter queens (secondary polygyny, common). Stuart *et al.* (1993) showed that in *Leptothorax curvispinosus*, daughter queens are adopted in 60% of laboratory colonies. In a substantial number of species, the original queen is replaced by secondary reproductives that differ in morphology. For instance, in *Technomyrmex brunneus*, many ergatoid queens mate with brothers and reproduce after the death of the founding (dealate) queen (Yamauchi *et al.* 1991). In the relatively few species in which workers have retained the capacity for sexual reproduction, one or several gamergates reproduce in the colony after the founding queen dies (Monnin and Peeters 2008).

9.6 Causes and consequences of the shift from ICF to DCF

Irrespective of whether ICF is claustral or non-claustral, natural selection led to its replacement by DCF in many unrelated ant taxa. Evidence for this shift comes from comparing congeneric species (e.g. in *Mystrium* and *Odontomachus*, Molet *et al.* 2007a, 2009). In a minority of species, both strategies occur as alternatives (e.g. Briese 1983), but in others, DCF has become obligate. Indeed, selection has frequently eliminated winged queens because they give no benefits if they disperse together with wingless workers. Loss of dispersal has often led to the evolution of flightlessness in solitary insects as well (Roff 1990).

9.6.1 What ecological pressures favour DCF?

Several environmental factors cause difficulties for solitary foundresses (Heinze and Tsuji 1995). First, there can be a strong cost associated with long-range dispersal; for example, habitat patchiness limits the chances for a flying queen to find a suitable habitat. Second, limitation of food or nesting sites increases competition between solitary foundresses and established colonies. This competition is decreased in claustral species because incipient colonies are 'hidden' until the first workers begin to forage. However, in non-claustral species, foundresses come in direct contact with the foragers of established colonies. Last, if incipient nests are disturbed before there are adult workers, it is difficult for a lone foundress to carry all of her first brood to a new nest site. All these parameters lead to a high failure rate of ICF: Less than 1% of queens are thought to succeed (e.g. *Pogonomyrmex*; Gordon and Kulig 1996). Alternative ICF strategies such as pleometrosis or parasitism can limit this failure. However, there are also selective pressures against DCF. In species with obligate DCF, the fate of colonies is entirely dependent on the quality of the local habitat. When environments become patchy or unstable, there can be no 'escape' by flight, since winged queens are generally absent. Lack of aerial dispersal also increases local competition for resources when a habitat is saturated. Accordingly, even if obligate DCF facilitates the colonization of local habitats over a short period, it may be costly over a longer timescale because of the loss of long-range dispersal by flight, raising the question of whether this strategy turns out to be an evolutionary dead end in many taxa. This could explain why most species exhibiting obligate DCF are nested among clusters of species that retain ICF. DCF only seems to be evolutionarily stable in taxa with a very specific life history (e.g. nomadic behaviour). It should be noted that dispersal on foot does not limit geographic range (e.g. the wide distribution of *Neivamyrmex nigrescens*) unless there are barriers; the time needed to colonize a habitat is simply increased.

These opposing selective pressures on ICF and DCF are probably responsible for the coexistence of both strategies in various species. When winged queens perform both ICF and DCF, there is often morphological variability in these queens, i.e., normal winged queens specialized in ICF coexist with microgynes (dwarf winged queens found for instance in *Ectatomma ruidum*; Lachaud *et al.* 1999) or brachypterous queens (queens with short wings unsuitable for flying found for instance in *Temnothorax longispinosus*; Howard 2006). In all other species, either winged or permanently wingless reproductives are specialized for ICF and DCF, respectively. In the *Rhytidoponera impressa*-group, the relative frequency of colonies reproducing by ICF decreases over a 3,000 km gradient from tropical rainforests to temperate forests (Molet *et al.* 2008). Founding queens need to hunt outside, so they have decreased success in a harsher and less predictable temperate environment. Instead, gamergate colonies exhibiting DCF become the predominant colonial strategy once the long-range dispersal benefit of ICF is exceeded by the enhanced survival rate of incipient DCF colonies.

There is a paucity of field studies that investigate the processes of either ICF or DCF, as well as their relative benefits. Monitoring of solitary foundresses over several weeks is laborious, and human observers can have a negative impact on success itself. DCF is often an unpredictable event in time, thus even harder to monitor. Hence, the assessment of success rates is difficult (but see Wiernasz and Cole 2003). In addition, empirical measurements of the link between environmental parameters and colony reproduction can only lead to correlative results. For a direct assessment of the effect of environment on colonial reproduction, one should transfer colonies from one habitat to another, but such a manipulation would interfere with colony performance and only give an insight into phenotypic plasticity, not evolution. Alternatively, mathematical modelling can reveal causal relationships between environmental conditions and colonial reproductive strategy. Environmental parameters together with individual and colonial traits of ants can be measured experimentally, and models can be developed to explore the underlying mechanisms that could connect them together. Using this approach, Molet *et al.* (2008) developed a population dynamical model to predict the effect of environment on the coexistence between ICF and DCF in *Rhytidoponera*. A system of two differential equations describes the dynamics of the two types of colonies: (*a*) queenright colonies reproduce by ICF, and following the queen's death, they turn into (*b*) gamergate colonies that reproduce by DCF. This model showed that fluctuations in environmental parameters (notably food availability) are responsible for the shift from ICF to DCF along a gradient from tropical to temperate forests. Empirical field data not only confirmed this prediction, but also indicated that ICF can be retained in harsh conditions due to gynes being provisioned with more fat reserves. Since the model did not take into account the benefits of long-range dispersal, the increased quality (and thus cost) of gynes suggests that retention of aerial dispersal is strongly selected. Geographic information system (GIS) maps of environmental variables can be used in the near future to study the causes of transitions in colonial reproductive strategies in ants (Kozak *et al.* 2008).

9.6.2 Investment in queen or worker castes in species with DCF

In contrast to the high mortality of winged queens during ICF, mortality of queens engaging in DCF is much lower. Accordingly, investment in queens can be much reduced, especially since there can be only one or a few DCF propagules each year. In *Myrmica rubra*, a large polydomous colony produced over 2,000 males and only five new queens in one year (Walin *et al.* 2001). Thus, in DCF, colonial resources are redirected from sexuals towards workers (Pamilo 1991). Indeed workers contribute to colonial reproduction by helping nestmate queens found daughter colonies. Their performance in non-reproductive labour determines the success of DCF. Hence, it is very difficult to quantify reproductive investment in species reproducing by DCF, unlike species that reproduce by ICF.

9.6.3 Loss of pre-mating dispersal and shift to DCF

Aerial dispersal prior to mating (see Section 9.2) dictates the subsequent behaviour of founding queens; it is very unlikely that they can return to their natal nest after mating away from it (Bourke and Franks 1995), so DCF is not an option. We suggest that changes in mating behaviour precede any evolutionary shift to DCF. In several species having winged queens, mating flights have never been observed (e.g. *Lasius neglectus*), and intra-nidal mating is probably the rule. It is then a short step for such queens to be selected to forgo ICF. We conceive that winged queens are initially involved in DCF, but that they do not fly to find sexual partners. Since mating close to the natal nest leads to a sharp drop in mortality, fewer gynes need to be produced annually, and this is entirely consistent with DCF strategy (see Section 9.6.2). A further morphological adaptation can then follow in many species (see Section 9.6.4).

9.6.4 Evolution of wingless reproductive phenotypes

The phenotype of winged queens is mostly shaped by the strong selective pressures linked to ICF. In particular, the bulky flight muscles in many formicoid ants represent essential metabolic reserves used to feed the first generation of workers. Constraints on flying ability are relaxed with the shift to DCF, except in species where it continues to alternate with ICF (Heinze and Tsuji 1995). When DCF is obligate, flight is no longer needed since queens walk together with workers; similarly, storage of metabolic reserves becomes useless since queens are never alone, and workers feed all offspring. Therefore, in thousands of species belonging to 66 genera at least, winged queens have been replaced by wingless reproductives (Plate 7) that have a simplified thorax as in workers (i.e. flight muscles are lost, and sclerites become fused). Ergatoid queens always have functional ovaries and spermatheca; they are a caste distinct from workers, which are unable to mate and store sperm in most ants. It is only in 200–300 species (mostly Ponerinae) that ergatoid queens did not need to evolve; the workers have a functional spermatheca, hence gamergates can reproduce instead of winged queens. Intermediate stages in the elimination of winged queens still exist. In *Cardiocondyla batesii*, winged queens exhibit a large variation in body volume, one in five of them having smaller thoraces and short wings (=brachypterous) ineffective for flying, whereas the others retain normal wings, although they do not actively fly (Heinze *et al.* 2002a). This suggests that once flying is no longer adaptive, stabilizing selection on the flight thorax stops.

The evolution of ergatoid queens corresponds to a strategy of colonial economy, because per capita costs of gynes are reduced at two levels: before and after adult emergence. Thus, (*a*) modifications in larval development result in gynes emerging without wing muscles and with a simplified thorax; (*b*) adult gynes no longer need to accumulate additional metabolic reserves (fat, glucose, and storage proteins) prior to dispersal. Indeed, the continuous presence of workers ensures that there is food available to sustain egg production and feed larvae. This reduction in energy investment associated with DCF has been shown in species with winged queens (Keller and Passera 1989), but it is even more pronounced in species with ergatoid queens. Ergatoid queens appear incapable of performing claustral ICF, because the lack of wing muscles as a metabolic store strongly reduces their chance of succeeding as would-be foundresses. Nonetheless, if ergatoid queens can predictably obtain nourishment outside, they succeed in non-claustral ICF. In *Pogonomyrmex cunicularius* (R. Johnson, personal communication), ergatoid foundresses are able to forage for seeds similarly to dealate queens of congeneric species, and thus be independent of nestmate workers. Similarly in *Plectroctena mandibularis*, ergatoid foundresses can hunt insect prey like their winged counterparts in other species (Villet 1991).

Ergatoid queens exhibit an extremely wide range of phenotypes across ants, making them much more diverse than winged queens (Plate 7). This great heterogeneity is due to independent evolution from numerous unrelated ancestors having winged queens. In many species, ergatoid queens are morphologically closer to the winged queens of congeneric species than to their workers. They are produced in very small numbers, and colonies are monogynous. Such 'sole-purpose' ergatoid queens are found across taxonomic groups (e.g. *Leptogenys, Myrmecia, Monomorium, Dolichoderus cuspidatus*-group; references in Table 9.2). In army ants, they are very fertile and called 'dichthadiiform'; their gaster becomes physogastric to accommodate extensive ovarian development. However, in other species (mostly Myrmicinae) 'multi-purpose' ergatoid queens are morphologically closer to the workers than to winged queens of congeneric species, and they are produced in greater numbers than sole-purpose ergatoid queens. Only some reproduce, while the others perform worker tasks. Infertile ergatoid queens do not disperse from the natal nest, and can be involved in brood care, grooming of nestmates, nest maintenance, and even foraging (e.g. *Ocymyrmex, Myrmecina*; references in Table 9.2).

9.6.5 DCF leads to changes in colonial life cycle

ICF species exhibit high seasonality (Figure 9.2) because sexuals of both sexes must disperse at the same time for mating to succeed. Moreover, mating is immediately followed by colony founding. Indeed, after long-range dispersal by flight, queens

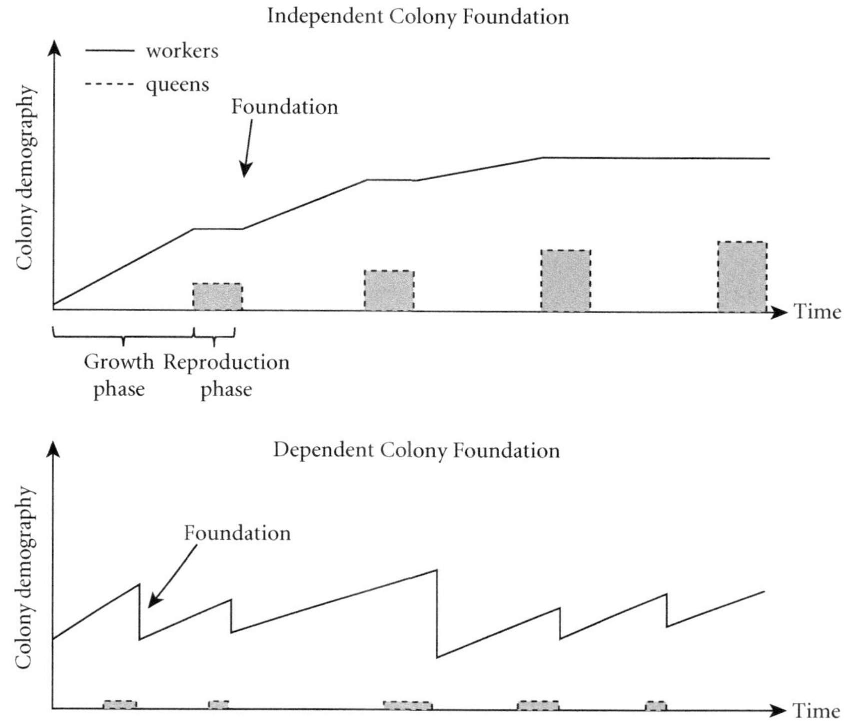

Figure 9.2 Changes in colony demography over time relative to the reproductive strategy. In ICF (*top*), colonies start with a single queen and need to grow for some time before being able to produce new gynes. These disperse during a nuptial flight. In contrast, colonies started by DCF (*bottom*) consist of a group of nestmate workers that help the new queen. Such colonies can start producing sexuals earlier; seasonality is not always selected for since mating and colony foundation can be disconnected. Only few gynes are produced as only one or a few propagules are possible, but their success rate is high.

cannot return to their natal colony and wait. In contrast, in DCF species, seasonality is less adaptive because mating and founding of new colonies can be disconnected (Figure 9.2). Males can be produced during many months of the year, not only in gamergate species, but also in species where ergatoid gynes are reared (e.g. in *Aphaenogaster senilis*, from May to January with a peak in July to August; R. Boulay, personal communication). Once mated, young queens can wait until conditions are optimal to leave the maternal nest with a group of workers. When seasons are marked, there are temporal constraints on excavation of new nests and survival of small colonies. When gynes emerge seasonally, this may provide a trigger for DCF if only one functional reproductive is tolerated per colony (e.g. *Cataglyphis cursor*; Lenoir *et al.* 1988; *A. senilis*; Boulay *et al.* 2007b).

Another important difference between the two modes of reproduction is the size of daughter colonies. ICF results in colonies that can take several years to produce enough workers, before they can rear the first pulse of sexuals (Figure 9.2). In contrast, new colonies founded by DCF start with a worker force that is already sufficient, and they can start producing sexuals much earlier. This shorter generation time is a strong advantage in growing populations (Roff 1992).

9.7 Future directions

The prevalence of DCF has been underestimated in ants. Although very few studies have described it in the field, the existence of wingless reproductives in a large number of species is strong evidence that ICF is very unlikely. Morphological adaptations in ergatoid queens need to be studied in a wider taxonomic range of species, and followed up by focusing on colony-founding behaviour. The ability of ants to readily produce reproductive phenotypes that are wingless and hence cheaper (impossible in social wasps and bees) probably contributed to the evolution of DCF as an efficient alternative to ICF. There are also unanswered questions about the evolutionary factors selecting for different modes of ICF, and these are best studied in genera (e.g. *Pogonomyrmex)* that exhibit variations in the levels of foraging by founding queens (i.e. obligate, facultative, or absent).

9.7.1 Finding more evidence for DCF

Field studies are hindered by the unpredictability in time of DCF events in various species (especially in the tropics, see Section 9.6.5), unlike ICF which always immediately follows seasonal mating flights. Moreover, many species that exhibit DCF are prone to regular nest relocation (discussed for Ponerinae in Peeters 1997), so emigrations must be distinguished from DCF events. Brood and adult exchanges between nests sometimes need to be monitored over a few days. In polydomous species, new nests can either remain connected with an existing colony, or they can become autonomous. Accordingly, describing DCF not only implies studying the behaviours during the division itself, but also the fate of the two nests over several days. It is important to excavate both nests involved to compare their demography and reproductive structure (using ovarian dissections and/or DNA markers).

It is commonly thought that molecular markers are powerful tools to confirm the occurrence of DCF. For instance, Liautard and Keller (2001) showed that nests of *Formica exacta* are very similar genetically within pastures, but different among pastures, suggesting that queens do not disperse far from their natal nest, probably because colonies are founded dependently. Ross *et al.* (1997) found that in the native range of *Solenopsis invicta*, queens produced by polygynous colonies do not disperse far (hence DCF is likely to occur), whereas queens from monogynous colonies perform long-range dispersal by flight (hence ICF). Sympatric monogynous and polygynous forms were genetically differentiated, so there was no queen exchange between colonies of different types and assortative mating may occur. In other species, complex patterns of intra-colonial relatedness linked to polygyny and/or polyandry can complicate analyses. Furthermore, a very fine-scale approach is required, i.e., analysis of neighbouring nests for which recent DCF is suspected. Nevertheless, population genetic data have shown extremely short ranges of female dispersal, as opposed to the contributions of flying males (e.g. Doums *et al.* 2002; Giraud *et al.* 2000). Such data corroborate the occurrence of DCF, but cannot unambiguously exclude ICF close to the nest of origin.

9.7.2 Understanding the behavioural processes in DCF

Diverse questions remain about the mechanisms of DCF. First, a few gynes are produced annually, but these may vary in size, potential fertility, and level of relatedness with respect to workers. Conflicts for the choice of the new queen are expected when more gynes are produced than the number of future daughter colonies (in *Aphaenogaster senilis*, Chéron *et al.* 2008). Queens compete for survival, and all workers may not prefer the same queen. Individual interactions must be studied, and relatedness between individuals as well as queen quality (e.g. cuticular hydrocarbons signaling levels of fertility) can be assessed. In a polyandrous and polygynous *Proformica* species, daughter colonies produced by DCF exhibited a higher intra-colonial relatedness than the mother colonies from which they originated (Seppä *et al.* 2008). Since workers are carried from the mother nest to the daughter nest, kin discrimination can be involved at this step. However, evidence for kinship-based DCF has not been found in other species. This may be linked to the need for an efficient daughter colony to include workers of all ages (i.e. nurses as well as foragers). Age is probably a factor that overrides genetic relatedness during the organization of DCF. In addition, workers may be unable to distinguish which nestmates they are more related to.

Second, contrary to ICF, where the winged queens land in an unknown environment and must quickly find a nesting site, DCF allows for preliminary exploration by the workers of the neighbourhood of the mother colony in order to find a suitable new nest. Some recent studies have focused on nest emigration to assess which characteristics of the nests are most important to workers (reviewed by Visscher 2007). It must be investigated whether such choices are also made before DCF in ants and involve similar mechanisms as those found in swarming honeybees, e.g., existence of a worker quorum that triggers emigration behaviour.

Third, the dynamics of DCF are unknown. Is division triggered once colonies exceed a certain size threshold? Or simply because physical space becomes limiting in the nest? Across species, is there an effect of colony size on dispersal distance? Are behavioural mechanisms affected by colony size across species? Group decisions may be hindered below a critical colony size, and individual decisions may then be more important. During nest emigration, the queen and brood are transported at specific times, and one should study whether these parameters are similar during DCF. Testing as to whether solitary ergatoid queens can succeed in ICF by foraging outside, and how the help of workers increases founding success, is also necessary. Although this has been done under laboratory conditions (e.g. Schrempf and Heinze 2006), experiments in the field are needed and must be generalized to more taxa.

Finally, we do not know how many propagules are produced at each DCF event (this is likely to vary across species), and how frequent DCF events are. The role of parasites also warrants more attention because in infected colonies parasites could be a strong selective pressure against DCF (assuming that founding queens dispersing for ICF are not infected).

9.7.3 Morphological adaptations for more efficient DCF

Future studies need to quantify per capita costs of either winged or wingless reproductives that occur in the same species or genera. This will test the selective advantages involved in replacing the ancestral winged queens. In *Mystrium*, interspecific comparisons revealed that ergatoid queens (5.95 mg) are cheaper to produce than winged queens (14.5 mg) (Molet *et al.* 2007b). Similar data are needed across all subfamilies of ants.

DCF is associated with a substantial increase in the success rate of incipient colonies, because it eliminates the dangerous solitary stage. Hence, the advantages of social life are retained throughout the colonial life cycle. One could almost expect that ICF would be completely lost in ants, were it not that dispersal on the ground introduces severe constraints, i.e. no colonization of new territories, no escape from deteriorating habitats. ICF continues to exist in a majority of ants, despite DCF being an efficient strategy. This is a powerful evidence for the benefits of aerial dispersal (Hamilton and May 1977).

9.8 Summary

All ants live in perennial colonies that exhibit three phases: foundation (initiation of new colonies), growth (production of more workers), and reproduction (production of sexuals). Colony foundation can be independent or dependent, and this dichotomy has important consequences on all phases of the life history. During independent colony foundation (ICF), winged queens are alone while they disperse by flight, mate, and raise the first generation of workers. Queens feed the larvae using energy provided by degradation of their wing muscles, but the queens of various species also need to forage. Both morphological (queens become much bigger than workers and no longer forage) and behavioural (cooperation with other queens; invasion of conspecific or heterospecific colonies; mutualism with different organisms or ant species) adaptations have evolved to improve the success rate of solitary foundresses. However mortality of foundresses often remains high (especially during aerial dispersal and mating), and colonies must invest a large proportion of their resources in the annual production of numerous winged queens. Hence, colonies must grow to a large size before being able to reproduce, which may take several years. Sexuals generally mate in aggregations that require their synchronous release from many colonies, but high predation can select against these aggregations.

Dependent colony foundation (DCF, also termed 'fission' or 'budding') is a fundamentally different strategy because young queens are continuously helped by a group of nestmate workers. Hence, reproductive investment is redirected towards the production of more workers, inasmuch as these determine the success rate of daughter colonies. Only a few gynes are reared annually, because only one or few propagules are possible. Given that colonies start at a bigger size, sexuals can be produced sooner. Males search for foreign sexual partners, who remain inside or near their natal nest. Depending on the species, mating of new gynes occurs before or after colony division. All ant workers are wingless, and thus queen dispersal occurs on the ground over short distances. Accordingly wingless 'ergatoid' queens evolved in many species (at least 66 genera, both poneroid and formicoid), and aerial dispersal is then restricted to males.

Contrary to widespread thinking, DCF occurs irrespective of monogyny or polygyny. In species founding independently, colonies can go extinct after the death of the original foundress, although queens have exceptional longevities in some species. Colony lifespan can be extended with secondary reproductives, and these are sometimes morphologically different from the normal winged queens (e.g. gynes of reduced size, mated workers). In polygynous species, the continuous turnover of queens also results in longer-lived colonies.

DCF is very efficient in ants as a result of adaptations in the morphology of reproductives, with ergatoid queens being cheaper to produce compared to the ancestral winged queens; ergatoid queens did not evolve in social wasps and bees. Shifts from ICF to DCF occurred many times in unrelated ant genera. Current research explores the causes and consequences of these shifts. It is likely that the loss of long-range dispersal by queens increases the chance of species extinction. DCF remains poorly known relative to ICF. Species where both strategies coexist are interesting material for future research.

Acknowledgements

We thank Adam Cronin, Claudie Doums, Jürgen Heinze, Bob Johnson, and Thibaud Monnin for their constructive comments, together with our editors. We acknowledge the financial support from ANR-06-BLAN-0268.

Chapter 10

Colony Structure

Florian M. Steiner, Ross H. Crozier, and Birgit C. Schlick-Steiner

10.1 Introduction: from ecology to evolution, and back

Ants live in colonies – societies of cooperating conspecific individuals. Colloquially, the terms colony and nest are synonymous, and indeed in many instances a colony can be allocated to one nest. However, to assess ant colony structure, we need to be precise: a colony is the society, a nest its vessel, and in many species, a colony has more than one nest.

A colony's structure is its caste, demographic, genealogical, and spatial makeup. This understanding of colony structure embraces a broad range of characters, albeit colony structure sometimes is used as a term for only one or a few of these. Colony structure is a key aspect of ant ecology and the organisms with which ants interact. Proximate causes shape colony structure, from queen pheromones to pathogen load, and from gene flow to nutrient availability. Ultimately, colony structure has resulted from millions of years of social evolution, under the constant influence of the ecosystem, which in turn is strongly influenced by colony structure.

The relevance of colony structure to ecology has long been recognized, but concepts have changed. In the first half of the twentieth century, the analogy of a single organism with many integrated organs led researchers to treat a colony as a superorganism (Hölldobler and Wilson 1990). From the 1960s onwards, understanding of the relatedness asymmetries, a result of male haploidy, with sons being more related than daughters to their mother, and sisters being more related to each other than to their brothers, spurred understanding of the intracolonial conflicts shaping colony structure (Crozier and Pamilo 1996). Recently, the influence of factors external to the colony have been emphasized as affecting sociality (Korb and Heinze 2008), but this emphasis does not negate the importance of kin selection (Crozier 2008).

Here we review the characters and character states of colony structure, the influences shaping the character states, and the ecological consequences of colony structure. We introduce supercoloniality as a paradox of colony structure and conclude with an outlook on future research. Throughout the chapter, we emphasize the usefulness of evolutionary concepts in ant ecology, and how ecology can contribute to unravelling evolutionary pathways.

10.2 The characters and character states

Colonies differ very greatly in their caste, demographic, genealogical, and spatial structure, best described by specific conditions (states) of recognizable attributes (characters). For example, colony size is a character and a particular size is a state. States are often associated across characters and form 'syndromes', such as large colony size being associated with more complex polyethism. Treating one character after the other is therefore somewhat artificial, although necessary before higher-order complexity can be examined. In Section 10.2, we introduce the characters, namely number and identity of reproductives, number and functionality of non-reproductives, and spatiality, as well as their most important states.

10.2.1 Number and identity of reproductives

The numbers and identities of female and male reproductives are two key characters of colony structure. They are connected to mating, dispersal, colony foundation, and colony growth (André *et al.* 2006; Bourke and Franks 1995; Crozier and Pamilo 1996; Hölldobler and Wilson 1990; see also Chapter 9).

Reproductively active females able to produce offspring of both sexes are called queens. But there are different approaches to defining the 'queen' caste, and thus distinguishing her from the other females, the worker caste. Under the morphological caste definition (e.g. André *et al.* 2006) a queen is morphologically distinguishable from a worker, usually by having wings at eclosure and by a larger mesosoma size (Plate 6). In contrast, according to the functional definition (e.g. Winter and Buschinger 1986), a queen is a mated individual laying eggs that develop into offspring of both sexes, irrespective of her morphology. In species without morphological differences between reproductive and non-reproductive females, the reproductives then are 'gamergates' according to the morphological definition, but 'ergatomorphic queens' according to the functional definition. 'Queens' in this review are queens under both definitions.

There is a broad range of states for the character 'number and identity of female reproductives'. A colony can either have a single (monogyny) or several to many queens (polygyny). Some colonies have more than one fertile, female sexual, but only one reproduces (functional monogyny), as in *Leptothorax* sp. A (Heinze and Buschinger 1989). Some species need several queens per colony (obligate polygyny), while others, like *Myrmecina graminicola* (Buschinger and Schreiber 2002), have both monogynous and polygynous colonies (facultative polygyny). A small and limited number of queens, intolerant to each other, can occupy different parts of the nest (oligogyny) as in *Camponotus ligniperda* (Gadau *et al.* 1998). This contrasts the more frequent true polygyny, where queens mix freely within the nest like in many *Formica* wood ants (Seifert 2007). Some monogynous ants replace the queen, after her death, by a daughter-queen (serial polygyny), to avoid colony dissolution, as in *Nothomyrmecia macrops* (Sanetra and Crozier 2002). Evolutionarily, the number of queens is flexible (Ross and Carpenter 1991), and highly variable in many species.

The same queen number can arise in different ways. Some colonies are founded by a single queen (haplometrosis) which remains alone over the colony's whole life (primary monogyny). Other colonies are founded cooperatively by several queens (pleometrosis), but one or multiple queens persist (secondary monogyny versus primary polygyny). Finally, some colonies are started by a single queen, but later other queens are adopted (secondary polygyny). These strategies can be species-specific, but in the extreme case, as in *Iridomyrmex purpureus* (Hölldobler and Carlin 1985), they can all be found in the same species.

In several species, there is more than one queen morph. Often this is a simple size dimorphism, with big (macrogynes) and small queens (microgynes), but it can extend to a range of phenotypes between worker-like and queen-like morphs of differing fecundity (Heinze and Tsuji 1995). An unusual case is *Crematogaster smithi*, in which individuals, termed 'large workers', are specialized for the production of unfertilized eggs, interpreted as live food provision (Heinze *et al.* 1999).

Unfertilized workers laying male-destined eggs are also reproductives (Crozier and Pamilo 1996). Worker reproduction is probably common in monogyny following the queen's death in species whose workers have ovaries, and was demonstrated in, for example, *Acromyrmex* species (Dijkstra *et al.* 2005). Worker reproduction is, however, incompletely understood in its quantitative relevance. Finally, in a few ant species (Keller 2007), females are produced from unfertilized eggs (thelytokous parthenogenesis).

The number of fathers contributing to a colony may also vary, due to variation in the mating frequency of queens (Crozier and Pamilo 1996). The queens of many species mate once (monandry), but multiple mating by queens (polyandry) can be extreme, for example, over ten times in *Pogonomyrmex badius* (Rheindt *et al.* 2004). In males, generally distinguishable from queens by having relatively tiny heads with thread-like antennae (queens and workers have elbowed antennae), polymorphism is generally rare and normally involves simple size variation (Sundström 1995). Strong dimorphism occurs in *Cardiocondyla*, however, with normal and worker like (wingless) males (Hölldobler and Wilson 1990).

The number and identity of mothers and fathers together define the degree of intra-colonial relatedness. Relatedness can vary from an average of 0.75 in monogyny–monandry (this high value resulting from male-haploidy; Crozier and Pamilo 1996), to little above zero in extreme polygyny–polyandry.

10.2.2 Number and functionality of non-reproductives

Workers make up most of the colony members (a notable exception being various social parasites in which they are rare or absent, Hölldobler and Wilson 1990), and are specialized for non-reproductive tasks. However, in many species, workers produce some or all of the males, and in quite a few scattered across several subfamilies, they possess a spermatheca and some mate and become reproductives (gamergates; see Chapter 9). Colony size in terms of number of workers ranges across six orders of magnitude, from about 10 in *Amblyopone pallipes* to over 20 million in *Dorylus wilverthi* (Hölldobler and Wilson 1990). Species with larger colony sizes tend to have more task specialists and those with smaller colonies tend to have workers which perform many tasks, that is, they are more generalist.

The tendency of workers to specialize in different tasks, namely to perform different roles, is an important and probably a universal aspect of ant social organization, and believed to contribute to their ecological success (Bourke and Franks 1995; Hölldobler and Wilson 1990; Oster and Wilson 1978). The range of tasks is diverse and varies a little between species, but includes brood care, nest construction, nest hygiene, foraging, and defence (Hölldobler and Wilson 1990). The occurrence of task specialization is termed polyethism, and in most genera is not associated with morphological differentiation (Oster and Wilson 1978). A common pattern is age polyethism, in which the tasks undertaken by a worker vary with her age, but there are also many instances where a minority became permanently assigned to a task, such as repletes (workers with enormously extended crops and functioning as stores of liquid foods, as in *Myrmecocystus mimicus*, Hölldobler and Wilson 1990; Plate 9). Brood care falls mostly to young workers; slightly older workers tend to other nest tasks, and the oldest forage and are first to engage in defence (Hölldobler and Wilson 1990; Sendova-Franks and Franks 1995; Tschinkel 2006).

In a minority of genera, task allocation shows an additional dimension, namely the occurrence of marked differences in worker morphology resulting from allometric changes in morphology. Workers of different morphology, falling into different subcastes, tend to have different arrays of tasks. Spectacular examples include phragmosis, the blocking of nest entrances with specially modified heads, as seen in various ants and characteristic of the subgenus *Camponotus* (*Colobopsis*) (see Wilson 1971), and the multiple subcastes of army ants (allowing the formation of teams for prey transport; Anderson and Franks 2001) and leaf-cutting ants such as *Atta laevigata*, with up to 50 different morphological task-specialists (Oster and Wilson 1978). The tendency for species with subcastes to be more often those with larger colonies leads to the plausible prediction that large colony size is a necessary condition for subcastes to occur, because in large colonies, it is easier to guarantee that all specialists are present, but this view does not withstand comparative analysis. Fjerdingstad and Crozier (2006) found that colony size effects on the likelihood of subcaste polymorphism are minimal, but that higher levels of genetic variation within colonies (as from polyandry and polygyny) and early divergence of queen from worker development favour such polymorphism. These findings, especially that of the effect of genetic variation, accord well with the changing picture of the importance of genetic differences in leading to subcaste and even caste differences (Section 10.3.1).

10.2.3 Spatiality

Nest number, size, and architecture define the spatial structure of a colony, which can vary within species and varies considerably across species (Hölldobler and Wilson 1990). Substrates used include soil, rock crevices, wood (both dead and living), and softer plant materials (including annual stems and seeds). Even when the same substrate is used, nest form may vary widely. Thus, there are soil nests with and without mounds, as seen in *Lasius flavus* (Seifert 2007) and *Cataglyphis bicolor* (Hölldobler and Wilson 1990), respectively. Nest size varies markedly; for ants inhabiting plant structures, the range extends from small acorn nests in some *Temnothorax* species to nests filling entire tree trunks, as in *Lasius fuliginosus* (Seifert 2007). Nest complexity varies, with entrance and chamber numbers ranging from one to hundreds. Possibly the most impressively complex nests are those of *Atta*, with potentially thousands of chambers, filling dozens of cubic metres (Hölldobler and Wilson 1990). Finally, chamber function also varies, from multipurpose to specialized, the latter including brood chambers, granaries, fungus gardens, and galleries for sap-sucking insects.

Among the more unusual nests, there is that of *Blepharidatta conops* where the queen uses her shield-shaped head to become a living gate to the brood chamber against the rest of the nest (intranidal phragmosis), which is interpreted as a measure against predators of the ants' larvae (Brandão *et al.* 2001). There also are nests inside clusters of living epiphytic plants that profit from the association (ant gardens; Hölldobler and Wilson 1990). These nests frequently are inhabited cooperatively by two ant species, e.g. *Crematogaster levior* and *Camponotus femoratus* (Vantaux *et al.* 2007) that, however, keep their brood separate (parabiosis, Hölldobler and Wilson 1990). Another example is the use of silk by a range of species that fasten together tree leaves using larval silk, the most widely known species probably being *Oecophylla smaragdina* (Hölldobler and Wilson 1990; Figure 10.1). Finally, the nest can completely lack materials as in some army ants that spend days or weeks as a mass of tightly locked individuals, or bivouac (Kronauer 2008).

A colony can live permanently in a single nest (monodomy) or maintain several nests, but live in only one at any given time (serial monodomy) as does *Aphaenogaster araneoides* (McGlynn 2007). Alternatively, a colony can live in more than one nest at a time (polydomy, Debout *et al.* 2007). Polydomy can be permanent, as in *Formica polyctena* (Rosengren and Pamilo 1983), or seasonal, as in *Cataglyphis iberica* (Cerdá *et al.* 1994). Polydomy can have various ontogenies. Monogynous polydomy, for example, can be a necessity due to colony growth, as is the case in the bark and twig dwelling *Dolichoderus quadripunctatus* (Seifert 2007). Polygynous polydomy can result from colony budding with incomplete separation as in *F. polyctena* (Rosengren and Pamilo 1983). Polygyny–polydomy is also known as supercoloniality (see Section 10.5).

a
b
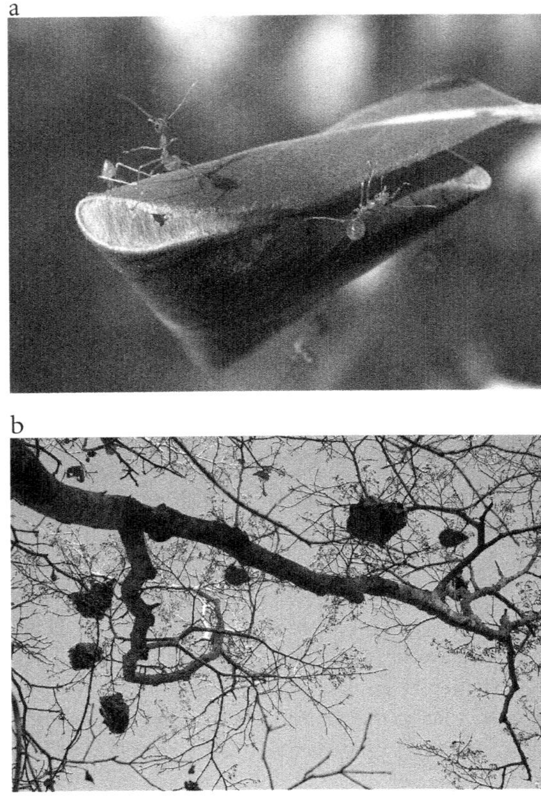

Figure 10.1 Several species use larval silk to construct nests: (*a*) the weaver ant *Oecophylla smaragdina* uses larval silk to fasten together tree leaves for nest construction; (*b*) the nests of a polydomous colony often are distributed over the tree's crown. (Photos: Alex Wild)

10.3 What shapes the character states?

The definitions of character states (Section 10.2) are descriptive; progress in research will not change them fundamentally. In contrast, explanations about the origins and interrelationships between characters and character states could well change markedly, as the system is complex, and includes interactions with the world outside the colony. Technical advances, for example in high-throughput genotyping, have discovered new interrelationships (see Section 10.3.1 for examples), and such progress is likely to continue, also through genome sequencing and gene expression studies (Goodisman *et al.* 2008). Furthermore, hypotheses are often regarded as strong when only a few supporting cases are known, a situation open to change with only a little future work (see Crozier and Pamilo 1996 on the evolution of polyandry). An example of radical change in such concepts is given by caste determination – until recently environmental factors were thought supreme, but now many cases of strong genetic influence are known (Section 10.3.1).

The question as to how different character states arise can be approached at different levels, referable to the proximate–ultimate dichotomy. Thus, the colonies of an ant species may be characterized by being very large: at the proximate level this is explicable by the queens having high fecundity, but at the ultimate level we have to ask about the long-term selection for such a large size due to habitat variables and the mix of competing species. In Section 10.3, we survey the diversity of factors affecting character states: genetics and gene flow, morphology, signal chemistry, nutrition, habitat, pathogen and parasite load, cooperation and conflict in the colony, colony age, and chance.

10.3.1 Genetics and gene flow

Colony structure is central to ant life history, yet sometimes there are differences in life histories within species. Where such differences occur, they may indicate unrecognized cryptic species, genetic variation, or a sensitive developmental switch responding to environmental differences. We restrict our discussion to genetic variation mediating life history traits, and also the nature of selection affecting it. Social insects present an unusually rich array of levels of selection, namely individual, patriline, matriline, nest, colony, and population (Crozier and Pamilo 1996); selection is expected to interact and even be antagonistic between levels. Selection also influences the flow of genetic information at a particular level. For example, where there are alternate reproductive strategies, as in the case of queen morphs differing in size and colony-founding methods, selection may restrict gene flow between these diverging gene pools (Steiner *et al.* 2006b), possibly leading eventually to speciation.

To what extent can we implicate genetic variation as contributing to variation in any one life-history trait? There is a range of possibilities, ranging from selective maintenance of genetically encoded life-history strategies (Maynard Smith 1998) to strong purifying selection for an optimum (implying that any observed variation in a trait springs solely from environmental variation). Between these extremes is the possibility that the observed variation is strongly mediated genetically, but is selectively neutral. Furthermore, this equilibrium view may not hold, in that directional selection may be taking place to modify important aspects of an ant's biology, as shown in the reduction in size of workers of invasive populations (McGlynn 1999a). Both selective neutrality and a lack of genetic influence are implausible in some cases examined, e.g. attine ants in genetically mixed groups better resist infection (Hughes and Boomsma 2004; 2006) and that honeybee colonies with a mixture of genetically mediated task specialists increase their colony fitness as compared to colonies of one pure type or another (Oldroyd and Fewell 2007). For ants, empirical evidence of benefits of genetically mediated polyethism is scarce, although there is highly suggestive evidence from harvester ants relating genetic variation and fitness (Wiernasz *et al.* 2004). On theoretical grounds, though, Page and Mitchell (1991) found that genetic mediation of task specialization is likely to be a universal emergent property of social insect genetics. This finding could still be compatible with such variation being non-adaptive and opposed by selection, except that further modelling and experiments show that this variation is likely to increase colony efficiency (Oldroyd and Fewell 2007 and references therein).

Genetic influences on polyethism are most easily studied when there are major lineage differences within colonies, such as when there are many queens present, or when there is multiple mating (fathers, being haploid, pass on all their genes unrecombined to daughters), but cannot be excluded when such genetic windows into behaviour are not available. If minor predispositions to task specialization, as seen in genetic polyethism, are adaptive, then so will genetic mediation of subcaste development be, when subcastes are themselves adaptive.

Unfortunately, ants present special difficulties to the empirical study of life history evolution; life histories themselves are usually poorly known (Seifert 2007; Tschinkel 1991) and their extraordinarily long colony lifespans (Keller and Genoud 1997) render the experimental study of life history variants practically impossible. Nevertheless, an impressive array of cases is now known in which genetic factors affect polyethism (Beshers and Fewell 2001), and genetic influence on caste determination has been found in 16 genera (Anderson *et al.* 2008). Effects range from strong (queens can arise only from certain genotypes) to weak determination (matrilines vary in their tendency to produce workers of different sizes).

A strong genetic effect is illustrated by *Harpagoxenus sublaevis*. Winter and Buschinger (1986) used this species and its two queen forms (gynomorphs and intermorphs) for their pioneering proof of genetic queen-morph mediation. From a long series of cross-breeding experiments, they concluded that a hypothetical, dominant allele *E* prevents the formation of gynomorphs, which can develop from *ee* larvae only, whereas intermorphs may be *EE* or *Ee*. All three genotypes can yield workers. Crossing gynomorphs and intermorphs with *e* males (sons of gynomorphs) and *E* males (sons of *EE* or *Ee* intermorphs) frequently produced the expected results: a gynomorph (*ee*) mated with a son of a gynomorph (*e*) had gynomorphic daughters, heterozygous (*Ee*) intermorphs mated with an *e* male had both gynomorphic and intermorphic offspring. Practically identical results were obtained with *Leptothorax* sp. A and with *Myrmecina graminicola* (Buschinger 2005; Heinze and Buschinger 1989).

Social form is strongly genetically determined in the red imported fire ant, *Solenopsis invicta*. The Gp-9 locus has strong effects on both queen development and behaviour and on worker behaviour. Gp-9^{BB} queens develop as heavier individuals suited to independent colony foundation, Gp-9^{bb} ones die during development, and Gp-9^{Bb} queens are small and committed to dependent colony foundation. Colony behaviour towards the queens depends on the proportion of Gp-9^{Bb} as against Gp-9^{BB} workers: below a threshold Gp-9^{Bb} proportion, workers enforce monogamy, favouring Gp-9^{BB} queens, whereas above that proportion workers allow polygyny, but kill Gp-9^{BB} queens, either young or mated (Gotzek and Ross 2007).

An important twist of male haploidy, the genetic system of ants, is that if a queen mates with a male of another species, her daughters are hybrids, but her sons are not (because they stem from her unfertilized eggs). Hence, it would be possible for a species to include some colonies which benefit from the hybrid vigour of hybrid workers to produce an enhanced number of males. This model has been suggested to explain hybridization between *Lasius* (*Acanthomyops*) species by Umphrey (2006), who terms this phenomenon sperm parasitism. When queens are polyandrous, mating with both conspecific and heterospecific males, they produce a mix of hybrid and pure same-species offspring. This is believed to be the basis of bizarre zones, where *Pogonomyrmex* harvester ant species exist as colonies with pure-species queens and a predominance of hybrid workers, with only pure-species males and new queens produced (Anderson *et al.* 2008). While the *Pogonomyrmex* case is reciprocal across species, one fire ant species appears to parasitize another when their ranges overlap: colonies occur with *Solenopsis xyloni* queens but *S. xyloni* × *S. geminata* workers, and produce only *S. xyloni* reproductives (Helms Cahan and Vinson 2003).

Other bizarre (or at least highly unexpected) phenomena have started to abound in the ant genetical landscape, perhaps none more so than the devastatingly invasive *Wasmannia auropunctata*, in which a complex life cycle has been found – queens arise from thelytokously produced eggs, workers from normal sexual reproduction, and males from fertilized eggs from which the maternal chromosomes have been ejected (Anderson *et al.* 2008). Queen lineages thus retain their own integrity while benefiting from hybrid vigour of their worker progeny, and males propagate as clonal lineages.

10.3.2 Morphology

Morphology is both genetically and environmentally mediated, but deserves separate treatment, as it directly influences colony structure. This influence is evident in, for example, the derived queen morphology of *Blepharidatta conops* adapted to close the brood chamber (Section 10.2.3), thus influencing colony spatiality.

Morphological variation within species is important, as in worker subcastes with task specialties (see Section 10.2.2), whether strongly influenced by genetics or not (see Section 10.3.1). Queen morphology also shows adaptive variation: mesosoma size defines physiological capacities (Keller and Passera 1989; Stille 1996), via the amount of flight muscles facilitating long-distance flights, and, after being transformed into storage tissue, enabling production and rearing of brood before the first workers emerge. Finally, even internal anatomy affects colony structure. The number of ovarioles a queen has influences the number of eggs she produces and thus colony size, which then may mediate worker polyethism (Section 10.2.2). Ovariole number per queen also correlates negatively with queen number: for example among closely related *Formica* species, there are more ovarioles per queen in monogynous species (Schmidt 1974). In workers, the presence or absence of ovarioles determines whether or not they are capable of reproducing (Helanterä and Sundström 2007).

10.3.3 Signal chemistry

Much of the information flow in an ant colony is transmitted chemically. The chemical signals that an ant species uses are genetically encoded, but in some cases environmentally modified (Hefetz 2007). Below are three examples showing the relevance to colony structure of communication via chemical signals.

First, many ants exhibit social mixing mechanisms to maintain uniformity of the colony specific bouquet of cuticular hydrocarbons (colony odour), facilitating colony-member recognition (Hefetz 2007). Colony odour uniformity makes colony-member recognition (see Chapter 11) more precise, and thus helps protect the inaccessibility of a colony to non-members (colony closure), for example, against unrelated reproductives seeking adoption. Colony odour uniformity is often indirectly proportional to the genetic variation within the colony and thus influenced by the number and identity of reproductives; thus, colony odour can generally be expected to be most uniform in monogynous–monandrous colonies (Bourke and Franks 1995), and indeed queen adoption has been experimentally shown to be less likely under monogyny than polygyny (Sundström 1997). This first example also illustrates how factors in the colony can be reciprocally interrelated, in that the colony odour is partly influenced by colony structure, but in turn determines how colony structure develops.

Second, queen chemical fertility signals, including those on queen eggs (Endler *et al.* 2004), are detected by workers who then minimize their own reproduction (Hefetz 2007). The death of a queen of a monogynous colony often stimulates worker reproduction. Chemical signals of the queen also inhibit the rearing of female sexual brood. In monogynous species with polydomy, polydomy allows an easy response to differences in queen signal compared to monogynous colonies with single nests, where a more complex response to the queen's presence is necessary. Thus, in *Myrmica punctiventris*, more resources are allocated to producing reproductive females in queenless than in queenright nests (Debout *et al.* 2007). Torossian (1967) reports a more extreme strategy for *Dolichoderus quadripunctatus* in that the queen produces all the female offspring, and the workers in peripheral queenless nests produce all the males. Queen signals are also important in another aspect which influences colony structure, namely in that they can govern the adoption of reproductives by the colony by non-reproductives (Hartmann *et al.* 2005).

Third, age polyethism, as a functional property of colony structure, can be influenced by signal chemistry, as shown in *Pogonomyrmex* by Greene and Gordon (2003). These authors found that cuticular hydrocarbons signal a worker's task to other workers, indicating which tasks have been and which have not been performed.

10.3.4 Nutrition

The processes around gathering and processing food as well as the food itself shape colony structure. In this section, we discuss effects that are governed by the ants, whereas food-related effects governed by the habitat a colony inhabits are treated in Section 10.3.5. Worker task allocation is, among others, a function of the specific diet an ant species has evolved; less readily accessible and more specialized nutrition such as on seeds and fungal tissue entails a higher degree of task specialization (Hölldobler and Wilson 1990; Oster and Wilson 1978). This relation is also reflected in worker polymorphism (Section 10.2.2) (see Plate 10). Spatiality of the colony is influenced by the species specific diet in that, for example, a centralized food source can be included in the nest, for example in species tending plant-sap sucking Hemiptera in chambers of their nest, like *S. invicta* (Tschinkel 2006) and *Lasius austriacus* (Steiner *et al.* 2007). In contrast, in nutrition-mediated plant–ant mutualism, the nest can be integrated into the living plant (Hölldobler and Wilson 1990). Last but not least, the complex developmental processes governing caste determination are influenced by many interacting factors including, besides genetic factors, food differentially fed to different female brood (Anderson *et al.* 2008).

10.3.5 Habitat

Habitat affects colony structure in various ways, for example through availability and quality of nutrients and nest sites as well as through temperature and humidity. Nutrient availability is important, for example, by limiting queen reproduction (Wheeler 1996) and thus affecting colony size. Experimentally, nutrient availability was shown to influence sex allocation, with more of the costly female sexuals being produced when there is more food (Deslippe and Savolainen 1995). Severe food shortage can lead to shifts in worker–queen ratios due to uneven mortality rates, with workers dying prior to queens (Rueppell and Kirkman 2005). Experimentally increasing the food supply, on the other hand, induced a shift from monogyny to polygyny in *Myrmica punctiventris*, as a consequence of the complete breakdown of colony integrity (Herbers 1993), maybe indicating that the abundant food rendered territoriality cost-ineffective. Reflecting 'you are what you eat', the specific quality of food available to the colony can influence the colony odour, and thus colony-member recognition. Such feeding-dependent change of colony odour occurs in *Linepithema humile* under laboratory conditions (Buczkowski *et al.* 2005), but influences on colony structure in wild populations await clarification. Spatiality is affected by nutrient availability and, for example, when food is patchily distributed in a specific habitat, this can increase the degree of polydomy in *Iridomyrmex purpureus* (van Wilgenburg *et al.* 2007). Finally, how food is distributed in a habitat influences the operations of workers, in extreme cases necessitating mixed strategies for finding food, as for swimming *Polyrhachis sokolova* workers (Box 10.1).

Different habitats offer different numbers of nest sites, in different spatial distribution, for different species of ants. Nest site availability influences number and identity of reproductives. When nest sites are limited, the degree of polygyny in facultatively polygynous species can increase through a rise in queen adoptions by established colonies (Hannonen *et al.* 2004), or through an increase in pleometrosis (Feldhaar *et al.* 2005). On the other hand, the identity of the colony queen, relevant to the relatedness structure within the colony, can be affected when nest site limitation enforces fusion of unrelated colonies, followed by termination of one of the queens (Foitzik and Heinze 1998). As a factor connected to nest site availability, nest material is also influential. Similar nest material can diminish the differences in colony odour of different colonies, thus reducing colony closure (Heinze *et al.* 1996), although the strength of the impact remains unclear.

Habitat temperature and humidity affect the depth of soil nests, ranging from a few centimetres in rainforests to over 4 m in deserts (Mikheyev and Tschinkel 2004 and references therein). Habitat temperature can trigger seasonal architectural adaptations aboveground to increase nest temperature from insolation, as in *Myrmica* species, and

Box 10.1 Ants in the intertidal zone: colony and behavioural adaptations for survival
Simon Robson

The nests of *Polyrhachis sokolova* are restricted to the mangrove mud of northern Australia and nearby tropical countries, where colonies survive despite seemingly inhospitable conditions (Kohout 1988; Nielsen 1997). The intertidal zone of mangroves, situated at the interface between land and ocean, constantly changes with the tidal cycle. Nests are frequently covered by incoming tides, can remain submerged for up to 3.5 h during a normal tidal cycle (Nielsen 1997) and during extreme spring tides can be covered by as much as two metres of water (Shuetrim 2001). Foraging areas can therefore rapidly change from sand and mud to pools of hot salty water, and the nests themselves face a constant risk of invasion via the burrowing activity of numerous mangrove organisms such as crabs and mudskippers.

Ongoing studies are revealing how *P. sokolova* manage to survive these conditions, but a great deal remains unknown. The nests appear to be of a typical structure for subterranean ants, based on a series of chambers and interconnecting tunnels. There are no obvious structural modifications that might reflect their intertidal location. The nest entrance collapses when the tide comes in and possibly forms a fairly water tight plug against further flooding, though the lower nest chambers become inundated with the rising water table. The colony itself is thought to survive submersion by individuals collecting in those chambers that continue to hold pockets of air (Nielsen 1997). Carbon dioxide levels become elevated during tidal submersion and reach some of the highest levels known for ant colonies. The processes enabling ants to survive these high CO_2 concentrations (up to 11%) must be impressive, but the actual physiological mechanisms involved remain unknown (Nielsen *et al.* 2003). Colonies can be polydomous and may consist of up to four nests, but there appears to be no relationship between the degree of polydomy and ecological factors such as the frequency with which individual nests are inundated (Shuetrim 2001).

Perhaps the most obvious and novel response to these inhospitable conditions is the ability of ants to swim. Individual foragers encountering a body of water simply stretch themselves out on the surface of the water and swim across (Kohout 1988; Nielsen 1997). Recent high-speed digital imaging and ultrastructural studies reveal that swimming is achieved through behavioural rather than morphological modifications. Individuals do not possess the elongated hairs and appendages or flattened smooth bodies often found in swimming arthropods (*P. sokolova* looks like a typical ant!), but they do display a modified gait when swimming: they do not simply continue 'walking' when they reach the water. Power when swimming is provided by the first two pairs of legs, which break the water surface and move in a rhythm similar to that used when walking. Movement of the hind legs, however, is suppressed. These legs are extended straight out across the surface of the water where they appear to maintain surface tension, prevent the ant from sinking, and may even act as a rudder (Fig. 10.1.1). Individuals do not swim in random directions, but are capable of changing course while swimming if need be, and can successfully navigate back to their colonies even when forced to both walk and swim (Raj 2007).

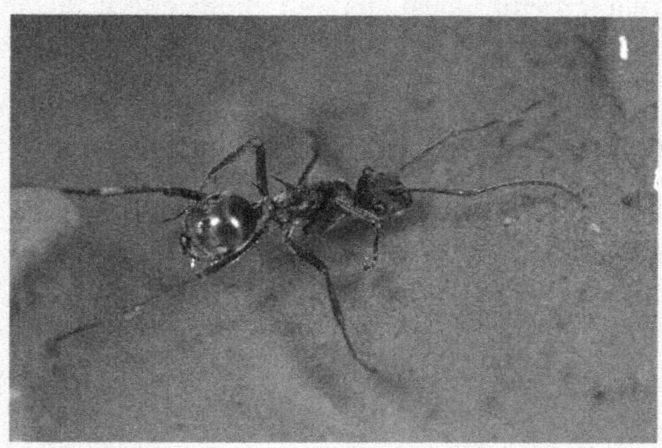

Figure 10.1.1. A *Polyrhachis sokolova* worker swimming across the surface of the water. (Photo: Ajay Narenda)

Active swimming is an extremely unusual behaviour for ants. Their relatively small size and the physical dynamics of the air—water

continues

Box 10.1 continued

interface mean that individuals falling into water typically stick to the surface or eventually sink, but a few species have derived novel solutions. Colonies of some species that become inundated with flooding (such as *Wasmannia auropunctata* and *Solenopsis invicta*) are able to form large rafts of interlocked individuals that float away until dry land is reached, and an unusual species of *Camponotus* is able to not only enter and walk through the liquid within a pitcher plant to forage for dead insects, but is able to extricate itself and its prey as well (Clarke and Kitching 1995). But as far as we are aware, *P. sokolova* is the only ant species that has successfully bridged the land—water interface.

permanent adaptations, for example in *Formica* wood ants (Seifert 2007). Polydomy can be favoured as a feature increasing the number of solaria, as in *Myrmica sulcinodis* (Pedersen and Boomsma 1999). In cold winters, thermoregulation and desiccation prevention are critical to reduce losses of colony members (Heinze and Hölldobler 1994). Probably as a strategy to avoid losses due to desiccation, and triggered by a decline in temperature in autumn, several monogynous colonies can aggregate into a larger polygynous one for winter, with subsequent fragmentation in spring, as in *Leptothorax acervorum* (Seifert 2007 and references therein). Habitat temperature probably also influences queen number (Elmes and Petal 1990), the investment in reproductives of either sex (sex allocation) and caste determination (through effects on larval growth rates), though these effects await quantification (Liautard *et al.* 2003).

10.3.6 Pathogen and parasite load

As ubiquitous and eternal plagues of ant colonies, pathogens and parasites (henceforth just 'pathogens') influence colony structure under ecological time scales and likely influenced its evolution. Interest in ecological immunology of social insects has increased in recent years (Cremer *et al.* 2007).

For species that are at a high risk of infection, colony spatiality is expected to evolve to yield polydomy and gallery systems, both of which adaptations decrease pathogen transmission (Cremer *et al.* 2007; Schmid-Hempel 1998). Leaf-cutting ants provide two examples. Founding queens use platforms to cultivate their incipient fungus gardens, reducing infection of the fungus by soil-borne diseases (Fernández-Marín *et al.* 2007), and parasitic fungi are deposited far from the fungus garden in waste dumps (Hart and Ratnieks 2001). Pathogen load also influences worker task allocation. Once a leaf-cutting worker begins work at the waste dump, she remains there and is thus confined to the task of garbage worker (Hart and Ratnieks 2001). The previous example might suggest that pathogen load also accelerates the evolution of worker polymorphism. However, Schmid-Hempel (1998) tentatively suggests the reverse – namely, that polymorphism imposes a cost that opens species to increased numbers of parasite species, perhaps by providing an increased number of niches within the colony for the parasites.

Pathogens also are of interest when we consider factors that shape mating frequency. Multiple mating provides more opportunities for a founding queen to contract infection (Cremer *et al.* 2007). On the other hand, herd immunity, the higher resistance of genetically variable groups compared to less variable ones (Serfling 1952), should be increased by queen multiple mating, so that selection by disease is also plausible as a selective force in favour of polyandry. These competing hypotheses await further testing.

10.3.7 Cooperation and conflict in the colony

Cooperation within the colony is the undisputed foundation of the ecological success of eusocial insects (Bourke and Franks 1995; Crozier and Pamilo 1996; Hölldobler and Wilson 1990). However, there is another reason why this topic deserves special attention here. Ever since Hamilton (1963; 1964) emphasized selection involving interactions among relatives (kin selection) in the evolution of social behaviour, kin selection has been considered the key factor leading to the evolution of sterility in eusocial species, under appropriate ecological conditions. In fact, kin selection has been seen as the key not only to the evolution of eusociality, but to multicellularity as well (Queller 2000). Recently, this view has been challenged, with the claim that group selection should have been more important in eusociality evolution (e.g. Wilson and Wilson 2007). However, the attacks against kin selection have been refuted by numerous commentators (e.g. Crozier 2008; Ratnieks *et al.* 2006). Also, kin selection predictions indeed are consistent with observations of sex allocation and caste ratio across many species (Bourke 2005; Heinze 2008).

Two kinds of conflict have received much study: queen–worker and worker–worker conflict. Both result from selection by the workers to maximize the number of gene copies similar to their own genes transmitted to the next generation (inclusive fitness), and hence are ultimate factors shaping colony structure. The queen–worker conflict is one over sex-allocation, i.e. over the investment in reproductives of either sex reared from queen eggs by workers: male haploidy produces relatedness asymmetries leading to the expectation that workers favour female production more than queens do, because sisters are more related to them than are brothers (see Crozier and Pamilo 1996). In a monogynous–monandrous colony with worker control, the proportion of investment in female production is expected to be three times that in male production, because it is at this ratio that the values of sisters and brothers become equal (Crozier and Pamilo 1996). Note that the investment ratio does not necessarily translate into a numerically identical sex ratio, because of the production of female reproductives being costlier. A 3:1 ratio of investment is compatible with a 1:1 sex ratio if three times as much investment is made per queen as per male. In many, but not all, monogynous–monandrous species, investment ratios consistent with 3:1 have been found (Bourke 2005), with exceptions suggested to result from proximate and perhaps evolutionarily transient environmental influences. Tying also into this conflict, males – having no sons – have been viewed as the sex-allocation allies of the workers in that both profit from more allocation to female production (Boomsma *et al.* 2005a; Tsuji 1996).

Worker reproduction is widespread, although not universal among ants (Crozier and Pamilo 1996), and leads to the potential for worker–worker conflict when selection for personal reproduction in one worker entails a fitness cost to others. The prevention of a worker's reproduction by other workers is termed worker policing, and is predicted to occur more readily in colonies with polyandrous than in ones with monandrous queens (because a worker's son is related to her by 1.00, that of a full sister by 0.75, that of the queen by 0.5, and that of a half-sister by 0.25). Especially when the queen mates more than twice, reproduction by a worker is not in the best interests of other workers, and they are expected to suppress her efforts. Indeed, worker policing trends occur (Wenseleers and Ratnieks 2006, but see Hammond and Keller 2004, who advocate more proximate causes). Worker-laid eggs for reproduction have to be viewed separately from those serving as food (especially for the queen, as in *Myrmecia forceps*, where queens feed only on worker-laid eggs; Freeland 1958), the latter being a result of cooperation and not conflict. In all, the findings and arguments around the queen–worker and worker–worker conflicts highlight the intertwining of conflict and cooperation in the operation of ant societies, and that both proximate and ultimate explanations must be considered in future research with the paramount need to place the former in the context of the latter, though the latter are often harder to pinpoint (see also Section 10.5 on the factors triggering supercolony formation).

In contrast to the above relationships, pleometrosis (Section 10.2.1) is queen–queen cooperation. In some species, conflict among queens arises when worker production begins. Queens then either fight until just one remains or all but one are expelled by the workers, resulting in monogyny; or, queens retreat to different territories within the nest, resulting in oligogyny (Hölldobler and Wilson 1990; Section 10.2.1). Polydomy can also be viewed as cooperation between separated parts of a colony, whenever there is at least one queen present in every nest. There is potential for conflict among the single nests, and some features of polydomy, including exchange of individuals and signal chemicals, can be viewed as selected to avoid conflict across nests (Debout *et al.* 2007).

A completely different and gene-centred view of cooperation and conflict arises when intragenomic conflict is considered, as between paternally and maternally derived alleles (Queller and Strassmann 2002). Epigenetic processes, involving the labelling of alleles as paternal or maternal (genetic imprinting), might inform individuals about relatedness patterns in the colony. Understanding such intra-individual conflict may elucidate hybridization-mediated caste determination (Section 10.3.1), as in explaining the unidirectionality of some hybridization, and hybrid systems may yield better understanding of intra-self conflict than single-species studies. The suggestion of epigenetic processes may resolve the paradox that selection imposed by kin recognition tends to destroy the necessary variation at cue loci (Crozier 1986), leading to suggestions of kin recognition being an unselected by-product of loci maintained variable by other forms of selection or by high mutation rates (Crozier 1989; Rousset and Roze 2007).

10.3.8 Colony age

Colony longevity differs greatly across colony types and species. In monogyny, colony age follows queen longevity, ranging from under 1 year to 3 decades (Hölldobler and Wilson 1990; Seifert 2007). In stark contrast, polygyny and serial polygyny make a colony potentially immortal. When studying colony age, awareness of its correlation with colony size is needed, the latter having been controlled for in some (e.g. Wagner and Gordon 1999), but by far not in all colony age studies.

Colony age influences task allocation. Very young colony age enforces broad worker flexibility and the level of task specialization increases with age until a species-specific maximum is reached. Older colonies are more persistent in task allocation when disturbed (Gordon 1987). In *Solenopsis invicta*, the proportion of larger workers increases continuously with increasing colony age, facilitating an increase of labour efficiency (Tschinkel 1988a). In general, the spatial extent of nest and colony increase with age (Gorosito *et al.* 2006). In *Pogonomyrmex badius*, the depth of the soil nest increases with colony age as a direct consequence of an increasing availability of older workers that do the digging (Tschinkel 2004).

10.3.9 Chance

Chance influences colony structure, both from inside and outside the colony. Sporadic ecological catastrophes such as floods lead to a range of outcomes according to species. In some species, colonies are destroyed by flooding, while in others the workers form floating rafts of ants in which are carried a queen and brood (Dietrich *et al.* 1998; Hölldobler and Wilson 1990). After the water has receded, the raft lands and the colony fragment can start a new, primarily monogynous colony (Dietrich *et al.* 1998; Lude *et al.* 1999). Although we are aware of some of the mechanisms by which ants survive unpredictable events, systematic studies are needed on the influence of stochasticity on the evolution of ant life history. The ecological-evolutionary framework of risk-spreading, the idea that unpredictably variable environments favour genotypes with lower variance in fitness at the cost of lower arithmetic mean fitness (Hopper 1999), could be applied to assess in how far selection for risk-spreading might influence intra- and interspecific variation in ant colony structure (see Section 10.2).

10.4 The ecological consequences . . .

Ant colony structure influences the ecosystem at many levels, although many effects still need detailed exploration. We present the ecological consequences of colony structure to the colony itself, to the population, to the ant community, to other organisms, and to the abiotic environment.

10.4.1 . . . to the colony itself

Many, probably all, features of colony structure have been shaped through selection by the ecological milieu experienced by the species, and in turn explain the place a species currently occupies in its community. Two examples suffice to show how such features can reflect adaptation under selection at the colony level to stable aspects of the environment, as well as to highly stochastic factors.

Selection by pathogens is an example of a highly predictable factor influencing colony structure. The behavioural exclusion of garbage workers from the rest of the nest reduces the backflow of pathogens (Hart and Ratnieks 2001). Clearly, selection for the appropriate genotypes does not follow from enhanced direct reproductive success, but through kin selection (Crozier and Pamilo 1996), with selection at the colony level favouring high genetic variation to guarantee the presence of such specialist genotypes. In contrast, accidental transport of colony fragments by people is an example of an unpredictable factor structuring colonies. Queen number influences survival rate, with increased probability that a queen is included in the colony fragment in polygynous colonies (Holway *et al.* 2002a). Human transport was not part of the selective regime under which queen number evolved, but such species may be adapted to colony fragmentation (Keller and Genoud 1997).

10.4.2 ... to the population

The population represents another level in multilevel selection (Crozier and Pamilo 1996) and ecological consequences of colony structure not discernible at lower levels are expected at population level, though little investigated. For example, the number and identity of reproductives define, through intra-colonial relatedness, the level of aggression and competition between colonies of the same species (Crozier and Pamilo 1996). This in turn affects the portion of resources allocated to activities increasing colony fitness. Along these lines, monogyny–monandry could be argued to reduce colony fitness, through the high costs of high-level aggression between colonies. We could, however, consider this life history as a K strategy (MacArthur and Wilson 1967), in that stronger territoriality might lead to lower density and reduced impact on the habitat. From this argument, a positive effect at population level in stable habitats might result, namely a reduced likelihood of exceeding the carrying capacity. This speculation is justified through the documentation of the opposite case, that of little competition within the population. Such reduced competition occurs in polygyny–polydomy (also see Section 10.5), and facilitates allocation of more resources to reproduction and brood rearing (Holway *et al.* 2002a) which results in increased biomass per surface area (Tschinkel 2006). This resembles an r strategy (MacArthur and Wilson 1967), and indeed a potential negative effect on population fitness can be expected: extreme levels of population density can result in severe damage of habitat structures crucial to the population, such as the death of trees that host honeydew-producing insects, as documented for *Lasius neglectus* (Espadaler and Rey 2001). However, in the evolutionary short term, supercolony populations of a species may outcompete populations of normal colonies (Holway *et al.* 2002a).

Relatedness across colonies is often influenced by the number and nature of reproductives and their dispersal strategy (Ross 2001). In polygynous ants, there is a tendency of related colonies to occur together within continuous populations (population viscosity, see Crozier and Pamilo 1996). Viscosity in turn should favour inbreeding, and there is evidence that polygyny yields inbreeding (Sundström *et al.* 2005). Is this phenomenon exerting a strong effect? Consider the following conservation issue in the age of anthropogenic habitat reduction. Increased inbreeding depression can lead to population decline. This risk is increased in ants and other Hymenoptera (Zayed and Packer 2005) because reduced variability at the sex locus increases the frequency of fertilized eggs, which yield not females but inviable or infertile diploid males. Some diploid males occur in all populations, but as inbreeding raises their number, population size decreases, more loss of diversity results, and a vicious cycle, termed diploid male vortex, potentially leads to the population's destruction (Seppä 2008; Zayed and Packer 2005). Conversely, low genetic variation due to founder events might lead to rapid evolution of traits that facilitate the success of invasive ants, especially in disturbed habitats (Seppä 2008). Future work is needed to elucidate the genetics of different responses of ants, with the same and with different colony structure characteristics, to landscape deterioration (Crist 2008).

Colony structure characters of non-reproductives are also relevant at the population level. For example, the number of workers, a function of colony age, in some species defines the degree of territoriality (Gordon and Kulig 1996), which could be interpreted as a colony-structure-based strategy regulating population growth, possibly as part of a K strategy. We are not aware of an ecological consequence of colony spatiality at the population level, but suspect that such effects await discovery.

10.4.3 ... to the ant community

Colony structure influences ant-community structure, for example through queen number and the number of nests per colony, and through colony size. Ant communities are structured hierarchically, often with three dominance levels: top = territorials, intermediate = encounterers and bottom = submissives (Savolainen and Vepsäläinen 1988; see also Chapter 5). Territorial species have the greatest influence on species of the other levels and even influence community composition (Hölldobler and Wilson 1990). This pattern is a function of the species present in a habitat. Prime factors determining the status of a given colony are behaviour and colony structure. The two are difficult to separate, but in cases of intra-specific variation, careful deductions may be attempted. Polydomy is considered to generally increase the ecological dominance of a colony (Debout *et al.* 2007). Queen number may also be influential, frequently in combination with the number of nests per colony. Several *Formica* species have both monogynous–monandrous populations as well as polygynous–polydomous ones, and the latter are usually higher in the community hierarchy than the former (B. Seifert, personal communication). This effect is illustrated in its extreme by polygynous–polydomous species such as *Linepithema humile*, which can severely disrupt ant communities outside of its native range (Sanders *et al.* 2003a). Extremely large colony size in monodomous *Myrmica* species elevates its position in the hierarchy (Seifert 2007), but it is unclear whether this is an effect of worker or queen number, the two being correlated.

Environmental factors also influence ant-community structure and composition in ecological timescales (LeBrun 2005). An extreme example is communities in periodically inundated habitats; species incapable of coping with the periodic catastrophes of inundations cannot persist, even if they represent the top level of the hierarchy in other habitats. Still, these communities are also shaped by colony structure, though indirectly. First, polygynous, ground-dwelling ants form floating rafts of workers and at least one queen, allowing recolonization after the water recedes (Section 10.3.9). Because of this polygyny-grounded flexibility they are the commonest species in these habitats (Lude *et al.* 1999; Schlick-Steiner *et al.* 2005). Second, the absence of top-level species in a community upgrades the position of the other species (Elmes *et al.* 1998; Vepsäläinen *et al.* 2000), the persistence of which is a function of queen number.

10.4.4 ... to other organisms

Ants have relationships with many other organisms (also see Chapter 6), from bacteria to fungi, and from vascular plants to vertebrates, in manifold ways, from hunters to prey, and from mutualists to parasites. Global numbers are lacking, but it may be illustrative that nests of the temperate *Lasius fuliginosus* sustain more than 100 other invertebrate species (Seifert 2007). Although seldom addressed, colony structure is likely to have an impact on all these relationships. Thus, the number and identity of reproductives define a colony's accessibility to non-ant social parasites, as argued, for example, for *Phengaris* butterflies (Elmes *et al.* 1998) and their *Myrmica* hosts, which are polygynous and thus with low levels of colony closure. Number and functionality of non-reproductives influence, in hunting ants, the activity of other arthropod predators (Laakso and Setälä 2000), as well as site selection by nesting birds (Haemig 1999), and in seed-dispersing ants, spatial properties of plant genetic diversity (Zhou *et al.* 2007). Nest structure shapes the structure and function of the soil-decomposer community living in *Formica* mounds through temperature and moisture regulation (Laakso and Setälä 1998).

10.4.5 ... to the abiotic environment

As major ecosystem engineers, ants influence, among other things, the chemistry, porosity, and water storage capacity of the soil, nutrient cycles,

Plate 1: *Adetomyrma* sp. mad01 has been found from only one location in Madagascar. These small, blind, hypogaeic ants have been termed 'dracula ants' for their habit of feeding on the haemolymph of their own larvae. (Photo: Alex Wild)

Plate 2: Ants often defend territories and food resources aggressively. (a) *Azteca alfari* (*Cecropia* ant) workers reknowned for their aggression immobilize an unfortunate *Odontomachus* sp., and (b) *Oecophylla longinoda* workers pin down a *Polyrhachis* ant that has strayed too far into their territory. (Photos: Alex Wild)

Plate 3: Ants are involved in an array of mutualistic interactions: (a) *Formica integroides* takes a droplet of honeydew excreted by an aphid, (b) a *Podomyrma* ant tends a lycaenid caterpillar, (c) *Pseudomyrmex* ants feed on special protein-rich food bodies provided by *Acacia* trees, (d) An *Ectatomma* ant feeds from the extrafloral nectary on an *Inga* plant. (Photos: Alex Wild)

Plate 4: (a) A healthy *Cephalotes atratus* worker and (b) one infected with the nematode *Myrmeconema neotropicum*. Effects of the parasite on the ant include the red gaster, erect posture, and nearly constant gaster flagging. (Photos: Stephen Yanoviak)

Plate 5: Some ants are specialized seed predators. This worker ant, *Pogonomyrmex desertorum*, is actively harvesting a seed still attached to the plant. (Photo: Alex Wild)

Plate 6: Members of an ant colony can vary reproductively, morphologically, and fuctionally. The image here shows the morphological distinctions among a queen (large winged), male (small winged), and worker of a *Camponotus discolor* colony. (Photo: Alex Wild)

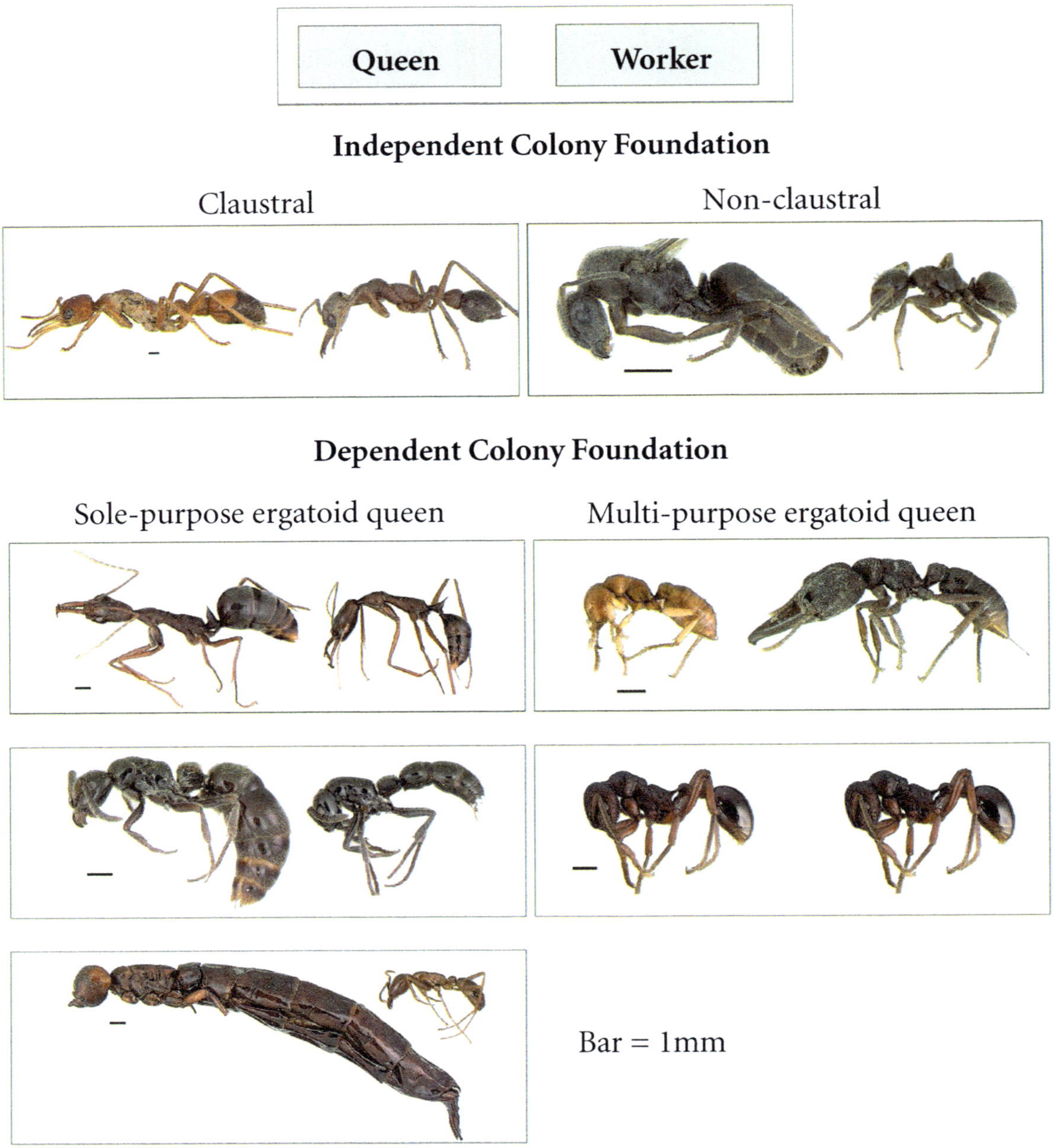

Plate 7: Comparison of queens and workers in relation to mode of colony founding. Species where queens perform non-claustral ICF exhibit low queen/worker size dimorphism (*Myrmecia gulosa*). In contrast, in species with claustral ICF, dimorphism is much larger and wing muscles are enlarged to function as reserves, resulting in a large mesosoma (*Lasius niger*). In species that perform DCF, the mesosoma of ergatoid (= permanently wingless) queens is simplified and closer to that of workers. The size of sole-purpose ergatoid queens varies a lot across species (from top to bottom: *Odontomachus coquereli*, *Cerapachys* sp. 1 from Madagascar, and *Dorylus molestus*). Multi-purpose ergatoid queens are more similar to workers (top: *Mystrium* 'red' from Madagascar, bottom: *Eutetramorium mocquerysi*). (Photos: www.AntWeb.org.)

Plate 8: Ants undergo complete metamorphosis in their life cycle. Life cycle stages of a twig ant, *Pseudomyrmex gracilis*, are shown: (left to right) an egg, three larval instars, pupa, and adult. (Photo: Alex Wild)

Plate 9: Replete workers of the honey pot ant, *Myrmecocystus mexicanus*, hang from the ceiling of a nest chamber. They are attended to regularly by workers and sometimes cover the ceiling of a nest chamber. The repletes' enormously extended crops are filled with liquid food for storage. (Photo: Alex Wild)

Plate 10: Some ant species exhibit polymorphism. (a) These African driver ants, *Dorylus helvolus*, vary in body size. (b) Workers can also differ in body proportions as well as size; the head of the largest worker of these *Camponotus sansabeanus* is much larger in proportion to its body than that of the smallest worker. (Photos: Alex Wild)

Plate 11: Two *Wasmannia auropunctata* foragers meet and assess each other. Information is communicated by touching antennae (antennation). (Photo: Alex Wild)

Plate 12: Worker ants can lead nestmates to food using tandem running where antennal tapping signals between the two ants control the speed and course of the run. Here, two *Pachycondyla* are tightly connected during a tandem run. (Photo: Alex Wild)

Plate 13: Ants possess a large range of morphological defences: (a) some *Meranoplus* species have a well-developed shield structure on their alitrunk while others possess dense hairs, (b) *Polyrhachis* species often have large petiolar and propodeal spines, (c) *Cataulacus brevisetosus* has heavily sculptured armour, and (d) *Acromyrmex versicolor* has prickle-like spines covering its head and alitrunk. (Photos: Alex Wild)

Plate 14: Some of the most notorious invasive ant species: (a) the Argentine ant (*Linepithema humile*), (b) the little fire ant (*Wasmannia auropunctata*), (c) the red imported fire ant (*Solenopsis invicta*), (d) the big-headed ant (*Pheidole megacephala*) (Photos: Alex Wild), and (e) the yellow crazy ant (*Anoplolepis gracilipes*) (Photo: Paul Zborowski)

Plate 15: Argentine ants (*Linepithema humile*) cooperatively attack a much larger Californian harvester ant worker (*Pogonomyrmex subdentatus*). Despite little overlap in resource use, harvester ants disappear from areas invaded by Argentine ants in California most likely as a result of aggressive colony raids. (Photo: Alex Wild)

the microrelief, and the carbon dioxide balance of the ecosystem (Cammeraat and Risch 2008; Frouz and Jilková 2008; Jouquet *et al.* 2006; Ohashi *et al.* 2007; Seifert 2007; see also Chapter 8). With few exceptions (e.g. Dauber *et al.* 2001), colony structure has not been considered in relation to the abiotic environment. This probably highlights a research gap and a lack of concerted efforts across different fields of ant research, rather than a lack of significance. We postulate some relationships, sketching a conservative minimum scenario. The number and identity of reproductives influence ant biomass per surface area, and thus all the effects mentioned earlier, because all relate directly to ant biomass. The functionality of workers matters in that intense foraging over a large area increases any effect. Spatiality plays a role, for example, through nest architecture, with soil nests reaching farther down affecting more soil horizons than others, and by monodomy *versus* polydomy through concentration *versus* dispersion of any effect.

10.5 A paradox of colony structure: supercolony formation

Whereas polygyny–polydomy was recognized more than a century ago (Debout *et al.* 2007), and received little publicity, 'supercoloniality' (see Gris and Cherix 1977 for the first use of 'supercolony'), which is essentially just another term for the polygyny plus polydomy social system when the number of nests becomes large, recently became a hotly debated biological topic. Whatever name one chooses, this social organization is ecologically relevant, in that supercolonial ants are among the ecologically most successful and invasive organisms.

An ant supercolony is a very large entity with very many queens, integrated harmoniously over a large area, from several square metres to many square kilometres, with individuals freely, though not necessarily evenly, mixing across spatially separate parts of the entity (Bourke and Franks 1995; Crozier and Pamilo 1996). The lack of aggression is advantageous; aggression is costly, involving direct and indirect losses and recognition errors (see Chapter 11). The complete disappearance of colony borders, on the other hand, tends to make relatedness among nestmates negligible. A paradox emerges, because reproductive altruism among unrelated individuals is not explicable by evolutionary theory that involves relatedness.

Supercolonial species have evolved many times (Crozier and Pamilo 1996) and finding a general principle to supercolony formation proves difficult (Bourke and Franks 1995). Three hypotheses for supercolony evolution have been proposed (Steiner *et al.* 2007 and references therein). First, reduced within-colony relatedness, as associated with polygyny, could ease the integration of unrelated individuals from other colonies, because relatedness would not be lowered by much. Species with both monogynous and supercolonial populations (e.g. *Solenopsis invicta*) make this pathway appear less universal though. Second, reduced ability to discriminate nestmates from non-nestmates, as in the case of reduced diversity in the relevant genes through a population bottleneck, could foster supercolony formation. However, in at least one species, *Formica paralugubris* (Holzer *et al.* 2006a), discrimination between nestmate and non-nestmate was shown to be upheld in a supercolony. Third, the adaptive value of avoiding the costs of aggression could be so strong, that under appropriate ecological conditions, it could promote the elimination of territorial aggression. This is the only hypothesis that views supercolonies as not a simple by-product of other processes – and otherwise huge supercolonies are unlikely to be evolutionarily stable (Crozier 1979) – but it awaits closer scrutiny.

There is an additional pair of terms to be considered in context with supercolonies. Wilson (1971) coined the terms 'multicoloniality' and 'unicoloniality' to distinguish what seemed then to be two distinct syndromes, with the former referring to populations of colonies each with one or a few queens showing independent colony foundation and outbreeding, and the latter referring to populations lacking colony boundaries and showing dependent colony foundation and inbreeding. The example par excellence of unicoloniality was seen to be *L. humile*, with vast populations showing free movement of workers. Since then, the distinction has become blurred, and indeed there is a continuum of social types. Ironically, the original impression of universal amity of the unicoloniality concept does not hold, with even *L. humile* occurring as mutually hostile 'supercolonies' which can be very large: Giraud *et al.* (2002) found one to extend over 6,000 km of coastline from Italy to the Spanish Atlantic. The other elements of Wilson's unicoloniality syndrome, dependent colony foundation and inbreeding, are most likely characteristic of supercolonial species. It seems likely that other 'unicolonial' species will be found on closer study to be better described as supercolonial. Invasive ant species tend to be supercolonial (Passera 1994).

The concept of supercolonies being large areas within which there is free and even distribution of workers fits one extreme of ant colony structure, although this has received particular attention because of its frequency in invasive species (see Chapter 14). Hierarchical population genetic studies of *Formica* (e.g. Chapuisat *et al.* 1997; Pamilo *et al.* 2005) show that, although the exchange of individuals can be deduced to occur between neighbouring nests, there is considerable genetic differentiation between distant nests of large supercolonies. Similar findings are known from supercolonial species in other genera, such as *Myrmica* (Pedersen and Boomsma 1999) and *Polyrhachis* (van Zweden *et al.* 2007). Hierarchical analyses also show the existence of a further type of organization, intermediate between supercoloniality and normal colony organization, such as when the strength of nestmate recognition varies seasonally (Katzerke *et al.* 2006; Pamilo *et al.* 1985; Peeters 1988). Thus, for *Rhytidoponera* sp.12, the seasonal variation in nestmate recognition (Pamilo *et al.* 1985; Peeters 1988) leads to exchange of individuals (Tay *et al.* 1997) which, in combination with the wingless nature of the reproductive females, may have led to an observed genetic similarity between neighbouring nests (Crozier *et al.* 1984).

There are probably several hundred species supercolonial in their native ranges – 11.6% of central European species are supercolonial (Seifert 2007). In contrast to non-native supercolonies, native supercolonies are not an ecological problem because they have coevolved with the ecosystem. Native and invasive supercolonies have been found to have similar organization (Pedersen *et al.* 2006) though invasive ones tend to be much larger (Suarez *et al.* 2008). However, there is at least one case, *Myrmica rubra*, in which the size of native supercolonies (Seifert 2007) can exceed that of invasive ones (Garnas *et al.* 2007). Recognition that variation in supercolonial organization may not be primarily a function of range makes supercoloniality a globally even hotter topic.

10.6 Future directions

Oster and Wilson (1978) noted that much of the theory concerning colony structure lacked ground truth. Since then, a range of startling structures such as thelytokous parthenogenesis and hybridization-mediated caste determination have come to light (Heinze 2008; Keller 2007), but how general these are remains to be established. Further work will clarify whether we are approaching good coverage of colony structure diversity.

We have identified some specific research needs throughout the chapter (Sections 10.3.1, 10.3.5, 10.3.6, 10.3.7–9, 10.4.2–5, 10.5), many of which reflect one or more of three effects. Firstly, evolutionary concepts have developed more rapidly than life history data could be collected for validation (Bourke and Franks 1995; Hopper 1999; Seifert 2007; Tschinkel 1991), impeding the universality of conclusions. Secondly, one strand of thinking in biology has been to study particular species ever more fully ('the mouse', 'the fruit fly', etc.). Without information on many species and analyses of the joint occurrence of traits of interest through comparative analyses of life-cycle evolution (Felsenstein 1985; Harvey and Pagel 1991), the evolution of the remarkable life histories that we are now finding would remain forever obscure. Thus, Keller (2007) called for a 'molecular naturalist' approach to survey many species to further find such life histories and to enable a search for ultimate causation, remarkably paralleling the remarks of MacIntyre (1985) about molecular evolutionists wandering in an 'enchanted forest'. Thirdly, genetic architecture is only one part of the explanation, which must also involve the ecological context and consideration of the various levels of selection identified earlier; some other evolutionists give insufficient attention to ecological conditions and their variability, while some ecologists neglect ultimate causation and evolutionary constraints. Fusing the two angles more frequently allows for a more comprehensive understanding of the causes and consequences of ant colony structure than ever before. Thus, remarkable species should be studied for more than just the feature that excited initial interest, an approach that will often demand teams of specialists (Vega and Blackwell 2005). Out of such collaborations new (inter-)disciplines may arise, much as molecular ecology arose from the conjunction of ecology, behaviour, and molecular evolution.

New frontiers open before us. Sociogenomics gives us powerful tools until recently unknown, and we can hope to eventually explain colony structure in terms of nucleotide sequences. Smaller steps will be feasible through cross-taxon application of advances in other insects, bigger steps following genome sequencing. Finally, we live in the era of rapid climate change, which will cause changes in distribution and may eliminate species that are unable to adapt or move. This may be the era to study rapid social evolution.

10.7 Summary

Ant colony structure is a colony's caste, demographic, genealogical, and spatial makeup. Characters of colony structure include queen number per colony, mating frequency of queens, worker number per colony, worker task allocation, and number, size, and architecture of nests. To capture the causes shaping character states in their complexity and interdependence and to reconcile ultimate and proximate viewpoints, it is necessary to call on divergent disciplines. In this chapter, nine causes working from inside and outside the colony are identified to shape colony structure: genetics and gene flow, morphology, signal chemistry, nutrition, habitat, pathogen and parasite load, cooperation and conflict in the colony, colony age, and chance. Colony structure in turn has ecological consequences at various levels: to the colony itself, to the population, to the ant community, to other organisms, and to the abiotic environment. Finally, supercolony formation is discussed as being a paradox of colony structure, in that reproductive altruism among unrelated individuals is not explicable by evolutionary theory that involves relatedness. For future research to achieve the most universal understanding of origin, maintenance, and consequences of colony structure, it will be necessary to combine evolutionary and ecological concepts to study a wide array of species by interdisciplinary approaches.

Acknowledgements

We thank Daniel Cherix, Abraham Hefetz, and Bernhard Seifert for information, and Alfred Buschinger, Simon Robson, three anonymous referees, and the editors for comments on the manuscript. BCS and FMS were supported by the Austrian Science Fund (J2639-B17, J2642-B17), RHC's research was supported by the Australian Research Council (DP0665890).

Chapter 11

Nestmate Recognition

Patrizia d'Ettorre and Alain Lenoir

11.1 Introduction

'... it begins to seem that some ability to recognize kin and to react accordingly will be found in any social animal if looked for carefully enough'.

—Hamilton (1987, p. 426)

Recognition of kin or group members is essential to the evolution of social behaviour, whether living in a small family group or in a society of millions of individuals, such as a mature *Atta* colony. Research on kin recognition has been prolific, and a good synthesis was achieved about 20 years ago, with the publication of two edited volumes, one by Fletcher and Michener (1987) – the source of the Hamilton quote above – and the other by Hepper (1991). Moreover, the contribution by Holmes and Sherman (1983), who investigated the *how* and *why* of kin recognition in one of the first models – the ground squirrel – deserves mentioning. The early history of Hamiltonian-based research on kin recognition, especially in vertebrates, has been nicely summarized more recently by Holmes (2004).

In this chapter, we review the recent literature on ant-recognition systems. We are aware that our approach is far from being comprehensive, but our aim here is to concisely highlight what we believe is the essential knowledge gained so far, with the hope of generating further studies aimed at filling some of the research gaps and answering what we think are important, but still unresolved questions.

Since terminology is often an issue that could easily shift the focus from a biological problem to a semantic one, we begin with defining a few key terms, and classifying some of the fundamental features of recognition systems. A minimum of two participants is required to play the recognition game: a *cue-bearer*, which shows the cues correlating with some significant factor, and an *evaluator*, which identifies and then assesses these cues by comparing them with some kind of template (see Liebert and Starks 2004 for a review of the alternative terminology used in recognition research). When this process takes place, we can usually observe an act of discrimination, for example, aggression. However, the absence of detectable discrimination does not necessarily mean that recognition did not occur, since recognition is defined as the internal neural or cognitive process that can also happen without producing any observable discrimination. Unfortunately, our knowledge of the strictly internal neural processes underlying recognition is still in its infancy, and thus the two terms are often used as functional synonyms.

An efficient way of studying recognition systems is to disentangle them by analysing three distinct components: the *expression* (also called production), the *perception*, and the *action* component (Gamboa *et al.* 1991; Sherman *et al.* 1997; Starks 2004). The expression component refers to all the processes involved to produce or acquire recognition cues (labels) by the cue-bearer; the perception concerns the evaluator and it is the process by which the evaluator detects, identifies the labels, and compare these with some kind of template; the action component is the response of the evaluator, usually a discriminating behaviour that we can somehow observe and quantify.

Later in this chapter, we specifically address the expression of recognition cues in ants, whereas here we give an essential, but a more general overview of the possible mechanisms underlying recognition, all of which might be relevant for ants. Several different classifications have been proposed by different authors (reviewed by Liebert and Starks 2004; Mateo 2004) and there has been a long debate on what should be defined as 'true' kin recognition (e.g. Grafen 1990; Sherman *et al.* 1997). We believe that the distinction originally made by Waldman (1987) has indeed some general heuristic value. Recognition is defined as *indirect* when the evaluator relies on some contextual cues, such as spatial location. In some particular circumstances, any individual encountered in a closed nest is reliably a group member, and the evaluator does not need to assess cues that are actually on the putative cue-bearer. In contrast, recognition is *direct* when it is based on phenotypic cues that are actually borne by other individuals (cue-bearers).

11.2 Mechanisms of recognition

The following recognition mechanisms (see also Figure 11.1) have all found theoretical and, most significantly, empirical support in different groups of organisms, from amoebae to insects and vertebrates. This is not to be viewed as a hierarchical classification; there is no 'best' mechanism of recognition, and the proposed ones are not necessarily mutually exclusive. The underlying forces that have favoured the selection of one mechanism in a particular social species can be successfully investigated only by taking into account ecological constraints and life history trade-offs.

(a) Prior association: During its development or early stage in life, the focal individual (evaluator) learns cues from the other individuals that are most frequently encountered. These become 'familiar' individuals, who are thus treated as kin (or fellows; see Section 11.3), while individuals who are not familiar are always

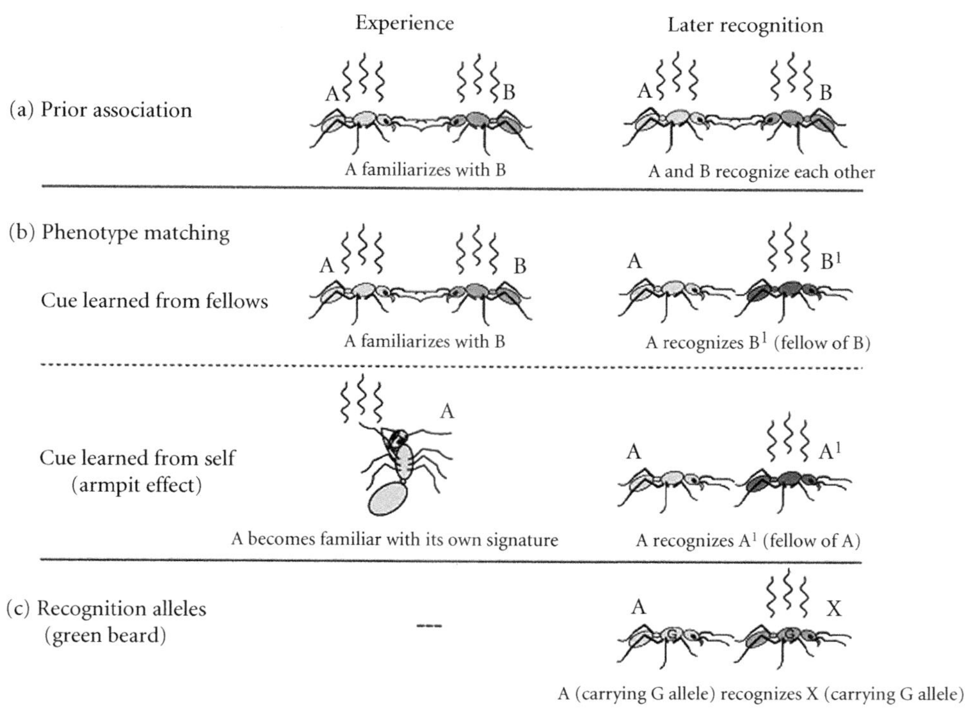

Figure 11.1 There are a range of possible mechanisms of direct recognition (see text Section 11.2). Inspired by Wyatt (2003).

treated as non-kin, independently of their relatedness with the evaluator.

(b) Phenotype matching: The focal individual learns cues to construct an internal, neural template. Once the template is in place, every encountered cue-bearer is compared with the evaluator's template, and recognition is based on the degree of similarity between label and template. The source of cues to be learned in order to form the template could come from other individuals (e.g. nestmates) or from the focal individual itself. The latter case is called self-referent phenotype matching (or armpit effect; cf. Dawkins 1982).

The difference between 'prior association' and 'phenotype matching' is that with the first mechanism only individuals that have been already encountered (familiar) can be recognized as kin (or fellow), whereas the second mechanism allows recognition of never-encountered individuals as kin, if they match the evaluator's *Gestalt* template (see Section 11.3.3).

(c) Recognition alleles (green beard): This is a concept proposed by Hamilton (1964) and then named by Dawkins (1976), the 'green-beard effect'. An allele at a single locus – or closely linked genes – could cause the expression of (*a*) a detectable phenotypic cue (a green beard), (*b*) the ability to recognize this same cue in other individuals independently of relatedness and (*c*) the preferential treatment of individuals expressing the cue. The same gene should encode all three functions (cue, recognition, and altruism), and thus this mechanism is not expected to occur frequently (see Grafen 1998). However, the green-beard effect has been shown in the red imported fire ant, *Solenopsis invicta* (Keller and Ross 1998) and has recently found additional theoretical support (Jansen and van Baalen 2006).

This recognition mechanism does not require any form of learning, contrary to the previous other mechanisms, which are based on cue-learning. However, it is very difficult to experimentally distinguish between self-referent phenotype matching and recognition alleles (cf. Crozier 1987; Mateo 2004).

11.3 Kin and nestmate recognition

When studying ants and social insects in general, a clear distinction should be made between kin and nestmate recognition. These two phenomena are essentially different. Efficient discrimination between colony members and aliens (nestmate recognition) is crucial for the organization of insect societies, since it prevents robbery and parasitism from outside (cf. Hölldobler and Wilson 1990). Nestmate recognition is typically manifested by rejecting alien intruders, thus it occurs between colonies and among unrelated individuals. Kin recognition, in contrast, could take place at a different level, within the colony. Whilst insect societies are usually composed of related individuals, the degrees of relatedness among nestmates within the same colony can vary. If the queen mates with more than one male (polyandry), the colony will contain workers from different patrilines, a mixture of full-sisters and half-sisters; the obvious example being the honeybee (cf. Tarpy *et al.* 2004). Obligate multiple mating is also the rule in some ant species, such as army ants (Kronauer *et al.* 2007a), leaf-cutting ants (Villesen *et al.* 2002), harvester ants *Pogonomyrmex badius* (Rheindt *et al.* 2004) and *Cataglyphis cursor* (Pearcy *et al.* 2004). Another possible complex scenario is the presence of multiple queens in the same colony (polygyny), which gives rise to the coexistence of several matrilines. Multiple queens with multiple matings (e.g. Kellner *et al.* 2007) can yield several patrilines and matrilines in the same colony.

11.3.1 Is kin recognition expected in social insects?

Kin and nestmate recognition coincide in ants only when colonies are headed by a singly mated single queen, and there is no queen turnover. In this case, workers are all full-sisters and there is no need to discriminate among different kinds of kin. By contrast, when different patrilines or matrilines coexist in the same colony, discriminating full-sisters versus half sisters could be advantageous for the single worker, which would benefit from favouring its full-sisters. However, kin recognition leading to nepotistic behaviour is expected to be selected

Box 11.1 Recognition between different species: natural and artificial mixed colonies of ants

Christine Errard

Social parasitism is the coexistence in the same nest of two species of social insects, one of which profits (the parasite), and the other of which generally suffers (the host). Many ant species are known to be engaged in some form of parasitic association with other ants (xenobiosis, temporary parasitism, dulosis or slavery, permanent parasitism, or inquilinism). Social parasitism in ants is a relatively rare form of parasitism, with about 220 cases described (~2% of all described ant species), though new parasitic species continue to be discovered. Social parasitism is not equally spread among the subfamilies; it is absent in the primitive subfamilies Ponerinae (with one exception) and Nothomyrmeciinae, and it is concentrated in certain genera in the Myrmicinae and Formicinae (Hölldobler and Wilson, 1990).

Social parasites have evolved to overcome the host nestmate recognition code, penetrate the host colony, and achieve social integration with their new colony. Newly eclosed social parasites, like all callow ants, are characterized by cuticular 'chemical insignificance' (odourlessness), which allows them to bypass the colony odour barrier (Schmid-Hempel 1998) at the time of usurpation of the host's nest. This is followed by a 'chemical integration' period when they acquire the specific chemical cues ('labels') of the host colony and incorporate them into their 'template' (internal representation of the environment chemical cues) by a learning process. Chemical integration is achieved by camouflage, in which the parasite gets cues from the host via contact with nest material and via allogrooming, and trophallaxis with the host (for reviews, see Dettner and Liepert 1994; Lenoir *et al.* 2001). For example, the xenobiotic ant *Formicoxenus provancheri* acquires the odour of its host, *Myrmica incompleta*, in the first days of its adult life and maintains the camouflage by intense host grooming (Lenoir *et al.* 1997). The slave-making ant *Polyergus rufescens* has not only evolved a species odour (chemical profile) that matches closely that of its most important and usual host species, *Formica cunicularia*, but it has also evolved the ability to modify its chemical profile should it penetrate any other *Formica* host species (e.g. *F. gagates*, rare host; *F. selysi*, non-natural host) to obtain social integration into host colonies. This chemical flexibility, possible only with the young parasite (newly emerged callows), could facilitate the change to different host species, if the main host species becomes rare (d'Ettorre *et al.* 2002a).

Artificially mixed-species groups composed of two non-chemically related ant species (Figure 11.1.1) provide a good tool for testing the chemical insignificance and chemical integration phases of alien-ant adoption and isolating the different parameters affecting recognition (e.g. label and template formation and plasticity). To achieve mixing, callow workers of each species are selected and removed from their mother colonies within 5 h of emergence and before they can interact with other colony members. Ten to fifteen workers of each species are combined and kept queenless and without brood, for at least three months before conducting the bioassays (dyadic encounters).

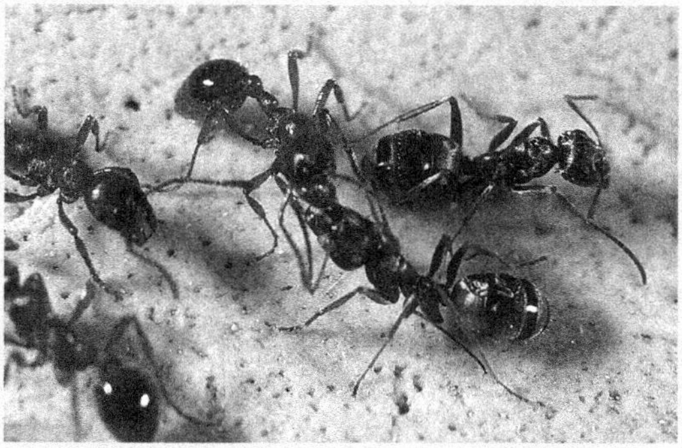

Figure 11.1.1 Food exchange between workers of *Manica rubida* (Myrmicinae) (left) and *Formica selysi* (Formicinae) (right) reared in an artificial mixed-species group created five hours after their emergence. (Photo: Y. Leclerc)

In mixed groups of *F. selysi* and *Manica rubida* callow ants, individuals acquire chemical cues characteristic of their allospecific nestmates via social interactions, thus progressively

continues

Box 11.1 continued

achieving a unified chemical profile (*Gestalt* i.e. mixture of the odours of the two associated species), that permits the two species to inhabit the same nest without displaying aggression (Errard 1994a). The reference cues are learned by the young imago shortly after emergence, the first interactions with their nurses (homospecific as well as heterospecific) being decisive. So, during their sensitive or critical period, the young ants are able to learn the odour of their nearest social environment, which strongly influences the recognition of colonial memberships during all their adult life (Errard 1994b). However, experimental mixed-species groups of *Manica rubida* with either *Myrmica rubra, Tetramorium bicarinatum*, or *F. selysi* show that the process of cue learning (see Chapter 11) during the sensitive period varies according to the specific chemical cues of the associated species. The post-imaginal learning, template reforming, and decision-making seem to be more precisely tuned (higher potential to discriminate between profiles) when the two species' chemical complexes are similar (Errard *et al.* 2006). The use of mixed-species groups of *F. selysi* and *Ma. rubida* also enables the exploration of the possible role of the volatile chemical cues within the nest that may affect the template formation during the early social experience of the ants. For example, *Ma. rubida* workers that were imprinted on *F. selysi* Dufour's gland constituents were always amicable towards the non-familiar *F. selysi* workers, indicating that undecane, the major product of *F. selysi* Dufour's gland, affects template formation in *Ma. rubida* workers. These results support the hypothesis that the perception of learned volatile cues permits a general recognition process that precedes the identification of cuticular chemical cues by contacts (Errard *et al.* 2008).

against at the colony level (Keller 1997; Boomsma *et al.* 2003).

Indeed, there is very limited – and controversial – evidence for nepotism in social insects (cf. Wenseleers 2007). In honeybees, several studies investigated the possible occurrence of nepotistic queen-rearing, but results have been equivocal or negative (review in Breed *et al.* 1994; see also Tilley and Oldroyd 1997; Moritz *et al.* 2005). In ants, only one study, conducted on the polygynous species *Formica fusca*, clearly suggested that workers indeed favour their own close kin when rearing eggs and larvae (Hannonen and Sundström 2003). However, a different study on another polygynous species, *Formica exsecta*, showed that workers do not discriminate between highly related and unrelated brood, but that brood viability differs between queens and this difference in viability could be sufficient to explain a relatedness pattern that could be interpreted as evidence for nepotism (Holzer *et al.* 2006b). This is in accordance with other studies that also failed to demonstrate nepotism in multiple queen colonies (e.g. De Heer and Ross 1997; Clémencet *et al.* 2007). Thus, the occurrence of nepotism remains controversial in ants, and we agree with previous authors who have suggested that recognition studies in ants usually deal with nestmate rather than kin recognition (e.g. Vander Meer and Morel 1998).

This does not mean that kin selection has to be discharged as one of the crucial forces for developing recognition systems in social insects. Kin selection has likely been very important for the evolution of eusociality in insects, but ecological pressures have contributed in shaping more complex societies where nestmate recognition conveyed higher advantages, and therefore kinship has been largely replaced by 'nestmateship' (cf. Lenoir *et al.* 1999). Thus, in social insect colonies, especially in the complex ant societies, individuals cooperate on the basis of familiarity and not necessarily on the basis of genetic relatedness. This familiarity has been termed 'fellowship' by Jaisson (1991, and references therein), and its strength has been elegantly shown by forming experimentally mixed colonies of phylogenetically distant ant species (e.g. Errard *et al.* 2006; Box 11.1). This is not a mere laboratory artefact, since natural mixed colonies do occur in the case of social parasitism (Lenoir *et al.* 2001).

11.3.2 How can recognition systems be stable?

For recognition to be possible, individuals must be somehow different: a polymorphism of labels is required. This cue diversity is supposed to be the expression of an underlying genetic variation: a polymorphism of genetic markers. When there is a high cost for being rejected, as in the case of a conspecific alien intruder that is attacked when attempting to enter an ant colony, individuals bearing rare labels will suffer costs at high probability. In contrast, individuals bearing common labels will very often match the template of evaluators and will suffer the cost of rejection only in few cases. The expected evolutionary scenario would result in rare labels being selected against with the consequent loss of the original genetic polymorphism. Eventually, all individuals in a population will be carrying the same genetic markers and recognition of friends and enemies would be impossible. Yet, polymorphic cue systems can be continuously observed. This paradox has been addressed for the first time in mathematical terms by Ross Crozier (review in Crozier 1987) and is known as the Crozier paradox (Tsutsui 2004). The subsequent debate has generated an array of verbal and mathematical models (discussed by Gardner and West 2007), and it now appears that the original suggestion by Crozier, that genetic marker diversity allowing recognition must be maintained by selection for something else, such as balancing selection imposed by host–parasite interactions, is indeed very likely (Rousset and Roze 2007).

In ants, nestmate recognition may be important, not only in competition between species and colonies, but also in mate choice. However, the phenomenon has not been extensively studied and, to our knowledge, only the following example is known. In *Leptothorax gredleri*, cuticular hydrocarbons of males and reproductive females are colony-specific and might thus act as a chemical cue (or signal) to avoid mating with sibs (Oppelt *et al.* 2008).

11.3.3 In search of the nestmate recognition cues

We have seen how in the context of nestmate recognition, which is vital for colony defence and is typically expressed by the action of rejecting alien intruders, recognition cues need not be directly correlated with genetic relatedness, since this would allow disruptive nepotism within the colony. One way of achieving nestmate recognition without allowing kin recognition is simply to mix things up by forming a cocktail of recognition cues. Cues produced by individuals may be combined to create a common colony odour (the *Gestalt* model; Crozier and Dix 1979). We have also seen that polymorphic cues are needed to discriminate between nestmates and non-nestmates. Insects live in a world of odours, thus we expect to find chemical cues that vary among colonies and are relatively uniform within a colony. Cuticular hydrocarbons appear to fit all the requirements needed to act as labels in the process of nestmate recognition in social insects. Insect cuticles are covered by waxy substances (mostly long-chain hydrocarbons from 20 to 35 carbon atoms) that probably evolved originally to avoid desiccation and were later used as recognition cues (Blomquist *et al.* 1998). Ants and other social insects show a complex pattern of cuticular hydrocarbons, which varies in quality among species and quantity (relative amount) within species, thus representing an ideal multi-component signal with the level of polymorphism required for recognition to be effective (Figure 11.2). These substances can be both genetically and environmentally determined and are permanently mixed to form a uniform blend. Trophallaxis and allogrooming are the main ways to obtain this uniform colony odour (Boulay *et al.* 2000; Chapuisat *et al.* 2005). It has been confirmed that the postpharyngeal gland (a head gland specific to the Formicidae) serves as a reservoir to concentrate and mix the hydrocarbons (review in Lenoir *et al.* 1999) that are transported by a lipophorin protein (Lucas *et al.* 2004).

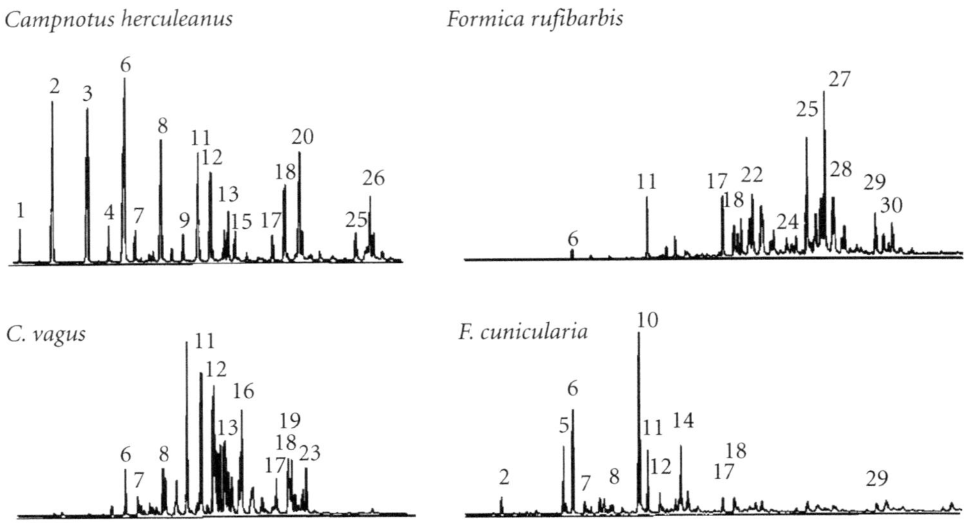

Figure 11.2 Gas-chromatograms showing the cuticular profiles of four different ant species. Some of the identified peaks are indicated as a reference: (1) n-C_{22}; (2) n-C_{23}; (3) n-C_{24}; (4) 2-meC_{24}; (5)$C_{25:1}$; (6) n-C_{25}; (7) 9- + 11-meC_{25}; (8) n-C_{26}; (9) 2-meC_{26}; (10) $C_{27:1}$; (11) n-C_{27}; (12) 11- + 13-meC_{27}; (13) 5-meC_{27}; (14) $C_{28:1}$; (15) n-C_{28}; (16) 10-meC_{28}; (17) n-C_{29}; (18) 11- + 13-meC_{29}; (19) 7-meC_{29}; (20) 7,13-dimeC_{29}; (21) 5,11-dimeC_{29}; (22) 12-meC_{30}; (23) n-C_{31}; (24) 13- + 15-meC_{31}; (25) 7-meC_{31}; (26) 5,13-dimeC_{31}; (27) 3,11-dimeC_{31}; (28) 15- + 17-meC_{33}; (29) 5-meC_{33}.

11.4 What do we know about recognition cues in ants?

Apart from the edited volumes on kin recognition cited earlier (see Section 11.1), which contain important chapters on social insects, there have been a number of comprehensive reviews more focused on the role of cuticular hydrocarbons, especially in ants (Lenoir *et al.* 1999; Singer 1998; Vander Meer and Morel 1998). These have stimulated an impressive body of research aimed at understanding the recognition code of ants. Our review here focuses on recent literature, and we encourage the reader to go back to these reviews for the basic knowledge.

11.4.1 'Bar-coding' and single compound recognition

Cuticular hydrocarbon (CHC) profiles are used by social insects to discriminate nestmates from non-nestmates according various levels: species, colonial, intra-colonial (castes, subcastes, reproductive status), and sometimes inter-individual. Recognition could occur through a process similar to reading a bar-code. Humans use bar-coding as a new promising tool for species identification via the mitochondrial gene cytochrome oxidase (COI). In modern taxonomy, it is especially helpful to discriminate cryptic species (Hebert *et al.* 2003; Hebert and Gregory 2005). In ants, bar-coding using cuticular hydrocarbons for species identification is theoretically possible and cheaper, but only few data are currently available, for instance on the genus *Cataglyphis* (cf. Dahbi *et al.* 1996; Oldham *et al.* 1999), the *Pachycondyla villosa* complex (Lucas *et al.* 2002) and the *Tetramorium caespitum/impurum* complex with six chemotypes (Steiner *et al.* 2002). In a recent study, the two sympatric colour morphs (red and black) of *Camponotus rufifemur* appeared to be chemically different with almost no hydrocarbons in common (Menzel *et al.* 2008). They may be two different species. Another study investigated 13 species of the genus *Formica* (Martin *et al.* 2008b) and a large program of species identification using CHCs is certainly an interesting challenge to be pursued. In termites, chemosystematics seems to be more advanced as phylogenetic analyses with cytochrome oxidase or microsatellites corroborate results obtained with chemical characterization (Copren *et al.* 2005; Dronnet *et al.* 2006).

New techniques have been used to discriminate between different species, colonies, and castes by measuring cuticular hydrocarbons levels with infrared photoacoustic spectroscopy, for example, for *Ectatomma* (Antonialli *et al.* 2007; Antonialli *et al.* 2008) and for *Oecophylla smaragdina* (Newey *et al.* 2008). Using spectroscopy may be a faster and less-expensive alternative to the analysis of cuticular hydrocarbons with gas chromatography or mass-spectrometry.

How many compounds are necessary for bar-code recognition among ants? Generally, bar-code discrimination cannot be based on one or a few compounds, but requires a complex blend of non-volatile compounds (Boomsma and Franks 2006). However, single compounds might play a role in within-colony discrimination, for example by characterizing the queen or reproductive individuals. In *Pachycondyla inversa*, the hydrocarbon 3,11-dimethylheptacosane is very abundant only on the cuticle of the queen and dominant egg-laying workers in queen-less colonies (Heinze *et al.* 2002b). Electro-antennography showed that workers react preferentially to this compound, which is correlated with ovarian activity and is likely to assume the role of a fertility signal (d'Ettorre *et al.* 2004). In *Dinoponera quadriceps*, this function is attributed to 9-hentriacontene (Monnin *et al.* 1998, 2002). However, evidence, although strong, remains correlative, and the role of these substances has not yet been demonstrated experimentally.

Other questions involve the respective role of the different hydrocarbon classes. The saturated *n*-alkanes have been considered to be important mainly for protection against desiccation, while branched alkanes appear to play a major role in nestmate recognition. In *Pachycondyla* species, internally branched methyl- and dimethyl-alkanes are indeed involved in recognition (Lucas *et al.* 2005). In *Camponotus cruentatus*, the colonial specificity is very important (see later) and trimethyl-alkanes, which are unusually abundant in this species, could play a central role (Boulay *et al.* 2007a). The picture is certainly more complex than previously thought, since recent data reveal different – and sometimes contrasting – results. In *Linepithema humile* and *Aphaenogaster cockerelli*, by using inert support such as glass beads or pieces of cotton coated with different hydrocarbon mixtures, it has been shown that a combination of at least two CHC classes is necessary to elicit an aggressive response. However, interestingly, no single class is more important than the others in eliciting the response (Greene and Gordon 2007b). In *Formica*, experiments involving glass beads and synthetic hydrocarbons showed that alkenes may have a more important role. In *F. japonica*, both *n*-alkanes and 9-alkenes are necessary to discriminate nestmates from aliens (Akino *et al.* 2004). But in *F. exsecta*, despite the cuticular profile being composed of alkanes and Z9-alkenes, aggression is elicited only by the alkenes (Martin *et al.* 2008b). Thus, more experiments are necessary to elucidate the roles of the various hydrocarbons classes, which appear to differ among species.

11.4.2 Cuticular hydrocarbons and task specificity

A correlation between the task an ant worker is performing and its CHCs is well known. For instance, ants modify their CHC profile when they become older and begin to forage. The role of juvenile hormone (JH) in temporal polyethism was first discovered in bees and wasps (Giray *et al.* 2005; Robinson 1985) and it has been recently confirmed in ants. The topical application of JH accelerates CHC modifications in the transition from brood-tender to forager in *Myrmicaria eumenoides* (Lengyel *et al.* 2007). Juvenile hormone has also been shown to be involved in the expression of possible fertility signals. Topical applications of a JH analogue (Cuvillier-Hot *et al.* 2004) could induce a decrease in fertility and a change of the cuticular profile in the monogynous queenless ant *Streblognathus peetersi*. Thus, cuticular hydrocarbons could inform nestmates about the hormonal state connected to dominance and fertility in a particular individual. Indeed, in *S. peetersi*, alpha workers are characterized by low levels of JH (Brent *et al.* 2006).

Despite much correlative evidence, experiments directly testing synthetic hydrocarbons and the role of different hydrocarbon classes are only few. The harvester ant *Pogonomyrmex barbatus* provides an interesting example. The CHCs are used for nestmate recognition in this species (Wagner *et al.* 2000), but the relative abundance of *n*-alkanes is 20% higher in foragers than in workers performing colony maintenance activities (Wagner *et al.* 1998). This might help prevent desiccation since these workers are foraging in a desert environment. Moreover, a particular group of workers, called patrollers, can stimulate foraging activity when a new seed source is discovered. These patrollers have a distinct CHC profile, and a series of elegant experiments showed that dropping glass beads coated with patroller CHC extracts inside the nest could mimic returning patrollers and induce foraging activity (Greene and Gordon 2003).

11.4.3 The genetic basis of cuticular hydrocarbon profile

Direct genetic control over CHCs is well known in *Drosophila* (Ferveur 2005). In social insects, there is also evidence that hydrocarbon composition, and therefore nestmate recognition cues, can be in part genetically determined. Nestmate recognition appears to be genetically based in *Formica polyctena*. Field experiments conducted on nests in pine forests of Germany showed the existence of a strong relationship between genetic distance and aggressive behaviour. This can be pictured as a sort of 'genetic gestalt': genetically related nests tend to show little aggressive behaviour (genetically determined recognition cues, namely CHCs), but there is no correlation between physical nest distance and aggression (Beye *et al.* 1997). Likewise, aggression increases with genetic distance between nests in *Formica pratensis*. But here dispersion often occurs by fission (a form of 'dependent colony foundation', see Chapter 9) and thus neighbouring nests tend to be more closely related than distant nests and are less aggressive to them (Beye *et al.* 1998). Moreover, the relative importance of environmentally and genetically determined cues in this species can vary according to the social structure (monodomy or polydomy (cf. Pirk *et al.* 2001). Similarly, the variation observed in the cuticular compounds of 12 populations of *Petalomyrmex phylax* from Cameroon could be explained by a combination of both genetic and social factors (number of queens), and by the spatial distribution of populations (Dalecky *et al.* 2007).

By contrast, there is no correlation between genetic distance and nestmate discrimination in *Plagiolepis pygmaea* (Thurin and Aron 2008) and in *Formica selysi* (Rosset *et al.* 2007). It is worth noting that in *F. selysi* these authors observed that there is no difference in nestmate recognition ability between workers of single- and multiple-queen colonies. Rosset *et al.* (2007) also suggested that workers might be able to detect a signal that is characteristic of the social structure (monogyny versus polygyny). However, this signal is not known, and it would be necessary to compare the odour profiles of the two types of colonies. In the super-colonies of *Formica paralugubris*, whereby individuals mix freely among separated nests, the ability to discriminate between nestmates and non-nestmates is maintained between populations, as indicated by longer antennation bouts, and aggression increases with geographic and genetic distance (Holzer *et al.* 2006a; see Plate 11 for more on antennation).

In conclusion, general patterns are difficult to find since the relative importance of genetic and environmental factors in shaping nestmate recognition cues seems to be linked to the particular life history of the different species. We discuss possible environmental factors in Section 11.5.

11.4.4 The discovery of very long chain hydrocarbons and other compounds

The recent use of high temperature gas-chromatography columns allowed the identification of new hydrocarbons with longer chains on the ant cuticle, which have remained undetected with the commonly used columns. The discovery of these long-chain hydrocarbons opens new avenues for research in some fields such as host–parasite interactions. Usually, social parasites mimic their host CHCs (chemical mimicry, cf.Lenoir *et al.* 2001), but *Acromyrmex insinuator* do not mimic their host. Instead, this social parasite is chemically insignificant in the 'normal' C29–C35 range, where it has a very low total amount of CHCs, but it possesses large quantities of unsaturated C43–C45 hydrocarbons. The role of these CHCs is not known; it has been suggested that they are difficult to perceive, and hence may support, the chemical insignificance hypothesis. They may also function as a 'sponge' and absorb traces of lighter hydrocarbons that are used as nestmate recognition cues so to blur them

(Lambardi *et al.* 2007). However, it is too early to draw any conclusion, and this discovery calls for re-investigating the chemical profile of all the ant species with high temperature GC-columns. For example, *Formica truncorum* was supposed to have a very simple CHC composition, with few compounds and not heavier than C31 (Boomsma *et al.* 2003). In fact, they have long-chain hydrocarbons, from 34 to 45 carbon atoms, accounting for 55% of the total CHC profile (Akino 2006). Apparently, this does not change the colonial identity, and the chemical signature of the colony is maintained with or without these long-chain hydrocarbons. Some ants will be shown not to have long-chain CHCs anyway, like *Formica japonica* (Akino 2006), whereas all the *Formica s. str.* species have C25–C37 chains (Martin *et al.* 2008a). *Pachycondyla villosa* has also very long chains CHCs, up to C45 (Lucas *et al.* 2004). The hydrocarbon profile of the tropical *Camponotus rufifemur* consists almost exclusively of methyl-branched alkenes from C35 up to C49 (Menzel *et al.* 2008). Interestingly, *Petalomyrmex phylax* from Cameroon has a long set of C32–C42 alkenes, and there is a geographical south bias towards substances that have a higher molecular weight (Dalecky *et al.* 2007), suggesting a role of environmental factors.

Other classes of compounds may be involved in ant nestmate discrimination. It is long known that free fatty acids and esters also exist on the insect cuticle, and steroids have been recently discovered (see parabiosis, Section 11.5.2). Cholesterol has been found in large quantities in males of *Leptothorax gredleri* (Oppelt *et al.* 2008). If these compounds have a role in recognition, it needs to be further investigated.

11.4.5 The possible role of volatiles

Cuticular hydrocarbons, which are not very volatile, have long been considered responsible for nestmate recognition as this occurs generally at very short distance between individuals: a few millimetres or maximum 1 cm (Brandstaetter *et al.* 2008; Cuvillier-Hot *et al.* 2005). Nevertheless, more volatile substances might also play a role, and nestmate recognition perhaps does not always rely only on CHCs. In two *Atta* species both inter- and intraspecific recognition seem to be mediated by alarm pheromone constituents as well as by substances from abdominal exocrine secretions (Hernandez *et al.* 2006). Akino and Yamaoka (2000) suggested that in *Lasius fuliginosus*, volatiles could act as a transient cue at short distances, while non-volatiles would serve as definitive signals for recognition of nestmates. Volatiles from the Dufour's gland are implicated in *Camponotus fellah* nestmate recognition (Katzav-Gozansky *et al.* 2004, 2008). However, here the chemical nature of the volatile cues still remains uncertain. Some simple alkanes may play a role, for example callow *Manica rubida* workers in mixed-species groups with *Formica selysi* can imprint on volatile alkanes (in particular undecane) from the *F. selysi* Dufour's gland and incorporate them into their own template. Since undecane is not present in the glandular secretion of *M. rubida*, it is learned from the *Formica* group-mates (Errard *et al.* 2008). We suggest that the role of volatiles has probably been largely underestimated, and thus requires more attention in future studies.

11.5 The ecological context

One of the principal reasons of the ecological success of social insects is their ability to exploit and monopolize food sources at the colony level. For this, they need to discriminate and exclude competitors. However, nestmate recognition plays different roles in different ecological contexts.

11.5.1 Inside the nest: role of nest material and food

Inside the nest, ants do not need to discriminate nestmates from non-nestmates as they are all supposed to be fellows (Jaisson 1991). The nest entrance is usually patrolled by very efficient guards, and aliens are not admitted into the nest. Thus, it is frequently observed that when an intruder manages to enter the nest, it is accepted. This is illustrated by the case of alate females of *Cardiocondyla elegans*, which are transported by workers into unrelated nests: outside the nest they are subject to aggression, but aggression ceases once they are inside (Lenoir *et al.* 2006). It is generally considered that nurses inside the nest are less aggressive than foragers. The walls of the nest chambers are probably saturated with hydrocarbons and other substances secreted by the ants, but their chemical identity is not known. Various myrmecophile beetles and crickets rub against the inner walls of the nest to obtain the colony odour and be tolerated by passive chemical mimicry (Lenoir *et al.* 2001).

The nest odour is important when the colony needs to emigrate. Explorers use it to mark the new nest, and when ants are given a choice, they will prefer a nest marked with colony odour over an unmarked one (e.g. *Lasius niger*; Depickère *et al.* 2004; *Temnothorax albipennis*; Franks *et al.* 2007a). On the contrary, workers of *Aphaenogaster araneoides*, which frequently migrate to a new nest, strongly avoid nests marked with colony extracts. This absence of nest marking might prevent detection by predaceous army ants (McGlynn 2007). The colony marks laid by minor workers of *Pheidole pallidula* are also used in ant clustering (Sempo *et al.* 2006). The aggregative role of cuticular hydrocarbons has also been documented in other insects, such as gregarious cockroaches. Recently, an elegant study showed that cockroaches indeed aggregate with robots impregnated with the CHCs of congeners (Halloy *et al.* 2007).

In wasps and honeybees, the nest is made with paper and/or wax that captures and retains odours and produces some key components used in recognition. Leaf-cutting ants can be compared to wasps and honeybees because the garden fungus is composed of degrading leaves and emits numerous substances influencing the nest odour. In *Acromyrmex*, the fungus absorbs the cuticular hydrocarbons of the ants and its odour is thus colony-specific (Bot *et al.* 2001b; Viana *et al.* 2001). Logically, the nest odour is influenced by the nature of the leaf used as substrate for the fungus, as was demonstrated in *Acromyrmex* long ago (Jutsum *et al.* 1979). That the diet (privet, roses, or bramble) can influence nestmate recognition has been confirmed more recently (Richard *et al.* 2004). The variation in chemical profiles of *Acromyrmex echiniator* and *A. octospinosus* ants is at least partly explained by the genetic differences in amides, aldehydes, and methyl esters, originating probably from the fungus. The fungus garden is therefore an important independent source of chemicals contributing more to the *Gestalt* than the innate chemicals of ants (Richard *et al.* 2007). In ant–plant interactions (see later) we do not know how the colony odour is influenced by the host plant, thus this is a promising field of research. *Allomerus* ants, for instance, build in their *Hirtella* host plants galleries pierced with numerous holes serving as traps to capture insect prey. A fungus that has not yet been identified is associated with the nest and probably produces compounds that are included in the ant colony odour (Dejean *et al.* 2005b).

The colony odour appears to be particularly influenced by environmental factors, such as food, in tramp species. For example, in *Linepithema humile*, the diet can significantly modify both CHCs and nestmate recognition (Liang and Silverman 2000). Similar behavioural results have been obtained in the crazy ant *Paratrechina longicornis*, although the chemical profiles were not analyzed and the study is based on laboratory observations only (Say-Piau and Chow-Yang 2003).

11.5.2 Outside the nest: territory-marking and foraging trails

Markings outside the nest may take different forms depending on the ecological context and the role of the species in the community. Workers mark the nest entrance with colony-specific chemicals to prevent intrusions. Thus, nest marking can play an important role in nestmate recognition. In *Myrmica rubra* and *Pheidole pallidula*, workers probably use their legs to transfer secretions onto the ground (Cammaerts and Cammaerts 1998; 2000b). In *Messor capitatus*, territorial marking near the nest entrance is made colony-specific by faecal spots containing hydrocarbons identical to those of the cuticle (Grasso *et al.* 2005). Wenseleers *et al.* (2002) showed that the readiness to fight in the desert ant *Cataglyphis fortis* was high for ants near the nest entrance and declined at 5–20 m. This clearly indicates that aggression is displayed in the context of the nest, and thus has more to do with nestmate recognition than with territory defence.

In some species, nestmate recognition is less strict, as shown by low inter-colonial aggression. Non-nestmate intruders are able to enter the nest in 60% of trials in *Myrmecia nigriceps* and up to 50% in *Cataglyphis cursor* (Lenoir *et al.* 1988; van

Wilgenburg *et al.* 2007). Here foragers can share food sources, and tolerance is probably favoured by high relatedness between neighbouring colonies since *C. cursor* reproduces by fission (Lenoir *et al.* 1990; Lenoir *et al.* 1988; Mayade *et al.* 1993). Another example is *Ectatomma tuberculatum*, which forms open colonies and forages in trees that are not defended as territories (Zinck *et al.* 2008). In confrontation tests between ants of different aggressive colonies, the 'winner' is generally the ant that is on its own territory (Fresneau and Errard 1994). This 'bourgeois strategy' has been investigated in *Cataglyphis niger* where the chemical cue advertising the ownership comes from the cloacal gland (Wenseleers *et al.* 2002). Some ant species might mark their entire home range with colony-specific chemicals. However, various species apparently mark their home range in a way that is not colony-specific, as has been observed in two *Tetramorium* species (Cammaerts and Cammaerts 2000a) and in *Lasius niger* (Devigne and Detrain 2002). In general, the identity of the marking substances is not known. The differences in all these data on 'territorial pheromones' may be explained by the different contexts used in the studies: for example, in *Myrmica* it refers to walking speed, while in others it refers to fighting advantages.

The famous wood-ant *Formica* wars with hundreds of corpses at the frontiers of colonies in spring and the ants' cannibalistic behaviour have impressed the human imagination (Mabelis 1979). Some ant species do have a real territory, which is actively defended and marked chemically according to the strict definition of Hölldobler and Wilson (1990). Territorial ants, such as *Pogonomyrmex* (Hölldobler 1974), usually form large over-dispersed colonies where the nest distribution allows foraging on non-overlapping areas or trunk trails thus reducing the number of aggressive interactions. Generally, these ants learn the colonial identity of their neighbours and consequently are less aggressive towards these known neighbours than towards complete strangers, a phenomenon called 'dear enemy' (see review in Knaden and Wehner 2003). On the contrary, in *Camponotus cruentatus*, where colonies have very different CHC profiles, the territories can overlap by 40%, but workers fiercely defend food sources against neighbouring colonies without any 'dear enemy' effect (Boulay *et al.* 2007a).

In the tropical rainforests, many ant species have evolved an arboreal life, some species are dominant and form very large colonies with absolute territories defended against neighbouring colonies of their own or other species. They are distributed in a mosaic pattern (Blüthgen and Stork 2007; Dejean *et al.* 2007a; see also Chapter 5). Weaver ants *Oecophylla* are a typical example of territorial arboreal ants, which mark the leaves with rectal pheromones that can persist for more than nine months under the tropical rains (Dejean and Beugnon 1991). These marks are used by other ant species to avoid the *Oecophylla* territories (Offenberg 2007). Herbivorous beetles are also able to detect these pheromones and avoid feeding on *Oecophylla* ant trees (Offenberg *et al.* 2004). Here again, we do not know the identity of the chemical signals.

Plant-ants are obligate associates of specialized plants called myrmecophytes (i.e. plants offering to their guest ants special structures called domatia; see Chapter 6). In these ant species, one colony generally occupies one tree for nesting and foraging, the tree being a real territory (Dejean *et al.* 2007a). When the distribution of trees is over-dispersed, the colonies tend to be isolated. In two *Allomerus* species in Guyana, it has been observed that intra-specific aggressiveness is very low, while interspecific conflicts between different species are very violent. This does not mean that the ants have lost nestmate recognition, but the strictly arboreal life of these ants and the distance between trees, which make the encounters almost impossible, may explain the loss of intra-specific aggression (Grangier *et al.* 2008).

Territorial ants are dominant in the ant community, and defend their territory not only against conspecific, but also against allospecific intruders (see Chapter 5). *Camponotus cruentatus* is a good example of ecologically dominant ant in the Mediterranean region, whereas *Aphaenogaster senilis*, which is not territorial, is subordinate (Figure 11.3). Subordinate ants use several strategies to avoid conflicts with the dominant ones, for example foraging in a different time-window leading to temporal partitioning (Cerdá *et al.* 1997). Are ants able to recognize the other species? Evidence suggests

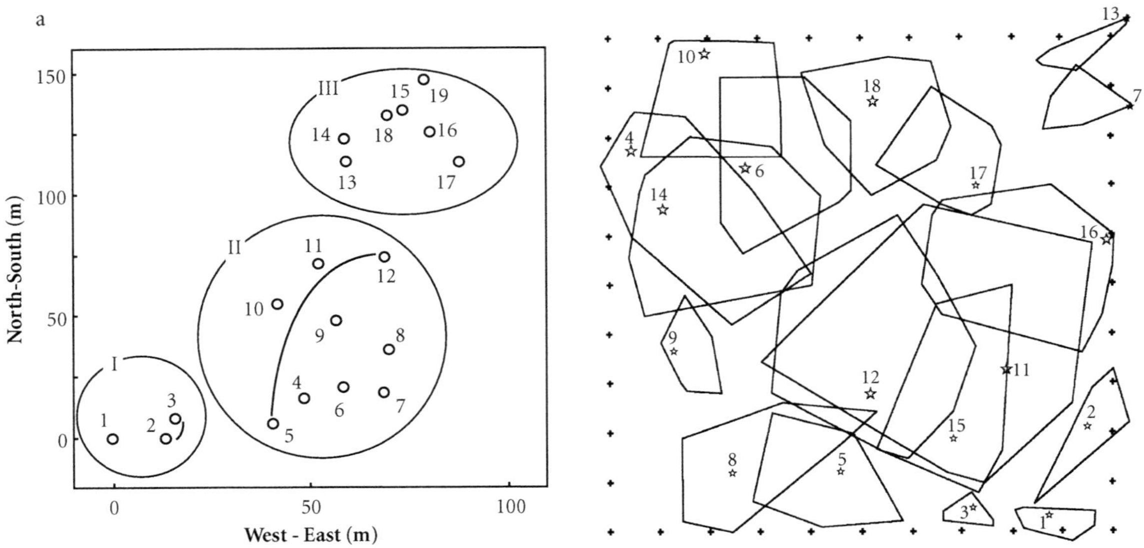

Figure 11.3 (a) Distribution map of *Aphaenogaster senilis* nests in Doñana National Park (Andalousia, South Spain, sea level). This ant species reproduces by dependent colony foundation and inter-nest aggression is low. Nests are presented in three groups according to their behavioural indices of aggression and chemical distances. Intra-group aggression is low, indicating a possible common ancestor fissioning group. Nests 2–3 and 5–12 have probably recently been founded. (Modified from Ichinose *et al.* 2005) (b) Map of 18 major nests of *Camponotus cruentatus* localised on or near the 50 x 50 studied plot in Sierra de Cazorla (South Spain, 1400 m asl). The polygons delimit the area within which 95% of the workers of a given nest forage. The overlap between the different areas is 44%. Nevertheless, food sources are fiercely defended against any other neighbour. Hydrocarbon profiles of the colonies are strictly different. (Modified from Boulay *et al.* 2007a)

that they do, for instance, *Camponotus foreli* workers always attack *Cataglyphis iberica*, whose colonies are then eliminated, while they tolerate *A. senilis* (Cerdá and Retana 1998).

Trails can contain colony-specific components also outside the territory, on the non-defended home range. In *Lasius nipponensis* (cf. *L. fuliginosus*) and *L. japonicus* (cf *L. niger*) trails are used by one colony only. The trail pheromone is not colony-specific, but the specificity is given by footprint hydrocarbons that are almost identical to CHCs (Akino and Yamaoka 2005a,b). This prevents the exploitation of trails by neighbouring colonies. A more elaborate association is parabiosis, where two (or more) species share the same nest and use the same trails. This phenomenon is frequent in Neotropical ant gardens (reviewed by Menzel *et al.* 2008). Since parabiotic species need to tolerate heterospecific ants as nestmates, they must have modified their recognition system. Habituation to the others' odour seems to be the mechanism. In the association between *Odontomachus mayi* and *Crematogaster limata*, the ants have completely different chemical profiles, and the learning is limited to the partner colony only (Orivel *et al.* 1997). In the rainforest of Borneo, there is the interesting case of parabiotic association between *Crematogaster modigliani* and *Camponotus rufifemur*. The latter is tolerant towards any colony of *Cr. modiglianii*, but not towards other *Crematogaster* species (Menzel *et al.* 2008). This might be explained by the unusual cuticular profiles of these species, which are covered by a set of steroids that have not yet been identified. The composition of these steroids differs between colonies, but is more similar for the two species of the same parabiotic nest. Whether steroids play a role in nestmate recognition is under investigation. The reduced discrimination of heterospecific nestmates might be caused by transfer of *Ca. rufifemur* hydrocarbons to the *Cr. modiglianii* profile.

Extremely long-chain hydrocarbons may be difficult to detect by antennal receptors, and hence result in chemical insignificance (Menzel *et al.* 2008). This species-specific, but not colony-specific tolerance contrasts with the above results, indicating that recognition in parabiosis may be due to a different learning process and different templates.

11.6 Concluding remarks

The amazing ecological success of ants is due in part to their ability to discriminate nestmates from non-nestmates, not only individuals belonging to colonies of the same species, but also to other species. Inside the colony, ants, for example, *Pogonomyrmex barbatus*, recognize individuals performing different tasks and can also discriminate social status (Sections 11.4.1 and 11.4.2). Outside the colony, ants know their surroundings, their nest entrance, and home range. The current body of evidence suggests that nestmate recognition in ants is mostly based on a mechanism of phenotype matching, even if other mechanisms cannot be excluded. Early in its development, an individual worker would learn the relevant cues from its fellows and build a template representing the colony odour profile. This is similar to the process of imprinting, and has been shown in several ant species (Jaisson 1991). In many cases, we have clear proof that cuticular hydrocarbons are among the relevant recognition cues, and that the colony odour is formed by mixing together the cues of basically all the colony members via social interactions (allogrooming and trophallaxis with the involvement of the post-pharyngeal gland (cf. Lenoir *et al.* 2001; Lenoir *et al.* 1999). Thus, the colony odour is not the simple sum of cues of the different individuals, but it is a new configuration, a pattern of elements resulting into a unified whole (*Gestalt*).

Depending on the life histories and the ecological and evolutionary constraints of the different ant species, the proportion of cues that are genetically and environmentally determined will vary (Sections 11.4.3 and 11.5.1). When the environmental component of the cue-expression is significant, the internal template of each individual needs to be flexible to adapt constantly to the changes in the local environment. Since nestmates and non-nestmates may have overlapping cues, the discriminating response of ants – similarly to other social organisms – cannot be perfect, and is likely regulated according to an acceptance–rejection threshold. Indeed, the acceptance threshold model (Reeve 1989) predicts that recognition systems are not fixed, but context-dependent, and the threshold should vary according to the cost and benefits of accepting non-nestmates and rejecting nestmates (recognition errors). The model has been tested in a host–social parasite system and has been supported by the observation of a significant adaptive behavioural flexibility (level of aggression) of the host species linked to the seasonal dynamics of the social parasite (d'Ettorre *et al.* 2004). According to the threshold model, aggression as a result of non-nestmate discrimination is an 'all-or-none' response: either there is aggression or not, but the threshold as well as the template can vary (Liebert and Starks 2004). Alternatively, the graded model proposes that ants progressively vary their level of aggression according to the difference between the template and the pattern of cues borne by the encountered individual (Lenoir *et al.* 1999). Evidence for a graded model in nestmate discrimination is given by the observation that longer antennation time is required when the chemical signature (cues) differs slightly from the template (Dahbi and Lenoir 1998; Holzer *et al.* 2006a). However, these two models are by no means mutually exclusive.

In some particular circumstances, ants have been shown to have unexpectedly sophisticated recognition abilities. This is the case of co-founding queens of *Pachycondyla villosa* and *P. inversa*, which are capable of individual recognition (d'Ettorre and Heinze 2005; Dreier *et al.* 2007). Unrelated queens found new colonies together, but when they first meet they aggressively establish a dominance hierarchy that later controls the partitioning of work and reproduction. Individual recognition in these small societies is advantageous because it facilitates the maintenance of stable dominance hierarchies and avoids the cost of repeated aggressive encounters. We know that individual recognition in *Pachycondyla* ant queens is based on the long-term memory of chemical cues, but there is no direct proof that these cues are indeed cuticular hydrocarbons, although the cuticular chemical profiles of queens are neither associated with dominance nor with fertility, and nestmate queens do not share a common odour.

Recent results suggest that ant workers of *Cataglyphis niger* can also discriminate different individuals. By using a habituation–discrimination paradigm (Nowbahari 2007) showed that adult workers learn the cues of individual ants that they have encountered and recognize them in subsequent encounters. Workers are less aggressive towards familiar non-nestmates than towards unfamiliar ones.

Finally, ants are apparently capable of a sort of 'latent learning'; for example they can learn what to do or not to do when they are confronted with the choice of a new nest using both pheromones and landmark cues. Thus, ants are possibly able to make plans for the future (Franks *et al.* 2007b).

11.7 Future directions

Despite the recent advances in analytical technology and the flourishing of studies in the last decades, the recognition code of ants and other social insects is far from being ultimately deciphered. As a usual occurrence in science – and this is one of the reasons why it is so fascinating – while investigating old questions, researchers find new questions instead of clear answers. We would like to draw attention to some issues that need to be considered and therefore constitute the ground for promising future studies.

- Is there something other than cuticular hydrocarbons acting as recognition labels?

Cuticular hydrocarbons have long been considered as the best candidates for recognition cues (cf. Howse 1975); however, although their importance has been confirmed in many cases, recent findings beg for exploring the potential role of other compounds by following new research directions. We have already discussed the possible implication of volatile chemicals from exocrine glands in Section 11.4.5. Here we point out a recent study that looked at a different category of substances. Paper wasps hibernate in particular safe locations that can be used by subsequent generations of foundresses. Turillazzi *et al.* (2006) experimentally showed that these hibernation sites are marked with venom secretions and cuticular peptides. A proteinaceous pheromone has also been recently identified as having a role in termite egg recognition (Matsuura *et al.* 2007). Thus, the unexplored world of proteins and peptides opens its doors to social insect recognition. With their complex tridimensional structure, cuticular peptides could contain essential information themselves, but could also somehow embed cuticular hydrocarbons and thus change their physical and chemical properties. This might explain why isolated hydrocarbons do not always elicit a behavioural response when used alone in experimental designs.

- Does recognition always need long-term memory and integrated information processing?

The label-template matching model discussed earlier (Section 11.6) requires learning the recognition cues and forming an internal neural template that is stored somewhere in the memory and can possibly be updated. This process implies information processing at high brain centres (e.g. mushroom bodies). Is there any other parsimonious alternative? The idea of habituation, which is the simplest form of learning not necessarily requiring high brain centres, is usually dismissed (Vander Meer and Morel 1998). However, a recent study suggests that even a simpler process, receptor adaptation, which does not involve any learning, could account for recognition of non-nestmates. Ozaki *et al.* (2005) described a sensory sensillum on the antennae of *Camponotus japonicus* with a surprising function. This sensillum responds specifically to non-nestmate CHC blends and does not react to nestmates' CHC extracts. A peripheral recognition mechanism in detecting colony-specific chemical signals is thus possible (but see Leonhardt *et al.* 2007). Such a mechanism cannot account for within-colony discrimination, and we know that ants are able to detect different classes of nestmates, but these results cannot be ignored, and more studies are needed to disentangle the different mechanisms that might intervene at different levels of recognition.

• Is there an ant queen pheromone?

The ant queen pheromone is like the Metastasian Arabian phoenix: everyone says it exists, but no one knows where it is. Three recent reviews have addressed this enigma from different angles (Hefetz 2007; Le Conte and Hefetz 2008; Peeters and Liebig 2009), thus we hope that it will be solved soon, at least in some ant species. There is evidence that cuticular hydrocarbons are involved in signalling queen fertility, but there is no direct proof so far. In *Aphaenogaster senilis*, the queen signal may involve the Dufour's gland secretion more than CHCs, and this ant is probably a good experimental model system because a simple biological test can be exploited: when the queen is removed, the workers immediately reorient the developing of worker larvae to produce gynes (Boulay *et al.* 2007b).

• Can recognition be studied in the laboratory?

Most of our current knowledge on recognition systems in ants is derived from laboratory assays. How much do these reflect the natural situation? This question has rarely been addressed and we believe it is an important one. Roulston *et al.* (2003) used the Argentine ant *Linepithema humile* as a model to compare four different laboratory aggression bioassays largely used to study nestmate discrimination in ants. The assays included interactions between one live and one dead ant, two live ants, five against five live ants, and one ant introduced to a foreign colony. All assays using live ants gave comparable results, independently of the scoring method used, but pairing a live and a dead produced inconsistent results and lowered aggression levels. Nevertheless, isolated aggressive acts did not necessarily predict whole colony interactions, as some colonies that fought in bioassays merged when the entire colonies were allowed to interact. Thus, aggression tests may give only limited information about interactions between colonies. This does not mean that we should stop working in the laboratory. Some particular questions can only be addressed under controlled conditions, as when trying to disentangle behavioral and chemical cues underlying recognition (cf. Guerrieri and d'Ettorre 2008; Lucas *et al.* 2005). Nevertheless, we would like to stress that laboratory results should be interpreted with caution, and we encourage researchers to trust their doubts and to go back to the field as much as possible.

11.8 Summary

The ability to recognize group members is a key characteristic of social life. Ants are typically very efficient in recognizing non-group members, and they aggressively reject them in order to protect their colonies from robbery and parasitism. There is a range of different recognition mechanisms including prior association, phenotype matching, and recognition alleles. The concept of kin recognition should be considered different from that of nestmate recognition in ants and other social insects. Most of the available studies address the nestmate recognition level, namely the discrimination of nestmates (colony members) from non-nestmates (strangers), independently of actual relatedness. Indirect and direct evidence identify long-chain cuticular hydrocarbons as the best candidates to act as recognition cues in ants, even if other chemical substances could also play a role, at least in some ant species. The relative importance of genetic and environmental factors on the expression and variation of the cuticular hydrocarbon profile is then analyzed in connection with ecological factors and life history characterizing the diversity of ant species. There are many ongoing debates and unanswered questions about recognition cues and mechanisms. The recognition systems of ants are extremely complex.

Acknowledgements

We would like to thank the editors, Kirsti Abbott, Lori Lach, and Catherine Parr, for inviting us to contribute to this volume. We thank Dominique Fresneau, Xim Cerdá, Raphael Boulay, and Claire Detrain for their useful comments and discussions.

Chapter 12

Foraging and Defence Strategies

Anna Dornhaus and Scott Powell

12.1 Introduction

In many habitats, the first animal that a visitor is likely to notice is an ant forager. Ant foraging trails can stretch for hundreds of metres, like pseudopodia from the central body of the colony, searching for and retrieving food. Attempts to interfere with these foraging trails may prompt a rapid and aggressive response from the ants, with individuals readily sacrificing their life in defence of the harvested resources. Such conspicuous activity is, however, only representative of a small subset of the striking diversity of foraging and defence strategies that have evolved in the ants. The goal of this chapter is to discuss the diversity of individual and collective strategies used by ants to find, retrieve, and defend resources. More specifically, we review how ants decide when and where to forage, what individual and collective strategies are used during foraging, and how ants communicate about food sources. Any foraging strategy is, however, only as effective as the defensive strategies that have evolved to safeguard harvested resources. Consequently, we also explore the defensive strategies used in the acquisition and retrieval of resources from the environment, and in retaining them at the nest. While the study of foraging strategies has a rich history, much is still to be learnt, and defence strategies remain a relatively understudied topic. We therefore identify numerous open questions in the study of foraging and defence strategies, and further discuss general approaches for advancing and integrating research in this area in the future.

12.2 Acquisition of resources

A resource acquisition strategy used by ants has to solve several specific problems. Most obviously, a relatively small number of foragers have to retrieve enough food to feed the entire colony. Frequently, less than 10% of workers participate in foraging (the other workers perform brood care and other in-nest tasks, or are inactive, e.g. Dornhaus *et al.* 2008; Robson and Traniello 2002). Contrary to many solitary insects, ants therefore perform 'central place foraging': all food must be brought back to a relatively immobile central place, the nest. A large body of theory covers the specific constraints and optimal strategies in central place foraging (Orians and Pearson 1979; Ydenberg and Schmid-Hempel 1994). There is also a discrepancy between the size of the individual foragers and the size of the colony, the 'superorganism' that needs to be supplied with food. In other animals, many physiological and ecological characteristics are tightly linked to body size, such as home range or territory (Adams 2001; Jetz *et al.* 2004), metabolic rate, and lifespan (West *et al.* 1999). In ants, both the body size of individual ants and the mass of their colonies may be important. An ant forager will typically forage much farther from its nest than a terrestrial solitary arthropod of the same size, because ant foragers have to cover a foraging range large enough to yield enough food for a much larger organism, the colony. For example, a leaf-cutting ant may travel several hundred metres to a foraging site (Roces 2002). In relation to body size, this is the equivalent of a human travelling 50–100 km (one way) to collect food.

It comes as no surprise that foraging is one of the most costly activities performed by an ant colony. Costs arise not only in terms of energy used, but also in the time spent and the mortality risks faced by workers as they leave the nest and travel great distances. Interestingly, the energetic costs of foraging, relative to the energy gained by this activity, vary widely among species. In harvester ants, the energetic costs of foraging are negligible compared to the energy collected (which is over 1,000 times more than the cost per trip), but in some nectar foraging ants, the ratio of energy gained to energy expended for the trip is only 3.9 (Fewell *et al.* 1996). Some ants may thus need to tightly optimize their energetic efficiency, whereas for others, minimizing the time- and mortality-costs of foraging may be more relevant (Nonacs and Dill 1990).

Mortality among ants that leave the nest is much higher than mortality of workers inside the nest; foragers may face dangers such as predation, parasitism, adverse weather, and simply losing their way (Hölldobler and Wilson 1990; Nonacs and Dill 1990). This has led to the hypothesis that older workers are allocated to foraging because their loss is less costly to the colony than that of younger individuals, which have many work-hours yet to live (Moron *et al.* 2008; see Section 12.3.1). Facing these difficulties of large foraging distances, high energy and other costs, and high mortality outside the nest, ants have evolved a number of strategies to improve their success at finding and retaining resources (see also reviews in Carroll 1973; Detrain *et al.* 1999; Gordon *et al.* 2008; Hölldobler and Wilson 1990; Roces 2002; Traniello 1989; Tschinkel 2006).

12.3 Individual foraging strategies

An ant leaving the nest to forage has to first search for a suitable food patch. Upon discovery, the forager may then have to overcome the prey item or dissect it into manageable pieces and decide what load size to carry. Foragers have to navigate back to the nest, either in a straight line or by retracing their steps; foragers may also memorize the location of a food patch to be able to return there or recruit to it. In addition, before even leaving the nest, forager ants may have to make decisions about the best time to forage. Solving all these problems is likely to require a mix of innate, species-specific behavioural rules, and learning abilities to be used by worker ants. We review each of these aspects of foraging in the subsequent paragraphs.

12.3.1 Deciding when to forage

Before specific foraging strategies come into play, workers have to decide when to initiate foraging. This can be studied at the individual level (how do individual workers decide when to forage) and at the level of the colony (how many workers are allocated to the foraging task). The mechanisms of task allocation at the collective level are relevant in the study of division of labour, which is not discussed here (but see Beshers and Fewell 2001; Gordon 1996; Tschinkel 2006 for reviews, and also Chapter 10). The decision to become a forager involves trade-offs between exploration and exploitation, and between flexibility and specialization (see Biesmeijer and de Vries 2001; Blanchard *et al.* 2000; Dechaume-Moncharmont *et al.* 2005; Detrain *et al.* 1999; Jaffé and Deneubourg 1992; Robson and Traniello 2002; Tripet and Nonacs 2004). Age (Hölldobler and Wilson 1990), genetic background (Robinson *et al.* 2005a), 'corpulence' (i.e. fat content, Blanchard *et al.* 2000), body size (Wilson 1980), or dominance status (Powell and Tschinkel 1999) may affect the probability that a worker will start foraging, as will cues and signals from the colony that food is available (Table 12.1) or needed (Burd and Howard 2005). Classic studies often claim that age and body size are the main determinants of task allocation; however, the fact that foragers are, for example, on average older than in-nest workers does not prove that task allocation is based on age *per se*; experience or disease may also play a role (Moron *et al.* 2008; Tripet and Nonacs 2004; Woyciechowski and Kozlowski 1998).

At the individual level, the decision to leave the nest to forage in many ant species is related to environmental conditions, in particular external temperature (Hölldobler and Wilson 1990; Traniello 1989; Tschinkel 2006; Table 12.1). Temperature may influence the expected foraging success, and thus the likelihood that a foraging trip will recoup its costs in a number of ways. As largely

Table 12.1 Collective strategies used in food retrieval by ants. See Hölldobler and Wilson (1990: Table 7–8, p. 280) for a more comprehensive list of genera.

Recruitment strategy	Function	Examplary genera	References
Solitary foraging: Foragers leave the nest individually and do not appear to interact while searching for or retrieving prey.	Presumably this strategy is used if prey are distributed, unpredictable, and can be carried back by a single forager. This strategy is often seen in predatory ants preying on other arthropods.	*Harpegnathos, Pachycondyla, Cataglyphis*	Hölldobler and Wilson (1990); Maschwitz and Steghaus-Kovac (1991); Wehner (1987)
Tandem running: a successful forager recruits and leads a single ant back to the food source.	Slow recruitment of individual ants to potentially hard-to-find sites, such as new nest sites. Also used in recruitment to food, although possibly less frequently.	*Temnothorax, Pachycondyla* (Plate 12)	Hölldobler and Wilson (1990: Table 7–7, p. 273); Möglich and Hölldobler (1974)
Group recruitment: a successful forager recruits a group of several nestmates to the resource	May be an evolutionarily intermediate form of recruitment between tandem running and mass recruitment by pheromone trails; often involves both motor and chemical signals from the recruiting forager.	*Camponotus*	Hölldobler and Wilson (1990: p. 276)
'Mass recruitment' by pheromone trail: successful foragers leave a pheromone trail back to the nest, which is reinforced in a positive feedback and quickly attracts large numbers of nestmates.	Useful if a food source is rich but short-lived, or if it has to be defended from competitors. Depending on the volatility of the pheromone, a certain minimum number of ants are necessary to maintain the trail.	*Solenopsis, Monomorium*	Beekman *et al.* (2001); Evison *et al.* 2008 Hölldobler and Wilson (1990); Tschinkel (2006)
Stable trunk trails: relatively wide, stable trails lead out from the nest and branch into smaller trails to cover a	Often used for stable resources such as plants (where leaves or extrafloral nectaries are harvested); may also be	*Atta, Forelius, Pogonomyrmex*	Hölldobler and Wilson (1990)

foraging area. Trunk trails may also be cleared of vegetation.	used where a foraging area is systematically covered by a large colony, or as connections between nests of a polydomous colony.		
Army ant 'raids': these include large numbers of ants moving in a unified foraging front that sweeps a tract of forest. A trail network behind the swarm consolidates into a single column of traffic for returning prey to the nest and allows foragers to return to the raid.	Using this strategy, army ants have the strength in numbers to overpower other social insect colonies, or, in a few species, to 'flush out' a wider variety of arthropod prey from the leaf litter.	*Eciton, Dorylus, Neivamyrmex*	Couzin and Franks (2003); Franks *et al.* (1991); Hölldobler and Wilson (1990)
Team transport: two or more ants cooperate in transporting items from the resource back to the nest.	Used for arthropod prey that is too large or cumbersome to be effectively transported by a single forager.	*Eciton*	Anderson and Franks (2001); Franks (1987); Hölldobler and Wilson (1990: Table 10–2, p. 389f)
'Bucket brigades': foragers only transport items a short distance to a 'cache station', where it is picked up by other ants that transport it to the next station.	This strategy is only likely to be used where large colonies harvest large numbers of items that are unlikely to be 'stolen' from caches; it has been observed particularly in fungus-growing ants harvesting leaves.	*Atta, Acromyrmex*	Anderson *et al.* (2002); Röschard and Roces (2003)

poikilothermic organisms, ants have an increased metabolic rate and thus not only a higher rate of energy use, but also higher running speed at higher temperatures (Hurlbert *et al.* 2008; Traniello 1989; Tschinkel 2006). However, especially in very hot and dry environments, too high a temperature may cause desiccation and death (Cerdá 1998; Hölldobler and Wilson 1990). In addition, temperature and weather conditions may change the availability of food items; for example, prey arthropods may be inactive at low or very high temperatures, and plant nectaries may produce low yields in cold or rainy weather, or may dry up during the middle of the day. Most importantly, temperature-dependent foraging by ants provides some of the best evidence for interspecific competition (although this has been debated, see Dunn *et al.* 2007c and Chapter 5). Species occurring in the same habitat often differ in the temperatures, and thus time of day, during which they forage. Competitively dominant ants typically forage in the morning, and other ant species forage at staggered times over the course of the day (Hölldobler and Wilson 1990; Traniello 1989; but see Dunn 2007 and references therein).

12.3.2 Search behaviour

Once a forager has decided to leave the nest, it can start searching for a food source. Most modelling studies of food-search algorithms assume a correlated random walk (a random walk in which the direction of each step correlates with the previous direction of movement, Harkness and Maroudas 1985) rather than a systematic search strategy (such as searching in an expanding spiral; Figure 12.1). This random walk pattern may be modified to become straighter (smaller turning angles) in response to encounters with other ants or lack of encounters with food sources, leading the forager away from the area, or more tortuous (larger turning angles), keeping the forager in the vicinity (Gordon 2002). The optimal tortuosity of the search path is dependent not only on the probable distribution of food sources (Fourcassié and Traniello 1993), but also on the number of cooperative searchers; if many workers from the same colony are foraging in the same area, foragers should use straighter search paths to minimize overlap. Single foragers, on the other hand, perform optimally if they use a tortuous search path to maximize area coverage around the nest without moving too far away, which would increase the costs of the return journey (A. Schmolke and A. Dornhaus, unpublished data). These results from modelling studies predict that ants from larger colonies may differ from those coming from smaller colonies in their search algorithms, a prediction that has yet to be tested empirically.

12.3.3 Orientation mechanisms

While searching, ant foragers have to use strategies to track their location relative to the nest entrance, in order to be able to return there. The ant *Cataglyphis bicolor* in particular has become a standard model system for studies of orientation and navigation (e.g. Müller and Wehner 2007; Wehner and Menzel 1969; Wittlinger *et al.* 2006). *Cataglyphis* ants use a path integration algorithm to keep track of their homing vector. Path integration means that foragers continuously update their memory of the vector (direction and distance) to the nest by measuring the directions and distances that they walk. In many flying insects, such as bees, distance moved is measured by visual perception of rate of movement (called optic flow, Srinivasan *et al.* 2000). This may play some role in distance measurement by ants, but in the ant *Cataglyphis* it was recently demonstrated that a 'pedometer' is used; distance is estimated from the number of strides made (Wittlinger *et al.* 2006, 2007). Direction can be determined using a variety of compass systems, such as the position of the sun or moon, canopy patterns, prominent landmarks, the direction of wind, or the polarization pattern of the blue sky (Collett and Graham 2004; Hölldobler and Wilson 1990; Müller and Wehner 2007). Wood ants (*Formica rufa*), on the other hand, memorize snapshots of landmarks at multiple positions along a route (Collett and Graham 2004). If the ants arrive at a site that matches their stored view, they can retrieve information on the next landmark, and thus follow their learned

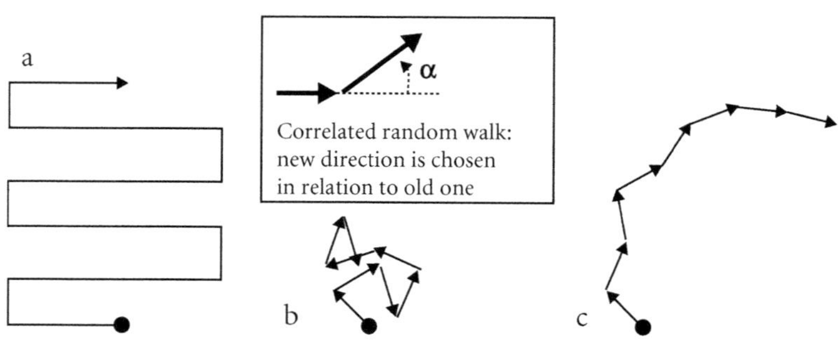

Figure 12.1 Possible search strategies: (*a*) systematic search, (*b*) correlated random walk with large turning angles, and (*c*) correlated random walk with small turning angles. In a regular random walk, the direction of a step is chosen randomly, independently of the direction of previous movement; in a correlated random walk, step directions are chosen from a distribution centred around the previous direction of movement.

route back to the nest from any point along it. Maintaining a library of snapshot landmark memories may be cognitively more costly (in terms of neural tissue or brain capacity required) than a path integration mechanism, but it is also more robust to errors in measurements of distance and direction, and it can be used more flexibly if the ant is displaced from its location by water, wind, or researchers (a condition under which a path integration mechanism fails completely). An even more costly mechanism of orientation may be to use a 'cognitive map', a map-like representation of the relative positions of landmarks, the goal (nest or food source), and the forager's own position (Collett and Graham 2004). It has been debated for a long time whether animals in general, and insects in particular, are capable of using cognitive maps, but new evidence from honeybees suggests that they can (Menzel *et al.* 2006). Such cognitive maps would enable foragers to take novel shortcuts between sites represented on the mental map, something that is impossible if only path integration or route memories are used. Whether ants use cognitive maps for orientation or learn routes based on landmarks, it is clear that learning and memory are important aspects of foraging (Dornhaus and Franks 2008). Learning also enables workers to return to profitable areas at the appropriate time (Schatz *et al.* 1999) or to remember previously found sites for later use or avoidance (Franks *et al.* 2007b).

12.3.4 Load size

Once a food source is located, a forager may have to choose the amount of food that it will bring back to the nest. It may seem that a forager should always carry as much as possible to make the search trip worthwhile, and indeed some species seem to tightly match their loads to forager body size (Powell and Franks 2005), whereas others may use tools to increase their load capacity (pellets of sand functioning as sponges, reviewed in Dornhaus and Franks 2008). However, it has been observed in several social insects that 'partial loads' are sometimes carried back to the nest, in spite of the fact that more food was available at the food source (Roces 2002). This phenomenon has caused some degree of controversy among researchers. There are several possible adaptive explanations for such partial loads, which either focus on the energetic costs of transport (Ydenberg and Schmid-Hempel 1994; reviewed in Raine *et al.* 2006), on the benefits of recruiting nestmates even at the expense of individual foraging success (Dornhaus *et al.* 2006a; Roces 2002), or on the limitations of the food processing chain at the nest (Burd and Howard 2005).

Many of the predictions of optimal forager behaviour have been derived from modelling studies, particularly in optimal foraging theory (Raine *et al.* 2006; Ydenberg and Schmid-Hempel 1994). In addition to determining the mathematically optimal behaviour, however, it is important to quantify the actual selection pressure or at least the degree to which such optimization can increase foraging success (Raine *et al.* 2006). In many cases, cognitively simpler rules of thumb may work almost equally well. For example, modelling studies predict that foragers may optimally return to the nest with a partial rather than a full load in order to collect information at the nest about possible new, superior food sources (Dornhaus *et al.* 2006a). However, within a parameter range that is biologically plausible for leaf-cutting ants or honeybees, such partial loads would increase foraging success on an average by a mere 0.000002% (ants) or 1–3% (bees) through this mechanism (Dornhaus *et al.* 2006a). It is therefore likely that if partial loads are adaptive, they evolved for another reason than quicker information collection by foragers (e.g. faster recruitment of other foragers or energetic efficiency, Roces 2002). An alternative, non-adaptive explanation is that partial loads simply result from the lack of strong selection for maximizing load size. This may be the case in taxa that do not incur significant energetic costs from foraging (see earlier), or if colony nutrition is limited by factors other than food delivery rates.

12.3.5 Morphological adaptations to foraging

In addition to potentially finely tuned behavioural strategies, ants possess a variety of morphological adaptations for capturing and processing food (see Box 12.1) (Carroll 1973; Hölldobler and Wilson

Box 12.1 Trap-jaw ants
Andrew V. Suarez and Joseph C. Spagna

The success of ants is often attributed to their remarkable social behaviour and cooperation. Group foraging species such as army ants can recruit hundreds or thousands of individuals to defend, divide, and retrieve resources such as a large insect or even a small vertebrate. However, not all ants are social hunters — some of the most successful predatory ants are solitary hunters. How do they compete with the social recruiters for resources? One way is by having some of the fastest jaws in the animal kingdom.

Of the many remarkable cases of extreme feeding ecology in the family Formicidae, few rival that of trap-jaw ants (Figure 12.1.1a). These ants use their oversized jaws and associated catapult-like muscle-firing ability to strike prey with extreme speeds and forces — exceeding 60 m/s and 500 times their own body weight, respectively (Gronenberg *et al.* 1993; Patek *et al.* 2006). These strikes are typically used for crushing, impaling, de-limbing, trapping, or ejecting prey or competitors. However, some trap-jaw ants in the genus *Odontomachus* can also use their high-powered strikes as an escape mechanism; by triggering their mandibles against the ground, they can launch *themselves* several centimetres into the air in response to threats (Patek *et al.* 2006) (Figure 12.1.1b).

Evolution and Ecology

The term 'trap-jaw ants' neither describes a single taxon nor a single clade. In a fascinating example of convergent evolution, trap-jaw morphology has evolved independently at least four times in ants, occurring in at least seven genera from three different subfamilies (Ponerinae, Myrmicinae, and Formicinae) (Figure 12.1.1c). The repeated evolution of this feeding syndrome makes it an ideal system for

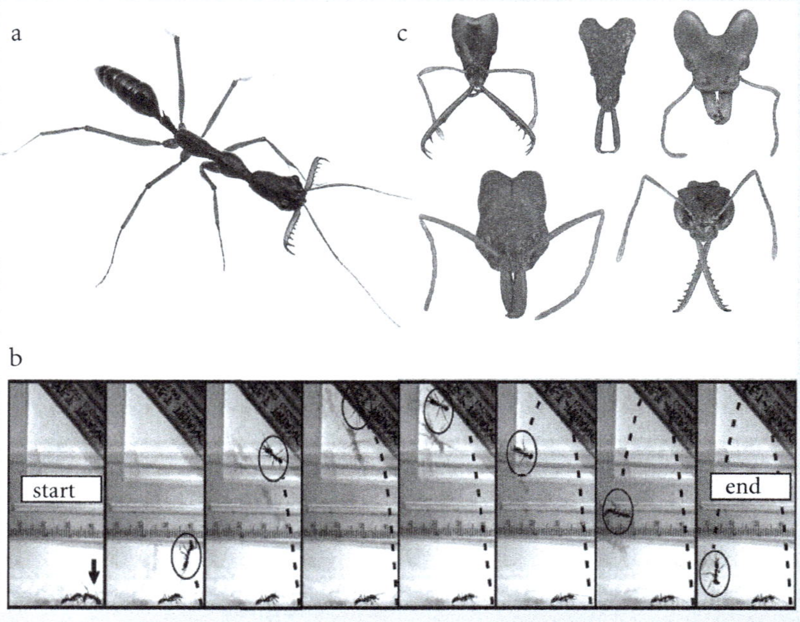

Figure 12.1.1 (a) The trap-jaw ant *Odontamachus coquereli* from Madagascar with its jaws "locked open" and ready to strike. (Photo: Alex Wild). (b) An image sequence of an *Odontomachus bauri* worker "jumping" with her jaws to escape an attacker. (Image: Patek et al. (2006) Copyright 2006 National Academy of Sciences, USA). (c) Examples of variation in trap-jaw ant morphology. Top row, left to right: *Acanthognathus*, *Strumigenys*, and *Daceton* (subfamily Myrmicinae); bottom row, left to right: *Odontomachus* (subfamily Ponerinae) and *Myrmoteras* (subfamily Formicinae). (Photos: www.AntWeb.org.)

continues

Box 12.1 continued

understanding the evolutionary prerequisites for, and ecological correlates of, feeding morphology and mechanics. Trap-jaw ants vary in size from a couple of millimetres (e.g. many *Strumigenys*) to over a centimetre (e.g. some *Odontomachus*). They are found in most biomes, but are particularly diverse in tropical and subtropical regions. Trap-jaw ants are commonly found in litter habitats (such as the genera *Anochetus* and *Strumigenys*), but will also nest arboreally (*Daceton* and some *Odontomachus*), and many are ground dwellers living in a variety of habitats including under stones, in rotten logs, and in termite mounds. Trap-jaw ants are highly predatory and most are dietary generalists, preying upon and scavenging a variety of arthropods. However, some species in the genus *Strumigenys* appear to specialize on springtails, while others in the genus *Odontomachus* prey predominately on termites, and a few have even been observed harvesting seeds (Brown and Wilson 1959a; Ehmer and Hölldobler 1995). However, for most trap-jaw ant species, little is known about their natural history.

Morphological and Mechanical Variation

Trap-jaw ants show considerable variation both within and among taxa in terms of mandible size, shape, and the mechanics of storing and generating force (Figure 12.1.1c). The mandibles are elongated and project from the head (anteriorly when relaxed, laterally when cocked), and store energy using a latch or 'click' mechanism. Across trap-jaw species, this mechanism is built using different anatomical structures, such as modifications of the jaw insertion points in *Odontomachus*, a modified labrum in *Strumigenys* and *Daceton,* which blocks the mandibles from closing, and by interlocking mandibular processes in *Acanthognathus* (Gronenberg 1995; 1996; Gronenberg and Ehmer 1996). In addition to the variation in locking mechanisms, the relative size, orientation, and attachments of mandible opener and closer muscles vary dramatically across trap-jaw ants. Across taxa, the mandible closer muscles can occupy over 60% of the head volume, and are often contained in visible oversized lobes extending the posterior margins of the head. Furthermore, variation in muscle volume is accompanied by variation in the relative composition of muscle fibre types. For example, in some trap-jaw ants, the small trigger muscles that release the strike are among the fastest muscles known in animals (Gronenberg *et al.* 1993). Finally, the shapes and surfaces of the jaws themselves vary considerably among species (Figure 12.1.1c). The jaws may be long or short, narrow or broad. Many trap-jaws are capped by large medially oriented terminal teeth, and the leading edges of the mandibles may also be lined with teeth, which may be sharp or blunt; the leading edges may also lack teeth and have a wedge- or scissor-like surface more suitable for cutting. Whether or not these variations are optimized for capture of certain prey-types, colony defence, or jumping ability is largely unknown and is a rich area for more research.

1990; Powell and Franks 2005, 2006). Aspects of body shape such as relative leg length are likely to be adaptations to the specific mode of foraging employed by the respective species; surface-running ants have longer legs, whereas ants that live in an interstitial environment such as leaf litter have short legs (Kaspari and Weiser 1999; Kronauer *et al.* 2007b; Schöning *et al.* 2005; Weiser and Kaspari 2006). Some species also display worker polymorphism, variation in worker sizes and body shape, producing a worker-caste particularly adapted to a foraging task (see Plate 10). For example, the leaf-cutting ants in the genus *Atta* produce a range of worker sizes: the largest workers with their strong mandibles cut leaves, other large workers walk fast to transport them, whereas smaller workers tend fungus inside the nest (Wilson 1980). Morphological differences may also occur interspecifically, predicting the diversity and toughness of leaves harvested (Wetterer 1995). Sometimes, very small workers ride on leaves carried by large workers, which may have two important defensive functions (see Section 12.6.3). Several species also possess 'soldiers', that is, workers morphologically specialized for defence (Section 12.6.3).

12.4 Collective strategies in foraging

Ants are eusocial, and collected food is shared among all members of the colony. Therefore, foragers do not operate alone, but are part of a colony-level foraging strategy. There are numerous collective strategies that are used by different species of ants (Table 12.1). These may include recruitment to the food source by pheromone trails and a wide variety of other communication signals.

12.4.1 Recruitment by pheromone trails

Pheromone trails are the most conspicuous, and therefore the best-studied mode of collective foraging in ants. Successful ant foragers in many species lay a trail on their way back to the nest, often by intermittent touching or dragging of the gaster on the substrate. Different ant species use different glands as sources of recruitment pheromone (Hölldobler and Wilson 1990), suggesting possibly many independent origins of this behaviour. Pheromone trails may serve many functions, and not all trails serve to find food and retrieve it; other resources, for example nesting sites (Dornhaus *et al.* 2004), building material (Aleksiev *et al.* 2007), or mutualist plants (Webber *et al.* 2007), are also sought by ants. Furthermore, individual-specific trails may aid in orientation (Hölldobler and Wilson 1990; Maschwitz *et al.* 1986), or be used to measure area of a nest site (Mallon and Franks 2000). Sometimes permanent trunk trails are maintained by ant colonies (Edelstein-Keshet *et al.* 1995), often cleared of vegetation or even reinforced with built structures (Anderson and McShea 2001b). In the most extreme case of the army ants, pheromone trails provide the foundation for an obligate group foraging strategy (Rettenmeyer 1963). Ant traffic on trunk trails can self-organize into separate traffic lanes, which increases running speed (Couzin and Franks 2003); interactions among ants on the trail may also have other functions (Dussutour *et al.* 2007; Gordon *et al.* 2008; see Burd 2006 for a review). In some ant species, worker ants improve their trail surface or width by plugging 'potholes' with their bodies (Powell and Franks 2007) or, more rarely, by forming living bridges (Hölldobler and Wilson 1990). Recent research has shown that in Pharaoh's ants (*Monomorium pharaonis*), foragers can use the trail branching angles to determine which direction on the trail will lead them back to the nest, and which direction will lead out to the food source (trail polarity; Jackson *et al.* 2004). In the same species, it was also found that ants cannot only recruit using pheromones, but also repel other ants from certain routes, for example to indicate unprofitable areas (Robinson *et al.* 2005a). Subtle differences in the chemical composition of pheromones may in addition indicate who laid them (Jackson *et al.* 2007). The shape of the trail system can be species-specific because of differences in these behaviours, or can be an emergent phenomenon resulting from particular resource distributions (Crist and Haefner 1994; Edelstein-Keshet *et al.* 1995; Franks *et al.* 1991).

12.4.2 Recruitment by tandem running

Pheromone trails, however, are not the only mode of communication available to ants. For example, a well-studied recruitment behaviour that involves both pheromone and mechanical signals is tandem running (Möglich and Hölldobler 1974; Plate 12). During a tandem run, an ant worker leads a single recruit to a resource; the recruit follows the leader closely, frequently touching its antennae to the gaster of the leading ant. If this touching is interrupted, the leading ant will remain in place and wait until the recruit catches up. Because of this feedback between leader and follower, tandem running may fulfil the criteria for 'teaching' (Richardson *et al.* 2007). To initiate a tandem run, the scout ant will use both antennation and a 'tandem calling pheromone' in the nest (Hölldobler and Wilson 1990; Möglich and Hölldobler 1974).

12.4.3 Other communication

As these examples show, collective foraging has mostly been studied in the context of signals informing recruits about the *location* of food sources or nest sites. However, foragers may communicate

Table 12.2 Categories of morphological defences seen in ant foragers. See also Plate 13 and Hölldobler and Wilson (1990: Table 10–3, p. 393f).

Morphological defence	Function	Associated foraging strategies	Exemplary genera	References
Armour: Thickened exoskeleton, and sometimes associated with a capsule-like gaster (expanded first gasteral segment)	Provides mechanical barriers against crushing, cutting, and puncturing forces exerted by aggressive prey or arthropod enemies	Common in predatory species. Also seen in subordinate omnivores that forage within the territories of aggressive species	*Cerapachys*, *Nomamyrmex*, *Paraponera*, *Cataulacus*, *Procryptocerus*, *Cephalotes* (Plate 13)	Buschinger and Maschwitz (1984); Hölldobler and Wilson (1990); Powell and Clark (2004)
Antennal scrobes: Depressions or cavities on the head that receive the antennae when folded	Provides sheltered protection for the antennae (primary sense organs) when attacking or under attack	Some predatory and slave-making species, as well as subordinate omnivores	*Aconthoponera*, *Harpagoxenus*, *Cataulacus*, *Procryptocerus*	Hölldobler and Wilson (1990); http://www.AntWeb.org
Shield: Lateral, membranous expansions of the exoskeleton on the head, mesosoma, or gaster	Provides expendable structures that can be seized, chewed, and damaged by arthropod enemies without injury to the ant	Slow-moving omnivores, potentially foraging within the territories of aggressive, territorial species	*Cephalotes*, *Meranoplus* (Plate 13)	Andersen (2006); de Andrade and Baroni-Urbani (1999)
Erect teeth and spines: Sharp triangular and elongate exoskeletal projections orientated outwards	May provide mechanical protection against vertebrate predators (spines potentially make the ants painful to capture and eat)	Common in relatively large taxa that forage on vegetation in the understory or canopy in the tropics	*Acromyrmex*, *Atta*, *Polyrhachis*, *Cephalotes* (*atratus* clade) (Plate 13)	Buschinger and Maschwitz (1984); de Andrade and Baroni-Urbani (1999) http://www.AntWeb.org
Decumbent teeth and spines: Lobed, triangular and elongate exoskeletal projections orientated along the plane of the body and usually over the main articulation points	May provide protection against arthropod enemies by blocking direct attacks on the main articulation points of the body	Widespread, including taxa that have foraging strategies with high-level interactions with arthropod enemies and low-level interactions with vertebrates	*Aconthoponera*, *Eciton*, *Phrynoponera*, *Cephalotes*, *Meranoplus*, *Harpagoxenus*, *Polyrhachis* (Plate 13)	Andersen (2006); Buschinger and Maschwitz (1984); de Andrade and Baroni-Urbani (1999); http://www.AntWeb.org
Pubescence: Entire body with a dense covering of long hairs	May provide mechanical barrier to biting attack by arthropod enemies	Appears to be associated with slow-moving, non-predatory species	*Apterostigma*, *Echinopla*, *Procryptocerus*, *Meranoplus* (Plate 13)	Andersen (2006); http://www.AntWeb.org
Crypsis: Structures that accumulate	Organic debris conceals the ants.	Debris camouflage associated with	*Basiceros*, *Stegomyrmex*,	de Andrade and Baroni-Urbani

(*Continued*)

Table 12.2 Continued

Morphological defence	Function	Associated foraging strategies	Exemplary genera	References
camouflage material or create a disrupted body outline	Exoskeletal structures, colouration and hairs may disrupt body outlines when ants are motionless	ambush predators. Outline disrupting morphology associated with slow-moving omnivores	*Cyphomyrmex, Cephalotes*	(1999); Diniz and Brandão (1993); Hölldobler and Wilson (1990); http://www.AntWeb.org

other information, such as *quality* of the resource (McCabe *et al.* 2006; Roces 2002). Communication with nestmates also influences the decision to start foraging: patroller ants may indicate the possible costs (Greene and Gordon 2007a), and successful foraging by others may indicate potential benefits of leaving the nest (e.g. McCabe *et al.* 2006). In addition, several workers may cooperate to transport heavy or unwieldy items to the nest (e.g. Franks 1987; Traniello and Beshers 1991). These 'teams' sometimes comprise workers of different sizes to maximize transport efficiency (Anderson and Franks 2001; Franks 1987; Powell and Franks 2005). Ants may also improve their foraging success by building new nests or moving existing nests into the proximity of stable resources (van Wilgenburg and Elgar 2007).

12.4.4 Ecology and evolution of different foraging strategies

The two factors whose influence on the evolution of foraging strategies has been studied most are spatial resource distribution and colony size. Ants that prey on solitary arthropods, which may be widely distributed across the foraging range and not occur in patches, may have no need for a mass recruitment system (Hölldobler and Wilson 1990). Ants that raid termite colonies, other ant colonies (i.e. predators, like army ants, or social parasites, like slave-making ants), or that exploit other highly profitable, stable resources (e.g. trees in leaf-cutting ants) on the other hand, are likely to benefit from the ability to recruit nestmates in large numbers. In these species, recruitment trails are common. Pheromone trails can generate a steep increase in the number of ants recruited through positive feedback (if recruits add to the pheromone trail and thus recruit more ants in turn). However, this effect is dependent on a sufficient colony size supplying a large number of potential recruits. In small colonies, pheromone trails may be slow to develop or unstable (Beekman *et al.* 2001; Edelstein-Keshet *et al.* 1995). Colony size may thus predict the complexity of the communication strategy used where pheromone trails are concerned (Beckers *et al.* 1989; Herbers and Choiniere 1996; Mailleux *et al.* 2003), but it is less clear whether colony size has an influence on the benefits of other modes of communication. This is because colony size (i.e. the number of potential recruits at the nest) is irrelevant to the success of a recruitment strategy when the rate of recruitment is not limited by the number of potential recruits. This is the case whenever each forager can only recruit a fixed number of nestmates at a time (Dornhaus *et al.* 2006b), such as in tandem running or small-group recruitment (incidentally, this is also true for the honeybee waggle dance). The evolution of group recruitment and tandem running are thus likely to be determined by factors other than colony size.

12.5 Individual defence strategies

During the act of foraging, individuals use a range of strategies to defend both themselves and any resources they are handling. Individual defence strategies can thus be broadly defined as any morphological, chemical or behavioural characteristic used against enemies in such a way that it improves the survival and resource acquisition of individual ants. These defences may, in some cases, involve characteristics used in the foraging process, while others appear to depend on adaptations to the special dangers of the taxon's ecology.

12.5.1 Morphological defences

Ant mandibles serve as the primary manipulation-tools in colony life, and they are typically robust and capable of exerting considerable force. The mandibles are therefore of universal importance as defensive weapons, and they are particularly effective in taxa with foraging strategies that have been selected for powerful cutting, crushing, or striking mandible morphologies. Good examples include the scissor-like mandibles of *Atta* leaf-cutting ants, the powerful chewing mandibles of *Camponotus* carpenter ants and the snapping mandibles of predatory genera like *Odontomachus* (Hölldobler and Wilson 1990; see Box 12.1).

In addition to morphological weapons, numerous protective morphological structures have evolved (Table 12.2; Plate 13). While the general function of these structures seems clear, their effectiveness under particular ecological conditions, including against specific types of enemies, have not been explored in detail. For instance, spines, a common morphological defence, are often thought to provide mechanical protection against vertebrate predation (Hunt 1983). Direct support for this hypothesis is lacking, but it seems reasonable for erect spines, particularly as they are common in larger taxa that forage on vegetation (Table 12.2), where predation pressure from birds and mammals is high. Other taxa, however, have lobe-like extensions to the cuticle, teeth, or spines that extend backwards over the main articulation points. These structures may protect against arthropod enemies by providing barriers that block biting attacks on these weak areas. This may be the case in army ants, for instance, where backwards-projecting vertexal lobes or teeth on the head prevent direct strikes by their ant prey (S. Powell, personal observation). Likewise, dense hairs are another common, but poorly understood morphological defence. While hair can be important for trapping camouflage material in some ants, other ant taxa have dense covering of long hair that is kept clean (Table 12.2). These may serve as a mechanical barrier against arthropod enemies, in much the same way that plant trichomes provide a mechanical barrier against insect herbivores (e.g. Johnson 1975).

Figure 12.2 Feeding at a rich food resource in *Crematogaster* and *Cephalotes*: *Crematogaster* sp. foragers vigorously defend the food and exude chemical repellents from the tip of their gasters while a heavily armoured *Cephalotes persimilis* forager pushes past the *Crematogaster* sp. defences, apparently without injury. (Photo: Scott Powell)

In addition to morphological structures that provide direct defence, warning colouration (aposematism) may serve as a deterrent to some enemies. In taxa like *Pseudomyrmex* and *Myrmecia*, bright and often patterned colouration would appear to be the advertisement of their potent sting (see Section 12.5.2; Hölldobler and Wilson 1990). In other taxa, the bright colouration may warn of distastefulness. For instance, the gynes and soldiers of numerous *Cephalotes* species have brightly coloured gastral eyespots (de Andrade and Baroni-Urbani 1999). *Cephalotes* lack a functional sting, but some evidence suggests that many species produce strongly distasteful chemicals (Coyle 1966; de Andrade and Baroni-Urbani 1999).

12.5.2 Chemical defences

A sting mechanism for venom injection is ancestral (plesiomorphic) in the ants, and remains a formidable chemical weapon in many extant taxa. The sting is formed by the modification of abdominal segments 8–10, and it is fed defensive chemicals from associated glands, with the venom typically produced in the poison gland (Buschinger and Maschwitz 1984). The sting is a particularly conspicuous weapon in predatory species that use it to kill prey, like many poneroid taxa that hunt solitarily (Buschinger and Maschwitz 1984), and group predators like the 'New World army ants' (Ecitoninae; Powell and Clark 2004; Powell and Franks 2005). When threatened, these ants also use the sting as a defensive weapon.

A functional sting has, however, been lost a number of times independently, and the degree to which it has been modified, reduced, or co-opted for other functions is also highly varied (Buschinger and Maschwitz 1984). For example, *Crematogaster* produce venom that does not need to be injected. Instead, it is effective when deposited on an enemy, with the sting little more than an applicator, and it also acts as a repellent (Buschinger and Maschwitz 1984; Marlier *et al.* 2004; Figure 12.2). In other taxa, the sting mechanism and associated glands have been co-opted for non-defensive functions, as in *Atta*, where the poison gland produces only trail pheromones (Buschinger and Maschwitz 1984; Hölldobler and Wilson 1990). In the more extreme case of the subfamily Formicinae, the sting mechanism has been lost entirely and formic acid, still produced in the poison gland, is deposited or sprayed directly onto enemies (Buschinger and Maschwitz 1984). Finally, in the Dolichoderinae, the sting and poison gland are greatly reduced, and a cocktail of defensive chemicals is instead produced in the pygidial gland (also known as Janet's gland; Buschinger and Maschwitz 1984). Numerous other glands produce defensive chemicals, but these tend to supplement other fighting strategies. Examples include foul-smelling and repellent secretions produced by the paired mandibular glands of many taxa, and sticky metapleural gland sections in some *Crematogaster* species (Buschinger and Maschwitz 1984). It is interesting to note that taxa with potent chemical defences often lack strong morphological defences, and the most morphologically defended ants (Table 12.2) tend to have limited chemical weaponry. As Hunt (1983) suggested, morphological and chemical defences may therefore represent alternative evolutionary defence strategies, but this idea has yet to be tested. Similarly, robust analyses of the selective pressures, and particularly the foraging strategies, that favour sting loss have not been conducted. Possible reasons for sting loss include selection for other types of chemical weapons that are more effective against other ants (Buschinger and Maschwitz 1984) and an evolutionary shift in diet away from predation on live arthropods (Kugler 1979). However, exceptions exist for both of these proposed scenarios. Sting loss is therefore likely to result from a number of interacting selective pressures.

12.5.3 Behavioural defences

Behaviour can provide unique defence strategies, and it can also modify and improve the effectiveness of morphological and chemical defences. Bursts of speed are a common behavioural strategy for avoiding danger in the foraging arena, and path complexity or tortuosity may also improve the likelihood of escape (Angilletta *et al.* 2008). These avoidance strategies are likely to be of the greatest importance in taxa where the foraging strategy has selected for high running speeds, such as *Cataglyphis* and *Ocymyrmex* (Hurlbert *et al.* 2008). Diametrically opposite to running is the defence strategy of freezing or 'playing dead', and it is typically seen in species with foraging strategies that are associated with slow movement and morphological defences. Examples include *Cephalotes* (de Andrade and Baroni-Urbani 1999), *Meranoplus* (Andersen 2006), and *Cyphomyrmex* (Hölldobler and Wilson 1990). In the case of the arboreal genus *Cephalotes*, body sculpturing and cryptic colouration may help disrupt the outline of the ants, making them even harder to see when they are motionless (Table 12.2; de Andrade and Baroni-Urbani 1999). However, when they visually detect that they have been spotted, or an enemy strikes at them, they use a radically different strategy: they jump (S. Powell, personal observation). Recent work has found that once these falling ants reach a critical speed, they are capable of

Box 12.2 The directed aerial descent of arboreal ants
Stephen P. Yanoviak

The evolution of winged flight in insects is unresolved due to a lack of fossil intermediate forms, but was likely preceded by directed aerial descent (i.e. gliding) in an arboreal setting (Dudley *et al.* 2007). Whereas a variety of vertebrates exhibit aerial gliding, the behaviour was unknown in wingless arthropods until it was recently documented in arboreal ants (Yanoviak *et al.* 2005, 2008a). Given that ants are a derived group among insects and are secondarily wingless, their gliding behaviour, while interesting and unexpected, is not directly relevant to the origins of winged flight in insects. However, their abundance in tropical forest canopies, their large variation in body size and morphology, and recent improvements to their phylogenetic resolution make ants an excellent focal group for investigating the selective pressures and aerodynamic mechanisms associated with this remarkable behaviour.

Arboreal ants forage in a relatively exposed physical setting. They may be accidentally dislodged from trees (e.g. Haemig 1997), or may voluntarily drop from branches when provoked (Yanoviak and Dudley 2006; Yanoviak *et al.* 2008a). In preliminary studies in Peru, worker ants composed 66% of wingless arthropods collected in ten passive funnel traps suspended in the forest canopy (Yanoviak, unpublished data). Thus, significant numbers of workers fall as 'ant rain' in tropical forests.

Lost workers are costly to ant colonies, and landing in the unfamiliar understory may have grave consequences for arboreal ants. Seasonally flooded forests are common in the tropics and present the most extreme circumstances — fallen insects are immediately consumed by surface-feeding fish. But even dry understory litter may pose a significant hazard. For example, up to 100% of arboreal ants released in the litter were attacked, and up to 40% were killed by the resident litter fauna in preliminary trials conducted in Peruvian terra firme forest (Yanoviak, unpublished data). Thus, the likelihood of a fallen arboreal ant returning to its point of origin after landing in the understory is presumably low, and gliding reduces this loss (Yanoviak *et al.* 2005).

Most research on gliding ants to date has focused on the myrmicine genus *Cephalotes*, especially the common Neotropical species *C. atratus* (Figure 12.2.1). However, at least six other ant genera include gliding species: *Camponotus*, *Cataulacus*, *Daceton*, *Nesomyrmex*, *Procryptocerus*, and *Pseudomyrmex*. Glide performance is generally size-dependent within and among species (Yanoviak *et al.* 2005, 2008a). Specifically, smaller workers within colonies, and smaller species within genera, tend to have larger glide indices (glide index = the horizontal distance travelled per unit vertical drop distance). The consistency of these size-

a

b

Figure 12.1.2 The Neotropical ant *Cephalotes atratus* (a) is a common inhabitant of rainforest canopies across South America. If a worker of this species is dislodged from the tree trunk it is able to (b) direct its aerial descent back to the tree trunk. (Photos: Alex Wild)

Box 12.2 continued

based patterns largely results from basic physics (i.e. smaller ants reach terminal velocity earlier in a fall). In contrast, mechanisms of aerodynamic stability and glide control are predominately behavioural and differ markedly among taxa. For example, *Camponotus* workers glide toward tree trunks head-first, whereas *Cephalotes* and *Cataulacus* glide abdomen-first (Yanoviak *et al.* 2005, 2008a). The aerodynamic relevance of different appendages during a fall also differs among taxa. Experiments with *C. atratus* suggest that the hind legs are necessary for aerodynamic stability in a fall, whereas field observations indicate that the forelimbs may serve this function in some *Camponotus* (Dudley and Yanoviak, unpubl. data).

Given that not all arboreal ants glide, and not all ants glide in the same manner, what traits can be associated with this behaviour? Gliding taxa share four characteristics: (*a*) costly workers (relatively small colony size and large per-worker investment); (*b*) arboreal nesting (ground-nesting arboreal ants like *Atta* and *Paraponera* do not glide); (*c*) good vision; and (*d*) diurnal activity (Yanoviak *et al.* 2005). The latter two characteristics are necessary for targeting during a fall. No nocturnal ants are known to glide, and falling *C. atratus* depend on visual cues to orient to light-coloured objects (e.g. lichen-covered tree trunks; Yanoviak and Dudley 2006). All available evidence indicates that gliding has multiple independent origins in ants. Comparative phylogenetic analyses and more information regarding the ecology, natural history, and morphology of arboreal ants will clarify the selection pressures associated with the behaviour.

In sum, ants provide an excellent model for studying gliding in small, wingless arthropods because they are abundant and experimentally tractable. However, ongoing research shows that gliding is not limited to ants — indeed, directed aerial descent is widespread among arboreal arthropods, some of which may support hypothesized terrestrial origins of insect flight (Dudley *et al.* 2007). Uncovering the mechanisms and constraints associated with the behaviour in ants will facilitate research on less common taxa that are relevant to understanding the evolution of winged flight in insects.

directed descent, returning them to their home tree or nearby vegetation with great reliability (Yanoviak *et al.* 2005; Box 12.2).

12.6 Group defence strategies

While individual defences may improve the survival and resource acquisition of individual ants, they must be put in the broader context of the defensive strategies of the colony as a whole. These group defences, while benefiting from individual defences, can be defined as those that require coordination of more than one individual for success, often at the cost of some of the individuals involved. The coordinated nature of these collective actions achieves defensive effectiveness well beyond the sum of the capabilities of the participating individuals.

12.6.1 Coordinated group defence at the nest

Coordinated group defences are defined here as strategies that use recruitment (pheromonal, tactile, or acoustic) to mobilize a mass defensive response to a specific threat at a specific location. Most ant species display coordinated group defence when an enemy is detected at the nest, but the strength of the response and the degree to which a colony relies on fight or flight depends on the species, life stage of the colony, and the enemy. The universal self-sacrificing behaviour of ant workers is the key in coordinated fight responses because it can increase their overall potency. In many taxa, a fight response is very general, involving widely broadcast alarm recruitment that releases excitement and aggression (Hölldobler and Wilson 1990). However, these responses can be more sophisticated, involving directed aggression, enemy-specific responses, and cooperative combat strategies. For example, in *Oecophylla longinoda,* multi-component mandibular gland secretions act together to alarm workers, attract them to a point location, and elicit aggressive behaviour (Bradshaw *et al.* 1975). In *Pheidole dentata,* workers recruit nestmates to locations where *Solenopsis* workers have been detected, but show limited or no response to other potential ant enemies (Wilson 1976). Likewise, *Atta* species immediately recruit a massive defensive force when their main invertebrate enemy, the army ant *Nomamyrmex esenbeckii,* is detected, but show no response to other army ants (Powell and Clark 2004). In both *P. dentata* and *Atta,* the defensive response involves a special role for soldiers (see Section 12.6.3), and *Atta* also use cooperative combat strategies to immobilize army ant raiders. In ant–ant fighting, theory suggests that individual fighting prowess and numerical superiority both play key roles in the outcome of a battle, and that their relative importance depends on the type of combat (i.e. a series of one-on-one battles, or an all-against-all war of attrition) that the battlefield allows (Adams and Mesterton-Gibbons 2006; Franks and Partridge 1993). These predictions have been largely supported by observational data from natural large-scale battles (Powell and Clark 2004) and from experimental studies (McGlynn 2000; Plowes and Adams 2005). Further work in this area is likely to prove valuable in understanding the evolution of coordinated group fighting strategies in ants, both at the nest and in the foraging arena.

Flight, or coordinated nest evacuation, is also a common defensive response in ants. Evacuation involves all colony members abandoning the nest with whatever brood they can carry, and seeking temporary shelter until the threat has passed. In some taxa, evacuation is a secondary or simultaneous strategy to fighting; in others, often small colonies or evacuation specialists, it is the first response. In all cases, evacuation appears to be particularly important when ants are under attack by army ants, which may occur frequently for ground-nesting taxa in tropical regions (O'Donnell *et al.* 2007). Thus, while mature *Atta* colonies fight *N. esenbeckii* raiders (Powell and Clark 2004), they resort to evacuation when the army ants enter the nest, and young colonies immediately evacuate without fighting (S. Powell, unpubl. data). *Aphaenogaster* colonies, on the other hand, tend to take flight as soon as a threat is detected, initiating rapid and highly coordinated evacuation of the entire nest (Hölldobler and Wilson 1990; Smith and Haight 2008).

12.6.2 Coordinated group defence when foraging

In the foraging arena, species with a large workforce of aggressive workers often recruit a strong defensive force to resources that cannot be harvested by individuals. These species are often referred to as 'extirpators' (Hölldobler and Wilson 1990) and the resources that they defend may be depleteable, like large prey items, or renewable resources like honeydew-producing herbivores and nectar-producing plants. Examples include members of the genera *Azteca, Camponotus, Crematogaster, Pheidole, Solenopsis,* and *Wasmannia* (Hölldobler and Wilson 1990; McGlynn 2000), with soldiers playing an important role in resource defence in *Azteca, Solenopsis,* and *Pheidole* (see Section 12.6.3). Collective use of chemical weapons can also aid resource defence, both in direct combat with enemies and as repellents around or on a resource (Buschinger and Maschwitz 1984). Interestingly, however, resource defence is mostly directed against other resource-defending species, and is generally ineffective against morphologically well-defended foragers or so-called insinuators. Species of this guild use a combination of stealth and bulldozing tactics to reach and feed at guarded food (Figure 12.2). It is not known how much food insinuators can steal from extirpator species in this way, or whether the lost food has any significant fitness consequences for the extirpators.

In a relatively small number of highly aggressive taxa, a territory is established to provide the colony with exclusive access to resources contained within the territorial border (reviewed in Hölldobler and Wilson 1990; Traniello 1989). Conspecifics, and other potential competitors are excluded from the territory with a number of coordinated border defence strategies. The first is simply the recruitment of a large number of individuals to defend a border

Table 12.3 Ant genera with at least one species that has a morphologically and behaviourally specialized soldier caste. List compiled by cross-referencing Baroni-Urbani (1998), Bolton *et al.* (2006), and Formis 2008 database (Wojcik and Porter 2008).

Ant subfamily	Genera with a soldier caste
Aneuretinae	*Aneuretus*
Dolichoderinae	*Azteca, Philidris, Tapinoma*
Dorylinae	*Dorylus*
Ecitoninae	*Cheliomyrmex, Eciton, Labidus*
Formicinae	*Camponotus, Cataglyphis, Cladomyrma, Gesomyrmex, Melophorus, Myrmecorhynchus, Notostigma, Pseudolasius*
Myrmecinae	*Acanthomyrmex, Adlerzia, Anisopheidole, Atta, Cephalotes, Carebara, Machomyrma, Monomorium, Orectognatus, Perissomyrmex, Pheidole, Pheidologeton, Solenopsis*
Pseudomyrmecinae	*Tetraponera*

when a threat is detected, as is the case in *Atta* (Whitehouse and Jaffe 1996). In species like *Azteca trigona*, a substantial presence is permanently maintained at the border, with reinforcements recruited when a threat escalates (Albrecht and Gotelli 2001). However, full territorial battles can result in massive mortality on both sides and even colony death (reviewed in Hölldobler and Wilson 1990). In some taxa, this has apparently selected for mechanisms that allow peaceful resolution to territorial disputes. For instance, border marking establishes chemical signals at the territory perimeter that can be honoured by neighbours to prevent unnecessary border skirmishes and escalated aggression (reviewed in Buschinger and Maschwitz 1984; Hölldobler and Wilson 1990). Ritualized combat, on the other hand, allows disputes to be resolved by way of a non-destructive show of force: the colony that displays its numerical superiority wins. Ritualized combat can take many forms, but includes non-escalating fights (e.g. *Cataglyphis niger*, Werner 1976), ritualized fighting behaviours like front-leg boxing (e.g. *Camponotus gigas*, Pfeiffer and Linsenmair 2001) and display tournaments (reviewed in Hölldobler and Wilson 1990).

In discussing coordinated group combat strategies in ants, the special case of repellent and propaganda pheromones that have evolved in arms races between ant predators and their ant prey is worth noting. These chemical weapons, used in a coordinated group context, neutralize the defensive strategies of the victim species by sending them into misdirected chaos and, in some cases, fighting each other instead of the enemy (Lenoir *et al.* 2001). Examples include slave-making species like *Formica subintegra* and *Harpagoxenus sublaevis* (Lenoir *et al.* 2001), and agro-predators like *Gnamptogenys hartmani* (Dijkstra and Boomsma 2003).

12.6.3 Defence, soldiers, and self-sacrifice

Colony defence appears to be a powerful selective force in the evolution of polymorphism in ants, with numerous independent origins of a soldier caste within the family. A soldier is defined here as a worker phenotype that is morphologically specialized for the role of colony defence. These individuals all display specialized defensive behaviours, but they can differ significantly in the degree to which their full behavioural repertoire is reduced from that of a standard worker. Building on the list compiled by Baroni–Urbani (1998), and adjusting for recent changes in ant taxonomy, good evidence can be found for a soldier caste in 30 genera from 7 subfamilies (Table 12.3). Among these taxa, soldiers can be categorized as specialized for (*a*) active defence against vertebrates, (*b*) active defence against arthropods, and (*c*) passive defence (i.e. entrance-blocking function) against arthropods (Hölldobler and Wilson 1990). The army ant genus *Eciton* provides the clearest examples of soldiers specialized for defence against vertebrates. These soldiers have long mandibles with recurved tips (Figure 12.3a) that easily

a
b

Figure 12.3 Soldiers exhibit a range of defensive behaviours: (*a*) Guarding of a foraging trail by a soldier in the army ant *Eciton hamatum*. The soldier (*right*) stands alert as foragers retrieve prey in the background. When a threat is detected, soldiers run in loops at the site of the alarm and immediately bite any vertebrate they encounter. The re-curved tips of the mandibles prevent these soldiers from releasing once they have punctured the flesh of a vertebrate enemy; (*b*) Entrance blocking by a *Cephalotes persimilis* soldier. These soldiers use their heavily armoured and specialised head-discs to prevent enemies from entering the multiple nests of their colony. Nests are made in pre-existing wood cavities, originally made by wood-boring insects, providing highly defensible structures once soldiers have blocked the entrances. (Photos: Scott Powell)

puncture vertebrate flesh, but the mandible blades lack grasping surfaces for seizing and cutting arthropods (Rettenmeyer 1963). In *Eciton* species that forage diurnally, like *Eciton burchellii and Eciton hamatum*, soldiers are deployed at large caches of prey and along foraging trails. In all *Eciton* species that have a soldier, members of this caste travel in a large group around the queen during emigrations, and remain in significant numbers at the nest at all times (Rettenmeyer 1963; Schneirla 1971). Some of the best examples of soldiers that are effective in active defence against arthropod enemies are seen in *Atta* and many *Pheidole* species (Hölldobler and Wilson 1990). A common strategy for soldiers of this type is to dismember their arthropod enemies, which are often other ants, and sometimes this is achieved with the help of smaller colony members (e.g. Detrain and Pasteels 1992; Powell and Clark 2004; Wilson 1976). Finally, the most striking examples of a soldier caste specialized for passive blocking defence at the nest are seen in *Cephalotes*. In the most derived species, the soldier caste has an elaborate disc-like structure on the head that is used to seal the entrances of the pre-existing arboreal cavities inhabited by these ants (Figure 12.3b; de Andrade and Baroni-Urbani 1999; Powell 2008). The basic function of a soldier caste, regardless of type, is clear in most taxa that have one, and one study has clearly demonstrated that the loss of soldiers can dramatically affect colony survivorship (Hasegawa 1993b). However, remarkably little is understood about how the fitness contribution of soldiers differs with environmental conditions and their temporal and spatial availability. Studies that address these types of issues will likely be of great value in understanding why soldiers are a convergent pattern in the evolution of defence strategies in ants.

While soldiers represent the common specialized defensive caste, the so-called hitchhikers provide a unique kind of caste-based defence in *Atta*. Here, members of the smallest caste (often referred to as 'minims') ride on leaf fragments being transported by larger individuals, removing fungal contaminants from the harvested leaves (Vieira-Neto *et al.* 2006) and possibly defending transporters against parasitic flies (Feener and Moss 1990; Vieira-Neto *et al.* 2006). Adaptations that increase the effectiveness of self-sacrificial defence are also not limited to soldier evolution. The most notable example is seen in the 'exploding' workers of some species in the *Camponotus cylindricus* clade. These ants have greatly hypertrophied mandibular glands that stretch the length of the body. Under conditions of an extreme threat, muscular contractions rupture the glands and the intersegmental membranes so that sticky secretions are released explosively onto an enemy. This defence immobilizes the enemy effectively, but kills the ant (Davidson *et al.* 2007).

12.6.4 Structural defences

In addition to the direct defence strategies employed by ants, constructed barriers or fortification can play an important role in fending off enemies. One of the most basic but common strategies is to seal nest entrances with soil or solid objects like stones or twigs. This is seen in a wide range of taxa, including *Atta, Camponotus, Cataglyphis, Harpegnathus, Messor, and Pogonomyrmex* (Buschinger and Maschwitz 1984; Hölldobler and Wilson 1990; Powell and Clark 2004). Redundancy may also be built into the nesting behaviour of a species by way of polydomy (two or more spatially separated nests used by one colony). Thus, when one nest is attacked, the colony may survive because predators do not locate all nests, or because additional nests provide a secure destination following evacuation of the attacked nest (Debout *et al.* 2007). In species that have the ability to construct their nest in soil or using other material, selection is likely to favour some level of defensive architecture. However, quantitative studies of ant nest architecture are scarce (Tschinkel 2004), and how architectural components may aid defence and increase colony fitness is essentially unknown. A few peculiar defensive nest architectures have been identified, however, with one of the strangest seen in the Neotropical species *Blepharidatta conops*. These ants construct a highly defensible internal chamber that is separated from the other parts of the nest by a wall built from insect carcasses. A single entrance hole is shaped in the defensive wall, and the queen seals the hole with her head, which is morphologically specialized for the task (Brandão *et al.* 2001). Structural defences can also play a role during foraging. For instance, foragers of *Labidus* army ants bury large food items (Rettenmeyer 1963), and many army ant species build soil tunnels over the stable foraging trail that is used to deliver prey to the nest (Gotwald 1995). In territorial species, so-called barrack nests housing large numbers of soldiers or large workers are built near to the border, with examples seen in *Oecophylla* (reviewed in Hölldobler and Wilson 1990) and *Camponotus gigas* (Pfeiffer and Linsenmair 2001).

12.7 Future directions

In exploring the diversity of foraging and defence strategies in ants, we have identified a number of specific questions that remain unaddressed. However, in considering future work on these topics, we would like to highlight two general directions that we feel are likely to be particularly fruitful. The first is a more explicit integration of ecological context and environmental variation into empirical studies of the efficiency and fitness benefits of foraging and defence strategies. The second is the use of modern comparative analyses to explore the evolutionary relationships between ecology and different strategies.

As discussed, the overall efficiency of foraging strategies in ants can be influenced by numerous components, including when to forage, search behaviour, orientation mechanisms, load selection, and morphology. The evolution of each component in any particular taxon may have been influenced by energy budgets (Fewell *et al.* 1996; Tschinkel 2006; Ydenberg and Schmid-Hempel 1994), mortality risk (Herbers 1981; Nonacs and Dill 1990), competitive environment (see references in Dunn *et al.* 2007c and Chapter 5), spatial and temporal distribution of resources (e.g. Dornhaus *et al.* 2006b; Sundström 1993; see Chapters 7 and 8), and colony size (Anderson and McShea 2001a; Beckers *et al.* 1989; Herbers and Choiniere 1996; Hölldobler and Wilson 1990). However, few, if any, empirical studies have explored how the effectiveness of any particular foraging strategy component varies under a range of realistic ecological conditions that a species might face. For instance, theory suggests that spatial and temporal distributions of resources have an important impact on foraging strategies (Dornhaus *et al.* 2006b; Raine *et al.* 2006). Yet, we have very little understanding of how ant foraging strategies perform under natural seasonal shifts in resource distributions and availability or across resource gradients within a population. We also have a very limited understanding of if and how foraging strategies undergo adaptive shifts among populations that face distinctly different environmental conditions. Do decisions on when to forage shift adaptively across temperature gradients? Do orientation mechanisms change with habitat structure? Does the relative use of solitary foraging and recruitment change predictably with different resource distributions found at different locations within a species' range? Studies that address these kinds of questions would be ideal for testing theory that explores how foraging strategies should shift with resource distributions. Similarly, the benefits of particular defence strategies relative to particular ranges of environmental conditions, and particularly different kinds of enemies, are very poorly known. Integrated understanding of how the interactions and benefits of foraging and defence strategies change under different conditions is equally rare.

We have also highlighted a number of hypotheses that seek to explain the evolutionary relationships between environmental conditions and particular foraging and defence strategies, and the relationships among strategies. For instance, it has been suggested that dietary shifts may explain the repeated loss of the sting in ants (Kugler 1979), and that morphological and chemical defences may be alternative trajectories in the evolution of defence strategies (Hunt 1983). These ideas, including those new ones suggested by us, are largely based on informal surveys of the diversity of foraging and defence strategies in ants, and have thus not been tested. Modern comparative analyses, which incorporate information about the evolutionary relationships among taxa, provide powerful and robust methods for exploring the relationships among ecological and phenotypic characters, while controlling for the relationships among taxa. The most commonly used methods are those that reconstruct character change over evolutionary time and test for evolutionary correlations among characters. To date, these methods have not been used extensively to study the evolution of foraging and defence strategies. However, examples include comparative studies of the evolution of recruitment systems in ants (Baroni-Urbani 1993), the evolution of slave-making and the associated propaganda pheromones in the tribe Formicoxenini (Beibl *et al.* 2005; Brandt *et al.* 2006), strata use and its role in the evolution of forager morphology in *Dorylus* (Kronauer *et al.* 2007b; Schöning *et al.* 2005), the role of diet in the evolution of a specialized prey-transport caste in *Eciton* (Powell and Franks 2006), and the role of nesting ecology in the evolution of a specialized soldier caste in *Cephalotes* (Powell 2008). As knowledge of the evolutionary relationships among ant taxa increases (see Chapter 1), the power of these kinds of analyses to test new and old hypotheses on the evolution of foraging and defence strategies will only increase. The patterns identified in these studies also have the potential to inform and focus the kinds of empirical studies of the ecology of foraging and defence strategies that we propose here.

12.8 Summary

Ants and other social insects are model systems for the study of foraging strategies, and have stimulated the development of theories in optimal foraging, central place foraging, risk and robustness, and search algorithms. This is partially due to their abundance and accessibility, and the ease with which colonies can be studied in the field and in the laboratory; but also due to the fascination exerted by their intricate social behaviours. Ant workers may communicate with their nestmates or adversaries using a variety of chemical, mechanical, visual, auditory, or vibration signals, or multimodal signals and cues, which may be specifically directed to single individuals or used to broadcast information and initiate 'mass-recruitment'. This diversity in communication strategies is mirrored by the diversity of food sources used, and in the array of defence strategies that have evolved to improve the acquisition and retention of resources. Many ants are generalists, not only preying on and scavenging a variety of arthropods, but also using nectar or honeydew sources where available. However, there are also many taxa that have become dietary specialists, feeding on seeds, fungus grown in their own gardens, honeydew produced by tended hemipterans or nectar produced by plants, and even feeding on other ants. Such differences in diet, as well as differences in habitat structure, competitive environment, and colony size, have likely led to the broad diversity of foraging and defence strategies in ants. Yet, much is still to be learnt about the process and patterns that underlie this diversity.

PART IV

INVASIVE ANTS

Invasive species, those species that demonstrate ecological, environmental, or economic impacts (Colautti and Richardson 2009), are now widely recognized as one of the primary threats to biodiversity and the functioning of ecosystems, and invasion ecology has become a discipline in its own right. Though invasive ants currently comprise <1% of all known ant species, the contribution of these species to understanding ant ecology is disproportional to their number.

Invasive ants represent an opportunity to apply everything we currently know about ants to a rapidly growing problem. Fundamental to understanding invasive ants is a comprehensive appreciation of the life cycles, social structures, and ecological roles of invasive ants in the context of the environment and evolutionary processes. Indeed, the study of invasive ants has led to opportunities for exploring new (or at least newly rediscovered) lines of inquiry that apply to ant ecology more broadly: How is social structure related to dominance? What ecological or environmental factors keep dominant ants in check? And how important is evolutionary history in shaping interactions with other organisms? Moreover, invasive ants grab the spotlight of public attention like few native insects; handled in the right manner, public concern about pest species can be a stepping stone for educating the public more broadly about the importance of insects and human impacts on them.

The chapters here build on the foundations of global diversity and biogeography, and community and population ecology provided earlier in the book. The first two chapters delve into processes mediating invasion success. Chapter 13 describes the variation in modes of dispersal observed in invasive ants. Suarez, McGlynn, and Tsutsui then take a close look at the biogeography and phylogeny of introduced ants in a search for patterns of which ants become invasive. Chapter 14 explores the question of invasion success at the community and population ecology levels. Krushelnycky, Holway, and LeBrun first examine factors that affect local spread, including biotic interactions, abiotic conditions, and habitat disturbance, before discussing causes of success including competitive displacement and predation, abundance, unicoloniality, generalist habits, ecological release, and genetic changes.

The final two chapters concern the impacts and management of invasive ants. Chapter 15 summarizes the consequences of ant invasions on native ants and other invertebrates, vertebrates, plants, and soil. Lach and Hooper-Bùi call for long-term and experimental approaches, and more attention paid to indirect effects, evolutionary adaptations for coexistence, and the potential for manipulating invasions to mitigate their harm. Finally, Chapter 16 provides a brief historical overview of the techniques used in ant management programmes, and highlights lessons learned from successful and unsuccessful eradication attempts. Hoffmann, Abbott, and Davis outline a management approach incorporating pest risk analysis, public education, data collection, cost–benefit analysis, decision-making, treatment, and monitoring, and call for a more proactive approach to invasive ant management.

Throughout Part IV, authors adhere to the definitions of Holway *et al.* (2002a) in distinguishing the term 'invasive' denoting impact, from terms that refer to origin (alien, exotic, non-native, and tramp). However, today's tramp ant may be tomorrow's invasive ant. As noted across the chapters, much of the information we have to date about the ecology of invasive ants derives from just two species: the red imported fire ant, *Solenopsis invicta* and the Argentine ant, *Linepithema humile*. Other ants may follow a different pathway to becoming invasive and researchers should remain open to finding new patterns to explain their success and impacts. We still have much to learn.

Chapter 13

Biogeographic and Taxonomic Patterns of Introduced Ants

Andrew V. Suarez, Terrence P. McGlynn, and Neil D. Tsutsui

13.1 Introduction

One of the many goals of invasion biology is the identification of factors that determine whether a species will establish successfully in a new environment and that predict whether a newly introduced species has the potential to become highly invasive. Most research in this area has focused on species-specific attributes that may convey ecological success. However, insights into these processes can also come from examining broad biogeographic and taxonomic patterns of introduction.

Introduced ants are an appropriate group for such an approach. Of the approximately 12,500 described species of ants, over 200 have established populations outside of their native ranges (McGlynn 1999b). Of these, many have become highly destructive invaders, and five (the Argentine ant *Linepithema humile*, the big-headed ant *Pheidole megacephala*, the yellow crazy ant *Anoplolepis gracilipes*, the little fire ant *Wasmannia auropunctata*, and the red imported fire ant, *Solenopsis invicta*; see Plate 14) are currently listed among the world's 100 worst invasive species (Lowe 2000). In addition to being economically costly in both urban and agricultural areas, invasive ants can greatly modify ecosystems by reducing native ant diversity, displacing other arthropods, negatively affecting vertebrate populations, and disrupting ant–plant mutualisms (see Chapter 15). Although hundreds of ant species have become established outside of their native ranges, most research has concentrated on the biology of only a few species (including those mentioned above). Despite clear insights gained from this approach, single-species treatments do not provide information on (*a*) patterns of ant species movements into new areas, (*b*) biogeographic patterns of invasion including the identification of regions that may either produce many invaders or be particularly prone to invasion, and (*c*) systematic perspectives on the evolutionary correlates of success. In this chapter, we address these three issues by first comparing natural versus anthropogenic means of dispersal in ants. We then examine broad-scale biogeographic and taxonomic patterns in ant invasions, and conclude by suggesting some areas of future research.

13.2 Patterns of dispersal in ants

13.2.1 Natural patterns of dispersal

Dispersal is a fundamentally important life-history characteristic of all organisms. The frequency, timing, and distance of successful dispersal determine the ability of species to colonize new areas and, in some cases, the survival of geographically isolated populations. Moreover, gene flow among populations via dispersal is a key determinant of their underlying genetic structures. In most organisms, dispersal distance distributions are 'leptokurtic' in shape – most new individuals settle relatively near where they originated but a few may disperse longer distances (Case 2000; Kot *et al.* 1996). However, patterns of dispersal can vary substantially among organisms, depending on a multitude of intrinsic and extrinsic factors including the mode of dispersal, nutrition, social structure, competition, and predation (Shigesada and Kawasaki 1997).

Normally, ant colony structure and social organization are intimately related to patterns of dispersal (reviewed in Bourke and Franks 1995; Hölldobler and Wilson 1990; Wilson 1971). In many ants, reproduction occurs via the production of winged alates that mate away from their natal nests. In contrast to species that undergo mating flights, virgin queens of other species mate within their natal nest with males that have been produced locally, from within the same colony, or from nearby colonies. This type of reproductive behaviour is particularly prominent in invasive ants (Passera 1994; Chapter 14), and ants with intranidal mating are expected to be genetically homogeneous within colonies, but highly genetically differentiated across colony boundaries, even at relatively small spatial scales (Pedersen *et al.* 2006).

In many cases, colony reproduction in species with intranidal mating can occur via dependent colony foundation, or colony budding (see Chapter 9). Budding occurs when a subset of a colony's population (tens to thousands of workers, one to hundreds of queens) disperses away on foot to a new location and becomes established as a separate colony elsewhere. Examples of species that reproduce by colony budding include many of the most widespread and damaging invasive ants, such as *A. gracilipes, L. humile, P. megacephala, W. auropunctata, Monomorium floricola, M. pharaonis, Tapinoma melanocephalum*, and the polygyne form of *S. invicta* (reviewed in Holway *et al.* 2002a and Wilson 1971).

During the spring, male and female alates are produced within *L. humile* colonies, as occurs in many species of ants. Although the female alates possess well-formed wings, curiously, they cannot, or do not fly and mate in their natal nest almost immediately after eclosion. Males, on the other hand, are clearly able to fly and, under appropriate conditions, large numbers of males can be observed aggregating at nest entrances and departing on the wing. These observations are confirmed by the presence of male Argentine ants at lights and in pitfall traps placed hundreds of metres from the nearest Argentine ant colony (A. Suarez, unpublished data). However, it is unknown if these dispersing males ever successfully enter other colonies and mate. Colony budding in *M. pharaonis* is reviewed in detail by Wilson (1971), and this description closely matches the reproductive patterns displayed by *L. humile*.

13.2.2 Variation in modes of dispersal, an example with *S. invicta*

While many introduced ants are polygynous and reproduce primarily by dependent colony foundation (see Chapter 14), this is not the case with monogyne populations of the red imported fire ant, *Solenopsis invicta* (reviewed in Tschinkel 2006). Under suitable conditions (warmth, high humidity, low wind), most often in the spring, mature monogyne colonies of *S. invicta* release reproductive males and, a short time later, the reproductive females. The reproductives fly to great heights of 50–100m (Markin *et al.* 1971), find each other via behaviours that are unknown, and mate. Newly mated queens may then descend immediately to the ground or fly some distance, typically downwind, before landing. New colonies may be founded by individual queens (haplometrotically) or by groups of queens (pleometrotically) (see Chapter 9), but the latter revert to the monogyne form when the first workers emerge and kill all but one of the foundresses (Tschinkel and Howard 1983). Various studies by George Markin and colleagues (reviewed in Tschinkel 2006) have estimated that dispersal distances are typically on the order of hundreds of meters or, occasionally, a few kilometres. Although nuptial flights can also occur in the introduced polygyne form of *S. invicta*, they typically occur closer to the ground (2–3 m elevation) and the newly mated queens do not seem to disperse as far as monogyne queens do. Moreover, polygyne *S. invicta* colonies may also disperse by dependent foundation.

These differences in mating and dispersal between monogyne and polygyne forms of *S. invicta* are reflected in the genetic structure of these ants in their introduced range (Shoemaker *et al.* 2006). Because newly mated monogyne queens travel much farther during nuptial flights, local genetic differentiation at maternally inherited mitochondrial loci is low or absent. Polygyne populations, on the other hand, are characterized by high levels of mitochondrial differentiation at relatively small spatial scales. At nuclear loci, however, substantial genetic differentiation is absent at a local scale for both forms because males may disperse more widely, and males from monogyne colonies can mate with queens from polygyne colonies, thus eroding genetic differences between the social forms. These population genetic differences between social forms are also evident at large spatial scales – pronounced mitochondrial differentiation, but at levels equivalent to that seen between sympatric populations, and very low levels of nuclear genetic differentiation (Shoemaker *et al.* 2006). At this larger spatial scale, genetic homogeneity, particularly at nuclear loci, likely arises in large part from the anthropogenic transport of *S. invicta* among spatially distant sites. These transfers increase the overall levels of realized gene flow among regions, thus eroding genetic differences among them.

13.2.3 Anthropogenic patterns of dispersal

Left to their own devices, ants are surprisingly poor long-distance dispersers and colonizers. As noted by Wilson (1971), prior to the advent of human-assisted dispersal, ants (as well as many other taxa) were unable to colonize many mid-Atlantic and distant Pacific Islands (see Chapter 2). Similarly, the pioneering island biogeographic studies of Simberloff and Wilson (1969) found that following the removal of arthropods from mangrove islands in the Florida Keys, ants were among the last arthropod taxa to recolonize. For invasive ants, the limited long-range (e.g. intercontinental or transoceanic) dispersal and colonization abilities of ants have been overcome by inadvertent human-assisted transport. In Hawaii, for example, there are no native ants, but about 50 ant species currently occur there, all introduced by human activities primarily in the mid-twentieth century (Krushelnycky *et al.* 2005b).

The large-scale anthropogenic dispersal of invasive ants is, in many cases, facilitated by their nesting habits and reproductive strategies. Many of these species do not construct elaborately excavated nests, but nest in superficial and/or ephemeral sites such as root masses or leaf litter, under stones, and in logs and plant debris. Moreover, many of these species are highly peripatetic, quickly absconding from sites as they become abiotically unsuitable (too dry, too wet, too hot) and colonizing new nesting sites. When coupled with human commerce and habitat alteration, these nesting habits translate into frequent human-mediated dispersal, as propagules colonize objects destined for transport to a new location. For example, ants transported to the United States and New Zealand are associated with a wide variety of commerce, but a majority of ants have been detected on plant material (Table 13.1; Suarez *et al.* 2005a; Ward *et al.* 2006). This likely reflects both the nesting habits of ants transported by humans, as well as the inspection policies of regulatory agencies that are biased towards plants and plant products.

Because many ant invaders are highly polygynous, individual colony fragments have a high probability of containing at least one fertile queen, thus increasing the likelihood of propagule viability. For example, laboratory colony fragments consisting of a single queen with as few as ten workers can be reproductively viable for *L. humile* and *P. megacephala* (Chang 1985; Hee *et al.* 2000). It is also possible that queens are not even necessary for establishment of some invasive ants. In the Argentine ant, for example, haploid eggs are present within colonies year-round; propagules containing only workers and brood can potentially establish new populations through the production of new queens and males followed by intranidal mating (Aron 2001).

New populations of invasive ants may be introduced from a single native source, multiple native sources, or from other previously established populations in other parts of the introduced range. Disentangling the history of ant invasions can therefore be quite difficult, but insight has been attained for some species by using genetic data (e.g. *L. humile*, Corin *et al.* 2007b; Tsutsui *et al.* 2001; *W. auropunctata*, Mikheyev and Mueller 2007; *S. invicta*, Caldera *et al.* 2008; *A. gracilipes*, Abbott *et al.* 2007). One useful example is the invasive little fire ant (*W. auropunctata*), which is native to large

Table 13.1 Commerce on which ants are transported to New Zealand (NZ) and the United States (USA). Data from Suarez *et al.* (2005a) and Ward *et al.* (2006). It should be noted that these data are heavily influenced by inspection policies that may favour the quarantine of plant material.

	% of Records	
Commodity	NZ	USA
Ornamental Plants		60
Fresh Produce	47	14
Shipping Containers	22	3
Personal Items	16	1
Vehicle	11	1
Timber	4	4
Other*		17

*Including soil and other plant materials.

portions of Central and South America. Phylogenetic analyses show that native populations fall into two major clades (Mikheyev and Mueller 2007); one is primarily distributed throughout Central America and the Caribbean, while the other is distributed throughout mainland South America (although some Brazilian populations fall in both clades). Introduced populations are also distributed across these two clades, indicating that multiple propagules have originated from different parts of the native range and have become established in different parts of the introduced range (Mikheyev and Mueller 2007).

The Argentine ant is another well-studied example. The spread of this invader occurs via two discrete modes: (*a*) dependent colony foundation, by which colonies advance at the rate of tens to a few hundred metres per year (Holway 1998b; Sanders *et al.* 2001) and (*b*) human-mediated jump-dispersal, which results in rates of dispersal three to four orders of magnitudes higher than by budding (Suarez *et al.* 2001). Current evidence suggests that some introduced populations may trace their ancestry back to different source populations in the native range. For example, some populations of *L. humile* from South Africa, Hawaii, and California are genetically distinct from other introduced populations, in terms of both allele identity and frequency at microsatellite loci (Tsutsui *et al.* 2001). However, the immediate source of most introduced populations is likely another previously established population (Corin *et al.* 2007b). In the southeastern United States, for example, almost every steamboat arriving at New Orleans and Baton Rouge in 1916 was infested with Argentine ants, as were many ports and rail stations (Barber 1916). Although Argentine ants are inherently slow dispersers (Holway 1998b; Sanders *et al.* 2001), through close association with humans they spread rapidly and discontinuously throughout the twentieth century, establishing themselves in at least 23 states in the United States and 35 countries worldwide (Roura-Pascual 2004; Suarez *et al.* 2001; Wetterer and Wetterer 2006).

13.3 Biogeographic patterns of ant invasions

13.3.1 Origins and exchange of introduced ants

Records of faunal exchange and invasions of ants date back to the European colonial era (Wilson 2005). Though many species of ants have been moved around the world, and new species continue to become established, only a small fraction of transported ants have become invasive (McGlynn 1999b; Suarez *et al.* 2005a). Biogeographic patterns in ant introductions show a number of clear trends (Figure 13.1). Like many other taxa, ants typically become established when introduced to a region different than from where they originated (Williamson 1996). Oceanic islands, with depauperate or absent ant faunas, have received the greatest number of introduced ant species by far. The tropical and subtropical

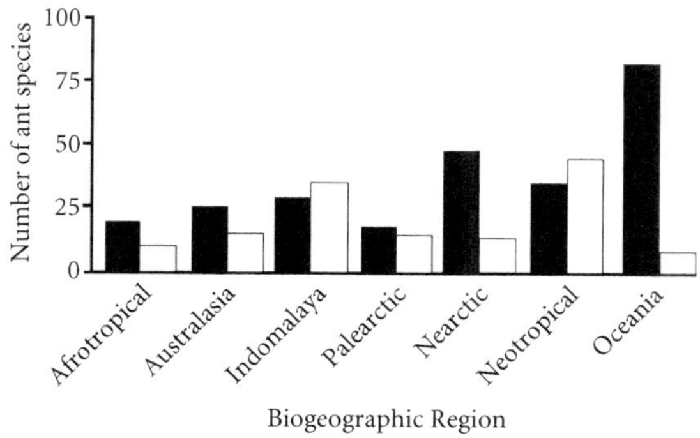

Figure 13.1 The number of introduced ants that have either become established in (black bars) or have originated from (open bars) each major biogeographic region (data from McGlynn 1999b). Notably the Oriental and Neotropical regions have exported more species than they have received. However, it should be noted that this pattern could be biased due to variation in sampling effort and taxonomic knowledge of the ant faunas among regions (Rackzkowski and Wenzel 2008).

regions of Australia host a number of invasions, but few invaders have been derived from the exceptionally diverse native ant fauna of this region. Likewise, North America has received many foreign ants, though ants originating from North America have apparently not established populations anywhere other than in Hawaii (McGlynn 1999b). The single region that has produced the greatest number of invaders appears to be South America. Palearctic and Indomalaya regions, according to existing records, have produced and received equivalent numbers of invaders. Currently, too little is known about the status of ant invasions in Africa (particularly north of South Africa) to meaningfully comment.

Because ant taxonomy and sampling are biased across regions (e.g. Dunn *et al.* 2007d; Raczkowski and Wenzel 2008), it is not clear how much inference can be made about these patterns at this point. The origin of multiple invasive species in central South America suggests either patterns of trade that are conducive for transporting ants, or that environmental conditions there may favour the evolution of species with a tendency for invasiveness. Behavioural experiments on interspecific competition in this region suggest high rates of general behavioural dominance and competitive interactions among ants (LeBrun *et al.* 2007). Moreover, a number of successfully introduced ants are native to and occur sympatrically in the seasonally inundated floodplains of the major rivers systems of northeastern Argentina, southern Paraguay, and Brazil including *L. humile*, *S. invicta*, *S. richteri*, *W. auropunctata*, *Brachymyrmex patagonicus*, and *Pheidole obscurithorax* (LeBrun *et al.* 2007; Wild 2007a; Table 13.2). It is possible that the adaptation to disturbance in the form of flooding, coupled with the species-rich and competitive environment may select for characteristics that allow species to do well in anthropogenically disturbed areas. The identification of other areas with similar characteristics (e.g. highly competitive environment (Andersen 1997) and natural disturbance regimes), coupled with better knowledge of the native ranges of introduced ants, is still needed to determine if specific regions can act as crucibles for creating potential invaders.

Since the latter half of the twentieth century, there are no examples of new, widespread (e.g. across many biogeographic regions), highly invasive ant species, even though this era has seen a major expansion in global trade that typically facilitates the emergence of invasions (Hulme 2003; Perrings 2005). However, it should be noted that many species not previously recognized as highly invasive (including *Monomorium pharaonis*, *M. destructor*, and *Paratrechina longicornis*) have become threats on many oceanic islands since the Second World War. The major invasive species established prior to the midpoint of the twentieth century predominately originated from tropical or subtropical climates (Table 13.2) (McGlynn 1999b). In contrast, it is possible that in the last 25 years or so, a new class of invasive ant is emerging with the potential to eventually become as widespread as some of the aforementioned species with tropical or subtropical origins. These include species that are temperate in origin, and that predominately invade other temperate climates (Table 13.2). For example, *Lasius neglectus*, likely native to eastern Europe, has emerged as an invader in temperate Europe. In North America, *Myrmica rubra* (also from Europe) is expanding its range in the northeastern United States. Two species from Japan, *Pachycondyla chinensis* and *Tetramorium tsushimae*, are spreading throughout the east coast of North America and the midwestern United States, respectively.

Table 13.2 The first six species listed below represent the most significant invasive ants, in terms of distribution, density, and damage to invaded environments, according to Holway *et al.* (2002a). All were invasive prior to 1900, except *S. invicta* which was first detected as an invader in the 1920s by W. S. Creighton (Wilson 1951). The bottom four species are more recently documented as exhibiting characteristics consistent with other highly invasive ant species and are largely from temperate climates.

Species	Origin
Anoplolepis gracilipes	Sub-Saharan Africa or tropical Asia
Linepithema humile	Central South America
Pheidole megacephala	Sub-Saharan Africa
Solenopsis geminata	Tropical Central and South America
Solenopsis invicta	Tropical and subtropical South America
Wasmannia auropunctata	Tropical Central and South America
Lasius neglectus	Probably Turkey (Seifert 2000, Cremer et al. 2008)
Myrmica rubra	Palearctic North Temperate Zone (Groden *et al.* 2005)
Pachycondyla chinensis	North Temperate to subtropical Asia (Nelder *et al.* 2006)
Tetramorium tsushimae	North Temperate China and Japan (Steiner *et al.* 2006)

Further evidence for this trend comes from the southern hemisphere, where approximately 20 species from temperate Australia have been detected in New Zealand (Ward *et al.* 2006). It remains to be seen how widespread these predominantly temperate species will become.

There is no overt explanation for the temporal disjunction between the emergence of tropical invasive ants and temperate invasive ants. Contemporary patterns of global commerce do not seem to be able to explain the emergence of temperate invaders. There is also a clearly documented time lag of several decades for two of the new invasive species: *P. chinensis* has been in North America for at least 70 years (Smith 1934) and *M. rubra* was introduced at least a century ago (Wheeler 1908). Furthermore, another temperate species, *Tetramorium caespitum*, which is native to Europe, has expanded its range throughout North America during the past 100 or more years, although it is not often considered 'invasive' (Steiner *et al.* 2007).

13.3.2 Using modeling approaches to understand biogeographic patterns of invasion

Species distribution models can be powerful tools for evaluating the potential for spread of invasive ants within and among geographic regions. Distribution models try to predict areas with suitable climate for the establishment or spread of species based either upon occurrence data from their native range, or other introduced populations (correlative approach), or by incorporating direct measurements of a species' response to environmental conditions (eco-physiological approach) (Pearson and Dawson 2003; Roura-Pascual and Suarez 2008).

Correlative approaches have been recently utilized for predicting the potential distribution for a number of introduced ant species including *S. invicta*, *L. humile*, and *T. caespitum* (Fitzpatrick *et al.* 2007; Hartley *et al.* 2006; Roura-Pascual 2006; Steiner *et al.* 2006b; Ward 2007a). Climate models have also been used to examine the potential distribution of ants in the face of global climate change. Scenarios that envision future increases in average global temperature generally predict an expansion of the ranges of invasive ants (Fitzpatrick *et al.* 2007; Roura-Pascual 2004; Zavaleta and Royval 2002).

In contrast to correlative (climate matching) approaches, eco-physiological distribution models base their predictions on direct measures of a species' response, in terms of growth, persistence, or reproduction, for example, to particular abiotic conditions, such as temperature and moisture (Helmuth *et al.* 2005). This approach has been applied to predict the distribution of red imported fire ants at a variety of spatial scales by examining temperature dependent colony growth and alate production (Morrison *et al.* 2004; Sutherst and Maywald

2005; Xiong *et al.* 2008). Similarly, degree-day models for colony growth of Argentine ants, based on development rates in relation to temperature, have been used to assess their potential for spread throughout New Zealand (Hartley and Lester 2003) and Hawaii (Krushelnycky *et al.* 2005a).

These modelling approaches do have some limitations. The geographic range of a species is not simply determined by abiotic suitability, but is determined by a variety of factors and processes that are often scale dependent (Wiens 1989). These are not often included in distribution models, and include fine-scale variation in abiotic conditions driven by anthropogenic disturbance (rather than climate), biotic interactions, dispersal capacity, and adaptability of the species (Roura-Pascual and Suarez 2008). Human-mediated disturbance and fine-scale changes in abiotic conditions due to land-use practices may be particularly important in determining the occurrence of introduced ants (King and Tschinkel 2006; Menke and Holway 2006) and need to be incorporated directly into modelling efforts (Menke *et al.* 2007). Despite these limitations, distribution modelling is a powerful tool for guiding monitoring efforts to help prevent the establishment of new populations of invasive species, and for identifying donor and recipient areas with similar climates that may be at high risk for new introductions.

13.4 Taxonomic patterns of invasion success

13.4.1 Taxonomic patterns of ant transportation across biogeographic regions

Most research on ant invasions has been conducted on a few widespread and highly invasive species. However, efforts to identify mechanisms of success or to establish a baseline of information that can be used to prevent future invaders hinges upon an informed understanding of the identity of species that are being transported – that is, the pool of species that have had the opportunity to become established. Only with this information can we make quantitative assessments of the characteristics that distinguish successful versus unsuccessful invaders.

A few studies have used historical records to identify the ant species that are most frequently transported by human commerce across biogeographic regions. Suarez *et al.* (2005a) found a high diversity of ants being transported by humans in the middle of the twentieth century. Identification of ants from roughly 400 separate interceptions in quarantine in the United States from 1920s to the 1970s revealed 232 species from 58 genera. Of this pool of potential invaders, only 28 species from 17 genera now occur as established non-native species in the continental United States. Moreover, the most commonly encountered ants in quarantine (species with five or more records: *Azteca* sp., *Cardiocondyla wroughtoni*, *Campanotus planatus*, *Linepithema iniquum*, *Monomorium pharaonis*, *Pseudomyrmex ferrugineus*, and *Tetramorium simillimum*) have either not become established, or are not considered among the most widespread and invasive ants in North America. Thus, these data suggest that the particular species that become the most successful invaders are not necessarily the same species that humans transport across biogeographic regions most frequently.

On the other hand, studies in different regions have reported a different pattern: the most frequently introduced species are also the most likely to establish. In New Zealand, for example, Lester (2005) found that a notoriously adept invader, *P. megacephala*, was the most frequently intercepted species, at an average of 17.67 interceptions per year.

There are several possible explanations for the different patterns reported in the United States and New Zealand. On one hand, differences in commerce and trade routes between the two countries may produce opportunities for different species to be sampled and transported (but see Table 13.1). Alternatively, the differences could reflect the different time periods examined by the studies. For example, between the periods of 1966–82 and 1997–2003, the interception rate for ants arriving in New Zealand differed across time: the interception rate for 13 species more than doubled while the interception rate for 12 other species was reduced by half. This suggests that the species pool may be homogenizing through time, perhaps as ports and other areas acting as sources for the

Table 13.3 Taxonomic composition of ants intercepted in ports of entry (POE) in New Zealand (from Ward *et al.* 2006) and the United States (from Suarez *et al.* 2005a), and of global ant species (from Bolton 1995b).

	POE records (%)		Established (%)		Ant species (%)
Subfamily	NZ	US	NZ	US	Global
Myrmecinae	36	47	54	52	48
Formicinae	30	22	7	22	27
Dolichoderinae	17	11	18	6	6
Ponerimorphs*	13	11	21	18	14
Pseudomyrmecinae	1	7	0	1	2
Dorylinae	1	1	0	0	1

*Includes taxa in the subfamilies Ponerinae and Ectatomminae.

transport of ants become dominated by fewer, ecologically dominant ant species. Future work on interception records of ants that examines temporal trends in occurrence data over long periods (e.g. 100+ years), or that contrast records within versus among regions, will be very valuable for addressing issues relating to opportunity and invasion success.

The taxonomic similarity (at the subfamily level) between ants intercepted at quarantine and ants generally suggests that ants with the opportunity to become established may be drawn from the species pool haphazardly (Table 13.3). More data are needed, however, to determine if taxonomic patterns will be evident at other levels (e.g. genera), and if taxonomic patterns are indicative of ecological characteristics that may promote success at other stages of invasions (e.g. establishment and spread). Using interception data as the basis for the examination of characteristics that promote invasion success or failure will be significant for illuminating mechanisms of success (Commonwealth of Australia 2006; Lester 2005; Suarez *et al.* 2005a). For example, many traits relating to diet, nesting habitats, and colony structure have been implicated in the success of invasive ants (see Chapter 14). How these traits interact with opportunity and propagule pressure to influence establishment remains unresolved, and provides a strong justification for the monitoring of interception data in a more scientific manner and also for the study of the basic natural history of ants generally.

13.4.2 Taxonomic/systematic patterns of success

Many characteristics, including high competitive ability, polygyny, dependent colony foundation, and a unicolonial colony structure have been suggested to promote invasion success in ants. A general description of these traits is covered in Chapter 14. However, in contrast to work conducted on invasive plants and birds (e.g. Grotkopp *et al.* 2002; Lockwood 1999), we know very little about taxonomic or phylogenetic patterns in the origin of invasiveness. While some biogeographic patterns are beginning to emerge, it is clear that success in terms of establishment and spread in new areas can emerge independent of phylogeny (McGlynn 1999b). For example, the five most invasive ant species come from the three most speciose subfamilies (Myrmecinae, Dolichoderinae, and Formicinae). Moreover, while the bulk of invasive species are in the subfamily Myrmicinae, this is not disproportionate to the species richness of this subfamily relative to other ants (McGlynn 1999b). When considering the taxonomic distribution of successfully introduced ants, we find a similar pattern to that described for ants being transported by humans above (McGlynn 1999b) (Table 13.4). Specifically, ants become established roughly proportional to how many species there are at the subfamily level. Whether this holds at other taxonomic levels (e.g. genera) remains to be seen.

A powerful approach for understanding taxonomic patterns of introduced species and uncovering the characteristics that promote invasiveness

Table 13.4 The number of ant species that have successfully established populations outside their native range and the total number of described species, by subfamily. Only subfamilies with at least one known introduced ant species are included. Data from McGlynn (1999a) and Bolton (1995b) and may not include recently discovered introduced species.

Subfamily	Number of known introduced species	Number of described species
Cerapachyinae	1 (0.7%)	198 (2.1%)
Dolichoderinae	11 (7.5%)	554 (5.8%)
Formicinae	39 (26.5%)	2,458 (25.8%)
Myrmeciinae	1 (0.7%)	89 (0.9%)
Myrmcinae	71 (48.3%)	4,377 (45.9%)
Ponerinae	23 (15.6%)	1,299 (13.6%)
Pseudomyrmicinae	1 (0.7%)	197 (2.1%)

will be to compare successful versus unsuccessful invaders in a phylogenetic context. For example, invasive ants are often unicolonial (see Chapters 10 and 14), and are also often among the smallest members of their genus (McGlynn 1999a). Mapping out these and other ecological characters onto generic level phylogenies will shed light as to which characteristics best predict the success of introduced species (see Grotkopp *et al.* 2002 for an example with pines). This approach will be particularly informative if done in conjunction with information gathered in interception data (Lester 2005; Suarez *et al.* 2005a; Ward *et al.* 2006). For example, using ants detected in quarantine in the United States, Suarez *et al.* (2005a) found 4 species of *Linepithema* and 13 species of *Pseudomyrmex*, yet only one from each genus (*humile* and *gracilis*, respectively) has become established outside of greenhouses. A phylogenic approach will shed light as to which characteristics may promote association with humans generally (if shared by all imported species) versus those that promote invasiveness (those unique to *L. humile* or *P. gracilis*).

Other genera for which a phylogenetic approach will be powerful for uncovering mechanisms of invasion success include those in which many species have become successfully introduced. For example, the Myrmecine genera *Cardiocondyla, Monomorium, Pheidole, Solenopsis*, and *Tetramorium* all have at least three successfully introduced species plus others that are moved around by humans but have not yet been recorded as established outside their native range (McGlynn 1999b; Suarez *et al.* 2005a). A careful, phylogenetically corrected comparison of natural history traits and biogeography of species within these genera will greatly increase our understanding of the factors that contribute to a species ability to become transported by humans, establish new populations, and spread.

13.5 Future directions

Two of the primary goals of invasion biology are predicting and preventing the future establishment of invasive species. Some of the recent research on invasive ants reviewed here suggests clear ways by which these goals may be accomplished. For example, continued and enhanced vigilance must be maintained as a first line of defence. This should involve a thorough examination of high-risk commercial items (such as living plants), and inspection programs should explicitly include sufficient flexibility to respond and adapt to changing situations. Although we recognize that the economic consequences of temporarily halting imports of problematic items are often unpalatable, federal and state agencies must be encouraged to adopt an assertive, forward-leaning stance when facing high-risk situations (see Box 13.1 for a description of New Zealand's risk management). Clearly, the economic consequences of introducing new invaders outweigh the short-term losses that may result from appropriate cautionary measures. This heightened vigilance should also be coupled with scientifically based data collection. For example, while a majority of ants detected in quarantine are intercepted on plant material, this may simply reflect the fact that plants are the most heavily scrutinized commodity.

Box 13.1 Preventing the international spread of invasive ants through implementation of offshore biosecurity risk management policy

Simon O'Connor and Grant P. Weston

Government biosecurity agencies are typically responsible for creating and enforcing policies to prevent the introduction or contain the spread of unwanted organisms. Invasive alien ants provide unique challenges to this process for many reasons. Ants contaminate an infinite range of commodities and are incredibly adaptable across a wide range of environments, which subsequently increases their chances of establishing in new places. Additionally, their reproductive strategies allow rapid nest maturity and efficient population expansion. Increased international trade has allowed the ranges of many invasive alien ant species to increase, and the sea cargo pathway presents a particularly high risk for the introduction of ants (Figure 13.1.1). For example, two of the three *Solenopis invicta* incursions eradicated in New Zealand were linked to contaminated cargo in imported sea containers. This dictates the need for biosecurity agencies to utilize innovative pre-border systems to prevent the entry of invasive ant species.

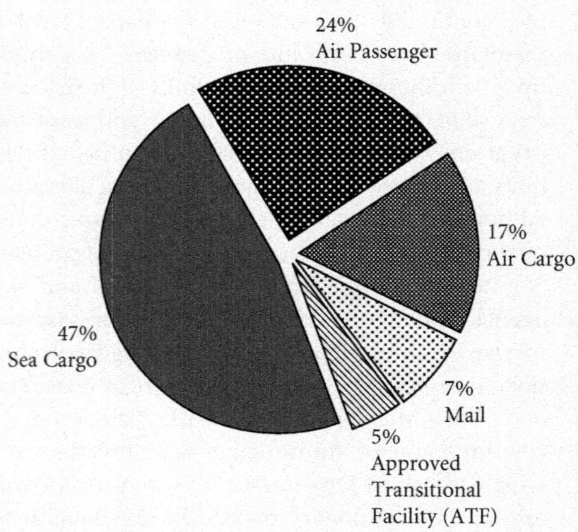

Figure 13.1.1 New Zealand intercept data demonstrating varied pathway use by ants. Approved Transitional Facilities are areas where imported containers and cargo are inspected by Quarantine Inspectors or importers, who have received basic biosecurity training. All ants detected were identified for a seven-month period. Data from Ministry of Agriculture and Forestry, Biosecurity New Zealand.

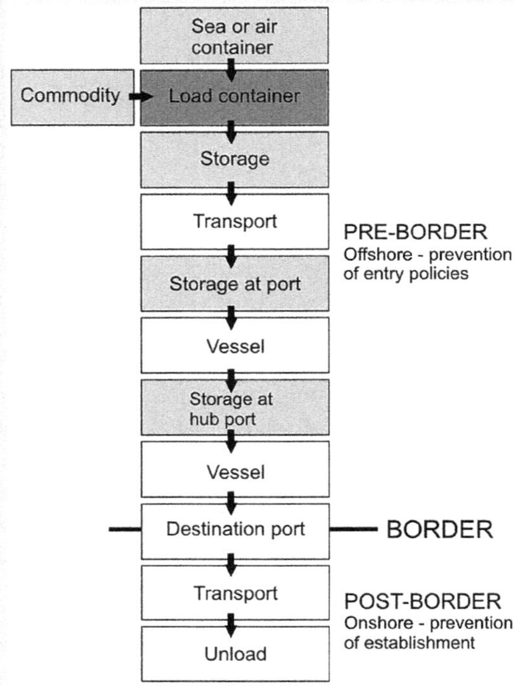

Figure 13.1.2 The supply-chain environment. Lightly shaded boxes are areas where initial contamination or re-contamination opportunities exist; the darkest shaded box represents the optimal control point for implementation of remedial and prophylactic biosecurity actions (i.e. critical control point).

Pathway contamination rates are likely to be a function of the number and length of contamination opportunities, the ant species and densities present at the source localities (e.g. container load and storage points), and attractants present within the container/commodity, such as suitable nest sites or food sources. A holistic supply-chain environment examination is therefore required, considering contamination likelihood at each opportunity (Figure 13.1.2). In reality, contamination can occur at any point on the supply chain, but is highest where containers remain undisturbed for extended periods and in direct contact with the ground. Effective offshore policies are documented in the International Standards for Phytosanitary Measures (ISPM, FAO 2006), but are rarely used for invasive ants. By utilizing the detailed guidelines in the appropriate ISPM

continues

Box 13.1 continued

standards and ant specific control methodologies, practical improvements in offshore risk management can be achieved as the following case study demonstrates.

Practical offshore risk management: a New Zealand case study

During 2003, New Zealand biosecurity officials became aware of multiple invasive alien ant incursions linked to a single import pathway: sea cargo. These incursions were traced back to three Melanesian ports utilized by one shipping line. Invasive ant contamination rates of empty sea containers on this pathway peaked at 17% of inbound containers. The high associated incursion response costs were being borne by the New Zealand taxpayer. Using site-based evaluations (ISPM No.6), it was determined that container origin ports had high densities of invasive ants and high species diversity (ISPM No. 8). There was also ample habitat and food to sustain populations, as well as multiple opportunities to enter sea containers prior to arrival in New Zealand. New Zealand's biosecurity agency entered into a collaborative risk-management agreement with the shipping line which saw the development and implementation of an offshore based sea container hygiene system (ISPM No. 24) at each of the three ports.

New Zealand biosecurity officials set maximum contamination limits for industry to meet, which provided the mechanism to measure system performance. Integrated pest-management programmes were designed and implemented by an ant-control specialist. The results dramatically reduced port-based ant populations via toxic baiting programmes and habitat reduction. Once containers were cleaned, re-contamination opportunities were also reduced to maintain cleanliness (ISPM No. 10) through the use of dedicated ant free storage areas. In addition, a prophylactic residual insecticide band was applied to all potential ant re-entry points on the cleaned containers to reduce the likelihood of recontamination of future supply chain points (ISPM No.14) where ant-control strategies were unlikely to be used. To gauge system performance, all containers were subsequently inspected by biosecurity staff in New Zealand (ISPM No. 23) over an initial 3-month period, and results fed back to industry partners. Following on from the initial successful three month trial period, a container hygiene system audit regime was implemented to ensure all components of the system were maintained. The resulting feedback systems (ISPM No. 13) facilitated continuous improvement, and ant contamination rates dropped from a recorded high of 17% to less than 0.2%, and commonly 0%. The success of this approach has resulted in the rollout of this system to other high risk import container ports in the Pacific.

Data should be collected at ports of entry to quantify the taxonomy, vectors of transport, and ports of departure for ants reaching international borders. At a minimum, long-term, well-curated sampling of quarantine specimens should be prioritized. These samples will provide invaluable insights into the dynamics of human-mediated transport through time and space.

Our ability to predict and prevent future invasions will be strengthened by a variety of predictive modelling approaches. Much good work has already been performed, but the development and application of new tools and approaches will provide increasingly precise and accurate insights into the future. For example, distribution models that examine the contribution of multiple mechanisms simultaneously (abiotic conditions, biotic factors including taxonomic similarity and species diversity, and exchange rates of commerce) are needed to identify geographic regions that are at a high risk of invasion, and to determine which species are most likely to be problematic in many regions. Climate matching can also be used to identify regions connected by trade that may be at a high risk for faunal exchange and determine if invaders have more commonality in geographic origin than in phylogenetic origin. This would suggest that invasion success may be tied to the evolution of characteristics related to geography (e.g. the presence of natural disturbances such as flooding) rather than life history characters constrained by phylogeny. In contrast, phylogenetic approaches can provide important insights into the mechanisms that contribute to why some species become invasive, yet related species do not.

Finally, our knowledge of the biology of introduced ants is restricted to a few 'well-studied' species. There are huge numbers of ants being moved around by humans that we know virtually nothing about, and it remains to be seen if generalizations about the spread and success of invasive ants will hold once others are studied in more detail. In fact, we have not even identified the native range for many invasive ants. Careful behavioural, ecological, and genetic studies of invasive ants in their native range are essential for understanding their success elsewhere. Are there particular selective regimes that groom some species to become invaders? What are the particular traits involved? How many different ways are there to become an invader? Studies that place this information in a comparative and explicitly phylogenetic context may be particularly insightful.

13.6 Summary

While research on biological invasions is often species-specific, insight into the mechanisms of invasion success can also come from broad biogeographic or taxonomic perspectives. Invasive ants have natural and anthropogenic modes of dispersal, which strongly influence genetic relatedness within a population. Ant introductions are taxonomically diverse; however, certain biogeographic regions are over-represented as either donors or recipients of introduced ants. While invasive ants have historically come from tropical and subtropical climates, contemporary invaders are also emerging from temperate regions. Future research needs to concentrate on the many rarely studied introduced ants to determine if current patterns are general, and to understand the relative importance of taxonomy versus biogeography in predicting future potential invaders and areas most likely to be invaded.

Acknowledgements

We would like to thank Lloyd Morrison and Darren Ward for their comments and insights on an earlier version of this chapter. This work was supported by the United States Department of Agriculture (NRI-CGP 2004-3502-14865; to NDT), the California Department of Consumer Affairs and the Structural Pest Control Board (NDT), and the National Science Foundation (DEB 0716966 to AVS).

Chapter 14

Invasion Processes and Causes of Success

Paul D. Krushelnycky, David A. Holway, and Edward G. LeBrun

14.1 Introduction

Biological invasions represent communities in flux. Although stasis is never the rule in nature, biological interactions in communities usually occur within a framework of shared ecological and evolutionary history. Consequently, biological invasions represent unique opportunities to study dynamics that can otherwise be difficult to observe (Elton 1958). Invasive ants are excellent organisms with which to pursue this goal, in part because ants as a group play a variety of important ecological roles within biological communities (Hölldobler and Wilson 1990). Ant invasions hold much potential for improving an understanding of ecological processes in general, as well as of interactions more specific to myrmecology. For instance, the dynamics that exist during ant invasions may reveal the traits that promote behavioural or ecological dominance. Furthermore, highly successful invasive ant species are often less conspicuous in their native ranges, so identification of the factors responsible for their greater prominence in introduced areas can provide insight into the more typical workings of ant communities, may highlight intrinsic differences between communities that have formed in separate biogeographic regions, and may suggest possible methods of control. Understanding ant invasion processes and causes of success thus has both basic and applied relevance.

Human activities have introduced many ant species to new biogeographic regions (McGlynn 1999b; Chapter 13). While most introduced ant species have limited success in spreading away from the human-modified habitats in which they usually first arrive, a subset can invade nearby, or even distant, undisturbed natural habitats. This distinction is not always hard and fast, as virtually all species exhibit different degrees of invasiveness and ecological dominance in different locales, and sometimes even under similar conditions in the same locale (e.g. Abbott *et al.* 2007). Notwithstanding, the following species are the most consistent in their ability to penetrate natural ecosystems and affect the composition or abundance of native species within them (Holway *et al.* 2002a): the Argentine ant (*Linepithema humile*), the red imported fire ant (*Solenopsis invicta*), the tropical fire ant (*S. geminata*), the big-headed ant (*Pheidole megacephala*), the little fire ant (*Wasmannia auropunctata*), and the yellow crazy ant (*Anoplolepis gracilipes*) (see Plate 14). We focus on these species because of their pronounced invasive tendencies, and also because they have been studied the most intensively, especially with regard to the factors that control spread and underlie dominance. In fact, the majority of our knowledge on these topics comes from studies on just two species: *S. invicta* and *L. humile*. This represents both a major weakness and a clear avenue for progress in the field, pointing to a need for more complete information on the ecologies of other invasive ant species, as well as their close relatives that fail to become invasive (Chapter 13; see also Section 14.4). This limitation must be acknowledged at the outset, as it impinges on our understanding of which causal factors are important only in particular cases, and which form more general patterns central to invasive ant success.

Chapter 13 described how ants are introduced to new areas: which species are likely to be transported, the pathways by which this occurs, and where they are likely to establish. In this chapter, we address patterns of invasion at the community scale, examining the invasion process after the point of establishment. This is the spatial scale at which invading colonies interact with a finite set of resident ant colonies and are affected by abiotic factors that vary over metres or tens of metres. This viewpoint complements the regional and global perspectives of Chapter 13, and together these two chapters provide an integrated picture of the invasion process and the forces that mediate it.

14.2 Invasion processes: the determinants of local patterns of spread

Once an invasive ant species establishes in a new area, what factors determine how quickly and in which directions it will spread? Embedded within this overarching question is a host of related questions. To what extent do biotic interactions influence patterns of spread? Do different invasive ants rely on a common set of environmental factors to spread into ecosystems? Do invasive ants require anthropogenic disturbance to invade, or are the frequent associations between disturbance and invader presence more often correlative?

There exist two broad modes of dispersal among ants (Chapter 9). The more common (independent colony foundation) is usually characterized by flighted dispersal of newly mated queens and males, and is typical of species that possess monogyne, discrete, and mutually antagonistic colonies. In contrast to this arrangement, in a minority of ant species generally, but in a majority of invasive ants (Hölldobler and Wilson 1977), colony foundation or expansion is dependent, and mated queens disperse from their natal nests on foot with retinues of workers and establish new nests nearby (Holway *et al.* 2002a; Passera 1994). Invasive ants thus typically form networks of mutually tolerant, polygynous colonies that are collectively termed supercolonies (see Box 14.1). The major exceptions to this rule are *Solenopsis invicta* and *Solenopsis geminata*, which possess both monogyne and polygyne forms (Tschinkel 2006). In *S. invicta*, the multicolonial monogyne form founds new colonies independently through mating flights, while the polygyne form can disperse both on the ground (dependently) and via mating flights (independently), although dependent dispersal appears to be the more successful mode among polygyne fire ants (Tschinkel 2006). In addition, the polygynous *A. gracilipes* engages in mating flights as well as dependent dispersal, but flying dispersal events and independent colony founding are believed to occur rarely (Abbott 2006).

Dispersal through mating flights and independent colony founding results in a different pattern of spread from that of dependent dispersal. For example, monogyne *S. invicta* queens can found new colonies 1.6 km or more from their natal colonies, although they usually fly much shorter distances (Markin *et al.* 1971), and this mode of dispersal can produce a highly scattered, discontinuous distribution. Similarly, small isolated supercolonies of *Anoplolepis gracilipes* on Christmas Island, Indian Ocean, may be formed by flying queens (Abbott 2006). Dependent dispersal, in contrast, usually proceeds at rates of tens to several hundreds of metres per year (Holway *et al.* 2002a; however, propagules may sometimes disperse long distances when moved by humans or water, Suarez *et al.* 2001; Walker 2006; Walsh *et al.* 2004), and often produces a clean separation of invaded from uninvaded areas. Because invasive ants generally spread locally through dependent dispersal, we focus here on spatially continuous expansion of supercolonies into surrounding uninvaded areas. Such patterns resemble classic examples of diffusive spread (e.g. Elton 1958) that have inspired the modelling efforts of mathematical biologists for decades (Shigesada and Kawasaki 1997). In this regard, ant invasions offer rare opportunities to examine the factors that control spatial spread.

Environmental factors interact in complex ways to influence patterns of ant invasion. While acknowledging that none of these factors act in isolation, we first discuss how biotic interactions can either repel or encourage the local spread of invasive ants. We then examine how physical

Box 14.1 The ecological consequences of cooperative behaviour among workers from different nests
Edward G. LeBrun

Unicoloniality, or the absence of behavioural boundaries among nests within a population, is considered a key trait of many invasive ant species (see Section 14.3.2.1). Unicoloniality and monodomous, multicoloniality, in which colonies occupy a single nest and are mutually intolerant, represent opposite extremes of a continuous gradient in social organization, characterized by increasing exchange of individuals among nests. Along this gradient, key transitions in social behaviour allow for heightened levels of internest cooperation and confer specific ecological advantages (see Chapter 10 for discussion on colony structure).

An ecologically critical transition occurs when intraspecific aggression declines to the point that neighbouring colonies within populations cease behaving aggressively towards one another. The potential evolutionary forces leading to this change may be diverse (Giraud *et al.* 2002; Steiner *et al.* 2007; Tsutsui *et al.* 2000), but the ecological consequences are similar. Colonies cease to engage in intraspecific interference competition and territorial contests. The greatly reduced costs of competition that result are thought to allow for higher local worker density and, as a result, to enhance interspecific competitive ability and ecological dominance (see citations in Section 14.3.2.1). *Myrmica rubra*, introduced into the northeastern United States, exhibits low-intensity intraspecific aggression that increases over short distances within local sites, but nevertheless forms large, continuous, and dense populations of colonies that closely resemble supercolonies (Garnas *et al.* 2007). Colonies within such populations do not engage in obvious intraspecific interference competition. Further along the gradient are populations of polygyne *Solenopsis invicta* in the southern United States, in which workers from different nests are mutually non-aggressive, but only limited exchange of workers or queens occurs among nests. These societies are best characterized as semi-permeable. Queens and workers within nests are unrelated, suggesting that inter-nest movement occurs. However, queen number, allele frequencies, and worker mass exhibit nest specific signatures, indicating the existence of boundaries preventing the free exchange of workers (summarized in Goodisman *et al.* 2007). These functional colony boundaries ensure that intraspecific competition is not entirely absent. In these populations, workers recruiting to a large food item tend to belong to only one nest in the neighbourhood, typically the closest. In fact, the degree to which resources are monopolized is significantly greater than would be expected, based upon the relative distances of the nests from the food resource (Weeks *et al.* 2004). Thus, a weak form of intraspecific exploitative competition may operate, and because nests do not commonly intermix, colonies may restrict each other's movements.

The next possible level of colony integration is the free exchange of workers, but not queens, among nests. *Formica exsecta*, a non-invasive ant, provides an example of a species in which adult workers mix freely among nests, but queens do not (Kümmerli and Keller 2007). Examining the genetics and behaviour of the workers, these populations appear unicolonial, but this social organization maintains a multicolonial genetic structure within the immature workers and queens. Emergent ecological benefits of this particular social structure have not been explored, but newly established nests could potentially suffer lower mortality rates given their ability to grow rapidly. This might confer advantages to new nests in environments where survival requires large numbers, thereby favouring populations exhibiting this trait. This form of social organization has not been documented for an invasive ant species.

A final critical shift in social organization occurs when queens and workers mix among nests: nest fusion. If nest fusion is sufficiently common, all genetic and functional boundaries between colonies disappear and populations become fully unicolonial (Suarez *et al.*

continues

Box 14.1 continued

2008; Vasquez and Silverman 2008). Known examples of invasive ants in this category include *Linepithema humile* and *Wasmannia auropunctata*, and this arrangement is also strongly suggested for some populations of *Anoplolepis gracilipes* and *Pheidole megacephala* (see Section 14.3.2.3). Ecologically, nest fusion allows for functional integration across large spatial scales. If nests can fuse, then space is no longer pre-empted, making fine-scale adjustments between worker density and resource availability or microhabitat suitability possible. During times of abiotic stress, nests can form large aggregations in potentially limited suitable microsites. Large aggregations of ants can maintain environmental homeostasis more easily, regulating temperature and humidity within the nest. During less stressful times of year, nests can disperse through the environment and track resource pulses.

The invasive *L. humile*, and non-invasive *Formica truncorum*, provide examples of species that excel at this opportunistic use of space (Elias *et al.* 2005; Heller and Gordon 2006). These two very distinct taxa both evolved in environments that select for the ability to exploit patches of space that are only ephemerally available. *F. truncorum*, a boreal species, migrates seasonally between its large winter nests and its dispersed summer nest sites on productive rocky outcrops, while *L. humile* evolved in the seasonally inundated floodplain of the Paraná River, and must be able to exploit dry ground as it appears and recedes. Many of the species with populations that lack intraspecific aggression but are not fully unicolonial, such as polygyne fire ants and some species of northern European *Formica*, require elaborate nest structures that are costly to construct. These may make colonies inherently less mobile, and thus reduce the opportunity for and ecological benefits of nest fusion.

conditions and disturbance influence invasibility, and how different factors can interact.

14.2.1 Biotic interactions

Given that 'the biggest enemies of ants are often other ants' (Hölldobler and Wilson 1990), interspecific competition from resident ants may repel invasive ants, or at least curb the rate at which invasive ants penetrate ecosystems. Few data are available to rigorously test this prediction. In northern California, the spread of Argentine ant supercolonies in riparian corridors over a 4-year period was independent of the number of native ant species present (Figure 14.1a; Holway 1998b), evidence inconsistent with the biotic resistance hypothesis (Elton 1958). Further research on another California system, however, revealed that under suitable abiotic conditions, Argentine ants invade sites to a greater extent when native ants are removed, compared to when they remain present (Menke *et al.* 2007). This latter result indicates that native ants can slow the advance of supercolonies. Extending this idea to a biogeographic scale, one might predict that regions with highly competitive ant faunas may be especially resistant to invasion. For instance, competition from Australia's dominant dolichoderines has been offered as a potential explanation for the apparent inability of invasive ants to easily spread beyond human-modified habitats on that continent (Andersen 1997; Majer 1994; see also Way *et al.* 1997 for a potential European example). Although no studies to date have tested whether the presence of dominant Australian ants, such as *Iridomyrmex* spp., can prevent the expansion of invasive ant supercolonies, several studies have recently examined their effects on Argentine ant activity and competitive ability (Rowles and O'Dowd 2007; Thomas and Holway 2005; Walters and Mackay 2005). Results of these studies have been mixed, but on balance suggest that biotic resistance interacts with abiotic conditions to limit Argentine ant competitive success in some habitats, and therefore possibly also its distribution in these areas. In mesic to wet forested ecosystems, to which it is well adapted, *Pheidole megacephala* invades in spite of competition from Australian ants (Hoffmann *et al.* 1999; Vanderwoude *et al.* 2000), but like the Argentine ant may be less able to do so in drier areas. More research is needed to clarify the importance of competitive resistance in influencing invasion patterns in regions with highly diverse ant faunas, like Australia.

While biotic resistance from competitors may discourage the spread of invasive ants, reciprocally positive interactions may act to encourage their spread. Ants participate in an astonishing array of mutualistic associations with other organisms (Hölldobler and Wilson 1990), but the commonly formed associations between invasive ants and honeydew-producing insects (especially Hemiptera) seem especially likely to contribute to ecological success. Although research suggests that these opportunistic mutualisms promote higher densities of invasive ants than would be possible if the ants were primarily acting as predators (see 14.3.2.2), it is unknown to what degree the distribution and density of honeydew-producing insects influences patterns of invasive ant spread (Helms and Vinson 2002). These interactions are of more general interest as well, because mutualisms may encourage the spread of a wide range of introduced species. Examples of potential facilitation and mutualism between two or more introduced species have been increasingly reported in recent years (Simberloff 2006).

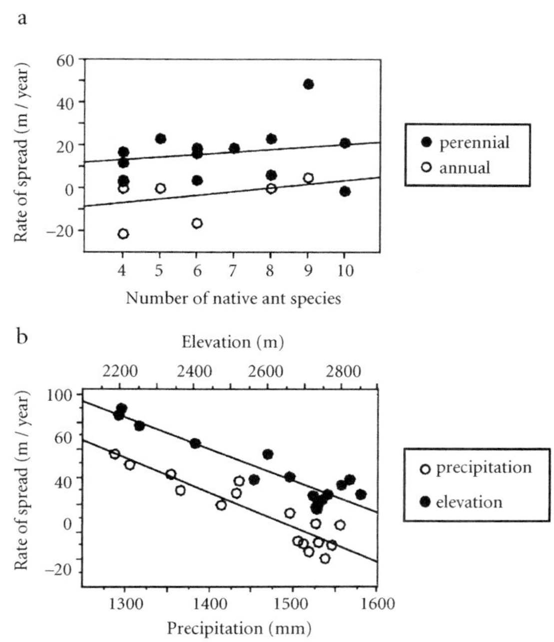

Figure 14.1 (*a*) Mean annual rate of spread of Argentine ants versus the number of native ant species in 20 riparian corridors in northern California. Over a 4-year period (1993–7), *Linepithema humile* spread more quickly at sites along perennial streams than it did at sites along intermittent streams (multiple regression: $t = 3.36$, $df = 17$, $p < 0.01$). Rate of spread was independent of native ant diversity 100 m ahead of the initial location of each invasion front ($t = 1.03$, $df = 17$, $p > 0.05$). (*b*) Mean annual rate of spread of Argentine ants versus elevation and precipitation in two spatially disjunct populations in Haleakala National Park, Maui, Hawaiian Islands. For the population in which spread is measured as a function of precipitation, the 16-data points represent net rates of outward spread between 1980 and 1997 along equally spaced radial lines (22.5° apart) emanating from a central point. For the population in which spread is measured as a function of elevation, the 16-data points are similar measures of outward spread between 1982 and 1997.

14.2.2 Abiotic conditions

The abiotic environment can strongly limit the activity and local abundance of invasive ants. As small-bodied insects, ants are highly vulnerable to temperature stress and desiccation (Hölldobler and Wilson 1990). Risks associated with exposure are especially acute for foraging workers, which must journey away from the safety of their nests, and which lack winged flight. With respect to ant invasions, abiotic factors can dictate patterns of invasion at multiple spatial scales (Menke *et al.* 2007). At small scales, the amoeba-like spread of ant supercolonies may be strongly influenced by abiotic suitability – with rainfall (or soil moisture) often invoked as a key factor. In seasonally dry California, for example, Argentine ants advance in riparian corridors along perennial streams, but not intermittent streams (Figure 14.1a; Holway 1998b), suggesting that the ability of this species to invade seasonally dry areas may be largely controlled by whether or not soils remain moist enough during the summer drought. This hypothesis is corroborated by observational studies that report positive correlations between rainfall and local patterns of Argentine ant abundance or rates of spread

(DiGirolamo and Fox 2006; Heller *et al*. 2008; Holway and Suarez 2006), field experiments that use irrigation to encourage the spread of *L. humile* (Menke *et al*. 2007; Menke and Holway 2006), and physiological studies showing that Argentine ants lose water more rapidly compared to native ants (Schilman *et al*. 2005). In arid locales, heat stress is also usually a factor during parts of the day, and Argentine ants are less capable of withstanding high temperatures compared to native ant species from arid ecosystems in both California (Holway *et al*. 2002b) and Australia (Walters and Mackay 2004). In both regions, a combination of hot and dry conditions appears important in preventing spread into xeric habitats from more suitable, and often anthropogenically modified, adjacent areas (Holway *et al*. 2002b; Thomas and Holway 2005; Walters and Mackay 2003a; 2004).

At the other end of the climatic spectrum, cold and wet conditions can also influence local patterns of ant invasion. In a montane shrubland habitat in Hawai'i, for example, both high elevation and abundant rainfall appear to have influenced long-term rates and spatial patterns of spread of *L. humile* (Krushelnycky *et al*. 2005a). In this system, areas with higher rainfall support a more dense vegetative ground cover, which reduces soil temperatures (S. Hartley and P. Krushelnycky, unpublished data). Lower temperatures thus appear to prevent Argentine ant spread into more windward, wetter habitat (Figure 14.1b). At the same time, the rate of spread of a nearby but separate unicolonial population of *L. humile* is strongly negatively correlated with elevation (Figure 14.1b), which also directly influences soil temperature. Similarly, low temperatures are likely responsible for limiting regional Argentine ant distributions in New Zealand (Hartley and Lester 2003). On a local scale, the distance of Argentine ant spread into New Zealand's natural areas from adjacent human-modified sites is much greater in open-canopy habitat than in closed-canopy habitat, probably because of higher ground temperatures in open sites (Ward and Harris 2005).

Seasonal patterns of invasion by Argentine ants are also largely tied to abiotic conditions. Nest clusters spread out as worker numbers increase during the warmer and drier parts of the year, at least in non-arid habitats (Heller *et al*. 2006; Newell and Barber 1913). This translates into overall expansion at the supercolony level during summer and fall, as well as some contraction in winter and spring in some locales (Heller *et al*. 2006; Krushelnycky *et al*. 2004; Sanders *et al*. 2001).

While rates and spatial patterns of spread of dependently dispersing supercolonies can be measured in a straightforward manner, identifying the factors that affect spread for species that disperse primarily by mating flights and independent colony founding is more difficult because of the complex spatial patterns that result. Nonetheless, abiotic factors can strongly influence patterns of establishment for invasive ants that colonize by mating flights, as in the case of monogyne *S. invicta* in Florida that proliferates in forest openings and in areas with moist soils (Tschinkel 1988b). In fact, *S. invicta* is generally absent from closed-canopy forests throughout the southeastern United States (e.g. Zettler *et al*. 2004), perhaps because thermal conditions in these environments are not consistently above the threshold for colony growth (Porter 1988), or because founding queens may have lower success in shaded forest habitats (McInnes 1994).

14.2.3 The role of disturbance

One factor likely to be of general relevance with respect to where ants invade is disturbance. Introduced ants commonly (but not exclusively) occur in anthropogenically disturbed environments (King and Porter 2007; Passera 1994). Although a distinction should be made between disturbances that directly result from human activity and those that do not, the causal relationship between disturbance and invasion success may be unrelated to the source of the perturbation. We view disturbance here as defined by Krebs (2000): any discrete event that disrupts community structure and changes available resources, substrate availability, or the physical environment. For ants, classes of disturbance that may often promote invasibility include flooding, soil perturbation, and the removal of trees. The clearest insights into how these factors affect invasibility will come from field manipulations that incorporate disturbance as an experimental treatment and that consider not only the effects of disturbance on the invader but also on resident ants. Disturbance may directly abet invaders that are predisposed to benefit from it. In addition, disturbance may indirectly promote invaders by negatively affecting native taxa, thus facilitating spread into recently impoverished communities (Diamond and Case 1986). The effects of disturbance, therefore, may commonly result from an interaction of biotic and abiotic factors.

The central role of anthropogenic disturbance in the success of monogyne red imported fire ants in the southeastern United States has been clarified in a series of studies by Walter Tschinkel and colleagues. Observational studies in northern Florida illustrate that *S. invicta* prefers open, recently disturbed areas (e.g. roadsides), whereas the putatively native *S. geminata* occupies less exposed sites with some canopy cover (McInnes 1994; Tschinkel 1988b, 2006). To identify the mechanisms underlying this pattern of invasion, King and Tschinkel (2006) and King and Tschinkel (2008) carried out manipulative experiments in Florida, adding or removing monogyne *S. invicta* colonies from study plots and physically altering some plots to emulate disturbances favoured by *S. invicta*. This body of work revealed – somewhat surprisingly – that the mixed assemblage of native and introduced ants present in the vicinity of plots in human-made pastures were unaffected by partial *S. invicta* removal (King and Tschinkel 2006). Conversely, the experimental establishment of *S. invicta* in natural forest plots was aided substantially by physical disturbance, while native ants were negatively affected by both the experimental additional of fire ants and disturbance (King and Tschinkel 2008). Taken together, these findings indicate that the spread of monogyne *S. invicta* in Florida may be driven largely by human-induced modifications to the landscape that simultaneously affect native ants.

However, disturbance may not be a necessary prerequisite for monogyne fire ant invasion in all systems. In a Texas savanna, monogyne *S. invicta* invaded a grassland habitat that had experienced no anthropogenic disturbance for at least 15 years prior to the arrival of red imported fire ants (Helms and Vinson 2001). In agreement with the results from Florida, there was no evidence that this invasion substantially affected native ant populations. In contrast, polygyne *S. invicta* invasions in Texas clearly depress native ant diversity (Morris and Steigman 1993; Porter and Savignano 1990), although in some systems local diversity has recovered (Morrison 2002b; see Section 14.4). Interestingly, Plowes *et al.* (2007) recently described a fine-scale distribution pattern similar to the disturbance-dependent invasion of Florida and the southeast United States in central Texas: in an urban landscape, *S. invicta* was common in open, recently cleared sites, whereas *S. geminata* persisted in historic neighbourhoods with mature shade trees. However, in the surrounding undeveloped matrix, all open habitats were invaded by polygyne red imported fire ants, regardless of disturbance history. Moreover, while polygyne *S. invicta* also thrives in deforested habitats, it has spread from these areas into naturally open and undisturbed habitats, such as prairies (Morris and Steigman 1993). This raises the question of whether a scarcity of unaltered naturally open and undisturbed habitats throughout the southeastern United States may be largely responsible for the close association of *S. invicta* with disturbed sites in this region.

The invasion of California by the Argentine ant has also been well studied, and offers an interesting comparison with that of the red imported fire ant. Like *S. invicta*, Argentine ants thrive in environments disturbed by humans. In southern California, for example, GIS analyses demonstrate the close association between *L. humile* and human-modified environments at the landscape scale (Menke *et al.* 2007). This positive relationship results in large part from two factors: (*a*) the Argentine ant's inherent dispersal limitations (Suarez *et al.* 2001), which keep this species from quickly spreading out from urban source populations, and (*b*) the tendency for it to benefit from irrigation (Menke and Holway 2006), which reflects its requirements for adequate levels of soil moisture. However, where moisture is naturally higher, Argentine ants readily invade, and in so doing displace a wide variety of native ants (Holway 1998a; b; Human and Gordon 1996; Ward 1987). Argentine ants become especially prolific in riparian woodlands of California (Holway 1998b, 2005) – perhaps unsurprising, given that they thrive in this habitat in Argentina.Thus, although *L. humile* benefits from certain aspects of human activity, it does not depend on these modifications to invade. Moreover, the kinds of disturbances (e.g. clearing of trees, soil disturbance) that promote the spread of *S. invicta* (Tschinkel 1988b; Zettler *et al.* 2004) fail to encourage *L. humile* in California unless they are accompanied by inputs of water. For example, anthropogenic degradation of riparian woodlands does not affect the rate at which Argentine ants spread in this habitat, nor does it influence native ant diversity or abundance (Holway 1998b).

One striking commonality of recent experimental work on *L. humile* and *S. invicta* concerns the importance of abiotic factors in the establishment and spread of these ants in North America. Native ants either interact weakly (monogyne *S. invicta*: King and Tschinkel 2006) or fail to repel (polygyne *S. invicta*: Porter and Savignano 1990; *L. humile*: Holway 1998b) invaders, and as a consequence the distribution of *L. humile* and *S. invicta* is largely dictated by physical conditions, which are altered to varying degrees by human activity. The local patterns of distribution of these two ant species in North America largely mirror their microhabitat selection in South America: Argentine ants most commonly inhabit floodplain woodlands or wetland habitats with emergent vegetation, whereas *S. invicta* is more common in open pastures and grasslands (LeBrun *et al.* 2007). These fine-scale habitat preferences are reflected in the spectacular success of *S. invicta* in the southeastern United States following landscape-level deforestation of this region (Tschinkel 2006). Likewise, the profligate use of water in seasonally dry coastal California opens up large areas to invasion that would otherwise be unsuitable to *L. humile* (Menke *et al.* 2007).

Despite the frequent connection between disturbance and ant invasions, it is apparent that, like *L. humile* and polygyne *S. invicta* in North America, most, if not all other invasive ant species are capable of spreading from human-modified habitats into largely undisturbed natural areas if the abiotic conditions are suitable. Examples of this phenomenon include *W. auropunctata* (Clark *et al.* 1982; Le Breton *et al.* 2003; Walker 2006), *A. gracilipes* (Abbott 2006; Hill *et al.* 2003), *P. megacephala* (Heterick 1997; Hoffmann and Parr 2008; Wetterer 2002), as well as *L. humile* in other parts of the world (Bond and Slingsby 1984; Krushelnycky *et al.* 2005a; Oliveras *et al.* 2005a; Ward and Harris 2005).

14.3 Causes of success

In cases where invasive ants do manage to spread into natural areas supporting native ant communities, they often displace a majority of native ant species in the process (Holway *et al.* 2002a; Chapter 15). This is a remarkable feat, in view of the fact that ant communities typically represent a diverse array of species employing a variety of life histories and competitive strategies. Monogyne populations of *S. invicta* in disturbed habitats of the southeastern United States may be notable exceptions to this pattern, in that their invasions in these areas may not greatly suppress colonies of many other ant species (Helms and Vinson 2001; King and Tschinkel 2006; Morrison and Porter 2003). But for other instances in which invasive ants have excluded the vast majority of native ant species, the question remains: how do they accomplish this and what characteristics account for this unusual ability?

14.3.1 Competitive displacement and predation

Coexistence among native ant species in intact communities is accomplished through a variety of mechanisms (Hölldobler and Wilson 1990; Chapter 5). For systems not structured by disturbance, coexistence among species that compete for similar food resources under similar abiotic conditions can be mediated, in part, by a foraging strategy trade-off between resource discovery and resource dominance (Davidson 1998; Feener 2000; Feener *et al.* 2008; Fellers 1987; Holway 1999; LeBrun and Feener 2007). Some species excel at rapidly finding and retrieving food (superior exploitation ability), while other species are slower at discovery but excel at displacing those species that arrived sooner through behavioural dominance (superior interference ability). There are few studies that have managed to elucidate the competitive dynamics between invasive ants and native ant communities at contact zones where displacement is actively occurring. Most information on this process comes from studies of Argentine ant invasions in California. In two different ecosystems, researchers found that Argentine ants were more likely to find baits, located and recruited to baits more quickly, and foraged over a greater part of the day than did native ants (Holway 1999; Human and Gordon 1996). Because of an omnivorous diet, Argentine ants also likely competed for food resources with a wide range of native species. Once they located baits, Argentine ants recruited more workers, persisted longer than native ants, and were more frequently aggressive towards native ants than vice versa. Thus, through pre-emption and displacement, *L. humile* controlled a majority of available baits in the contact zones. Similar results have recently been reported for an Argentine ant invasion front in Spain (Carpintero *et al.* 2007; Carpintero and Reyes-López 2008), although subtle differences appear to exist in the prevalence and effects of Argentine ant aggression in these three invaded communities.

These studies suggest that Argentine ants come to dominate the ant communities they invade by consistently excelling at both exploitation and interference competition (Figure 14.2). This presumably allows them to monopolize a majority of available resources at the leading edges of their expanding populations, and in the process to displace a majority of the resident native ant species. While such detailed studies of the competitive dynamics of invasions in progress are not available for most other invasive ant species, at least a few of them appear to possess similar abilities. For instance, *W. auropunctata* foraged 24 h a day and displaced other ants from a large majority of baits in contact zones at the edges of its range in the Galápagos (Clark *et al.* 1982). In areas where *S. invicta* still co-occurs with *S. geminata* in its native range in the southern United States, *S. invicta* has been found to retrieve more food (Morrison 1999) and to forage over a wider range of temperatures and humidities (Wuellner and Saunders 2003) compared to *S. geminata*. Similarly, in a recently invaded field in the southern United States, colonies of hybrid imported fire ants (*S. invicta* × *Solenopis richteri*) discovered and recruited to baits more quickly than did native ants, and also retained control of more baits after several hours (Gibbons and Simberloff 2005).

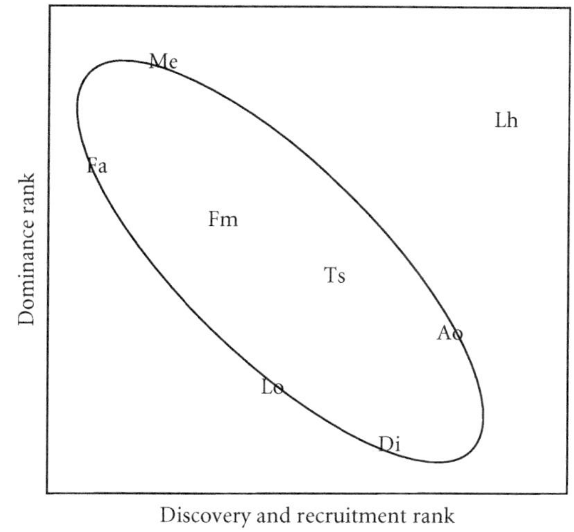

Figure 14.2 For ant assemblages of northern California riparian woodlands, an interspecific trade-off exists between the abilities of species to discover food resources versus their ability to dominate them. The invasive Argentine ant (Lh) does not conform to this relationship because it is highly successful at both discovery and dominance. Remaining ant species abbreviations are as follows: Ao = *Aphaenogaster occidentalis*, Di = *Dorymyrmex insanus*, Fa = *Formica aerata*, Fm = *Formica moki*, Lo = *Liometopum occidentale*, Me = *Monomorium ergatogyna*, Ts = *Tapinoma sessile*. (Modified from Holway 1999 and Feener 2000)

The aggressive behaviour of invasive ant species is not limited to the acquisition of food. Nest raiding of heterospecific colonies has been reported for *L. humile* (Carpintero and Reyes-López 2008; Fluker and Beardsley 1970; Rowles and O'Dowd 2007; Zee and Holway 2006), polygyne *S. invicta* (Hook and Porter 1990) and *P. megacephala* (Dejean *et al.* 2008), and although difficult to observe, this behaviour may be common. In fact, nest raiding may be the primary way in which colonies of some species are displaced (or destroyed). This is likely to be the case for species, such as harvester ants, that have little overlap with invasive ants in resource use but nevertheless disappear from invaded areas (Plate 15; Erickson 1971; Hook and Porter 1990; Human *et al.* 1998; Zee and Holway 2006). It is often unclear to what degree raided adults and brood are taken for prey, and therefore this behaviour may sometimes represent interference competition (Hook and Porter 1990), predation (Carpintero and Reyes-López 2008), or both (Zee and Holway 2006). Aggression directed at dispersing queen foundresses is likely important in preventing the re-establishment of native colonies within the territories of invasive ants (Human and Gordon 1996).

14.3.2 The advantages of numbers

But how, exactly, do invasive ant species excel in these competitive or predatory abilities? A number of studies have shown that invasive ants are not particularly successful in one-on-one competition with native ants (e.g. Buczkowski and Bennett 2007; Holway 1999; Holway and Case 2001). Instead, most evidence indicates that invasive ants derive their superior competitive abilities from higher numbers (Buczkowski and Bennett 2007; Carpintero and Reyes-López 2008; Holway 1999; Holway and Case 2001; Human and Gordon 1996; Human and Gordon 1999; Morrison 2000; Rowles and O'Dowd 2007; Tremper 1976; Walters and Mackay 2005). Numerical asymmetries commonly determine competitive outcomes in ants generally (Hölldobler and Wilson 1990), and these asymmetries can be taken to extraordinary levels in the case of invasive ants. Invasive ant colonies are not only larger than competing native colonies, they can sometimes attain such high abundances that they greatly exceed in number or biomass all of the native ant species, combined, in the communities they invade (Hoffmann and Parr 2008; Holway 1999; Human and Gordon 1997; Morris and Steigman 1993; Porter and Savignano 1990). These elevated densities are not only the key to their success against native ants, but are also the underlying cause for many of their other ecological and economic impacts (Holway *et al.* 2002a). Explanations for this phenomenon, reviewed here, have been the topic of much research.

14.3.2.1 Unicoloniality

Unicoloniality is shared among most (but not all) invasive ant species (Holway *et al.* 2002a; Passera 1994; see Box 14.1). While unicoloniality is not unique to invasive ant species, some *Formica*, for example, form expansive supercolonies (Bourke and Franks 1995; Hölldobler and Wilson 1990), unicoloniality has attained its most dramatic form among certain invasive ants. Introduced populations of *L. humile* and *W. auropunctata* can form vast supercolonies and occupy disjunct sites over hundreds of square kilometres (Corin *et al.* 2007a; Giraud *et al.* 2002; Le Breton *et al.* 2004; Tsutsui *et al.* 2000). Within such supercolonies, workers from different nests generally behave towards one another as if they were nestmates, even when the workers originate from distant locations. It is worth noting that this is not always the case, and in some situations multiple supercolonies of much smaller size occur in a region or locality (Abbott *et al.* 2007; Buczkowski *et al.* 2004; Sunamura *et al.* 2007). Although the sizes of supercolonies can vary, unicoloniality appears to consistently promote high densities of ants. Why is this the case?

At a local scale, the absence of intraspecific aggression can have profound consequences. Because aggression in ants is often strongest among conspecific colonies (Hölldobler and Wilson 1990), species that lack intraspecific aggression avoid large costs associated with territoriality. The polygyne forms of *S. invicta* and *S. geminata*, for example, while not considered strictly unicolonial, are believed to maintain higher nest densities and worker biomass compared to those of the monogyne forms (MacKay *et al.* 1990; Macom and Porter 1996) in part because of reduced intraspecific aggression. Argentine ants provide an additional example. Over most of their range in California, Argentine ants show little or no hostility to one another (Tsutsui *et al.* 2000), but where genetically differentiated supercolonies contact one another the story is entirely different. Battles between supercolonies at points of contact are frequent, prolonged and intense, and can result in considerable worker mortality (Thomas *et al.* 2006). The potential for this type of aggression to limit colony growth and competitive ability can be readily studied in the laboratory. Compared to non-aggressive pairs of experimental colonies collected from the same supercolonies, mutually antagonistic pairs have lower rates of worker foraging and food retrieval, higher worker mortality, and reduced productivity (Holway and Suarez 2004; Holway *et al.* 1998). In the laboratory, such costs also impinge on interspecific competitive ability. In a similar study, colonies of *Forelius mccooki* experienced greater worker mortality and lower egg production when reared with non-aggressive pairs of Argentine ants than they did when reared with intraspecifically aggressive pairs (Holway and Suarez 2004). Dramatically, two-thirds of *F. mccooki* colonies matched with cooperative *L. humile* colony pairs were killed outright, while none of those matched with mutually antagonistic *L. humile* colony pairs died. Although conducted in a simplified laboratory setting, these experiments on *L. humile* illustrate in principle not only how the absence of intraspecific aggression can lead to higher abundances of invasive ants, but also how it should translate into strong competitive advantages against native ants.

The breakdown of intraspecific territoriality within supercolonies can result in extremely high nest densities. After most heterospecific ant colonies have been displaced, invasive ants have all suitable nesting sites at their disposal. This allows them to saturate the environment and more thoroughly monopolize food resources, undoubtedly boosting densities further. Numerous studies have shown that within supercolonies (i.e. away from invasion fronts), invasive ant species are highly effective at food exploitation. Through higher local forager densities, often combined with a more rapid foraging tempo, they have been found to cover ground more quickly when foraging compared to assemblages of native ants (*L. humile*: Oliveras *et al.* 2005a), to find food faster compared to native ants (*L. humile*: Gómez and Oliveras 2003; Holway 1999; polygyne *S. invicta*: Porter and Savignano 1990; *P. megacephala* and *A. gracilipes*: Ward and Beggs 2007), and to find more of it (*L. humile*: Human and Gordon 1996; polygyne *S. invicta*: Porter and Savignano 1990; *W. auropunctata*: Le Breton *et al.* 2005; *P. megacephala*: Dejean *et al.* 2007b; Ward and Beggs 2007; *A. gracilipes*: Ward and Beggs 2007). In other situations, comparisons of performance with native ant species are unavailable, but data nevertheless indicate a thorough acquisition of offered food baits (Lester and Tavite 2004; Sarty *et al.* 2007; Vanderwoude *et al.* 2000).

This same absence of territoriality within supercolony boundaries also allows for flexibility in nesting arrangements. Instead of being subjected to the limitations of central-place foraging strategies, invasive ants that are unicolonial may more commonly employ dispersed central-place foraging as a result of their polydomy (Holway and Case 2000). Under this arrangement, food, workers, brood, and queens can be shifted among nests, and new nests can be constructed, to counteract spatial heterogeneity in available resources, and minimize energy expended in food retrieval and distribution (Holway and Case 2000; Newell and Barber 1913; Silverman and Nsimba 2000). Similarly, ants with this colony structure move their nests in order to take advantage of favourable abiotic conditions (Heller *et al.* 2006; Markin 1970b; Newell and Barber 1913). Such strategies are largely unavailable to multicolonial species, which would often need to usurp another colony in order to move. Despite the absence of territoriality in polygyne *S. invicta*, relatedness among workers within and across mounds indicates a general lack of exchange of workers or queens among nests (Goodisman *et al.* 2007; Ross 1993), revealing that it does not share this ecological advantage with fully unicolonial species (see Box 14.1).

14.3.2.2 Generalist habits

As a subset of the broader group of tramp species, invasive ant species typically have loose nesting requirements (Holway *et al.* 2002a). This allows them to establish nests in a wide variety of sites and reinforces the trend towards high nest density within supercolony boundaries (e.g. *P. megacephala* in Tonga, Wetterer 2002; *W. auropunctata* in New Caledonia, Le Breton *et al.* 2005; *L. humile* in California, Heller 2004; *A. gracilipes* in the Seychelles, Haines and Haines 1978a and on Christmas Island, Abbott 2005).

A broad diet also characterizes most invasive ant species (Holway *et al.* 2002a), and this trait enables a more complete use of available resources. However, it appears that a particular aspect of omnivory is especially important, namely the heavy use of liquid carbohydrate resources. Plant exudates and insect honeydew play a vital role in the energy budgets of many ant species, but may be of particular importance for ecologically dominant species that attain high densities and maintain high levels of activity (Davidson 1998; Davidson *et al.* 2003;

Chapters 6 and 7). It is therefore not surprising that strong associations with honeydew-producing insects have been reported for all of the major invasive ant species (Lach 2003; Ness and Bronstein 2004). For *S. invicta* in eastern Texas, up to half of the colony energy requirements may be supplied by hemipteran honeydew, most of which comes from an introduced mealybug (Helms and Vinson 2002). The exploitation of abundant and cheap carbohydrate-based fuel may be unsurpassed on Christmas Island in the Indian Ocean, where facultative mutualisms with possibly introduced honeydew-producing Hemiptera appear to contribute importantly to the extraordinary density and frenetic activity of *A. gracilipes* that currently exists there (Abbott 2005; Abbott and Green 2007; O'Dowd *et al.* 2003).

Besides boosting invasive ant abundances, liquid exudates may also influence behaviour. In the case of *L. humile*, carbohydrate deprivation reduces aggression and activity (Grover *et al.* 2007). It remains to be seen whether this is true of other invasive ants, or whether different species employ somewhat different strategies in achieving ecological dominance. For instance, despite heavy use of liquid exudates (Clark *et al.* 1982; Le Breton *et al.* 2005), *W. auropunctata* does not usually engage in high-tempo foraging and scouting activity. Instead, it seems to rely on an especially high saturation of the environment with its nests. This species therefore manages to locate and recruit to food quickly (Le Breton *et al.* 2005) despite the slow movement of its workers. In addition to disparities among ants, each invasive species may use plant-based exudates differently depending on site-specific variation and time since introduction. Such flexibility in resource use has recently been demonstrated for *L. humile* (Tillberg *et al.* 2007), and could be partially responsible for differences in density or behaviour among sites.

14.3.2.3 Ecological release and genetic changes

The characteristics discussed above explain how invasive ants achieve their remarkable dominance in introduced areas. However, they do not adequately address the important question of why invasive ants are able to employ these mechanisms to such greater effect in their introduced ranges than in their native ranges, where they typically coexist with many other ant species (Feener *et al.* 2008; Heller 2004; LeBrun *et al.* 2007; Porter *et al.* 1997; Suarez *et al.* 1999; Tennant 1994). For most ant invasions, ecological success likely results from the joint action of multiple causal factors, and a key challenge is to develop a clearer understanding of the relative importance of each factor and how they interact.

A common explanation to account for the success of invasive species involves escape from natural enemies (Mack *et al.* 2000), whereby an invader gains a competitive advantage over natives by leaving behind its suite of specialized pathogens, predators, and parasites in its native range, while its new competitors must still contend with theirs. *S. invicta* has until recently faced only a handful of pathogen and parasite species in North America, compared to several dozen in its South American homeland (Porter *et al.* 1997). Phorid flies, which parasitize adult worker ants, provide an example of one such enemy. Phorids modify interspecific competitive outcomes in ants through differential effects on worker behaviour (Feener 2000). Phorids are typically host-specific and often associate with dominant ants. When present, the disruption phorids create through their attempts to oviposit can substantially diminish the behavioural dominance and foraging efficiency of the beleaguered species, including *S. invicta* in its native range (Feener *et al.* 2008; Orr *et al.* 1995; Porter *et al.* 1995). In fact, the potential importance of these effects forms the basis of recent biological control programs that have introduced phorid flies into the southeastern United States in an attempt to suppress the dominance of *S. invicta* over native species. Where *S. invicta* co-occurs with *S. geminata*, for instance, it can forage unmolested day and night, while diurnal foraging by *S. geminata* is reduced when its own phorid parasitoids are present (Morrison 1999). However, the strength of phorid fly impacts on behavioural dominance appears to be community-dependent (Feener *et al.* 2008), and so far there is no evidence that introduced phorid populations have succeeded in suppressing populations of *S. invicta* (Morrison and Porter 2005). The importance of escape from pathogens, parasitoids, and predators is unknown in the case of other invasive ant species, because no such natural enemies have so far been identified. More attention should be directed towards this goal.

Ecological release, and thus greater dominance, of invasive ants in their introduced ranges could also result from a less competitive environment, mediated chiefly by the rigour of the new ant communities that are faced (Buren *et al.* 1974; Heller 2004). For *S. invicta, L. humile*, and *W. auropunctata*, ant communities in their native ranges in South America are typically more species rich than are those in various areas around the world where they have been introduced (e.g. compare: Clark *et al.* 1982; Tennant 1994; Heller 2004; Human and Gordon 1996; Le Breton *et al.* 2005; Suarez *et al.* 1999; Ward and Harris 2005; reviewed in Carpintero and Reyes-López 2008; Feener *et al.* 2008; LeBrun *et al.* 2007; Tschinkel 2006). This imbalance is most acute for oceanic islands, where many introduced ant species often occur and native ant faunas are often depauperate. Both *S. invicta* and *L. humile* experience strong interspecific competition within the diverse ant communities in their native ranges – including from each other where they are sympatric – and this competition shapes local distributions and patterns of dominance (LeBrun *et al.* 2007). Other species in these communities exhibit greater behavioural dominance, or are more efficient at resource location, and therefore attain some degree of ecological dominance even in the presence of *S. invicta* and *L. humile.* The lower-diversity ant communities in many invaded regions appear unable to offer this type of resistance. But what about cases where invasive ants have completely displaced highly diverse ant communities in undisturbed ecosystems, such as *W. auropunctata* in western equatorial African forests (Walker 2006) and *P. megacephala* in both open forests and monsoonal rain forests in Australia (Hoffmann *et al.* 1999; Vanderwoude *et al.* 2000)? Because the invaders are not entering species-poor communities in these examples, it would suggest that factors other than competitive release are operating. On the other hand, the competitive strength of an ant community may be only loosely tied to the number of species present, and could, for example, vary systematically among communities that evolved in different biogeographic regions (LeBrun *et al.* 2007; Suarez *et al.* 2008; Chapter 13). If true, simple comparisons between the diversity of an invader's source community and those of introduction sites may not accurately predict whether competitive release is occurring. Much remains to be learned in this area.

A final factor potentially involved in the greater ecological dominance and increased size of invasive ant supercolonies concerns genetic changes that take place during and after introduction. This idea has been studied in greatest detail with the Argentine ant and postulates that reduced genetic variation, resulting from bottlenecks during founding events, has led to a diminished ability to discriminate between nestmates and non-nestmates within introduced populations (Tsutsui *et al.* 2000). Giraud *et al.* (2002) suggest that losses in discriminatory ability arose primarily after establishment through differential selection against uncommon recognition alleles. In either case, the resultant reduction in intraspecific aggression may have contributed to the formation of the massive Argentine ant supercolonies found in a variety of introduced areas. In addition, reduced genetic diversity may play a role in maintaining or expanding the size of introduced supercolonies by mediating patterns of asymmetric aggression among different supercolonies, with lower diversity supercolonies exhibiting greater agonistic behaviour towards higher diversity supercolonies than vice versa (Tsutsui *et al.* 2003). It is clear, however, that the vast supercolonies in some introduced areas are also the product of widespread transportation of a small number of founder colonies. The relative importance of these and other factors is still uncertain. Colony-structure variation occurs in native populations of Argentine ants, and includes localized supercolonies, but these are typically orders of magnitude smaller than those in introduced areas (Heller 2004; Holway and Suarez 2004; Suarez *et al.* 1999; Tsutsui *et al.* 2000 Pedersen *et al.* 2006). Smaller supercolonies are more likely to experience intraspecific competition from neighbouring supercolonies, and this may be part of the reason that Argentine ants are less ecologically dominant in their native range (Suarez *et al.* 2008).

Differences in genetic diversity between native and introduced populations also exist for *S. invicta* and *W. auropunctata.* In the case of *S. invicta*, genetic bottlenecks during introduction to the United States have resulted in changes in the sex determination system, with the frequent production of sterile diploid males in polygyne populations (Ross 1993). However, this reduced genetic diversity is not linked to polygyny. Instead, polygyny appears to be determined at a single genomic region marked by the locus *Gp-9* in both introduced and native populations of *S. invicta* and its close relatives (Krieger and Ross 2002; Ross and Keller 1998), although differences exist between native and introduced populations in the number of queens per nest and patterns of relatedness within nests (Ross *et al.* 1996). Like some introduced populations of *L. humile, W. auropunctata* in New Caledonia has also passed through a severe genetic bottleneck, and there are large differences in the sizes of its supercolonies between native and introduced ranges (Foucaud *et al.* 2006). Unlike other ants, however, both introduced and some native populations of *W. auropunctata* reproduce predominantly clonally (in both queens and males; Fournier *et al.* 2005), and it appears that these clonal populations have arisen from normal, sexually reproducing populations in less disturbed parts of the native range (Foucaud *et al.* 2007). It remains to be determined how this variation in genetic diversity among populations relates to ecological dominance. In some cases, connections between genetic differences and ant density may be difficult to elucidate. For example, Abbott *et al.* (2007) documented two genetic haplotypes of *A. gracilipes* that are both unicolonial, but that attain strikingly different densities and levels of ecological dominance on the atolls of Tokelau. It is as yet unknown to what degree variation in the habitats occupied by each haplotype might contribute to this difference. In general, genetic differences between native and introduced populations strongly relate to patterns of human translocation, and genetic bottlenecks that occur from one introduced population to another or through human-mediated introduction events within the native range have the potential to confound the native versus introduced range dichotomy.

14.4 Future directions

Major gaps in our knowledge remain at each stage of the invasion process. At the earliest stage, the dynamics of establishment and initial persistence have received relatively little attention. How do small incipient colonies survive long enough and displace enough competing colonies to begin forming the supercolonies that subsequently make them so dominant? In some cases it appears that invasive ant supercolonies have managed to establish, survive, and grow in largely undisturbed natural habitats supporting diverse ant faunas (e.g. Vanderwoude *et al.* 2000), suggesting that they must possess fairly strong competitive abilities even at low densities. The most compelling evidence for such abilities comes from colony- introduction experiments. When single relatively small colonies (500 to 1,500 workers) of *L. humile* have been placed in the field among native Californian ants, these have often been found to be successful at winning control of baits (Holway 1999; Human and Gordon 1996). In Australia, however, results from similar experiments have been mixed. Small Argentine ant colonies were found to always win control of baits when facing colonies of *Iridomyrmex bicknelli* and two other native species in one community (Rowles and O'Dowd 2007), but even substantially larger colonies (5,000 workers) of *L. humile* were unable to displace *Iridomyrmex 'rufoniger'* from baits in another community (Walters and Mackay 2005). In cases where small Argentine ant colonies succeeded against resident ant colonies, they appeared to do so by recruiting higher numbers of workers to offered baits, frequently exhibiting aggression towards heterospecific workers, and raiding the nests of nearby competing colonies. Similar experiments have not been performed for other invasive ant species, but would seem to hold promise for illuminating key mechanisms involved in this early phase of the invasion process. Moreover, because they can remove much of the numerical imbalance that exists between native colonies and well-established invasive supercolonies, small colony-introduction experiments also have the potential to shed light on the question of possible asymmetries in inherent competitive abilities between species that originated in different biogeographic regions.

Important questions persist about dynamics that continue after establishment. For instance, while the heavy use of carbohydrate resources by invasive ants has long been recognized, it is unclear whether particular ant species require the presence of certain types of honeydew-producing mutualists to spread into some habitats, or to become dominant in some communities. Likewise, it would be useful to know whether invasive ants typically rely on introduced mutualists, or instead whether there are enough native mutualists in most communities to meet their carbohydrate needs. Another area of interest concerns mechanisms of coexistence with invasive ants on the part of native ants. It is often reported that a few resident ant species can persist with invaders, but the ways in which they do this need to be tested more explicitly (e.g. Sarty *et al.* 2006, 2007; Ward 1987).

Longer-term dynamics are also of great interest, but not well studied. For example, the remarkably high densities of nests and workers measured in some invasions may exist above carrying capacity and thus be unsustainable. In one case study, the initial invasion of Brackenridge Field Station in central Texas by polygyne *S. invicta* resulted in the local displacement of native ants and many non-ant arthropods, with the severity of these impacts positively related to fire ant density (Porter and Savignano 1990). Over the subsequent decade, however, the initially high numbers of *S. invicta* declined to the point that native ants re-colonised the site and effects on non-ant arthropods became difficult to discern (Morrison 2002). In other cases, long-term studies have documented that other invasive ant species can remain abundant over time and continue to suppress native ants (Heller *et al.* 2008; Hoffmann and Parr 2008; Holway 1995; Tillberg *et al.* 2007; Walker 2006). In yet another pattern, some ant species may be present at low densities for many decades before suddenly becoming very abundant and invasive (Groden *et al.* 2005; O'Dowd *et al.* 2003). As for invasions generally (Simberloff and Gibbons 2004), the mechanisms underlying long-term changes in density of invasive ants can be hard to pinpoint (Krushelnycky *et al.* 2005a; Morrison 2002b), but making progress in this area is important. Increasingly, global climate change needs to be considered in studies of long-term population dynamics (Heller *et al.* 2008; Morrison *et al.* 2005; Roura-Pascual 2004).

Finally, as mentioned in the introduction of this chapter, a major barrier to understanding general causes of success among invasive ants is the currently small sample size. We have accumulated considerable knowledge about a few ant species, but still know little about others. For example, *Technomyrmex albipes, Paratrechina fulva*, and *P. longicornis* can attain high densities in some situations (Wetterer *et al.* 1999; Zenner-Polania 1994 Holway *et al.* 2002a), but little is known about the dynamics of these cases. Several other species, including *Lasius neglectus, Myrmica rubra*, and *Tetramorium tsushimae* have been recognized as invasive, or have become the subjects of research only relatively recently (e.g. Espadaler and Rey 2001; Groden *et al.* 2005; Steiner *et al.* 2006b). Additionally, a number of species have long been regarded as urban pests (e.g. *Monomorium pharaonis, Monomorium destructor*, and *Tapinoma melanocephalum*, Vander Meer *et al.* 1990; Williams 1994), and while these often appear to share some of the biological traits possessed by the species discussed throughout this chapter, the processes and mechanisms by which they achieve their success have not been sufficiently studied.

Further examination of the most common and widespread invasive ants is needed, but research on these additional ant species should also be highly informative. For example, both *L. neglectus* and *M. rubra* exhibit many characteristics common to invasive ants, including polygyny, polydomy, and an apparent ability to form large, continuous, high density aggregations of nests that spread out from anthropogenic habitats into surrounding more natural habitats (Espadaler *et al.* 2007; Groden *et al.* 2005). However, *L. neglectus* is of interest because, while it is unicolonial and has rapidly spread via human transportation like other invasive ant species (Ugelvig *et al.* 2008), its queens are morphologically and physiologically intermediate between those of typically monogynous, claustral-founding species that undergo mating flights and those of most invasive ant species, which are typically polygynous and undergo dependent-colony founding (Espadaler and Rey 2001). Laboratory colonies of *L. neglectus* can be initiated by independent queens, and while there is currently little evidence that winged dispersal by solitary queens occurs under natural conditions, the retention of traits associated with independent colony founding may provide physiological advantages that enhance invasion success (Espadaler and Rey 2001). Meanwhile, *M. rubra* is unusual in that it forms large, dense populations of colonies that resemble supercolonies despite the presence of intraspecific aggression (Garnas *et al*. 2007). Apparently, aggression among nearby colonies is relatively subdued and does not prevent coexistence. These two more recently invasive species thus emphasize the fact that considerable variation exists among invasive ants. Their cases, as well as others, need to be further explored to gain a more complete picture of the factors that can lead to invasiveness. Moreover, these factors should be systematically compared between invasive ant species and their non-invasive relatives, as has been done for other taxa (e.g. Rejmánek and Richardson 1996), to strengthen inferences about their causal roles.

14.5 Summary

The nature of the current global economy all but ensures that introduced ants will continue to be transported to new regions of the world, and that they will initiate new invasions upon reaching some of these locales. A comprehensive understanding of the causal factors that promote invasiveness, as well as the mechanisms that mediate the invasion process, should assist in developing a predictive capacity about which species are likely to become invasive in which habitats. Comparisons among invasive ant species studied to date indicate that most appear to be pre-adapted to some degree for periodic disturbance, and most are either unicolonial or exhibit a reduction in inter-nest aggression. They can form relatively large colonies, even within their native ranges, and employ abundant active and generalist workers. When invasive ants meet suitable abiotic conditions within introduced ranges, native ant communities have shown little ability to stop their spread, although longer-term dynamics may follow various trajectories. The characteristics of several common invasive ant species in their native ranges are still not known, and much remains to be learned about a number of other species. Current conclusions, therefore, may be biased. Even if most invasive ant species appear to share a common group of behavioural traits, they may have arrived at this condition in different ways. As an example, recent research has revealed that the genetic mechanisms underlying unicoloniality, or at least polygyny and the breakdown of intraspecific territoriality, appear to be different for *L. humile, S. invicta* and *W. auropunctata*. Among other topics, future research should seek to understand how multiple strategies, genetic pathways, and ecological factors combine to create the emergent property of invasiveness in ants.

Acknowledgements

We would like to thank K. Abbott, B. Hoffmann, L. Keller, L. Lach, K. Parr, J. Silverman, D. Simberloff, A. Suarez, and N. Tsutsui for helpful comments on earlier versions of this chapter.

Chapter 15

Consequences of Ant Invasions

Lori Lach and Linda M. Hooper-Bùi

15.1 Introduction

As with other invasive species, many facets of the consequences of ant invasions have been observed and measured only in the last 20 years. However, because ants are virtually ubiquitous and play so many important roles in ecosystems, impacts of invasive ants have received much research attention. An analysis of over 400 primary research papers (1900–2007 inclusive) that have investigated the impacts of invasive insects or their underlying mechanisms found that though invasive ants comprised only 8 of the 72 species investigated, they were the subject of 41% of the studies (Kenis *et al.* 2009). The red imported fire ant (*Solenopsis invicta*) and the Argentine ant (*Linepithema humile*) accounted for 42% and 33% of these studies, or 18% and 14% of all studies, respectively.

Several reviews have summarized the impact of specific invasive ant species (e.g. Allen *et al.* 2004; Wetterer 2007; Wetterer and Porter 2003) or invasive ants in specific geographic locations (Krushelnycky *et al.* 2005b; Lach and Thomas 2008) or on specific interactions (Ness and Bronstein 2004; Suarez *et al.* 2005b). In this chapter, we focus on developments since the publication of the comprehensive review by Holway *et al.* (2002a). Chapter 14 considered mechanisms of invasive ant success. Here, we review the documented ecological consequences of invasive alien ants, including effects on invertebrates, vertebrates, plants and their associated arthropods, and soil, in both natural and agricultural systems. We include all ant species previously defined as invasive (see Part IV Introduction) for which there are published studies. We do not consider economic or public health effects, although we acknowledge that these are the primary drivers in funding research on invasive ants.

15.2 Effects on native ants

The displacement of native ants is the most commonly documented effect of ant invasions (Hölldobler and Wilson 1990; Holway *et al.* 2002a). Holway *et al.* (2002a) cited 37 studies that reported displacement of native ants or that demonstrated differences in competitive ability between native and invasive ants. Few of these studies used experimental approaches, and only a handful explored effects of ant species other than the red imported fire ant (*Solenopsis invicta*) and the Argentine ant (*Linepithema humile*).

Recent studies have continued to document native ant displacement by alien invasive ants (Table 15.1). Whereas *S. invicta* and *L. humile* continue to be popular topics of research, the long-legged, or yellow crazy ant (*Anoplolepis gracilipes*), the little fire ant, or electric ant (*Wasmannia auropunctata*), and to a lesser extent, the big-headed ant (*Pheidole megacephala*) are receiving more research attention in some parts of the world (Table 15.1). Other invasive ant species have been noted relatively recently and are only beginning to be studied, for example, *Lasius neglectus* (Cremer *et al.* 2006), *Myrmica rubra* (Morales *et al.* 2008b) *Tetramorium tsushimae* (Steiner *et al.* 2006a), and the Rasberry crazy ant, *Paratrechina* sp (nr. *pubens* [Meyers 2008] or *P. fulva* [J. Trager, personal communication]). Studies with repeated sampling are still rare, but confirm native ant displacement by Argentine ants and big-headed ants (Table 15.1, and see Plate 15). Short-term

Table 15.1 Summary of recently published studies (>2002) of invasive ant effects on ants and other invertebrates in their introduced range.

Location	Reported effect on other ants	Reported effect on other invertebrates	Reference
Anoplolepis gracilipes			
Christmas Island, Australia	*A. gracilipes* activity explained 46% of variation of other ant species richness; only one other ant species present in *A. gracilipes* supercolonies	*A. gracilipes* occupied red crab burrows and killed resident red crabs	Abbott (2006); O'Dowd *et al.* (2003); Davis *et al.* (2008)
Tokelau	50% of ant species recorded from the seven islands absent in areas inhabited by haplotype D		Abbott *et al.* (2007)
Sulawesi, Indonesia	Decreased species richness of forest ants in the presence of *A. gracilipes*; no effect on non-forest species		Bos *et al.* (2008)
Bird Island, Seychelles	Displaces large predatory native ant, *Odontomachus simillimus*; other ants positively associated with *A. gracilipes* abundance	Ant parasite increased in invaded areas, no other differences detected	Gerlach (2004)
Bird Island, Seychelles	No difference in abundance or richness of native ants between invaded and uninvaded plots	Crustacea and Dermaptera absent from invaded sites; one orthopteran species more abundant in invaded sites	Hill *et al.* (2003)
Tokelau	Negative correlation between *A. gracilipes* abundance and ant species richness; attack other ants at baits	Density of crabs one-fourth that in uninvaded forest; no significant differences in Diptera, Orthoptera, or isopods between invaded and uninvaded sites	Lester and Tavite (2004)
Tokelau	Reduced number of co-occurring ant species when highly abundant, but not when less abundant		Lester *et al.* (2009)
Tokelau		Hermit crabs competitively excluded from carrion and shift to lower trophic level in invaded sites	McNatty *et al.* (2009)

Tokelau	*A. gracilipes* discovered and dominated baits in the low rugose treatment but not in high rugose treatment		Sarty *et al.* (2006)
Tokelau	Invaded and non-invaded ant communities 89% dissimilar, 11 ant species able to coexist with *A. gracilipes*, 4 ant species only found in uninvaded plots		Sarty *et al.* (2007)
Fiji	*A. gracilipes* able to break discovery-dominance trade-off		Ward and Beggs (2007)
Lasius neglectus			
Spain	Showed higher chemical and physical aggression in laboratory encounters against 3 native *Lasius* species		Cremer *et al.* (2006)
Linepithema humile			
CA, USA		Negative correlation between *L. humile* and native spider abundance; positive correlation with non-native spider	Bolger *et al.* (2008)
Laboratory, USA	*L. humile* did not consistently win one-on-one interactions, but excelled at symmetrical group interactions and dominated food and nesting sites		Buczkowski and Bennett (2008)
Spain	*L. humile* was the only ant to displace other ants at tuna baits		Carpintero and Reyes-López (2008)
Spain	Ecologically similar species modified food retrieval strategy in the presence of *L. humile*; subordinate species or those with little temporal overlap not as strongly affected		Carpintero *et al.* (2007)
Spain	23 trees occupied by native ants in 1992 exclusively occupied by *L. humile* 8 years later		Carpintero *et al.* (2005)

(Continued)

Table 15.1 Continued

Location	Reported effect on other ants	Reported effect on other invertebrates	Reference
CA, USA	Lower abundance and richness of native ant species in invaded sites		DiGirolamo and Fox (2006)
CA, USA	Sites invaded by *L. humile* had 4 native ant species and were deficient in ant species larger than *L. humile*; 20 native ant species found in uninvaded areas	Ant lion larvae were heavier, had longer mandibles and grew more quickly when preying on *L. humile* than on native ants	Glenn and Holway (2008)
CA, USA	Fewer native ant species detected in invaded plots; native ant species richness increased with time since start of survey		Heller *et al.* (2008)
Spain		*L. humile* did not affect pine bark beetle ability to colonize logs and did not injure them	Henin and Paiva (2004)
CA, USA	Native ant abundance increased linearly with decreasing *L. humile* abundance (distance from riparian corridor)		Holway (2005)
CA, USA	*Forelius mccooki* reared with non-aggressive *L. humile* colonies produced fewer eggs, foraged less, and supported fewer workers; only *L. humile* lacking intraspecific aggression displaced *F. mccooki* from baits		Holway and Suarez (2004)
HI, USA		Arthropod community compositional change strongly correlated with ant density; compositional changes occurred more frequently among endemics	Krushelnycky and Gillespie (2008)
HI, USA		Fewer native carabid beetles in *L. humile* range; no difference in alien carabid abundance	Liebherr and Krushelnycky (2007)

Spain	Abundance 2× as high as native ant abundance in uninvaded zone; ant biomass 4× higher in uninvaded zone; 3 and 17 native ant species in uninvaded and invaded zones, respectively		Oliveras *et al*. (2005a)
VIC, Australia	*Iridomyrmex bicknelli, Pheidole* sp. 2, and *Rhytidoponera victoriae* displaced at baits within 20 min of *L. humile* introduction		Rowles and O'Dowd (2007)
VIC, Australia	Ants 14× more abundant and ant species richness greater at baits in invaded sites, ant abundance and richness in litter not different between invaded and uninvaded sites	No detectable effect on abundance or richness of non-ant invertebrates; Lepidoptera lower and Psocoptera higher in invaded sites	Rowles and O'Dowd (2009)
CA, USA	Invasion leads to aggregation, as opposed to segregation, of the ant community		Sanders *et al*. (2003a)
WA, Australia	Native ants, including dominant *Iridomyrmex* displaced from baits under warm conditions; *L. humile* displaced under hot conditions		Thomas and Holway (2005)
Hiroshima, Japan		Myrmecophagic spider more frequently found and more abundant in invaded than non-invaded sites	Touyama *et al*. (2008)
SA, Australia	9 ant genera only captured in non-invaded areas, 2 genera captured in lower abundance in invaded areas; 3 genera captured in larger numbers in invaded areas	No significant between site differences in abundances of earwigs, spiders, wasps, isopods, beetles, mites, or Hemiptera (excluding aphids)	Walters (2006)
SA, Australia	Native ants coexisted with *L. humile* at 3% of baits, *Monomorium* sp. 2 particularly common; ant density and	Invaded areas have more isopods and amphipods, no consistent differences in dipterans or millipedes	Walters and Mackay (2003b)

(Continued)

Table 15.1 Continued

Location	Reported effect on other ants	Reported effect on other invertebrates	Reference
	native ant abundance lower in invaded areas than non-invaded areas		
SA, Australia	Laboratory: only larger *L. humile* colonies able to displace *Iridomyrmex rufoniger* colonies; field: large *L. humile* colonies only able to displace *I. rufoniger* from baits in the first 5 min after introduction		Walters and Mackay (2005)
Madeira, Portugal	Mean number of native ants higher at sites without than with *L. humile*; limited spread 150 years after arrival		Wetterer *et al.* (2006)
CA, USA	*Pogonomyrmex subnitidus* decreased foraging activity and increased number of nest entrance workers in presence of *L. humile*		Zee and Holway (2006)
Myrmica rubra			
MA, USA	Significant decline in native ant diversity with presence and density of *M. rubra*		Morales *et al.* (2008a)
Pheidole megacephala			
Mexico		Termite nest density lower in invaded areas; *P. megacephala* more successful at capturing termites than native ant species	Dejean *et al.* (2007b)
Mexico	8 of 11 ant species unable to resist attack by *P. megacephala* in colony-colony interactions		Dejean *et al.* (2008)
NT, Australia	Abundance increased 20-fold in 9 years, biomass 4–18× greater than native ants in non-invaded sites, only one native	Abundance of macroinvertebrates lowest in oldest invaded sites, Coleoptera and Orthoptera less abundant in invaded sites	Hoffmann and Parr (2008)

	ant specimen collected in newly invaded area		
HI, USA		Difference in arthropod community composition between invaded and uninvaded sites	Krushelnycky and Gillespie (2008)
Fiji	*P. megacephala* able to break discovery-dominance trade-off		Ward and Beggs (2007)
Madeira, Portugal	Mean number of native ants higher at sites without than with *P. megacephala*; limited spread 150 years after arrival		Wetterer *et al.* (2006)
Solenopsis invicta			
FL, USA	60% reduction in abundance of *S. invicta* had no effect on the abundance or species richness of other ants; *S. invicta* positively correlated with abundance of other ants in control plots		King and Tschinkel (2006)
FL, USA	Habitat alteration and *S. invicta* introduction caused richness decline of 4 and 5 species, respectively		King and Tschinkel (2008)
FL, USA	Density positively correlated with ant species richness; 37 ant species coexisted with *S. invicta*	*S. invicta* positively correlated with non-ant species richness	Morrison and Porter (2003)
TX, USA	Native fire ants (*S. geminata*) able to persist in area with dense leaf cover and little disturbance		Plowes *et al.* (2007)
TX, USA	Colonies of 30–480 workers attacked and killed by 3 native ant species and 2 exotic ant species; brood abandonment and queen death dependent on *S. invicta* colony size		Rao and Vinson (2004)
GA, USA		Earwigs more abundant when fire ants suppressed, spiders more	Seagraves *et al.* (2004)

(*Continued*)

Table 15.1 Continued

Location	Reported effect on other ants	Reported effect on other invertebrates	Reference
		abundant on some dates in untreated plots; no differences for ground beetles, lesser cornstalk borers, crickets, or mole crickets	
FL, USA		*S. invicta* responsible for 29.5% of predation of dropped root weevil larvae; other ants responsible for 2.5–27.8%	Stuart *et al.* (2003)
Wasmannia auropunctata			
New Caledonia	Abundance and richness of native ants higher in non-invaded zone; 4 cryptic species of 23 native species found in presence of *W. auropunctata*		Le Breton *et al.* (2003)
New Caledonia	*W. auropunctata* displaced both tested native *Pheidole* species from baits; only major *Pheidole* workers effective at defense against *W. auropunctata*		Le Breton *et al.* (2007)
Gabon	Relative density of other ant species reduced to 0–10% at the introduction point and 80m beyond		Ndoutoume-Ndong and Mikissa (2007)
Gabon	Sharp decline in native ant richness with *W. auropunctata* presence across all forest types; lowest ant species richness and abundance in oldest invaded sites		Walker (2006)

introductions and baiting have been utilized to ascertain the competitive ability of *A. gracilipes*, *L. humile*, and *W. auropunctata* against specific native ants in various parts of their introduced ranges. Experimental microhabitat modification has further revealed abiotic conditions under which Argentine ants are superior competitors in Australia and California, and limitations to *A. gracilipes*' competitiveness in Tokelau (Table 15.1).

In contrast, a few recent studies have called into question whether *S. invicta*, at least the monogyne form, has any long-term detrimental effects on native ant assemblages. Previous studies throughout the southern United States (e.g. Camilo and Philips 1990; Vinson 1997) were consistent in their conclusions that *S. invicta* severely reduced native ant diversity (Holway *et al.* 2002a). Two recent studies (King and Tschinkel 2006; Morrison and Porter 2003; Table 15.1) report a positive correlation between monogyne *S. invicta* and the abundance of other ants, and one (Morrison and Porter 2003) also found a positive correlation between monogyne *S. invicta* density and ant species richness. These findings suggest that the same abiotic and biotic factors control monogyne *S. invicta* and native ant populations (Morrison and Porter 2003).

Habitat disturbance has long been recognized as a confounding factor in studies that compare invaded and non-invaded areas to assess the effects of invasive ants on native ants and other biota (Holway *et al.* 2002a; Table 15.1). In a recent large-scale factorial experiment, King and Tschinkel (2008) separately tested the effects of habitat disturbance (mowing or plowing) and the introduction of *S. invicta* on native ants in a Florida pine forest. They concluded that disturbance, not *S. invicta*, had the greater impact on structuring ant communities. However, it is noteworthy that in the absence of disturbance, *S. invicta* reduced native ant richness at least as much as plowing or mowing, and the addition of *S. invicta* to disturbed habitats resulted in a greater loss of native ant richness than disturbance alone.

Additional experiments and long-term rigorous studies may be helpful in teasing apart covariates and any independent effect of *S. invicta*. A follow-up study at Texas sites previously found to have severely reduced native ant diversity following invasion by polygne *S. invicta* revealed that 12 years later *S. invicta* abundance had declined by an order of magnitude, and the richness and abundance of native ants did not differ between invaded and non-invaded sites (Morrison 2002b). A 60% reduction in monogyne *S. invicta* abundance achieved by killing colonies with hot water did not result in a positive response in the abundance or species richness of other ants in a north Florida pasture (King and Tschinkel 2006), prompting the authors to conclude that prior disturbance and recruitment limitation affect native ant diversity more than fire ants do. However, with an average of 400 *S. invicta* remaining in each pitfall trap in this experiment, any colonizing native ant species may still have been outcompeted. Long-term sampling in additional locations and more complete removal experiments will help to discern whether population declines are typical, and whether there is some density of *S. invicta* at which native ants can recolonize.

15.2.1 Native ants able to coexist

Many recent studies have documented that some native ants are able to coexist with invasive ants. Because invasive ants frequently break the discovery-dominance trade-off (Davidson 1998; Chapter 5), that is, they both discover food resources more quickly, and recruit to food in higher numbers than their competitors (e.g. Ward and Beggs 2007; Chapter 14), ants that can coexist with invasive ants are often those that can utilize different resources (e.g. Carpintero *et al.* 2007; Sarty *et al.* 2007), use the same resources at different times (e.g. Carpintero *et al.* 2007; Witt and Giliomee 1999), or have potent chemical defences (Holway *et al.* 2002a). In other cases, native ants may persist because the invader is at the margin of its abiotic tolerance (Wetterer *et al.* 2006), or is not numerically dominant (Rao and Vinson 2004).

15.3 Effects on other ground-dwelling invertebrate taxa

Other invertebrates may also be directly or indirectly affected by invasive ants. Multiple studies of invasive ant effects on invertebrates cited by Holway and colleagues (2002a) reported a negative effect of the invader ranging from observed predation to reduced abundance in invaded areas, to complete absence in invaded areas. Gastropods, Crustacea, several orders of insects, and other arthropods were among those affected primarily by *S. invicta*. For many of these effects, however, evidence was either anecdotal or qualitative.

Some recent studies have confirmed that invasive ants prey on or are competitively superior to other invertebrate taxa, but others reveal inconsistencies in effects (Table 15.1). *A. gracilipes* has a strong negative effect on crabs on Christmas Island (Abbott 2006; O'Dowd *et al.* 2003; Box 15.1) and Tokelau atolls (Lester and Tavite 2004; McNatty *et al.* 2009), but findings on effects of Bird Island arthropods vary (Table 15.1). Though big-headed ants remain poorly studied, recent findings are consistent with previous conclusions (e.g. Zimmerman 1970) that this ant has superior predatory abilities (Dejean *et al.* 2007b) and continues to displace many invertebrate taxa years after initial invasion (Hoffmann and Parr 2008). In contrast, as shown by recent (Table 15.1) and past studies (e.g. Bolger *et al.* 2000; Cole *et al.* 1992; Holway 1998a; Human and Gordon 1997), effects of Argentine ants on other ground-dwelling arthropods appear to be strongly context-dependent. New (Morrison and Porter 2003) and follow-up studies on *S. invicta* (Morrison 2002b) have cast doubt on the ability of *S. invicta* to sustain long-term effects on ground-dwelling arthropod communities, but these findings need to be considered against the large body of literature that concludes otherwise (see Holway *et al.* 2002a).

The mechanisms by which invasive ants affect other invertebrates are rarely elucidated. Most invasive ants possess a suite of traits that make them both formidable predators and interference competitors, and few studies documenting displacement have attempted to determine which mechanism was responsible (Holway *et al.* 2002a). Even for interactions involving taxonomically similar taxa, the mechanism may vary. For example, yellow crazy ants directly kill red land crabs on Christmas Island (Abbott 2006; O'Dowd et al. 2003), but competitively exclude hermit crabs from Tokelau islands, as revealed by ant exclusion experiments and stable isotope analysis (McNatty *et al.* 2009). Stable isotope analysis (see Chapter 7 and Box 7.1) will be a useful tool for future studies that attempt to discern predation from competition and explore changes in trophic structures of communities associated with ant invasions.

Clear patterns in factors associated with vulnerability to displacement also remain to be uncovered. Arthropods that have never encountered ants (for example in Hawai'i), appear to be more vulnerable than those that have evolved behaviour or morphology to coexist with ants (Liebherr and Krushelnycky 2007). Carnivores may be especially vulnerable because they can suffer from direct predation as well as competition for prey (P. Krushelnycky and R. Gillespie, unpublished data). However, elsewhere, evidence for effects of invasive ants on spiders and other carnivores is mixed (Holway *et al.* 2002a; Table 15.1). Several early studies showed increases in scavenger abundance in invaded areas, but the study designs precluded ruling out disturbance to the sites as the primary cause (Human and Gordon 1997 and references therein). Species-level analyses will be necessary to elucidate the species that are truly vulnerable and the traits or contexts that inure native species to displacement by invasive ants.

Some ground-dwelling invertebrates that prey on ants have benefited from ant invasions. Myrmecophagic spiders in Japan (Touyama *et al.* 2008) and ant-lions in California (Glenn and Holway 2008) have responded positively to Argentine ant invasions (Table 15.1). The higher abundance of invasive ants relative to the displaced native ants appears to be a key feature driving the benefits to these organisms.

Given the array of defensive mechanisms of both the ants and other invertebrates and the numerous ways in which they may interact, as well as the range of methods employed to investigate invasive ant effects, variation in the reported consequences of their interactions is expected. As with effects on native ants, long-term and experimental studies will be most helpful in discerning the effects of the ants, and covarying abiotic and biotic factors, and the underlying mechanisms for vulnerability or resilience of invertebrates in the face of invasions.

Box 15.1 Invasional meltdown: do invasive ants facilitate secondary invasions?

Dennis J. O'Dowd and Peter T. Green

Biological invasions can fundamentally alter the structure, composition, dynamics, and function of natural ecosystems. Direct and indirect effects of some invaders can be so pervasive and strong that they reconfigure entire interaction networks and lead to state changes in ecosystems (Croll *et al.* 2005). Simberloff and von Holle (1999) go still further to suggest that they can lead to 'invasional meltdown' whereby invader–invader synergism amplifies and diversifies impacts so as to facilitate secondary invasions and further accelerate impacts. The invasional meltdown metaphor remains controversial: few studies have demonstrated conclusively that synergies between invaders pave the way for secondary invasions.

Invasive alien ants frequently have large and varied impacts on natural ecosystems. This may be especially so on islands where native species richness and functional redundancy are low, and propagule pressure can be high. Some invasive ants form expansive supercolonies with high, sustained densities of worker ants that extend from hectares to many square kilometres. The invasion and supercolony formation by the yellow crazy ant, *Anoplolepis gracilipes* (YCA hereafter) in rainforest on Christmas Island (Indian Ocean) is a notable example of the manifold impacts of a single invader on a natural ecosystem. On the forest floor, this ant attacks and kills the dominant native omnivore, the red land crab, *Gecarcoidea natalis* (O'Dowd *et al.* 2003). In the forest canopy, YCA forms new mutualistic associations with herbivorous, honeydew-secreting Hemiptera (Abbott and Green 2007). These effects change the network and strength of interactions among producers, herbivores, and detritivores, deregulating seedling recruitment, increasing tree mortality, reducing litter decomposition, and affecting higher-order consumers such as birds (Davis *et al.* 2008; O'Dowd *et al.* 2003).

Does this qualify as an invasional meltdown as defined by Simberloff (2006)? True meltdowns comprise two distinct but complementary components. First, invader–invader interactions should generate positive population-level feedbacks that amplify impacts. There seems little doubt that interactions between YCA and honeydew-secreting scale insects sustain elevated populations of both that go on to amplify impacts. Second, impacts resulting from this synergism should enhance secondary invasions. We evaluate this second criterion by considering whether YCA invasion promotes invasion by an alien snail and facilitates the broader assemblage of introduced ants on the island.

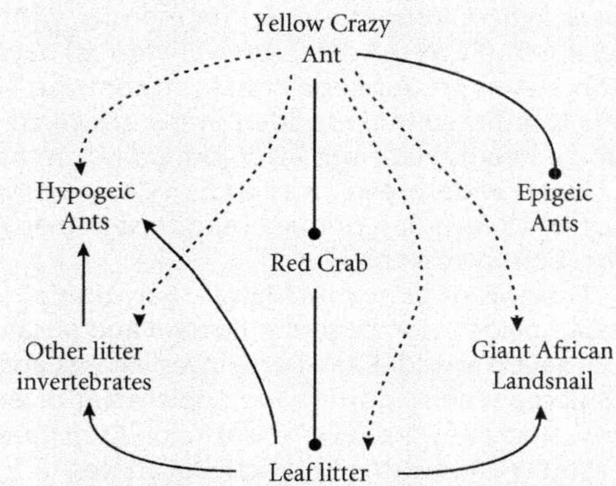

Figure 15.1.1 Interaction pathways by which invasion and supercolony formation by the yellow crazy ant, *Anoplolepis gracilipes* (YCA) facilitate other invaders on Christmas Island. Solid lines are direct effects, dashed lines are indirect effects; arrows are positive effects; knobs are negative effects. The YCA invasion facilitates secondary invasion of rainforest by the giant African landsnail (GALS) and hypogaeic ants through its impacts on omnivorous red land crabs. Red crabs are both predators of GALS and regulators of litter breakdown. By extirpating red crabs, YCA releases GALS from predation pressure, allowing entry into primary forest and the build-up of a key resource, litter. The abundance and diversity of hypogaeic ants are also facilitated by litter build-up that provides habitat, food resources, and a refugium from the epigaeic YCA. However, the YCA also has direct and adverse effects on some epigeic ants, probably through interference and exploitative competition.

The giant African land snail (*Achatina fulica*, GALS hereafter), a noteworthy invader throughout the tropics, has been present on

continues

Box 15.1 continued

Christmas Island for decades. However, it never managed to penetrate intact primary rainforest, despite its widespread distribution in disturbed habitats across the island. The reason is simple: predacious native red crabs are a formidable barrier to snail invasion of primary rainforest (Figure 15.1.1). Experiments show that red crabs rapidly discover tethered GALS and devour them within hours (Lake and O'Dowd 1991). The YCA, by extirpating red crabs, allow GALS to breach the barrier and establish in primary forest. Tethered snails persist in YCA supercolonies for months, with 53 ± 6 % (SE, $N = 3$ sites) survival after 60 days. Some even produce egg masses. In contrast, all snails tethered in uninvaded sites were killed and eaten by red crabs after just 6 days. In the wake of widespread invasion by YCA, we now see GALS invading primary rainforest in many locations across the island.

The ants of Christmas Island – a synthetic assemblage – comprise the flotsam and jetsam of the ant world. Elsewhere, invasive ants are typically seen to disrupt and deplete ant diversity. However, the reverse is true on Christmas Island: YCA invasion indirectly facilitates increased abundance and species density of non-native ants in island rainforest. On a per-area basis, ant abundance is three-fold greater in invaded sites (33.6 ± 9.2 ants m^{-2}) than in uninvaded sites (9.5 ± 2.0 ants m^{-2}; $F_{2,12} = 12.88$, $P = 0.001$). Likewise species density is twofold greater in invaded sites (2.4 ± 0.2 species m^{-2}) than in uninvaded sites (1.0 ± 0.1 ants m^{-2}; $F_{2,12} = 7.98$, $P = 0.006$). However, differences in species density but not abundance disappear when differences in litter mass between site types are considered. When expressed on a per kilogram litter basis, the abundance of other ants is still twofold higher in invaded sites ($F_{2,12} = 11.65$, $P = 0.002$), but there is no difference in species richness ($F_{2,12} = 0.14$, $P = 0.872$). Strong compositional differences also occur between ant assemblages in YCA-invaded and uninvaded sites (Global R = 0.521, ANOSIM $P = 0.008$).

These impacts of the YCA on the diversity of other ants are largely indirect and mediated by its effects on litter (Figure 15.1.1). By eliminating the red crab, which otherwise regulates leaf litter on the forest floor, YCA invasion increases litter biomass and, thus, habitat and food for other ants, especially small-litter foraging (i.e. hypogaeic) species. Nevertheless, the YCA does disrupt a few surface-foraging (i.e. epigaeic) ants so can also directly affect species composition. These results illustrate invasional meltdown whereby direct interaction between the YCA and the red crab, itself accelerated by YCA-scale mutualism, facilitates the rapid population increase and local species richness of other introduced ant species.

We argue that these data demonstrate invasional meltdown *sensu stricto* (Simberloff 2006). For us, invasional meltdown is a real phenomenon and a most fortunate metaphor (cf. Gurevitch 2006), evocative of complex changes that can be wrought by biological invaders in general, and invasive ants in particular.

15.3.1 Myrmecophiles

Higher abundance of invasive ants relative to native ants may not benefit all invertebrates that have associations with ants. Myrmecophiles are organisms that spend at least part of their life cycle with ant colonies as commensals, parasites, or mutualists (Hölldobler and Wilson 1990). There are some reports of myrmecophiles in invasive ant colonies (e.g. Dekoninck 2007; Kistner *et al.* 2007; Neece and Bartell 1981; Wojcik *et al.* 1991), but few studies report quantitative consequences of ant invasion on myrmecophiles. However, at least one study found an increase in ant-associated parasites (paussine beetles) following invasion by *A. gracilipes* (Gerlach 2004, see Table 15.1).

Lycaenid butterfly larvae can be parasites or mutualists and rely on chemical cues to blend into their host ant's nest. Those that are obligatorily associated with ants are less likely to be able to switch ant hosts (Eastwood and Fraser 1999). In lycaenid-rich Australia, 39 out of 56 obligate lycaenid myrmecophiles have distributions that overlap with *L. humile*, *P. megacephala*, or *A. gracilipes*. It is unknown whether these invasive ants tend or prey on the larvae (Lach and Thomas 2008). Some lycaenid species have more facultative relationships with ants and there have been reports of associations between their larvae and non-native *L. humile*, *P. megacephala*, *Tapinoma melanocephalum*, *Paratrechina bourbonica*, and *Pa. longicornis*, although it is unknown whether the introduced ants confer the same degree of benefit (Agrawal and Fordyce 2000; Lach and Thomas 2008; Saarinen and Daniels 2006).

15.4 Effect of invasive ants on vertebrates

Early findings of invasive ant impacts on vertebrates were largely anecdotal or correlative (Holway *et al.* 2002a). In contrast, recent studies have been more experimental and continue to show that invasive ants, particularly stinging *S. invicta*, can adversely affect birds, mammals, and herpetofauna (Allen *et al.* 2004; Table 15.2). Taxa that have altricial young, are oviparous, prefer open habitats, nest on the ground, and that hatch or birth when fire ants are more active or have their greatest protein needs, are most susceptible to attack by red imported fire ants (Allen *et al.* 2004). Indirect effects of invasive ants are also possible, including reduction in arthropod prey (Allen *et al.* 2001; Suarez and Case 2002) and changes in habitat use and foraging, nesting, and parenting behaviour (Allen *et al.* 2004; Suarez *et al.* 2005b).

15.4.1 Avian fauna

The majority of research on invasive ant interactions with birds has involved *S. invicta* (Suarez *et al.* 2005b). Multiple species in multiple habitats suffer from nesting disruption and/or significantly increased mortality in the presence of *S. invicta* (reviewed in Allen *et al.* 2004; Table 15.2). Most ground-nesting chicks are vulnerable because of the considerable time spent hatching. They are highly attractive as prey because they are large, moist, defenseless morsels. Reports of effects on northern bobwhites are particularly numerous and document nest and population reduction, correlations between bobwhite abundance and years since

Table 15.2 Summary of recent (>2002) published studies of invasive ant effects on vertebrates in their introduced ranges. Ag = *Anoplolepis gracilipes*, Lh = *Linepithema humile*, P = *Paratrechina* sp. (Rasberry crazy ant), Pm = *Pheidole megacephala*, Sg = *Solenopsis geminata*, Si = *S. invicta*, Wa = *Wasmannia auropunctata*

Ant	Taxa	Location	Reported Effect	Study
Birds				
Ag	Emerald dove (*Chalcophaps indica natalis*)	Christmas Island, Australia	Counts 9–14 times lower	Davis *et al.* (2008)
Ag	Island thrush (*Turdus poliocephalus erythropleurus*)	Christmas Island, Australia	Reduced nest success and juvenile counts	Davis *et al.* (2008)
Ag	Christmas Island white-eye (*Zosterops natalis*)	Christmas Island, Australia	Increased counts in invaded areas	Davis *et al.* (2008)
Ag	Sooty tern (*Sterna fruscata*)	Bird Island, Seychelles	Reduced nesting area, caused excess stress on birds and failure of nests	Gerlach (2004)
Lh	Dark-eyed junko (*Junco hyemalis*)	CA, USA	Recruit to experimentally placed eggs, but responsible for <2% of failed nests	Suarez *et al.* (2005b)

(*Continued*)

Table 15.2 Continued

Ant	Taxa	Location	Reported Effect	Study
Pm	Wedge-tailed shearwater (*Puffinus pacificus*)	HI, USA	No change in hatching success, growth, or fledging success following eradication of *P. megacephala*	Plentovich *et al.* (2009)
Sg	Wedge-tailed shearwater (*Puffinus pacificus*)	HI, USA	Chicks that lose more than 20% of skin on feet and/or suffer holes and tears in feet due to ant bites will not fledge	Plentovich *et al.* (2009)
Si	Least tern (*Sterna antillarum*)	TX, USA	Reduces nesting area, kills chicks	Campomizzi (2008)
Si	Colonial waterbirds	USA	Reduces births, 92% reduction in recruitment	Allen *et al.* (2004); Jetter *et al.* (2002)
Si	Northern bobwhite (*Colinus virginianus*)	southern USA	Attacks on chicks, changes in behaviour and fitness, attacks on nests, predation of intact and hatching eggs	Seymour (2007); Staller *et al.* (2005)
Si	Black-capped vireo (*Vireo atricapillus*)	TX, USA	Disrupts nest sites, kills chicks; 30% nest failure	Smith *et al.* (2004); Stake and Cimprich (2003)
Wa	Melanesian scrub fowl (*Megapodius eremita*)	northern Melanesia	Attacks hatchlings	Wetterer and Porter (2003)
Reptiles and Amphibians				
Si	Gopher tortoise (*Gopherus polyphemus*)	MS, USA	Predation of eggs and hatchlings	Epperson and Heise (2003)
Si	Fence lizard (*Sceloporus undulatus*)	southern USA	Hind leg length and body twitch behaviour increased with time since invasion	Langkilde (2009)
Si	Texas horned lizard (*Phrynosoma cornutum*)	TX, USA	Disappearing from area	Moeller *et al.* (2005)
Si	Loggerhead sea turtle (*Carreta carreta*)	FL, USA	Predation of eggs and hatchlings	Parris *et al.* (2002)
Si	Green turtle (*Chelonia myda*)	FL, USA	Predation of eggs and hatchlings	Parris *et al.* (2002)
Si	Slider turtle (*Trachemys scripta*)	FL, USA	Predation of eggs and hatchlings	Parris *et al.* (2002)
Si	Hognose snake (*Heterodon simus*)	southeast USA	Population decline	Todd *et al.* (2008)
Mammals				
P	Wildlife, domestic animals, livestock	TX, USA	Irritate and inflict bites	Texas A and M University (2008); Wynalda (2008)
Si	Old field mice (*Peromyscus polinotus*)	SC, USA	Reduced foraging in presence of ants	Orrock and Danielson (2004)
Si	Cotton rats (*Sigmodon hispidus*)	TX, USA	Reduced foraging in presence of ants	Pedersen *et al.* (2003)
Wa	Domestic dogs and cats, leopards, forest elephants (*Loxodonta cyclotis*), red river hogs (*Potamocherus porcus*)	Gabon	Cause blindness	Walsh *et al.* (2004)

S. invicta invasion, and increased bobwhite density when *S. invicta* is experimentally reduced (Allen *et al.* 2004). The stinging ability of *S. invicta* is purported to be a key feature of its ability to directly attack vertebrates. The paucity of reports of effects of other stinging ants such as *Solenopsis geminata* and *W. auropunctata* on avian fauna (Table 15.2) relative to *S. invicta* suggests that other attributes unique to *S. invicta*, such as its venom, may be more important than its stinging ability. Underreporting of effects, particularly for the less studied *W. auropunctata*, is another possibility.

As with observations of ecological interactions generally, it is important to note that observed interactions do not always translate to population-level effects. For example, despite observations of recruitment and attack by Argentine and big-headed ants on dark-eyed junkos and wedge-tailed shearwaters, respectively, experimental investigation revealed little or no effect of the ants on the birds (Plentovich *et al.* 2009; Suarez *et al.* 2005b; Table 15.2).

Indirect consequences mediated through nest or food availability are also possible and may involve multiple cascading effects. For example, emerald dove and island thrush counts are affected not only by interference by yellow crazy ants, but also by changes in the habitat structure and resource availability associated with the displacement of red crabs and the tending of scale insects by the ants (Davis *et al.* 2008; see Box 15.1). Similarly complex cascades may occur elsewhere, but even simpler indirect effects are likely underreported because of difficulty in detecting them.

15.4.2 Herpetofauna

As with birds, reports of adverse consequences of ant invasions on reptiles and amphibians primarily involve *S. invicta*. These ants are attracted to the disturbance, mucous, and moisture associated with nests of many herpetofauna species (Allen *et al.* 2001). Taxa, such as sea turtles, that display delayed hatchling emergence (whereby hatchlings do not emerge from the nest until all eggs have hatched) are especially vulnerable (Allen *et al.* 2004; Table 15.2). Moreover, *S. invicta* presence may preclude use of woody debris as shelters by reptiles and amphibians (Todd *et al.* 2008) and affects nest placement by sea turtles (Wetterer *et al.* 2007). *Solenopsis invicta* has been linked to the decline of the Texas horned lizard (*Phrynosoma cornutum*) via displacement of native harvester ants (*Pogonomrymex* spp.), its primary food source (Allen *et al.* 2004 and references therein). Similarly, experiments have revealed that Argentine ants decrease the growth and survival of coastal horned lizards (*Phrynosoma coronatum*) in California by displacing their native ant prey (Suarez and Case 2002; Suarez *et al.* 2000). However, documented effects, experimental or otherwise, on herpetofauna by other invasive ant species are limited (Table 15.2).

Red imported fire ants invaded parts of the southern United States as long as 70 years ago. The approximately 40 generations that this represents for fence lizards (*Sceloporus undulatus*) has been sufficient for the evolution of behavioural and morphological traits that facilitate their escape from *S. invicta* attack (Langkilde 2009). Undoubtedly, other fauna are also evolving in response to selection pressures to minimize their susceptibility to invasive ant effects.

15.4.3 Mammalian fauna

Documented effects of invasive ants on mammals are limited (Allen *et al.* 2004; Holway *et al.* 2002a; Table 15.2). This may reflect a lack of investigation or a lower susceptibility of mammals to direct attack because they are viviparous and covered with hair or fur. Most of the studies involve measures of foraging behaviour of small rodents, whose behaviour may change in response to both exploitative and interference competition (Orrock and Danielson 2004; Table 15.2). As with herpetofauna, mammals for which ants comprise a large part of the diet may be affected indirectly by the displacement of native ants, but to date there are no published studies.

15.5 Interactions on and with plants

As with native ants (see Chapter 6), invasive ants are typically attracted to plants by carbohydrate-rich resources either offered by the plant or by honeydew-producing herbivores. Ants in the dolichoderine and formicine subfamilies have modified physiologies for carrying liquid food (Davidson *et al.* 2004). *Pheidole megacephala* soldiers can carry liquid externally (Dejean *et al.* 2005a). Fire ants are not known to have these adaptations, but have been reported consuming nectar and honeydew (e.g. Helms and Vinson 2002; Lanza *et al.* 1993; Le Breton *et al.* 2005; Tennant and Porter 1991), although they also readily prey on arthropods or seeds (Ness and Bronstein 2004; Vogt *et al.* 2003 and references therein). Few interactions have been observed between invasive ants and myrmecophytes (but see Gaume *et al.* 2005; Krombein *et al.* 1999; Wetterer and Wetterer 2003); therefore we focus on interactions with EFN-bearing plants, flowers, hemipterans, and seeds.

15.5.1 Extrafloral nectaries

Ness and Bronstein (2004) reviewed 15 studies that investigated invasive ant relationships with EFN-bearing plants and concluded that in 11 of these the plants benefited from the invasive ants by having increased fruit or seed production, increased growth, decreased herbivory or herbivore abundance, and/or decreased incidence of pathogens. In contrast, one study found *S. invicta* abundance to be negatively correlated, and other ants to be positively correlated, with plant growth (Stiles and Jones 2001). Only 6 of the 15 studies compared invasive ant performance to that of native ants; EFN plants performed better with invasive ants than with the median ant in five of these (Ness and Bronstein 2004). The abundance and aggression of invasive ants may be key characters in making them effective plant guards (Lach 2003).

As with native ants (see Chapter 6), there are no data quantifying how invasive ants may benefit from EFNs. The composition of extrafloral nectar can vary widely, but consists primarily of sugars with small amounts of amino acids (Blüthgen *et al.* 2004a). Considering the importance of carbohydrate-rich resources to invasive ants (Holway *et al.* 2002a), we would expect that EFNs would be an attractive and readily harvested resource. In many cases, invasive ant visitation to plants is associated with EFNs (e.g. *S. invicta*: Agnew *et al.* 1982; Fleet and Young 2000; *S. geminata*: Díaz-Castelazo and Rico-Gray 1998; *L. humile*: Freitas *et al.* 2000; Koptur 1979; *W. auropunctata*: de la Fuente and Marquis 1999), but in others, invasive ants appear to ignore these resources (*S. invicta*: Ness 2003b; *L. humile*: Lach 2008a). Laboratory experiments revealed that *L. humile* and *P. megacephala* in Australia had similarly high worker survival as native *Iridomyrmex* when they had access to extrafloral nectar of a native *Acacia*, but none of the three ant species could induce extrafloral nectar production by the plant (Lach *et al.* 2009). Future studies that explore extrafloral nectar composition, abundance, and availability in relation to other available resources and ant colony needs may shed light on ant foraging patterns and the extent to and context in which ants benefit from EFNs.

15.5.2 Flowers

Ants in general are considered floral antagonists (Kerner 1878; Lach 2003) and some plants have evolved mechanisms to deter ants (Ghazoul 2001; Willmer and Stone 1997; Chapter 6). However, several studies show that invasive ants, especially *L. humile*, are able to consume floral nectar in situ with largely negative effects on other floral arthropod visitors, including pollinators (Table 15.3; Buys 1987; Visser *et al.* 1996). It is noteworthy that despite volumes of research, none of the fire ants has been reported as having tritrophic interactions that involve flowers, although *S. invicta* consumes nectar from several flower species ex situ (Koptur and Truong 1998).

The displacement of pollinators might be expected to have negative consequences for seed set, but only one published study has reported such an effect (Table 15.3). Plants that are not pollen-limited may not show any effects of pollinator loss (e.g. Lach 2007). The consistently negative effect of some invasive ants on other floral arthropod visitors (Table 15.2) suggests that pollen-limited, arthropod-pollinated plants, and any other plants that may be members of the same pollinator network, may be most affected by invasive ants. However, taking nectar without pollinating or 'nectar-thieving' can increase pollination in some contexts (Maloof and Inouye 2000). Hence, the possibility that invasive ants may have positive effects on seed set should not be overlooked.

15.5.3 Hemiptera

15.5.3.1 The ants and the bugs

Much of the early research on invasive ant–hemipteran interactions documented associations between invasive ants and outbreaks of these sap-sucking insects. Holway *et al.* (2002a) cited 20 studies in which invasive ants interfered with natural enemies or were otherwise positively associated with hemipteran abundance. Ness and Bronstein (2004) reviewed 41 studies involving invasive ants and trophobionts (including one study involving a lycaenid butterfly) and found that in 72% of the ant–trophobiont–host plant combinations, invasive ants had a positive effect on the trophobiont. Recent studies continue to support the pattern that invasive ants are associated with increased abundance of honeydew-producing hemipterans (Table 15.3).

Ants may increase hemipteran populations by removing honeydew that contributes to the growth of sooty mould, moving nymphs to better sites, and deterring parasites and predators (Way 1963; Chapter 6). Perhaps the most dramatic example, and one with multiple cascading effects, is the tending of scale insects by *A. gracilipes* on Christmas Island (O'Dowd *et al.* 2003, see Box 15.1). Several traits of invasive ants may explain how they excel at their role in the relationship. The high abundance of invasive ants relative to native ants (see Chapter 14) and the ability of many to carry a large amount of liquid food likely aid in collection of honeydew (Lach 2003). High abundance and aggression also may contribute to interference of parasitoid oviposition (Barzman and Daane 2001; Daane *et al.* 2007; Martinez-Ferrer *et al.* 2003). Moreover, there is some evidence that Argentine ants are pre-adapted to respond to aphid alarm pheromones in novel environments; release of the pheromone increases both the abundance and aggression of Argentine ants around the aphids (Mondor and Addicott 2007).

Despite these traits and the documentation of invasive ant-associated hemipteran outbreaks, there is not yet strong evidence that invasive ants are always superior tenders relative to native ants. Many of the more dramatic examples of invasive ant-associated hemipteran outbreaks occur on islands (e.g. Christmas Island, Palmyra Atoll, Seychelles; see Table 15.3) or in highly disturbed areas (e.g. agricultural systems) that may have had depauperate native ant faunas or hemipteran natural enemy populations prior to ant invasion. In a ranking of trophobiont partners described in several studies, invasive ants did better than the median ant in only 6 of 12 comparisons (Ness and Bronstein 2004). In particular, *S. invicta* is reported to have negative or neutral effects on hemipterans in a number of studies, although comparison to native ants is usually lacking (Harris *et al.* 2003; Ness and Bronstein 2004 and references therein). Argentine ants do not reduce mutilation or parasitism of California red scale by parasitoids as well as the native gray ant, *Formica aerata* (Martinez-Ferrer *et al.* 2003).

The converse questions: how much do invasive ants benefit from these associations, and do they benefit more than native ants, also require further investigation. Honeydew and sugary plant exudates are important resources in the fuelling, foraging, and defence activities of dominant ants and the structuring of tropical native ant assemblages (Blüthgen and Fiedler 2004a; Davidson 1998; see Chapters 6 and 7). Laboratory *S. invicta* colonies with access to honeydew and prey grow 50% larger than those with access to prey only (Helms and Vinson 2008). Other studies have documented density-dependent responses of ants to hemipteran abundance (*L. humile*: Grover *et al.* 2008; *S. invicta*: Kaplan and Eubanks 2005, but see Marti and Olson 2007), and positive associations with presence (*A. gracilipes*, Abbott and Green 2007; *S. invicta*, Helms and Vinson 2002). Quantifying the effect of these resources at the population level may contribute to the development of additional control methods. There is some evidence that *L. humile* is not able to achieve ecological dominance in the absence of honeydew resources (Addison and Samways 2000).

15.5.3.2 Effects on other herbivores and the plants

Whether attracted to a plant by a sugary resource or in the absence of a specific attractant, ants can have extensive follow-on effects on the plant-associated arthropod community and the plant. Indeed, the

Table 15.3 Summary of recent (>2002) studies of invasive ant interactions on plants. Abbreviations for ant species: Ag = *Anoplolepis gracilipes*, Lh = *Linepithema humile*, Pb = *Paratrechina bourbonica*, Pv = *Paratrechina vaga*, Pm = *Pheidole megacephala*, Sg = *Solenopsis geminata*, Si = *S. invicta*, Ta = *Technomyrmex 'albipes'*, Tb = *Tetramorium bicarinatum*, Tm = *Tapinoma melanocephalum*, Wa = *Wasmannia auropunctata*

Ant	Plant & location	Attractant: effect on	Effect on other plant-associated invertebrates	Effect on host plant	Reference(s)
Nectar and other plant-based rewards					
Ag	*Metrosideros polymorpha*, HI, USA	Floral nectar: single ant consumes up to 4.5% of nectar from an inflorescence			Lach (2005)
Lh	*Euphorbia characias*, Spain	Floral nectar	Decrease visitation time of dipteran pollinator and overall number of arthropod visitors	Reduced fruit and seed set in invaded area	Blancafort and Gomez (2005)
Lh	*Euphorbia characias*, *E. biumbellata*, Spain	Floral nectar	Displace native pollinating ants		Blancafort and Gómez (2006)
Lh	*Quercus lobata*, CA, USA	'Honeydew' from galls	Decreased total parasitism of galls and gall-maker emergence; changed composition of parasitoid community		Inouye and Agrawal (2004)
Lh	*Protea nitida*, South Africa	Floral nectar	Decreased floral arthropods	No effect on seed set or ovule predation	Lach (2007)
Lh	*Leucospermum conocarpodendron*, South Africa	Floral nectar	Decreased floral arthropods; decreased visitation time of native honeybee	No effect on seed set or flower predation	Lach (2008a)
Lh	*Metrosideros polymorpha*, HI, USA	Floral nectar: single ant consumes 0.27% of nectar from an inflorescence	No effect on frequency or length of honeybee or native *Hylaeus* spp. bee visits		Lach (2005, 2008b)
Pm	*Barteria nigritana*, Guinea	Domatia, EFNs occupied 13% of tree clumps, often with other ants	Preyed on nymphalid eggs, but not on larvae	More herbivore damage than trees occupied by other native ants, similar damage as unoccupied trees	Djiéto-Lordon *et al.* (2004)
Pm	*Metrosideros polymorpha*, HI, USA	Floral nectar: single ant consumes 0.13% of nectar from an inflorescence	Native *Hylaeus* spp. bees never visited inflorescences with ants		Lach (2005, 2008b)

Ta	*Humboldtia brunonis*, India	Domatia, EFNs: up to 50× greater *T. albipes* activity compared to other ants	No herbivorous insects observed on *T. albipes* patrolled leaves	Decreased leaf herbivory, increased fruit production compared to trees with other ants and no ants	Gaume *et al.* (2005)
Hemipterans					
Ag	Rainforest trees, Christmas Island, Australia	Scale: 100% decline when ants excluded		2.8-fold increase in tree dieback in invaded sites	Abbott and Green (2007); O'Dowd *et al.* (2003)
Ag	*Pisonia grandis*, Bird Island, Seychelles	*Pulvinaria urbicola*: 'encouraged' by ants	Other herbivores scarce in infested areas, but other invertebrates more abundant	Less insect chewing damage, but lower foliage density and mean leaf size in invaded areas	Hill *et al.* (2003), but see Gerlach (2004)
Ag	*Scaevola taccada*, Bird Island, Seychelles	Coccids: no effect	Lower invertebrate densities in invaded areas	More insect chewing damage and lower mean leaf size observed in ant-infested areas	Hill *et al.* (2003)
Ag	*Carica papaya*, Bird Island, Seychelles	Coccids: not distinguished from ant counts	Higher invertebrate abundance in invaded areas	No difference in mean leaf size	Hill *et al.* (2003)
Ag	*Phyllanthus pervilleanus*, Bird Island, Seychelles	Coccids: not distinguished from ant counts		No difference in mean leaf size	Hill *et al.* (2003)
Lh	*Baccharis halimifolia*, FL, USA	Aphid: greater density with high ant density treatment	Higher predator abundance on larger aggregations of aphids, and on small scale ant exclusion; no effects on leaf miners		Altfeld and Stiling (2006)
Lh	*Baccharis halimifolia*, FL, USA	Aphid: greater density when tended by ants	Reduced abundance of leafminers, stemborer; increased abundance of chyrsomelid and coccinellid; no effect on gall fly	Greater stemborer-induced mortality on trees without *L. humile*	Altfeld and Stiling (2009)
Lh	Vineyards, CA, USA	Obscure mealybug and grape mealybug: increased densities when tended by ants, even in the absence of natural enemies	Higher abundance of mealybug destroyer on ant-tended vines		Daane *et al.* (2007)
Lh	*Solanum nigrum*, greenhouse	Aphid	Ants found aphid predators 5× faster when aphids present		Grover *et al.* (2008)

(*Continued*)

Table 15.3 Continued

Ant	Plant & location	Attractant: effect on	Effect on other plant-associated invertebrates	Effect on host plant	Reference(s)
Lh	*Protea nitida*, South Africa	Membracid, floral nectar	Ant-membracid mutualism increased *L. humile* discovery of inflorescences; decreased floral arthropods	No effect on seed set or ovule predation	Lach (2007)
Lh	Lemon, laboratory	California red scale	59% and 79% reduction in parasitism by parasitoids		Martinez-Ferrer *et al.* (2003)
Pm Pv, Pb, Tb	*Pisonia grandis*, Palmyra Atoll	*Pulvinaria urbicola*: ants observed tending scale, high densities of scale observed		>50% leaf loss; dieback reported	Handler *et al.* (2007)
Sg, Tm	Corn, Philippines	Aphid	Ants prey on Asian corn borer eggs and larvae		Litsinger *et al.* (2007)
Si	Tomato, AL, USA	Aphid: alate abundance higher in high fire ant plots	No effect on non- aphid herbivore abundance; reduced abundance of natural enemies in high fire ant plots in second year		Coppler *et al.* (2007)
Si	Cotton, TX, USA	Aphid: up to 5.5 times more abundant when fire ants present	Bollworm eggs and beet armyworm egg mass disappearance greater when fire ants present		Diaz *et al.* (2004)
Si	Pecan, TX, USA	Aphid: did not affect aphid densities	Some aphid natural enemies increased on some sampling dates when ant populations were reduced		Harris *et al.* (2003)
Si	*Cynodon dactylon* and *Aristida oligantha*, TX, USA	Mealybug: frequency of occurrence decreases with distance from mound			Helms and Vinson (2003)
Si	Citrus, FL, USA	Aphid	6.5-fold decrease in parasitoid emergence when fire ants present		Hill and Hoy (2003)
Si	Cotton, AL, USA	Aphid: 69% higher average abundance in high fire ant density plots	Lady beetle larvae and lacewing larvae less abundant with higher fire ant density		Kaplan and Eubanks (2002)

Si	Cotton, greenhouse	Aphid	Reduced predator and herbivore survival when aphids present		Kaplan and Eubanks (2005)
Si	Cotton, AL, USA	Aphid	27–33% of herbivores and 40–54% of predators adversely affected by ant-aphid mutualism		Kaplan and Eubanks (2005)
Si	Citrus, FL, USA	Aphid	Fire ants preferentially prey on parasitized aphids		Persad and Hoy (2004)
Wa	*Basselinia pancheri*, New Caledonia	Margarodids: significantly higher density in invaded areas			Le Breton *et al.* (2005)
No attractants reported					
Si	Cotton, AL USA		Abundance of most natural enemies negatively correlated with fire ant abundance; lady beetles, spiders, big-eyed bugs increased when ant populations decreased; no effect on damsel bugs and hooded beetles		Eubanks *et al.* (2002)
Si	Cotton, greenhouse		50% lower lady beetle survival and 38% lower green lacewing survival; no effect on spiders		Eubanks *et al.* (2002)
Si	Collard, AL, USA		No effect on caterpillar parasitoid abundance or parasitism		Harvey and Eubanks (2005)
Si	Soybean, GA, USA		Increased predation on pest caterpillar eggs and pupae relative to plots with suppressed fire ant densities	No effect of fire ant suppression on soybean yield	Seagraves and McPherson (2006)
Si	Soybean, AL, USA		Lower noctuid larvae and leafhopper abundance; higher alfalfa hopper abundance in high fire ant plots	Lower caterpillar herbivory on plants with trichomes, but higher on glabrous plants under high ant densities	Styrsky *et al.* (2006)
Si	Sugarcane, LA, USA		Predated 6–24% of parasitoid cocoons despite suppression attempts		White *et al.* (2004)
Sg	Rice, Philippines		As effective as entire predator complex at eliminating brown planthopper adults; attack planthopper eggs and nymphs, and leaffolders		Way *et al.* (2002)

ant–hemipteran relationship has been termed a 'keystone interaction' because of its effect on ants as predators (Styrsky and Eubanks 2007). Whereas effects of *A. gracilipes* on plant-associated arthopods are mixed, most reported effects of *L. humile* and *S. invicta* are negative, with predation being the most commonly observed mechanism (Table 15.3). It is striking that all nine published studies of invasive ant effects in the absence of a carbohydrate-rich attractant on the plant involve *S. invicta* or *S. geminata* (Table 15.3), potentially reflecting these ants' preference for protein-rich prey.

For the plant, the outcome is dependent on the relative displacement or encouragement of antagonistic and beneficial arthropods by the ants. Even when ants tend sap-sucking bugs, positive outcomes for the plant are possible. For many ant–hemipteran interactions involving native ants, the plant benefits; the displacement of other herbivores more than compensates for any detrimental effects of the ant-tended herbivores (Styrsky and Eubanks 2007). Ants also may need to compensate for other negative effects, such as the displacement of other predators (i.e. intraguild predation) before they can have a positive effect on the plant. For interactions involving invasive ants, positive outcomes for the plant appear to be rare outside of agricultural systems with *S. invicta* (Table 15.3). Maximizing predation of yield-reducing herbivores by understanding and manipulating *S. invicta*, and to a lesser extent, *S. geminata*, foraging has been an active area of research for several decades (e.g. Table 15.2; Agnew *et al.* 1982; Ali and Reagan 1985; Dutcher *et al.* 1999; Stuart *et al.* 2003; Tillman *et al.* 2004; Wyckhuys and O'Neil 2007; Zappalà *et al.* 2007).

15.5.4 Seeds

Ants that interact with seeds can generally be divided into seed dispersers or seed harvesters, although there is some overlap. In addition to their direct effects on seeds, whether harvesting or dispersing, invasive ants can also have indirect effects via the displacement of native seed dispersers or harvesters.

15.5.4.1 Seed dispersal

Ant invasions generally have negative consequences for myrmecochorous plants (plants that rely on ants for seed dispersal) (Holway *et al.* 2002a; Ness 2006, but see Rowles and O'Dowd 2009). Ness and Bronstein (2004) reported that invasive ants had detrimental effects on myrmecochorous plants in 11 of 19 studies reviewed. They found five suboptimal interactions to be prevalent: invasive ants may collect fewer seeds per unit time compared to other ant species (*L. humile*), they may function as seed predators (*S. invicta*, *S. geminata* and possibly *Ph. megacephala*), may leave seeds exposed on the surface (*L. humile*, *S. invicta*, *S. geminata*, *W. auropunctata*), may ingest the elaiosome, but fail to move the seed (*S. invicta*, *S. geminata*, *W. auropunctata*, *Pa. longicornis*) or they may move the seed to shorter distances than the native ants they displace (*S. invicta*, *S. geminata*, *W. auropunctata*, *Pa. longicornis*, *L. humile*). The invasive ant dispersed the seed a shorter distance than the median ant in eight out of the nine studies (Ness 2006). Failed seed dispersal may mean that seeds are less likely to escape predation and the parent canopy and to arrive at an optimal germination site, such as a forest interior (Ness 2004). Incomplete elaiosome removal, such as by *L. humile*, can also impede seed germination (Gómez *et al.* 2003), though in some contexts it may protect the seed from predation (Christian and Stanton 2004).

Why are invasive ants typically poor seed dispersers? Seed dispersal distance is a function of ant size, and invasive ants tend to be smaller than native seed-dispersing ants (Ness *et al.* 2004). Small ants disperse seeds shorter distances and are less likely to move large seeds (Carney *et al.* 2003; Ness *et al.* 2004; Witt and Giliomee 2005) because of their smaller mandible gap (Oliveras *et al.* 2005b). Small native ants may be equally poor seed dispersers, but coexist with large native ants that are effective dispersers (Carney *et al.* 2003; Ness *et al.* 2004). As with native ants, seed dispersal by invasive ants may be a function of reward attraction and ease of handling. For example, an experiment manipulating reward and seed size revealed that Argentine ants disperse small diaspores independent of reward size, but large diaspores with large rewards elicit mass recruitment (Rowles and O'Dowd 2009).

As with other ant–plant interactions, benefits to ants are rarely assessed. Whereas carbohydrate-rich food sources that fuel large and active worker forces are likely to be especially relevant to attaining and maintaining ecological dominance, lipid-rich elaiosomes are more likely to benefit queen(s) and brood (Wheeler 1994), and therefore may facilitate greater brood production. Studies of *M. rubra* in its native range report that elaiosome-supplemented laboratory colonies produced more pupae than control colonies (Fokuhl *et al.* 2007), and that elaiosomes contributed 87% and 79% of daily nitrogen and carbon incorporation, respectively, to *M. rubra* larvae (Fischer *et al.* 2005). Comparisons in elaiosome use between native and invasive ants, and between the native and introduced ranges of invasive ants, as well as data on how any differences may affect ant invasibility are lacking.

15.5.4.2 Seed harvesting

Some ants, including some invasives such as *S. invicta* and *S. geminata*, regularly consume seeds. Seeds have been reported to comprise up to 17% and 29.6% of *S. invicta* (Vogt *et al.* 2002) and *S. geminata* (Tennant and Porter 1991) diets, respectively. Both species have been reported as seed pests in agricultural systems (Morrison *et al.* 1997a; Veeresh 1990). Invasion by seed-harvesting invasive ants into both agricultural and natural systems has the potential to change the composition of the seed bank because some seed species will be more palatable, and therefore more subject to harvesting than others (Vogt *et al.* 2003; Zettler *et al.* 2001). For example, *S. invicta* facilitated the success of ragweed in old fields by preferentially preying on the seeds of four other early successional plants, but increased predation on ragweed seeds as the other seeds became less abundant (Seaman and Marino 2003). The extent of the changes will depend on the behaviour of the invading granivorous ants relative to any displaced native granivores. *Solenopsis invicta* appears to prefer water-soaked seeds (Drees *et al.* 1991; Vogt *et al.* 2003) and under laboratory conditions, seed damage by the ant has been related to accessibility of the seed embryo, colony size, and the availability of other food (Drees *et al.* 1991). Displacement of native granivorous ants by invasive ants that do not readily consume seeds, such as *L. humile* (Oliveras *et al.* 2007; Zee and Holway 2006), can also have consequences for the plant community. Plants that have lighter or weaker seeds that would have been more susceptible to harvesting may benefit, although plants with seeds that were effectively dispersed by seed harvesters may suffer (Oliveras *et al.* 2007).

15.6 Soil

Ants can further indirectly affect plants by affecting the biogeochemical and physical properties of soil, as well as nutrient levels (e.g. Farji-Brener and Ghermandi 2008; Lafleur *et al.* 2005; Seaman and Marino 2003). All published studies on the consequences of ant invasions on soil to date have focused on *S. invicta*. Red imported fire ants have been termed 'ecosystem engineers' for the influence that their mound-building has on surrounding habitat (Bohlen 2006; DeFauw *et al.* 2008; Figure 15.1). DeFauw *et al.* (2008) summarized the effects of *S. invicta* on soil as reported in nine published studies as 'increased aeration and infiltration, altered soil pH, increased levels of available phosphorus and potassium, lower surface soil bulk density, reductions in organic matter, lowering of texture grade, and greater fungal abundance coupled with lower species richness and diversity.' Seasonal shifts in mound soil chemistry have been hypothesized to further influence microfaunal pathogens and parasites, including biological control agents (DeFauw *et al.* 2008). However, mound-building and tunnelling by a number of non-invasive ant species result in changes in soil properties, and studies that consider nest densities and patterns are necessary to understand effects at the landscape level (Cammeraat and Risch 2008). The shallow and sometimes ephemeral nests of species such as *L. humile*, *A. gracilipes*, and *W. auropunctata* may not result in the same soil turnover as nests of displaced native ants.

Figure 15.1 Many ant species are involved in soil bioturbation. Mound-building by the red imported fire ant (*Solenopsis invicta*) alters soil properties. This image shows a cross section of a *S. invicta* nest mound with chambers visible. (Photo: Alex Wild)

15.7 Future directions

Relative to some other fields of ant ecology, the study of the consequences of ant invasions is in its infancy. Whereas Argentine and red imported fire ants have received a lot of research attention, there is still much opportunity for further research to elucidate how these species interact with and affect their novel environments, especially in the long-term and over large spatial scales. We still need basic information for several other species: *A. gracilipes*, *P. megacephala*, and *W. auropuncata* have only been studied in limited parts of their introduced ranges, and we know little about the consequences of the relatively recent invasions by *M. rubra* in eastern North America, Rasberry crazy ants (*Paratrechina* sp.) in Texas, and *L. neglectus* in Europe. Several other species, such as *Pachycondyla chinensis* and *Vollenhovia emeryi*, have attributes consistent with known invaders (Kjar and Sunian 2007) (see Chapter 13), but are entirely unstudied in their adopted habitats. We know almost nothing about the direct or indirect effects of invasive ants other than *S. invicta* on soil.

Multiple and synergistic effects of ant invasions are possible and may lead to large-scale ecosystem-level consequences. Most indirect consequences of ant invasions (such as the effects of native ant displacement on seed dispersal, or the effects of floral arthropod displacement on seed set) have been explored in relative isolation from each other. However, these effects may be present in the same community and combine to change the community greater than either one by itself. Disparate effects can also combine, as has been documented on Christmas Island where displacement of red land crabs and tending of hemipterans by *A. gracilipes* have together resulted in dramatic changes in the rainforest structure (O'Dowd *et al.* 2003; Box 15.1). There may be many more cases where ecosystem level effects resulting from the combination of multiple interactions are more insidious and difficult to detect and occur over longer time scales. Invertebrate displacement and ant-forced shifts in trophic position (e.g. McNatty *et al.* 2009), the evolution of behavioural, morphological, or other traits (e.g. Langkilde 2009) that enable coexistence with an ant invader, or even changes in trophic position of the invasive ant over time (e.g. Tillberg *et al.* 2007) add to the virtually unstudied potential pathways through which an ecosystem may be affected by invasive ants. Considering the consequences of ant invasions in a community context may reveal that some traits that make communities susceptible to some consequences result in decreased vulnerability to others. For example, many myrmecochorous plants are associated with nutrient poor soils (Westoby *et al.* 1991a), and plants in nutrient poor soils often have foliage with a high carbon/nitrogen ratio, which protects them against herbivores (Orians and Milewski 2007). Thus, while myrmecochorus plants may have reduced seed dispersal, they also may not be as susceptible to hemipteran outbreaks following ant invasions.

Further research on any invasive ant species should aim to bring us closer to the ultimate goal of mitigating or preventing adverse consequences. It is increasingly important to document not only what effects an invader has, but to also determine what behavioural, physiological, or other characteristics allow it to have such effects, under what range of conditions, and for how long. For example, many consequences of ant invasions described in this chapter are linked to, or enhanced by, high invasive- ant density relative to the density of the native-ant fauna. The logical corollary to this observation, that low densities of an invasive ant would fail to have adverse consequences, has some support from observations (e.g. Abbott and Green 2007, but see Lach 2007; 2008a) and short-term introductions (e.g.Walters and Mackay 2005), but deserves further explicit attention, particularly for interactions with native species other than ants. For even the most-well studied species, global comparisons of their effects across their introduced ranges will be helpful in elucidating how the abiotic and biotic environments affect their biology and behaviour, and how this links to their impacts.

Invasion success is often linked to disturbed habitats (Chapter 14) and separating the effects of disturbance from those of ant invasion is an ongoing challenge. Researchers cannot ethically introduce invasive ants to previously uninvaded areas on a large scale; however, short-term introductions (e.g. Thomas and Holway 2005), targeted removals (e.g. King and Tschinkel 2006), and small-scale exclusions (e.g. Lach 2007) are examples of experimental approaches that can help discern the effects of confounding biotic and abiotic factors associated with invasion. Additional long-term or follow-up studies (e.g. Hoffmann and Parr 2008) that investigate the same areas before and after invasion will also be helpful, as well as studies of the same areas years after initial invasion (e.g. Heller *et al.* 2008; Morrison 2002b). Research that considers the effect of invasive ants relative to displaced native ants, rather than to the absence of all ants, may be especially informative.

Finally, the possibility that ant invasions may be managed to human benefit should be further explored. To date, such research is limited to agroecosystems infested with *S. invicta* in parts of the United States. However, *A. gracilipes*, *P. megacephala*, and *W. auropunctata* have all been exploited as natural biological control agents in some parts of their introduced ranges (Lach 2003, see Box 7.2). We do not advocate intentional spread of invasive ants. However, elucidating the means by which predacious behaviours can be manipulated to affect harmful herbivores rather than beneficial predators will mitigate some of their damage.

15.8 Summary

Elucidating the consequences of ant invasions continues to be an important component of invasive ant research. Red imported fire ants and Argentine ants are among the most studied invasive insects, but there are several other invasive ants with the capacity to have adverse ecological consequences. The displacement of native ants is the best documented consequence of ant invasions, though recent research calls into question the long-term effects of monogyne *S. invicta*. A range of other invertebrates is also affected by invasive ants, though interactions with the same taxa can vary across different parts of an invasive ant's introduced range. Some myrmecophagous species have been shown to benefit by the increased ant densities associated with invasions. Stinging invasive ants, particularly *S. invicta*, have the greatest documented effects on vertebrates. However, very few studies have examined indirect effects of invasive ants on vertebrate fauna. Carbohydrate-rich resources attract invasive ants to forage on foliage, though *S. invicta* will visit plants in the absence of such rewards. Most plants with extrafloral nectaries benefit from invasive ant tending, but invasive ants in flowers can deter pollinators. Effects on hemipteran herbivores are largely positive, but evidence that invasive ants are superior tenders is still lacking. In agroecosystems, *S. invicta* can be a beneficial predator of yield-reducing herbivores. Due to their relatively small sizes, invasive ants are often poor seed dispersers relative to native ants. Seed-harvesting invasive ants can be serious pests in agricultural systems and can alter plant communities, as can non-seed harvesters, such as *L. humile*, that displace native granivorous ants. *Solenopsis invicta* nesting habits alter many soil properties, though it is unclear how extensive these changes are relative to those caused by the displaced mound-building *S. geminata*. Future research directions include understanding basic consequences of invasion by lesser known species, further investigation of indirect effects and effects on soil, consideration of effects in a community context, and explicit investigation into the importance of high densities in effecting impacts. Experimental approaches that are adequately replicated in space and time will be most helpful in meeting the goal of mitigating or preventing adverse consequences, or manipulating the predacious characteristics to our benefit.

Acknowledgements

We thank Alexei Rowles and Paul Krushelnycky for their helpful comments on earlier drafts and Ashley Miller and Rachel Strecker for their assistance in gathering references.

Chapter 16

Invasive Ant Management

Benjamin D. Hoffmann, Kirsti L. Abbott, and Peter Davis

16.1 Introduction

The environmental, social, and economic impacts of ant invasions are well known and severe (Holway *et al.* 2002b; Chapter 15), and as such, invasive ant species have been the target of many management efforts for containment, control, or eradication. Yet, despite nearly a century of efforts managing invasive ants, there are very few publications that describe management attempts, regardless of success or failure. The lack of published literature on invasive ant management programmes is largely due to the demand for managers to do exactly that – manage, not communicate strategies, outcomes, or lessons to a global audience. It is also only in the last 10–15 years that management-oriented journals have sought to report the details of management programmes in an attempt to enhance our knowledge of invasive species management. Consequently, a vast amount of valuable information on this theme is buried in internal government reports, or has not even been recorded.

Whereas there have been recent publications that synthesize knowledge of invasive species eradications in order to elucidate key criteria that determine success or failure (Myers *et al.* 2000a; Simberloff 2009), none deal specifically with ants. This chapter brings together global knowledge of ant management, as well as personal experiences of the authors in conducting, coordinating, and overseeing invasive ant management programmes throughout Australasia and the Pacific. We provide a broad overview of modern management techniques, highlight some specific programmes that have provided valuable lessons for effective ant management, and present a general framework for invasive ant management detailing requirements for effective programmes. In addition, we highlight the three actions that we believe will provide the greatest immediate enhancement of exotic ant management. This chapter does not attempt to synthesize the ecological information in this book to inform management, but provides readers with an additional insight into the interface between ant ecology and invasive ant management.

16.2 Development of modern invasive ant management techniques

Modern invasive ant management aims for either eradication – the intentional extirpation of geographically discrete populations of a species, irrespective of whether other discrete populations still remain within the landscape; or control – management actions that aim to reduce populations of an invasive species, thereby reducing impacts, or preventing further dispersal. Both eradication and control currently rely predominantly on chemicals, but non-chemical options are also being developed.

16.2.1 Chemical control

The evolution, historical use, and consequences of sequential chemicals used for pest ant management have been reviewed many times (Lofgren *et al.* 1975; Tschinkel 2006; Williams 1983, 1984; Williams *et al.* 2001), and will not be discussed here. Rather, we focus on the lesser-reviewed topics of characteristics of ant treatment products and the challenges of continued development in this field. We further limit this discussion by ignoring dusts and products utilising inert carrier granules as these treatments have no potential for large-scale utilisation.

Treatment products consist of at least two components: an active constituent and a matrix. The active constituent is the chemical that mediates ant control (e.g. metabolic inhibitors, juvenile hormone analogues), and is supplied to the ants within a delivery matrix that the ant must ingest in order to receive a dose of the active constituent. Matrices are either liquid (predominantly water and possibly other attractive substances such as sugar) or solid (corn grit or fishmeal). The treatment products are termed according to their mode of delivery to the ant, being either a nest drench or spray for the liquid forms, and baits for the solid forms.

Pioneering work that can be considered the precursor to effective modern ant management using chemicals was undertaken by the United States Department of Agriculture (USDA) and the Federal Bureau of Entomology (Newell and Barber 1913). This work was initiated following the detection of the Argentine ant *Linepithema humile* in New Orleans, USA, in 1891, and its subsequent escalation to become a severe urban and agricultural pest. Most importantly, two biological factors affecting control were identified that are fundamental to the effective treatment of these social insects. First, only a small fraction of ants in a colony forage outside of the protection of the nest at any one time; thus killing foraging workers is not an effective means of control. Second, it is futile to use a product that only kills the sterile workers; the active constituent must have a delayed action so that the workers live sufficiently long to feed the queens and immature stages within the nest.

Following these initial biological breakthroughs, research has identified characteristics of active constituents and products that make them suitable for effective social insect management (Williams *et al.* 1980): (*a*) they must be non-repellent at the product concentration; (*b*) they must have a delayed action of at least 24 h at the product concentration, thereby allowing trophallaxis within ant colonies, whereby ants regurgitate liquid food to share with colony members, inadvertently distributing the toxicant to fellow workers, brood, and queens (Figure 16.1); (*c*) they should be fatal to the target ants at concentrations as little as 1/100th of the original dosage to account for dilution during trophallaxis; and (*d*) surviving ants should not avoid the product in subsequent treatments.

All original treatment products were liquids and were applied as a nest drench or a spray when the active constituent was a contact insecticide, or within aqueous sugar as a food source when the active constituent was a toxicant requiring ingestion. In general, except where solid baits are targeted at ant larvae, products requiring ingestion must present the active constituent in a liquid form because adult ants are liquid feeders, restricting the passage of solids into their gut within an infrabuccal pocket and only allowing liquid to pass into a crop, where it can be regurgitated to conduct trophallaxis (Hölldobler and Wilson 1990). However, only the sprays were suitable for large-scale treatments (i.e. bigger than a house block) and the effects of these treatments were far from focused on the target ant species (Carson 1962; Markin *et al.* 1974; Rosene 1958). A means had to be found that delivered to ants an active constituent in liquid form whilst having significantly fewer non-target impacts. The development of a solid matrix infused with the active constituent dissolved in soy oil provided an important breakthrough for invasive ant management. Solid matrices have two advantages over liquid treatments that provide reduced non-target impacts. First, solid matrices can consist of a potentially large number of combinations of protein and carbohydrates (e.g. dehydrated honey, peanut butter, fishmeal, corn grit, and sausage meat) that can be made to suit the dietary preferences of a target ant. This maximizes uptake by the target species over periods where their dietary preference changes between food types and reduces uptake by non-target species. Second, active constituents that require ingestion rather than contact also greatly reduce unnecessary non-target impacts because not all biota that come into contact with the bait will ingest it.

The greatest hindrance to successful ant control appears to be the lack of a universally attractive and effective treatment product. Species do not have equal preferences for protein or carbohydrates (Robinson *et al.* 2005b), and there can be marked seasonal differences in dietary preferences within a species (Cornelius and Grace 1997; Stein

Figure 16.1 Workers exchange liquid food, in which toxins are carried, by means of trophallaxis. (Photo: Alex Wild)

et al. 1990). Efficacy is further influenced by how food is managed within a colony (e.g. stored versus utilized immediately) and pathways by which nutrition is provided to queens (Broekhuysen 1948; Tschinkel 2006). Most bait development has targeted fire ants, *Solenopsis* spp., and unfortunately much lower efficacy has been recorded for use of these baits against other species (Rey and Espadaler 2004). Indeed, current products are largely considered inadequate to provide effective control of *L. humile* (Silverman and Brightwell 2008). This lack of product efficacy is greatest for species that prefer aqueous sugar matrices and are not greatly attracted to corn grit, or to the oil in other dehydrated solids (e.g. *Tapinoma melanocephalum*, *Ochetellus glaber*, and *Anoplolepis gracilipes*).

Insect growth regulators (IGRs) with juvenile hormone analogues as active constituents, including *s*-methoprene, pyriproxifen, and fenoxycarb, lead the field in treatments that do not aim to directly kill the target species. These active constituents reduce queen fecundity to prevent further reproduction, especially of sexuals, as well as prevent the transition from pupa to adult. The lack of reproduction results in colonies reducing in size and ultimately dying after the remaining workers complete their lifecycle and die naturally. Treatments using IGRs, however, have thus far proven to be variable in their efficacy, both within and among ant species (Banks *et al.* 1983; Fowler and Roberts 1983; Williams and Vail 1994).

Finally, there are relatively new chemicals that will further revolutionize ant management due to their environmental acceptability (i.e. low toxicity to non-ant fauna) and increased efficacy compared to historically used chemicals. For example, the naturally occurring bacterial toxin, spinosad (spinosyns A & D), has been commercialized as an active ingredient in a general fire ant bait, and has also been granted organic status by the USDA National Organic Program (www.ams.usda.gov). Another toxin, indoxacarb, has potential for large-scale use in invasive ant management as it is considered a 'reduced risk pesticide' (US Environmental Protection Agency 2000), with improved efficacy following metabolic degradation (i.e. the ants' metabolic processes produces a more potent active compound). Ideally, treatment products would attract the target ant species but repel non-target organisms, but no such options are yet available for ant management.

16.2.2 Non-chemical control

It is clear that chemical treatment products currently offer the best form of ant control. However, there are non-chemical control methods that show promise for effective ant management. In most cases, these non-chemical options alone are unlikely to eradicate an incursion or even prevent further spread. As such, for eradication campaigns they should be used as methods supporting chemical treatment within an Integrated Pest Management Program (IPM) framework (Oi and Drees 2009).

The use of pathogens, parasites, and parasitoids of invasive ants as control options remain almost completely limited to fire ants, predominantly *Solenopsis invicta*, because, unlike most other highly invasive ants, the native range of these species is known, allowing natural biological control options to be identified. Fortuitously, many of these natural enemies have high host specificity, which is a requirement for any organism to be used as a biocontrol agent. The biocontrol agents with the most advanced utilization are phorid flies. These dipteran parasitoids lay eggs in the thorax, after which the larva migrate to the head where they complete their development and emerge following decapitation of the host (Porter *et al.* 1995). However, because sometimes as few as

3% of ants from a colony are parasitized (Morrison *et al.* 1997b), it is the behavioural modification of individual workers that has the highest ecological impacts on the colony, particularly by reducing food-harvesting capabilities (Mehdiabadi and Gilbert 2002). Many phorid fly releases have been undertaken throughout the United States (Graham *et al.* 2003; Porter *et al.* 2004), and populations of two species, *Pseudacteon tricuspis* and *Pseudacteon curvatus*, have successfully established in the southern United States (LeBrun *et al.* 2008; Thead *et al.* 2005). However, although the flies induce stress in laboratory fire ant populations (Mehdiabadi and Gilbert 2002), their effectiveness in the field remains uncertain (Morrison and Porter 2005).

The microsporidian, *Kneallhazia* (=*Thelohania*) *solenopsae*, is a natural pathogen of multiple fire ant species throughout their home ranges in South America (Allen and Buren 1974; Williams *et al.* 1999) that was recently discovered in the southern United States (Williams *et al.* 1998). This pathogen infects all ant developmental stages, but does not necessarily significantly reduce the life expectancy of the host (Tschinkel 2006). It does, however, reduce brood production, which leads to reductions in both nest sizes and nest densities (Williams *et al.* 1999). Infection rates of *S. invicta* colonies within the United States have been recorded as high as 93% (Oi and Williams 2002), but the impact of this pathogen in the field is currently unclear.

Where possible, habitat modifications such as fire (Hoffmann and O'Connor 2004), scrub clearing, or drainage restriction (Holway and Suarez 2006) can be applied to complement chemical treatments. These modifications aim to increase stress on the invader or reduce the abiotic suitability for invasion; burning temporarily reduces food (carbohydrate) sources by destroying extrafloral nectaries and kills phytophagous insects that cannot escape the fire; draining restriction creates more open and drier environments that restrict the spread or distribution of some species such as *L. humile* (Holway *et al.* 2002b; Human *et al.* 1998; Menke and Holway 2006). These two techniques may simultaneously promote biotic resistance from some aggressive native ant species (Hoffmann and O'Connor 2004; Menke *et al.* 2007). However, such modifications are not always appropriate, considering that some invasive ant species are associated with habitat disturbance (e.g. Colby *et al.* 2008; Travis 1941), and not all vegetation is fire-resilient.

Controlling mutualistic honeydew-producing insects to naturally reduce ant populations is a research area and management strategy that is likely to develop in the near future. Carbohydrate supply in the form of honeydew is regarded as a key driver of ant population densities (Davidson 1997), including invasive ants (Addison and Samways 2000; Le Breton *et al.* 2005; O'Dowd *et al.* 2003; Chapters 7 and 14). Thus, reducing or eliminating populations of mutualistic species is likely to reduce invasive ant populations. Chemical control of phytophagous insects is currently possible by using broad-spectrum sprays or systemic insecticides. However, this is only an option within urban and agricultural settings, not within intact ecosystems due to unacceptable environmental impacts. The key for developing this technique appears to be the identification of biocontrol options for phytophagous insects that are effective in the presence of ants.

16.3 Key lessons from eradication programmes

Both the successes and failures of ant management programmes provide valuable lessons for effective ant management. Eradication attempts, rather than control programmes, provide the best opportunity for learning because the mistakes have a greater (possibly devastating) impact on project integrity, and there is a single clear end-point on which to gauge success. Here we briefly describe some eradication programmes and the key lessons they have provided that have proved fundamental to the success of all ant-eradication projects.

16.3.1 Programmes for *Solenopsis invicta*

The largest, best-known, and most debated ant control effort has been for fire ants (both native and introduced) in the United States, particularly *Solenopsis invicta*. These 'fire ant wars' have been reviewed numerous times (including Tschinkel 2006 Williams *et al.* 2001) and therefore will not be detailed here, but are noteworthy for the many lessons that were hard learnt by their failings. Most importantly, the campaigns against *S. invicta* failed to prevent further spread, as well as the reinfestation of effectively treated areas, and treatments ignored the biology of the ant, at least up until the banning of organochlorines in the 1980s. Moreover, treatments were originally conducted with such little regard for environmental issues that this management effort possibly has had the greatest adverse environmental impacts of any eradication project conducted for any species (Carson 1962). It is now recognized that nationwide eradication of *S. invicta* in the United States is not a feasible option, and management is now aimed at mitigating its impacts at local scales and preventing further spread (Oi and Drees 2009). Ironically, despite this being the largest and longest fought effort, few documents detailing results from any location exist within the publically available literature. We are aware of only seven isolated infestations of mostly unknown area in six eastern states (Thorvilson *et al.* 1992; Williams *et al.* 2001) that have been declared eradicated largely from what appears to be the northern-most range limits of *S. invicta* (Korzukhin *et al.* 2001).

In contrast, recent efforts to eradicate *S. invicta* in New Zealand and Australia have proven to be more successful. In less than a decade, New Zealand authorities have detected three separate incursions of *S. invicta*. The first, a single nest, was treated and declared eradicated two years later (Pascoe 2003). The second detection was of ants on a yacht recently arrived from Bermuda, and all ants were immediately extirpated (S. O'Connor, personal communication). The most recent incursion, three nests within a small area, was declared eradicated in April 2009. (Biosecurity New Zealand 2009). In each case, the incursions were detected so early on their arrival or establishment that no evidence of spread was found, demonstrating the value of investment in active surveillance and public engagement to achieve early detection.

The detection of two *S. invicta* infestations in Brisbane in 2001 initiated Australia's largest eradication programme of an invertebrate (Vanderwoude *et al.* 2003). The programme was initially planned to last five years with a cost of AU$123.9 million. While still ongoing after seven years and almost AU$200 million, the programme has had many noteworthy successes, and highlighted valuable lessons. For example, outlying infestations were detected early and subsequently eradicated, a direct result of a coordinated national surveillance programme operating externally to the known infested area. These include detections in Dandenong in Victoria, 1,400 km southwest of Brisbane, and at Gladstone, approximately 500 km north–northwest of Brisbane (P. Davis and B. Hoffmann, unpublished data). In addition, the treatment strategy appears to be capable of eradicating *S. invicta*. The strategy involves three to four applications of baits containing IGR's (either pyriproxyfen or *s*-methoprene) per year over three consecutive years followed by two years of intensive surveillance. The programme has also revealed clear differences in the efficacy of bait application methods: by hand (~40% of treated area), land vehicle (four-wheel motorbikes ~ 10%), and helicopter (~50%) (Figure 16.2). Aerial application has proven to be the most effective, whereas four-wheel motorcycle application has been the least reliable due to 'islands' of poor treatment around ground obstacles. Finally, novel techniques have been developed that greatly enhance programme performance and reduce costs (McNicol 2006). One such development is a 'Habitat Model' that uses LANDSAT imagery and known locations of *S. invicta* nests prior to treatment to create a probability map of *S. invicta* occurrence (R. George, unpublished data) which can then be used to prioritize the surveillance areas. The programme also has trained sniffer dogs to detect *S. invicta*. These dogs have proven to be cost-effective compared to visual surveillance by trained survey teams (McNicol 2006).

16.3.2 Programmes for other species

The management of *L. humile* has an extensive history throughout all southern Australian states. The largest and best documented of these ran from 1954–1988 in Western Australia. After trials proved that the organochlorine dieldrin applied as a broad-scale spray was capable of achieving extirpation of *L. humile* infestations within small trial plots of approximately 0.5 ha (Forte 1956), a large-scale eradication project commenced (Van Schagen *et al.* 1994). Over the duration of the program, the infestations were reduced from approximately 18,000 ha to 1,458 ha, but the successful treatments did not cover the entire infested area simultaneously. It

Figure 16.2 Granular bait distribution techniques can be via (*a*) hand, (*b*) land vehicle or (*c*) helicopter. (Photo: Benjamin D. Hoffmann)

took six years to provide a first round of treatment to the 18,000 ha (average of 2,900 ha treated each year), and on average only 500 ha were treated within each subsequent year as new or persistent populations were found (Western Australian Department of Agriculture 1988). Of the 1,458 ha that remained infested, 75% were within areas where treatment was not possible because of environmental or agricultural concerns about the products in use (Western Australian Department of Agriculture 1988). The project was terminated in 1988 when organochlorines were banned, and an acceptable and effective alternative treatment could not be found (Van schagen *et al*. 1994). Despite the failure of this programme to fully eradicate *L. humile* from Western Australia, the programme is noteworthy for successfully containing the ant for such a long period, for the large area from which the ant was eradicated, and for the lesson that the programme's success eventually undermined public support for the ultimate goal of total eradication, as most of the community had not experienced the negative aspects of *L. humile*.

A successful eradication programme was conducted for *Wasmannia auropunctata* over 2 ha of Santa Fe island in the Galápagos archipelago (Abedrabbo 1994). It was successful predominantly due to the persistence of land managers when initial attempts failed. Original treatments over several years using DDT, pyrethrin, and resmethrin applied as broadcast sprays achieved high levels of control, but not eradication. Eradication was finally achieved by broadcast use of hydramethylnon in the product form of Amdro® following the clearing of scrub to make access pathways, and burning. The success of this project led to a subsequent eradication attempt of *W. auropunctata* from 22 ha on Marchena island, also within the Galápagos archipelago (Causton *et al*. 2005). Similar to the attempt on Santa Fe, the first three treatments conducted over four years failed (Roque-Albelo *et al*. 2000), probably because the bait was not adequately dispersed and did not cover the entire infestation, leaving residual populations (Causton *et al*. 2005). A fourth attempt using a triple treatment of Amdro® within seven months, coupled with more frequent and intensive post-treatment surveys and scope for detailed follow-up treatments, proved successful. Both projects not only demonstrated that eradications of *W. auropunctata* are possible, but are fine examples of the fact that ongoing and increasing commitment to a programme by funders and land managers can be rewarded following the creation of more effective ant control products and improvements in treatment application protocols.

Hoffmann and O'Connnor (2004) reported the eradications of 24 infestations of *Pheidole megacephala* (ranging in size from a single nest to 10 ha and with a combined area of 30 ha), as well as two infestations of *Solenopsis geminata* (covering 3 ha) from areas associated with buildings within Kakadu National Park. Both species were treated with Amdro®, and some *S. geminata* colonies were also treated with a localized drench of diazinon. Most *P. megacephala* were killed by a single treatment that was conducted external to buildings, with only a few small locations requiring a second treatment where populations had survived within buildings. This project primarily demonstrated the relative ease by which *P. megacephala* could be eradicated when compared to other invasive ant species. Indeed, there are least a further 12 completed eradications of *P. megacephala* from locations throughout northern Australia (B. Hoffmann, unpublished data). The project also demonstrated that *S. geminata* could be eradicated while the infestation was young and confined to a small area.

The management of *W. auropunctata* in Hawai'i illustrates both success and failure of ant management. This species was found in the town of Puna, Hawai'i, in March 1999. It was detected soon thereafter in other nearby areas, and later that same year an outlier population was found on the island Kaua'i, having originated from plants transported from Hawai'i (Krushelnycky *et al*. 2005b). The outlier infestation was treated, which reduced populations to very low levels, and currently only a few small populations persist (P. Krushelnycky, personal communication). The significant reduction of *W. auropunctata* in this treated outlier population is a success in that it shows promise for achieving eradication over the whole infestation. However, the greater management programme for *W. auropunctata* in the Hawai'ian archipelago is failing dismally to prevent further spread from infested areas due to lack of funding and a dearth of public awareness of the ant (Krushelnycky *et al*. 2005b), and the expanding front of the infestation means that eradication over the entire archipelago is unlikely to occur.

Finally, a recent eradication of *P. megacephala* in northern Australia from 5 ha of intact vine thicket is noteworthy because a follow-up study has been the first to demonstrate ecological recovery of the native ant fauna following the elimination of an invasive ant (Hoffmann 2009). Ecological impacts of the target species are often documented within ant management programmes to establish the need for management, and non-target impacts are often monitored to ensure that there are no unacceptable treatment effects on fauna of concern. However, very few programmes have documented recovery dynamics of the native fauna post-treatment. At best, previous work has noted the persistence or rapid increases in the abundance of certain native ant species post-treatment (Abedrabbo 1994; Hoffmann and O'Connor 2004), or persisting differences in taxonomic group metrics between treated and untreated areas (Marr 2003). Given that invasive ant species are largely subject to management efforts within intact natural environments because of their deleterious environmental impacts, the full recovery of native biota within treated areas should be of great interest and importance.

16.4 A general framework for invasive ant management

The phases of management programmes are logical and generic for most invasive species; it is the detailed protocols within each phase that will have project specificity due to the great array of working environments and target species. Unlike phases, management frameworks can differ by taking either a reactive or a proactive approach. Historically, ant management has been reactive, only commencing following the detection of an incursion. We argue that invasive ant management must become proactive; preparedness should begin prior to the detection of an incursion. Here we describe issues and considerations within five phases of invasive ant management and place them in a proactive framework.

16.4.1 Pre-emptive phase

The pre-emptive phase enhances a jurisdiction's ability to prevent an incursion, as well as to rapidly detect an incursion and initiate on-ground measures, should an incursion occur. Neither of the two components of the pre-emptive phase is restricted to this phase; rather, they are integral components of all other phases.

16.4.1.1 Pest risk analysis

Pest risk analyses (PRAs) were originally developed to model the spread of disease for public health (Andersen *et al.* 2004b), but are now widely used for invasive species management (Andersen *et al.* 2004b; Leung *et al.* 2002). PRAs integrate and model two components: (*a*) species risk assessment; and (*b*) risk management assessment. An international standard provides a basic framework and guidelines for PRAs within the scope of the International Plant Protection Convention (FAO 1997).

In their simplest form, species risk assessments consolidate global knowledge of the biology, ecology, and impacts (beneficial and negative) of target species, and use this to assess the overall potential benefit or impact within a landscape, should it establish there. As such, they are a useful tool to rank species in relation to the risk that they pose and prioritize management actions accordingly. While there is an almost inexhaustible list of species that can potentially invade or be analyzed, species risk assessments should at the very least be conducted for the few (<50) ant species that are potential invaders (e.g. McGlynn 1999b).

Following the identification of an unacceptable risk to a region, a PRA combines the knowledge contained within the species risk assessment with a risk management assessment to develop on-ground procedures that actively prevent incursion, improve detection, allow rapid response to detection, or effectively manage established populations. In particular, for each species identified as a risk, risk management assessments (*a*) identify the most likely invasion pathways to prevent incursion or prevent further spread, (*b*) detail detection strategies that maximize the detection of invasive ants, (*c*) provide immediate treatment options for use upon detection of an incursion to prevent establishment or effectively manage established populations, (*d*) strategically plan outreach to all stakeholders and the public to improve awareness of the potential risks of invasive ants, and (*e*) detail strategies to ensure their own effective implementation and integration with other ant management efforts. Importantly, a PRA can highlight the potential lack of treatment options available in a location, such as a proven treatment product not being available because it is not registered for use in the jurisdiction or on the target species. This knowledge can subsequently be used to proactively obtain permits or registrations for product use and supply of treatment products prior to an incursion, thereby eliminating unnecessary delays in the commencement of treatments following detection. Although product registration is normally a commercial decision made by chemical supply companies, such pre-emptive registrations have been implemented by the New Zealand government for many invasive ants following the incursion of *S. invicta* in 2001 (S. O'Connor, personal communication).

PRAs can operate at different jurisdictional levels and may be broad or specific in focus. For example, the ant prevention plan for the state of Hawai'i applies to all non-established ant species, but focuses particularly on *S. invicta* and *W. auropunctata* (Hawai'i Ant Group 2001, 2007). At a national level, the Australian Tramp Ant Threat Abatement Plan (Commonwealth of Australia 2006) establishes a comprehensive framework to guide and coordinate Australia's response to exotic ants. At a regional level, the Pacific Ant Prevention Plan (Pacific Invasive Ant Group 2004) addresses the risk of exotic ants for multiple countries throughout the south Pacific.

16.4.1.2 Public education

In a proactive management environment, public education on key invasive ant species would occur prior to an incursion. Public education serves not only to enhance the use of public vigilance as an early detection tool, but aids in building a support base, should a programme be launched. Public education efforts aim to inform people of the pest, the risks it poses, and the need for management to eliminate or contain those risks. Public support is highly advantageous because it facilitates access to, and treatment of, property without the need for legal enforcement, as well as greater adherence to quarantine (measures to prevent further spread of an exotic ant) and biosecurity (inspection of goods and people entering a jurisdiction) measures. Public vigilance is also a useful tool to detect persisting colonies post-treatment, satellite populations, or new incursions. The usefulness of public education makes it an important requirement throughout all phases of ant management.

16.4.2 The scoping phase

The detection of an exotic ant species places authorities in a position of deciding whether to initiate an eradication programme, or control programme, or to not take any action. The scoping phase gathers all information required to make this critical decision.

16.4.2.1 Pest risk analysis

When a species is new to a landscape, knowledge-based decision making can be extremely difficult because information such as quantified impacts, documented rate of spread, biology, and proven control and detection techniques is non-existent, and must be inferred from other locations, if such information is available (Smythe *et al.* 1996; Williams *et al.* 2001). The use of consolidated knowledge within a PRA aids in removing as much of the uncertainty as possible. If a PRA was not conducted in the pre-emptive phase of management, one must be compiled at this stage, making this action the first step of a scoping phase. A well-prepared PRA is highly advantageous, as it can greatly accelerate the speed at which the scoping phase can be conducted, and ensure the highest integrity of on-ground data collection (see Section 16.4.2.2). Moreover, a pre-prepared PRA would have already identified whether a species poses a risk significant enough to warrant management, eliminating much of the need to quantify impacts to justify management (Harris *et al.* 2005; Simberloff 2003).

16.4.2.2 On-ground data collection

The most important question to answer in the scoping phase is: What is the area of infestation? Accurate determination of the infested area and subsequent treatment area (usually incorporating a buffer zone) is central to the decision-making process. Generally, the size of this area is inversely proportional to the likelihood of achieving control or eradication, and directly proportional to the cost. Effective delimiting (mapping) of the infestation boundaries is critical, considering that in most cases, failure to treat all populations simultaneously will either increase the duration and cost of a programme, or destine an eradication attempt to failure. The infested area, combined with a buffer zone, will become the treatment and quarantine area. It is no coincidence that many programmes that undertake to control or eradicate invasive ants are on oceanic islands (Causton *et al.* 2005; Green and O'Dowd 2009; Lewis *et al.* 1976) as an island's coastline provides a definite dispersal limit and effectively delimits the quarantine area.

While assessment of the area infested is the key quantifiable component of on-ground data collection required to make an informed decision about whether to proceed, other data and knowledge can be obtained at this point that will be helpful in setting baselines, justifying management actions, and aiding management decisions by increasing site-specific knowledge of the target species. Baseline data that provide population measures of the infestation allow quantifiable analyses of treatment success. For invasive ant management, these data are most commonly measures of ant forager abundance, nest density, or frequency of detections, rather than just the size of the area infested. The status of co-occurring biota prior to treatment, or ideally prior to invasion, is also helpful to assess the impacts of treatment on non-target organisms, and to document ecological recovery following control or eradication of the target species. For established invasions, quantitative measures of ecological impacts are useful for justifying or modifying project importance, as well as to aid scientific understanding of invasion dynamics, including the consequences of the invasion. Site-specific knowledge of the target species in its new environment is highly beneficial to improve project protocols (Box 16.1). For example, knowledge of reproductive phenology, bait preferences, and efficacy of proposed control techniques, can often be critical for treatment success (Myers *et al.* 2000a; Simberloff 2002), but such information may be unknown or differ at the new location. Likewise, methods to detect low-density populations that will be used for post-treatment monitoring are best tested while populations are known pre-treatment, rather than having to assess efficacy by the detection of often cryptic populations post-treatment. Information on the genetic structure of an invasive ant population is another example of data that can inform managers of details critical to decision making. For example, genetics can help explain invasion history (single or multiple incursions), the source of ancillary populations detected at later dates (originating from a known infestation or a new incursion), potentially the differences seen in biology and impacts among populations (Abbott *et al.* 2007), and control within and among programmes. The scoping phase is the best time to conduct all these studies, as outcomes can be used at the outset to influence management decisions and refine best practice protocols. Obtaining this information early can help avoid costly mistakes, or even worse, commencing projects that are unfeasible (Abdelkrim *et al.* 2007). However, if data collection prior to management initiation is not possible, programmes should still take an active adaptive management approach (Box 16.1) and where possible proactively incorporate outcomes of any studies into subsequent phases.

16.4.2.3 Cost–benefit analysis

The determination of the treatment area coupled with information of potential impacts and control strategies identified in a PRA supply the data required to conduct a cost–benefit analysis (CBA), which compares the calculated cost of one or multiple proposed management strategies to the projected cost of doing nothing (Antony *et al.* 2009). Inevitably, programmes that show a large financial benefit are more likely to be funded than those that do not.

The biggest challenge in a CBA is ascribing accurate monetary values to elements that will be affected by the ant species, but as yet have no standard cash value, including human aesthetics and environmental services. As a result, traditional CBAs are inadequate to underpin decisions to initiate or continue with management of invasive ants that primarily affect ecosystems, rather than for species that affect the more readily quantified interests of agriculture or human health. Various measures have been used as proxies to value the environment such as direct values of logging and tourism (Shrestha *et al.* 2007), as well as the cost of environmental services, such as freshwater regulation and supply, pollution treatment, and habitat provision (Ingraham & Foster 2008). Future CBAs that contain politically accepted values of natural environments and ecosystem services (see Costanza *et al.* 1997; Farley 2008) will be more likely to avert the 'no-action' option for species that primarily affect native ecosystems.

Box 16.1 Active adaptive management for invasive ant management
Benjamin D. Hoffmann and Kirsti L. Abbott

Adaptive management at its most basic level involves the continual incorporation of new knowledge into a decision-making process to refine and improve project actions and outcomes (Walters and Holling 1990; Figure 16.1.1). Knowledge accumulation in adaptive management programmes has traditionally been passive through the use of information gained either during quantification of programme progress towards specified goals, or obtained external to the programme. However, such knowledge can also be gained *actively* through the incorporation of programme-driven research that is targeted at the managed system (McCarthy and Possingham 2007). This active approach enhances management outcomes, as information gained is specific to the management programme and continually fed back to managers. As such, active adaptive management is now accepted as best practice (Hauser and Possingham 2008).

Importantly, adaptive management acknowledges knowledge gaps, and allows management to proceed despite uncertainty, with evaluation and learning explicitly incorporated within its framework (Figure 16.1.1). This is important because early intervention improves the chances of successful eradication (Lodge *et al*. 2006; Simberloff 2003). A risk assessment essentially starts the adaptive cycle by assessing the potential risks posed by an invasive species and directly linking management strategies to those risks, as well as clearly identifying knowledge gaps (see Section 16.4). However, just as early intervention has proven important to achieving eradication, so too has the removal of uncertainty. Addressing uncertainty in early programme phases can help prevent unwise decisions and costly mistakes. For example, the failed attempt at rat (*Rattus rattus*) eradication from the Sainte Anne islets in the French

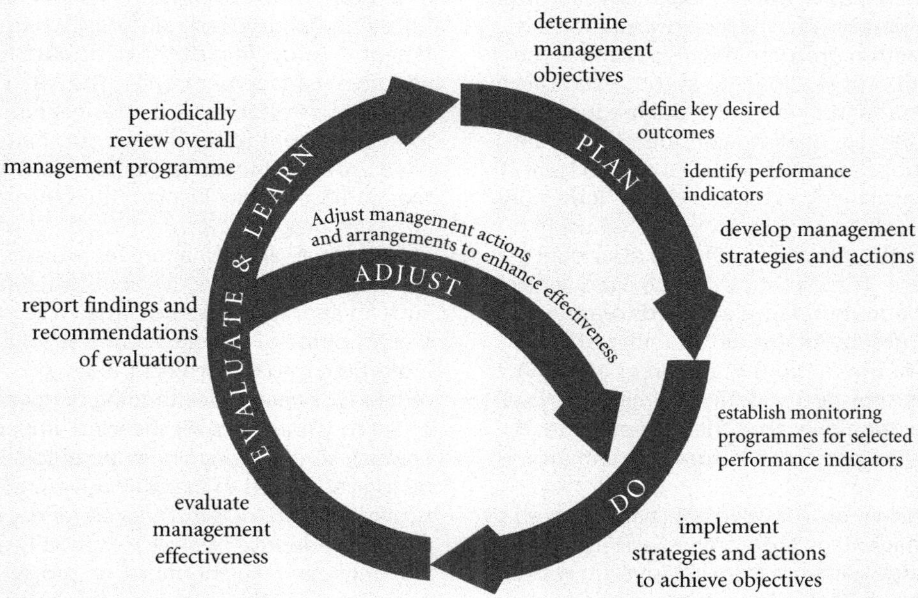

Figure 16.1.1 The adaptive management cycle (Jones 2005).

continues

Box 16.1 continued

Caribbean was attributed to constant recolonization from the nearby mainland (Abdelkrim *et al.* 2007). Had basic dispersal dynamics been investigated during the project's pre-emptive or scoping phases, the eradication would have been deemed unfeasible and not commenced.

Historically, invasive ant management has suffered from a lack of basic biological information of most target species, especially within new environments. Yet, management programmes have rarely addressed this uncertainty due to an under-appreciation of the importance of understanding ant biology to achieve effective management. The lack of an active approach to elucidate such biological information may well be a major contributing factor to relatively few ant management projects having achieved their objectives. Even when such information has been regarded as important, complete reliance on information obtained from locations elsewhere has been insufficient to achieve project goals. Instead, successful invasive ant management has often been reliant on site-specific information obtained within programmes. For example, an eradication of *Wasmannia auropunctata* from Marchena Island within the Galápagos archipelago had to take into consideration great fluctuations of ant abundance due to local climate, particularly extended dry periods, none of which could have been predicted using knowledge of *W. auropunctata* abundance elsewhere. The climatically controlled variations of ant abundance affected treatment efficacy and subsequent detectability of persistent populations (Causton *et al.* 2005). It was only site-specific data informing managers of these anomalies that then gave rise to refined and project-specific management actions.

Even when an adaptive approach is taken by ant management programmes, most new knowledge gained is from invasive ant research conducted external to such programmes, which creates three issues for potential users of such information. First, unless there is a dedicated researcher within a project, project managers and officers typically do not have the time or resources to access and interpret the vast amount of scientific literature. Consequently, knowledge uptake is slow, and adoption of new strategies even slower. Second, invasive ant ecological research is not predominately directed at improving management outcomes; thus the implications are not always obvious or relevant. Third, there will always be an element of uncertainty associated with the use of information from a different site, no matter how similar those sites are (e.g. potential variation in ant activity due to slight differences in rainfall and temperature). Active adaptive management provides the framework to resolve these problems by providing site-specific research outcomes that are targeted at improving ant management to managers in real time. Moreover, such research simultaneously advances broader ant invasion ecology. For example, Abbott (2006) provided information about *Anoplolepis gracilipes* supercolony expansion and boundary dynamics on Christmas Island to help determine baiting protocols. The research contributed to invasion ecology by aiding understanding of the nature of transition zones between infested and uninfested rainforest while simultaneously providing site-specific data into the adaptive management cycle. The rate of the spread of infestations is a clear example of a topic important to both theoretical and applied aspects of invasive species management that could be easily incorporated into all management programmes.

All invasive ant management projects can be undertaken with an active adaptive approach. Such an approach does not entail a separate or distinct phase of a project, but is wholly integrated into the project framework. The current lack of clear understanding of many aspects of ant invasions (e.g. mechanisms, impacts, dynamics), let alone any understanding of basic biological aspects of many invasive species, particularly within their introduced range, makes an active adaptive approach especially relevant and important. Furthermore, no project should operate in isolation. Greater knowledge sharing among invasive ant programmes, combined with active adaptive management will more rapidly increase our knowledge bank of invasive ants and their management, improve programme success, and aid in the adoption of active and adaptive approaches into invasive ant management globally.

16.4.2.4 Indirect effects of management

A consideration often overlooked in determining management options is the potential for the occurrence of undesirable indirect effects of management activities, such as subsequent invasion or population increase of another, possibly worse, invasive species following eradication or control of the target species. For example, eradications of *Pheidole megacephala* in northern Australia and Hawai'i have resulted in subsequent invasion by *Solenopsis geminata*, which is even harder to manage, and in these cases has had arguably worse social and environmental impacts (Plentovich *et al.* 2009; B. Hoffmann, unpubl. data). Such adverse indirect effects are by no means restricted to the management of ants. Most recently, the eradication of cats from Macquarie Island, Australia saw a dramatic population rise in rabbits, due to the loss of top-down control of rabbits by cats. It was initially assumed that rabbit populations would remain regulated by another control measure, the *Myxoma* virus. However, this was not the case, and the dramatic rise in the rabbit population has resulted in devastating ecosystem effects (Bergstrom *et al.* 2009). It is unclear how such adverse outcomes can be avoided except through having a thorough knowledge of the ecology of the affected ecosystem, vigilant post-treatment monitoring, and adapting management strategies as necessary.

16.4.2.5 Programme responsibility, authority, and structure

For a management programme to proceed, an entity (usually government) must be available to take responsibility for it, or a new entity must be formed. Expertise and legal authority usually determine which entity is given the responsibility for an ant management programme. In the United States and Australia, which have the longest histories of large-scale invasive ant management, management programmes aiming to contain biological organisms were first implemented to protect human health from disease and were later expanded to shield essential food-production industries (e.g. agriculture, horticulture) from serious invasive pests. As a result, the legislative and funding pathways are well established to protect human health and food production industries, but they still remain rudimentary for the protection of ecosystems from pests that have little or no impact in other sectors. Consequently, government departments responsible for food-production industries (e.g. USDA, California Department of Food and Agriculture, Western Australian Department of Agriculture, Queensland Department of Primary Industries) have, until recently, been the only government departments with the required legal powers and expertise to conduct eradication programmes, despite other government agencies having portfolios that are directly or more greatly affected by the ant species of concern. It is only in more recent times that non-agricultural government departments have been provided mandates and powers to manage threats that are primarily environmental (e.g. Biosecurity New Zealand, Biosecurity Queensland), and the primary management responsibility (i.e. not contract work) of new incursions has been given to non-government entities such as local land management organizations (Hoffmann 2004).

Because ants do not respect jurisdictional borders, invasive ant management inevitably involves more than one tier of government and/or political boundary. For these entities to cooperate effectively and enable effective on-ground operations, two factors are fundamental to success (Myers *et al.* 2000a). First, the programme must have legal authority that supersedes that of all affected jurisdictions so that necessary groundwork can be completed without delays. Second, a single line of authority must be determined within the programme structure to ensure the programme acts as a single entity with centralized coordination. Implementation of such a structure is considered a major contributing factor to the success of the current Brisbane *S. invicta* eradication campaign, while lack of centralized coordination is a hallmark of failed *S. invicta* eradication attempts in the United States (Oi and Drees 2009). This single line of authority can also extend past individual programmes in the form of an overseeing committee whose role is to effectively integrate biosecurity responses across jurisdictions and simplify reporting to respective stakeholders at all levels. For example, within Australia, a National Tramp Ant Committee provides coordination and decision-making for national exotic ant issues by utilizing advisory representatives of all state, territory, and federal governments.

16.4.2.6 The decision

The ultimate decision of whether to apply management actions or not is dependent upon five criteria, all of which are addressed in the scoping phase: (*a*) technical possibility; (*b*) practical feasibility; (*c*) environmental acceptability of treatments; (*d*) economic sensibility; and (*e*) political and social acceptability (Norton and Mumford 1993). Technical possibility is addressed in the PRA through a review of prior control efforts, products, and methods shown to be effective, as well as products and methods available in the newly invaded jurisdiction. Practical feasibility is based on the conclusions of technical possibility, the size of the required treatment area, the protocols determined to be required to achieve proposed project goals, and the resources available to conduct the project. Environmental acceptability of treatments is based on the balance of perceived treatment impacts versus invader impacts (both detailed within the PRA) in both the short and long term. Economic sensibility is determined by the CBA. While strictly quantitative, a limited or negative economic benefit should not necessarily declare management options void where there are clear advantageous non-monetary outcomes, such as aesthetics of 'natural' environments and reduction of pests inside buildings. Political and social acceptability are probably the least predictable criteria, because while all other criteria may indicate that management, especially eradication, is feasible, there may well be other external factors that influence the decision. For example, some weeds and vertebrate pests within Kakadu National Park are either useful or now hold some cultural significance to local traditional land owners, and as such are forbidden to be removed from this World Heritage Listed area (S. O'Connor, personal communication). All five criteria are important throughout the life of a programme as they are at the scoping phase and are the key issues to address within a programme's review process.

The decision to take no action against a new incursion may well be the right decision in some cases and should not necessarily be seen as a failure of authorities to act. However, a decision to take no action, and allow unrestricted spread should also be seen as potentially having consequences that are irreversible to both human interests and the environment. Indeed there are many examples where inaction by default or due to lack of concern about a new incursion that could have easily been managed or eradicated has ultimately resulted in serious deleterious impacts at a point when eradication is no longer feasible (Simberloff 2003). Prior to making a decision to do nothing due to some technical or feasibility issue, authorities should consider that an appropriately designed management programme may 'buy time' while advances in knowledge, control products, techniques, and technology are made (e.g. McNicol 2006) that may allow eradication to become a realistic future option.

16.4.3 Treatment phase

The treatment phase commences following a decision to manage a species. The primary goal of the treatment phase is not necessarily to treat everything as quickly as possible but rather to prevent further spread. No matter how successful treatments are, if ancillary populations continue to establish, thereby further increasing the treatment area and resources required, a management programme is unlikely to succeed. Conversely, unsuccessful treatments in the absence of further spread can always be reattempted without any detriment to project feasibility. This is not to say that treatments do not require urgency; rather, when containment is effective, treatments that eliminate populations can be conducted over longer timeframes.

Preventing further spread of all species requires effective containment strategies preventing human-aided dispersal (quarantine). For species dispersing by nuptial flight (e.g. monogyne *S. invicta*) this additionally requires the prevention of development of reproductive individuals, currently achieved predominantly by the use of IGRs (Section 16.2). The production of sexuals has little impact on the spread of truly unicolonial species (Passera 1994; see Chapter 14), particularly *P. megacephala*, *L. humile*, and *W. auropunctata*, as natural queen dispersal is restricted to only a few metres of the parent colony (Heller and

Table 16.1 Publications of ant management programmes targeting established populations where eradication was declared.

Source	Location	Target species	Size of infestation
Haines and Haines (1978b)	Praslin Island, Seychelles	*Anoplolepis gracilipes*	Area not stated
Thorvilson *et al.* (1992)	Lubbock, Texas, USA	*Solenopsis invicta*	3 ha
Abedrabbo (1994)	Santa Fe Island, Galápagos	*Wasmannia auropunctata*	2 ha
Williams *et al.* (2001)	Kentucky, Virginia, Maryland, Washington DC, Delaware, USA	*Solenopsis invicta*	At least 6 locations, area not stated
Pascoe (2003)	Auckland, New Zealand	*Solenopsis invicta*	1 nest
Hoffmann and O'Connor (2004)	Kakadu, Australia	*Pheidole megacephala*	24 locations, 10 ha
Hoffmann and O'Connor (2004)	Kakadu, Australia	*Solenopsis geminata*	3 ha
Causton *et al.* (2005)	Marchena Island, Galápagos	*Wasmannia auropunctata*	22 ha
Lester and Keall (2005)	Devonport, New Zealand	*Myrmecia brevinoda*	1 nest
Espadaler (2005)	Barcelona, Spain	*Monomorium destructor*	infested cargo
Hoffmann (2009)	Daly River, Australia	*Pheidole megacephala*	5 ha

Gordon 2006; Hoffmann *et al.* 1999). Indeed, numerous projects targeting these species that did not use treatment products that cease reproduction and did not treat all infested areas simultaneously have achieved eradication (Abedrabbo 1994; Causton *et al.* 2005; Hoffmann and O'Connor 2004; Van schagen *et al.* 1994).

Within a treatment programme, the presence of multiple invasive ant species can provide unexpected eradication opportunities, and (temporary) beneficial partnerships between jurisdictions. For example, the first incursions of *S. geminata* and *L. humile* within the state of Queensland were also detected within the *S. invicta* eradication programme in Brisbane, Australia (Vanderwoude *et al.* 2003), which prompted the treatment and probable elimination of both species (C. Vanderwoude and P. Davis, unpublished data). Moreover, one invasive species can also be used to contain another species. On the Tiwi islands, Australia, an infestation of *P. megacephala* that was destined for eradication was temporarily allowed to persist as it was restricting the dispersal of the much harder to eradicate species, *S. geminata* (B. Hoffmann, unpublished data).

Ultimately, the treatment phase is the implementation of the control strategies that were determined in the scoping phase from details within the PRA. We do not elaborate here on the options or methods that have been either successful or unsuccessful, but do note that projects are often faced with unique situations, effectively making them great 'testing arenas' for the trial of new products and techniques (Walters and Holling 1990). As such, we strongly recommend that ant management, and in particular the treatment phase, take an adaptive approach (Box 16.1). Indeed, only two of the nine eradication projects documented to have achieved eradication (Table 16.1) did not have to change treatment strategy (Hoffmann and O'Connor 2004; B. Hoffmann unpublished data).

16.4.4 Post-treatment monitoring phase

Post-treatment monitoring provides quantifiable measures to assess treatment impacts and progression towards achieving programme goals (i.e. eradication or control). The commencement of this monitoring phase is dependent upon the question that needs answering, for example 'Are there any non-target impacts of concern?' versus 'Are there any target ant species persisting?'. Obvious options include monitoring after each treatment, after a certain number of treatments or amount of time, when none of the target species has been noticed for, typically, up to two months (but depending on weather), or after treatment stops. As such, treatment and monitoring will often occur concurrently until treatment ceases.

The most difficult form of post-treatment assessment is monitoring for persistent populations of the target species when the goal is eradication. The difficulty lies not only with the intensity of on-ground labour requirements, but also in that the decision that eradication has been achieved is contestable. Just like any scientific research, the conclusions reached (eradication) should reflect the results (assessments), and the quality of the results is dependent upon the study design (assessment protocols). Yet, despite the fundamental role that assessments for persisting populations play in the completion of an eradication effort, there are no clear standards for how monitoring should be conducted, or when there are sufficient data to indicate that eradication has indeed occurred (FAO 1998). This issue stems solely from the very nature of declaring complete eradication, in that lack of detection does not prove that a species does not exist somewhere (Reed 1996; Regan *et al.* 2006). False negatives, the lack of detection when a species is actually present, can occur for numerous reasons, including inappropriate spatial and temporal sampling methods, too few sample locations, sampling when too hot or cold, incorrect identification, and insufficient techniques. All that can be done to avoid false negatives is to conduct surveys of highest quality that alleviate the issues mentioned here, and provide the greatest likelihood of detection if a species is present.

In attempts to remove the issue of false negatives, numerous mathematical approaches have been developed that can be used to infer eradication including detection probabilities for rare species (McArdle 1990), inferring extinction from observational data (Solow 1993) and optimal economics (Regan *et al.* 2006). With only one exception (Thorvilson *et al.* 1992), and excluding Haines and Haines (1978b) and Williams *et al.* (2001) where no details of the eradications are provided, all programmes that have declared ants eradicated (Table 16.1) have used the simple criterion of finding no individuals of the target species for at least two years. There is no problem with this minimum two-year criterion *per se*; however, assessments within this timeframe can vary greatly in intensity and frequency. Minimum standards are required for sampling or surveying within the two-year timeframe to be able to state that an area has been adequately assessed, and that a species can be declared eradicated. Scientific assessment directly addressing this issue remains non-existent, so development of minimum standards will remain elusive for quite some time. Obviously in the meantime, the greater the survey intensity and frequency and the greater the number of techniques used, the more confident is the declaration of eradication.

16.4.5 Programme completion phase

For programmes aiming at eradication, and where predetermined protocols to satisfy the criteria for a declaration of eradication are met, a species can be considered to be eradicated. The other extreme scenario is that a project's assessments have demonstrated that its management actions have failed, coupled with project administration deeming management to be unworthy of further implementation, resulting in project termination. Alternatively, a programme may have successfully managed a species, but the management of this species is no longer considered a priority, and the programme is terminated. For any of these scenarios, on-ground work can cease.

We argue, however, that these declarations are not where a programme should end. The information gained throughout all phases of the programme, especially of techniques used, results of monitoring treatment efficacy, mistakes made and project costs is highly useful to the establishment and improvement of ant management programmes globally, and as such should be published where it is easily accessible (i.e. not in the form of internal reports). While much literature can be produced during a project, we believe a publication thoroughly reviewing a programme at its conclusion, particularly describing factors influencing successes and failures, would be highly beneficial for others involved in ant management. As such, resources to publish articles should be a component of project budgets.

16.5 Future directions

Invasive ant management techniques will improve as new treatment products, methods, and technologies develop in the future, and as information is shared in the published literature. However, we believe that there are strategies that can be implemented now that require no further demonstration of efficacy and will greatly improve ant management globally. We see the greatest advances in ant management coming from a shift from reactive to proactive management. Proactive management will provide an increased ability to prevent further incursions, a high level of preparedness prior to an incursion being detected, and an increased likelihood of detection of an incursion at an early stage of its establishment before eradication becomes an unrealistic option. The development of a proactive PRA (Section 16.4.1) is only one part of this advancement; there are two additional on-ground management actions that can be conducted: biosecurity and surveillance. In addition, the advancement of knowledge-based protocols and decision-making frameworks equip managers with tools that remove uncertainty and improve programme outcomes. Here we describe what biosecurity and proactive surveillance can contribute to invasive ant management, and highlight what we believe is the current greatest knowledge-based need for ant management; standards for verifying eradication and programme completion.

16.5.1 Biosecurity

Australia has among the most stringent biosecurity protocols at ports-of-entry in the world, yet even these standards are inadequate to detect many incursions (Stanaway *et al.* 2001), and indeed incursions repeatedly occur. Surely the most effective approach to invasive ant management globally is active prevention at not only the location of entry of a country, region, or locale but also at the port-of-exit, to prevent contamination elsewhere. New Zealand is likely the most proactive jurisdiction preventing exotic species incursions with their biosecurity activities now extending into four ports in three surrounding nations (Nendick 2008; Chapter 13, Box 13.1). The logic is that if a port-of-exit is free of unwanted pests, and the goods are pretreated prior to arrival in New Zealand, there should be a much lower incidence of incursion, further reduced risk of a species becoming established, and a significant cost-saving in both quarantine requirements and potential invasive-species management. Indeed, this has proven to be correct with a 98.5% reduction in contamination rates of inbound goods within just 12 months of active management (Nendick 2008). While the New Zealand example may be relatively easy to achieve because the ports of the surrounding Pacific island nations are relatively small, financially dependent on New Zealand, and export relatively few goods, such a reduction in contamination rates could be just as achievable worldwide if ports-of-exit proactively ensured they were free of organisms declared as pests in trading destinations.

16.5.2 Proactive surveillance

Early detection of incursions is among the most critical factors that determine eradication success (Lodge *et al.* 2006; Simberloff 2003). Yet, proactive surveillance for new incursions has been historically rare because it is seen as a large expense. Instead, most governments rely solely on passive surveillance and the discovery and reporting of incursions by the public. Proactivity of governments to monitor high-risk areas would greatly enhance prospects of early detection. Primarily, ports-of-entry must be constantly monitored. This is now standard within New Zealand since the detection of *S. invicta* at Auckland International Airport in March 2001. But proactive surveillance of ports-of-entry alone will not detect those incursions that have established farther. For example, well-established populations of *A. gracilipes* have been inadvertently found in no fewer than 12 locations throughout the east coast of Australia since the first incursion was detected in 2004. Secondary locations regarded as 'high risk', such as imported timber yards, should also be identified and monitored.

16.5.3 Decision frameworks and knowledge-based protocols for declaring eradication

Protocols and decision-making frameworks based on knowledge, especially programme-specific knowledge, are clearly more effective, and more socially and politically acceptable than those utilizing 'gut feeling' or at best, hypothetical modelling. In moving towards a proactive and adaptive approach to ant management, there has been much research into facets that affect ant management, including ant biology, detection techniques, and greater target-specific treatment products, yet there has been no research into improving protocols for declaring eradication of ants. The evaluation and determination of eradication for ants currently lack standards due to the complete absence of knowledge-based measures and decision frameworks. This is in stark contrast to other invasive taxa, such as plants, that have well-developed knowledge-based protocols for determining eradication success (Regan *et al.* 2006 and references therein). There is a desperate need for research into this vital aspect of ant management, as well as an important need for continued knowledge improvement of all aspects affecting ant management programmes.

16.6 Summary

Invasive ant management has evolved greatly in nearly a century, from simply broadcast spraying general contact insecticides with no regard for the target species' biology, to the development of highly specialized chemicals and species-specific biological controls, as well as an increasing incorporation of IPM. The paucity of literature on ant management programmes is very likely to have resulted in lessons and advances being learnt and re-learnt in isolation, and may well have been a major contributor to the failure of past management attempts. We have no doubt that the success of invasive ant management will increase with time, especially if greater importance is given to publishing the details of management programmes, successful or otherwise. However, we believe the greatest improvement of ant management will come from a shift from reactive to proactive management. As such, we have presented a relatively generic five-phase management framework that can be used to guide decisions and actions to ensure success. The pre-emptive phase includes pest risk analyses that help either avoid an incursion or prepare for a rapid response should an incursion occur; the scoping phase brings together all information required to determine the best management option; the treatment phase undertakes tasks required to achieve the management goal; the post-treatment monitoring phase provides quantifiable measures that indicate progress towards the management goal; and the programme-completion phase provides the opportunity to inform land managers of lessons learnt in order to improve ant management globally. Improvements in invasive ant management will also be made from an increase in knowledge-based protocols and decision-making frameworks, especially for declaring eradication.

Acknowledgements

We thank Bart Drees, Richard Harris, and Dan Borg for commenting on the draft manuscript.

Synthesis and Perspectives

Lori Lach, Catherine L. Parr, and Kirsti L. Abbott

In 1866, Ernst Haeckel coined the term 'oecology' in reference to Darwin's description of 'the study of the multifaceted struggle for existence' (Kingsland 1991). The discipline of ecology was thus born and has flourished in the intervening century and a half. Ants as a group have several attributes that are attractive to ecologists. The ubiquity of ants in the terrestrial environment means that they can be studied just about anywhere; their diversity of character states allows specific insight into evolutionary adaptations to changing ecological conditions; and they are abundant and readily observed as they forage, attack, defend, tend, harvest, tunnel, and otherwise go about the tasks that have earned them their place among 'the little things that run the world' (Wilson 1987b).

Ant Ecology synopsis

This volume summarizes many of the advances in ant ecology over the last century, and introduces new findings, ideas, perspectives, and explanations. Understanding the diversity and distribution of ant species and the factors responsible for their success underlies much of this research. Ant ecologists are now equipped with a relatively stable generic phylogeny, largely monophyletic at the subfamily and tribe levels, with which they can address questions of evolution and biogeography (Chapter 1). Despite the high number of undescribed ant taxa, it is clear that biogeographic patterns of ant diversity largely correspond with other taxa (Chapter 2) and ant diversity declines with increasing latitude and elevation (Chapters 2 and 3). Unlike most other taxa, however, explaining patterns of ant diversity requires explicit consideration at both the colony and individual levels (Chapter 3). More data on the diversity and distribution of ants, as well as their phylogenies are especially important for the prioritization of conservation efforts (Chapter 4).

As a taxon, ants interact with an array of other organisms in the community, and in turn ant assemblages are shaped by these interactions as well as the abiotic environment in which they occur. Both as individual workers and as colonies, ants are formidable competitors for solitary invertebrates and other ants (Chapter 5). Many ant species have evolved morphological, behavioural, or physiological traits to facilitate the acquisition of food or shelter from their insect, plant, or microbial mutualists (Chapter 6). Whether from mutualist sources or not, the availability and accessibility of food and shelter strongly affect the structure of ant assemblages (Chapter 7). Habitat disturbance, particularly from anthropogenic sources, can disrupt and transform communities via their effect on food and nesting resources and microclimate (Chapter 8).

The social structure of ants underpins their interactions in the community. Uncovering the repeated shifts from independent to dependent colony foundation has revealed each mode's benefits, constraints, and consequences for other phases of ant life history (Chapter 9). Colony structure is shaped by, and has ecological consequences for, multiple levels including the population, the ant community within which the colony is embedded, and the broader abiotic environment (Chapter 10). Nestmate recognition is the basis upon which a colony acts as a single integrated unit (superorganism), defending itself from foreign intrusion (Chapter 11). As with many other characteristics of ants, the striking diversity of foraging and defence strategies that has evolved has no doubt contributed to their success as a taxon in nearly every type of habitat (Chapter 12).

The study of invasive ants has incorporated all of these aspects of ant ecology and in doing so, has reignited debates about the importance of competition and the evolution of social structure. To date, most invasive ants have their origin in tropical and subtropical climates and have traits that enable them to thrive in association with humans and in disturbed habitats (Chapter 13). Reduced inter-nest aggression and the formation of large colonies appear to be key traits underscoring the success of invasive ants, but the mechanisms by which they are achieved may differ across species (Chapter 14). Invasive ants have multiple direct and indirect consequences for the communities they invade including the displacement of native ants, other invertebrates, and vertebrates, and the alteration of mutualistic interactions (Chapter 15). Improved understanding of ants' social structure and feeding requirements have helped provide major breakthroughs in the development of chemical controls used in invasive ant management (Chapter 16).

Contributions of ant ecology to ecology

Ants, with their diverse morphological specializations, extensive behavioural repertoire, and ability to inhabit a variety of environments, are model organisms for testing ecological theory and studying evolutionary processes. The chapters in this book illustrate the many ways in which ant ecology has contributed, or has the potential to contribute, to the discipline of ecology. For example, because ants are widespread, diverse, easy to sample, and have been the subject of gradient studies for nearly half a century, they make suitable invertebrate candidates for testing some of the competing hypotheses to elucidate the mechanisms underlying diversity patterns (Chapter 3). At the level of communities, work on ants has informed our understanding of the structure and functioning of ecosystems, with studies exploring a range of species interactions and food web patterns. Competition, long considered a key process structuring local assemblages, has received particular attention in ant ecology. Work on ants has highlighted the difficulties of unequivocally demonstrating competition, and illustrated the many factors, both biotic and abiotic, that can modify competition (Chapter 5). Because ants are easily manipulated, monitored, and can remain stable over time, the study of mutualistic interactions involving ants has greatly expanded the approach to studying mutualisms, specifically in measuring the benefits, costs, outcomes, and context-dependency of these interactions and demonstrating their community-wide importance (Bronstein 1998; O'Dowd *et al.* 2003; Chapter 6). Ant–hemipteran relationships in particular have provided excellent opportunities for exploring 'bottom-up' trophic cascades and their role in community structure (e.g. Blüthgen *et al.* 2004b; Ohgushi 2008). Moreover, the extreme diversity of food resources used by ants, and their huge variation in dietary specialization and body size opens up opportunities for testing predictions of metabolic theory based on allometric relations (Chapter 7). Their abundance and accessibility and the ease with which they can be manipulated in the laboratory have made them model systems for studying foraging strategies, in particular stimulating the development of theories in optimal foraging, central place foraging, risk and robustness, and search algorithms (Chapter 12). Other characteristics of ants, including their sensitivity to environmental disturbance, their great functional importance, and the ease with which they can be sampled make them practical and powerful indicators of ecosystem health and change (Chapter 8, Box 8.1).

The trail ahead

Advances in ant ecology are being made all the time. Even in the final few months of compiling this book there have been several key studies that have shifted our perspective and understanding of ant ecology and have uncovered some important aspects of ant behaviour, biology, and evolution. For example, we have recently learned that acoustic information can play a greater role in information exchange than has been previously thought; parasitic butterflies (*Maculinea rebeli*) can use acoustic mimicry to infiltrate and achieve higher status within nests of *Myrmica schencki* (Barbero *et al.* 2009). And, the division of labour, long thought to be key to the ecological success of ants, has also received closer scrutiny. That the allocation of workers to tasks is not necessarily related to their ability to perform those tasks suggests there is still a lot to be learned about the adaptive benefits of the division of labour (Dornhaus 2008). A new subfamily of ant has been discovered (Martialinae), whose evolutionary position and morphology strengthens the dynastic succession hypothesis for ant evolution (Rabeling *et al.* 2008), which proposes that ants evolved and spread from the soil and ground litter to occupy other habitats (e.g. tree canopies), coinciding with the diversification of the angiosperms (Wilson and Hölldobler 2005). And, within the molecular realm, a recent study has revealed that social harmony and cohesion within the colony is maintained by molecular signals that constrain reproduction in worker ants (Khila and Abouheif 2008).

Undoubtedly, additional discoveries are on the horizon that will further advance our understanding of ant ecology. Several broad ecological themes have emerged from contributions to this book that offer inspiration for future research. Below, we explore these future research avenues, take a look at factors likely to play a role in facilitating research, and consider what a changing world means for ants and ant research.

What factors contribute to the ecological success of ants?

That ants are ecologically successful is seldom disputed. They have persisted as a clearly defined taxonomic family for 120 million years, number as many as 10,000,000,000,000,000 individuals alive at a time, are the most diverse of the social insects, and occupy a variety of habitats across a wide geographic range (Wilson 1987a; Hölldobler and Wilson 1994; Grimaldi and Engel 2005). Identifying the factors that have contributed to the ecological success of the Formicidae and their relative importance remains a long-standing goal, and one that has spawned research at every level of ecological organization. We restrict our focus here to what we consider to be some of the most important and promising research directions in the immediate future.

Resources and relationships

In ecology there has long been debate about whether 'top-down' versus 'bottom-up' factors control assemblages and ecosystems (e.g. Hunter and Price 1992; Power 1992; Worm *et al.* 2002). For ants, much attention has focused on bottom-up control, yet there is still much to be learned about how resources, and particularly certain nutrients, structure ant assemblages in space and time. The role of limiting nutrients for ants is especially unclear. We've recently learned that the availability of sodium influences the geographic distribution of ants (Kaspari *et al.* 2008); other potential nutrients, such as phosphorus, warrant a closer look. Future studies that span a range of scales across environmental gradients of productivity, rainfall, and habitat types are likely to be especially informative. Carbohydrate-rich resources, particularly honeydew, have been hypothesized to be key for fuelling dominant ants (Blüthgen *et al.* 2004b; Davidson 1998; Davidson *et al.* 2003), but this requires confirmation across regions, and a great deal remains to be explored regarding why some sources are preferred over others, how carbohydrates affect behaviour, and how the availability of carbohydrates affects colony performance. Furthermore, the role of endosymbiotic micro-organisms in, for example, facilitating nitrogen uptake, nutrient balance, and food detoxification, warrants greater consideration; for example, how does the ability to digest food affect niche differentiation? Relationships can also be investigated at broader scales; is the composition of gut flora reflected in phylogenetic relationships, or by biogeographic area? And, given the vast number of interactions ants engage in, it would be useful to look more broadly at how ants compete for resources not only among themselves, but with other taxa in their environment; under what situations can they compete successfully and when do they lose?

From the opposite perspective, the role of ants as resources for other organisms and the extent of top-down control for regulating ant colonies and shaping assemblages have received little attention. We know very little about predation pressure and its effects on either ant assemblages or colonies although the handful of studies on predation indicate that it can influence the distribution, abundance, and behaviour of ants (Gotelli 1996), and mortality via predation at the colony founding stage can be substantial (e.g. Nichols and Sites 1991). In addition, invasions by the apparently unpalatable *Solenopsis invicta* and *Linepithema humile* have demonstrated how changes in ant assemblages can have significant effects on lizards (Chapter 15); the potential for consequences to other myrmecophages deserves exploring. The importance of managing ant invasions further necessitates a better understanding of how parasitoids and disease can affect colony performance (Chapter 16).

Trade-offs

Life-history trade-offs are considered fundamental to understanding the structure of ecological communities and as such have been central to theories addressing coexistence (see Kneitel and Chase 2004, though see also Hubbell 2001 for discussion on neutral theory). Ants have many strategies for accomplishing the tasks necessary to maximize their fitness, and some of these strategies require traits that likely preclude the ability to efficiently perform another task. These trade-offs are evident at several levels of ant ecology and include those between discovery and dominance ability, benefits obtained from other organisms and the costs of acquiring them, proximity to a food resource and nest permanency, independent versus dependent colony founding, small versus large colony size, specialization versus flexibility, and investment in morphological versus chemical defence, to name a few. Still other trade-offs have probably yet to be recognized. Questions for future research include: How do behavioural and morphological traits interact? What trade-offs are important for structuring inter- and intra-colonial behaviour and division of labour? How frequently and under what circumstances can trade-offs be broken, what are the consequences to the species and the community of breaking the trade-off, and what trade-offs are important at different spatial scales? (Kneital and Chase 2004).

Underpinnings of colony structure and foundation

Though great strides have been made in understanding the varying ecological and evolutionary forces that underlie colony structure and modes of foundation, many other questions remain. For example, during dependent colony foundation, queens vary in size, potential fertility, and level of relatedness with respect to workers (Chapter 9). What are the mechanisms and constraints behind this variation, and what conflicts are expected to arise when more gynes are produced than the number of future daughter colonies? Given that colony foundation influences subsequent growth and reproduction, how tightly are colony foundation and colony structure linked? In elucidating the interdependency of the two, there is a need to integrate genetic, evolutionary, and ecological research, and test emerging hypotheses with empirical data, especially from the field. The advances made by seeking answers to these questions will undoubtedly help to resolve questions about the evolutionary origin of supercolony formation (see Chapter 10 and Box 14.1), and the evolution of eusociality itself.

What factors will facilitate our understanding of ant ecology?

The diverse and exciting range of research foci for the future offer plenty of challenges to aspiring and established ant ecologists. As with all ecology, rigorous experiments and appropriately framed and tested hypotheses will be the cornerstone of advancement. Given the longevity of ant colonies and the variety of niches and habitats ants occupy, long-term and large-scale studies have the greatest potential to make significant contributions to advancing the field. We see advances in ant research being significantly facilitated by the following:

Resolution of taxonomic uncertainties

That about half of all ant species are currently undescribed clearly demonstrates the need for an ongoing and sustained effort in species discovery and delimitation. Taxonomic uncertainty hampers identification of potentially harmful species, hinders comparisons across studies, and slows advances in species-level ecology. While some progress is being made with ant macroecological studies, improved species-level taxonomy would expand the scope and resolution of studies (presently limited to generic level). Without reasonably complete species lists, we cannot even begin to look more broadly at ecological questions relating to distribution and range sizes. Although the potential is tantalizingly close for North America and Europe, for the vast majority of the world it remains a distant possibility. Conservation efforts will also benefit from a more inclusive and refined formicid taxonomy. A species-level understanding of ant biology and behaviour will help to identify species of conservation concern and appropriate and realistic action plans.

Integration of taxonomy with robust phylogenies

Continued research into a well-supported phylogeny is critical for understanding the origin, dispersal, radiation, and spread of taxa across biogeographic regions. Such a phylogeny will inform hypotheses that attempt to explain gradients in animal communities and the evolution of mutualistic and other interactions, and will allow for more informative comparative analyses (e.g. the evolution of foraging and defence strategies and metabolic theory).

A focus on the neglected masses

Ant ecology has traditionally focused on the conspicuous ground-foraging members of the assemblage. Yet we currently have limited information and relatively little appreciation for what goes on below and high above ground, despite the many thousands of ant species and the substantial portion of ant biomass inhabiting these realms. Undoubtedly, hypogaeic, arboreal, and otherwise more cryptic species play important functional roles, shaping not only ant assemblages but the broader community. A closer look deep in the soil, within the humic layer, and high in the tree canopies is likely to reveal many novel behaviours and adaptations. The recent discovery of the new subfamily, Martialinae, illustrates the potential for surprises the neglected masses hold in store.

Information sharing

Information technology has already provided massive benefit to scientists of all disciplines, making instant communication and data sharing possible around the globe. Ease of communication among scientists is likely to facilitate collaboration across disparate fields resulting in new (inter)disciplines. Online databases, journals, ant collections, and photo libraries, and informal electronic 'coffee breaks' between scientists will continue to increase the rate at which information is shared, and the rate at which we build upon new discoveries.

Development and utilization of new tools and techniques

In addition to the internet, ant ecologists and other scientists have already benefited from advances in molecular ecology, population genetics, genome sequencing, stable isotope analysis, imaging techniques, and interactive identification keys. Few could have predicted such rapid development and widespread utilization of these tools even 15 years ago, and we anticipate similar advances and their adoption in the years to come.

Discovering new applications of ant ecology and communicating them to non-scientists

Discoveries from studying ant ecology and its evolution, such as the structure, function, and utility of antimicrobial agents produced by ants and their symbionts, could provide new tools for improving agriculture and human health. Communicating the realized and potential human benefits of this knowledge to funding bodies, politicians, and society-at-large may enhance future funding and research opportunities.

What does a changing world mean for ant ecology?

Although change is expected over evolutionary and ecological time, the speed and magnitude of global change we are currently experiencing is enormous. We see three main drivers of global change that yield questions for ant ecology research.

Climate change

Given that the abiotic environment affects so many aspects of ant ecology – the distribution and behaviour of ants, their life cycles, their relationships with other organisms – the anticipated (and already experienced) changes in temperature, precipitation, and sea level associated with global climate change are likely to have a large impact on ant assemblages. At a global scale, climate change is anticipated to increase temperatures (between 1.1°C and 6.4°C during the twenty-first century), and alter rainfall patterns and intensity (IPCC 2007). Shifting distributions and phenologies in response to climate change have already been observed for other taxa (e.g. Walther *et al.* 2002). Ants in montane habitats, on low-lying islands and lowland rainforest may be at particular risk (e.g. Colwell *et al.* 2008). Baseline information on ant species' physiological tolerances is urgently needed, particularly in potentially vulnerable habitats (e.g. mountain tops). Importantly, this needs to be linked to information on specific habitat and resource requirements and species interactions.

Biological invasions

Global trade will continue to provide ample opportunity for ants and other taxa to be transported from their native ranges to other suitable habitat. Invasive species, biotic homogenization, and the formation of novel ecosystems (e.g. McKinney and Lockwood 1999; Hobbs *et al.* 2005) will become increasingly common. How will native ants respond to incursions of exotic flora and fauna? Can we identify new potential ant invaders before they become invasive? How successfully can we apply our knowledge of ant ecology to curtail future pest outbreaks or mitigate their effects? How can we control species that have already become established? Or can we?

Increased human population

Human population, currently 6.7 billion, is anticipated to reach 9 billion by 2040 (U.S. Census Bureau, Population Division 2008). Concomitant urbanization, agricultural intensification, deforestation, and increased pollution will allow for complex research into landscape and local disturbance effects on ants. Improved understanding of the functioning of ecosystems and the mitigation of human-induced disturbances will be facilitated with the use of ants as indicators of ecosystem health, and the refinement of ant functional groups for use at regional and local scales. Given the array of functions ants perform, and diverse range of mutualisms they are involved in, ants are also likely to be increasingly recognized for their role in ecological restoration. The contributions ants make to ecosystem services requires more detailed quantification.

Conclusion

In tackling the many research opportunities and questions we have posed, multiple tools and approaches will be called upon, and both reductionist and holistic lines of inquiry will be needed. As these questions are answered, new questions will emerge and myrmecology will develop in new directions or revisit familiar ground with new lenses of interpretation. Whatever the research future holds, one thing is certain: ants will continue to fascinate, inspire, and frustrate, and will surely endure as one of the most successful and abundant animal groups to ever roam the planet.

Glossary

active constituent: A chemical in a bait matrix that mediates ant control; includes direct toxins and juvenile hormone analogues.

aculeate wasps: A group of wasps in which the egg-laying ovipositor has been modified to form a sting; most closely related group to ants.

adaptive management: A natural resource management approach that incorporates systematic experimentation and monitoring to compare alternative management actions and update these actions accordingly.

agricultural intensification: generally associated with crop specialization, increasing mechanization, and generalized use of agrochemicals and other external inputs in the crop field.

agroforestry: Agricultural systems incorporating trees.

alate: Winged reproductive; can refer to either queens or males.

allometry: A non-linear scaling relationship between the size of an organism and the size of any of its body parts. For example, head width increases more than leg length as body size increases. The opposite is isometry, where proportions between body parts stay constant across a range of body sizes.

altruism: Lifetime improvement of a beneficiary's reproductive success at the expense of the lifetime reproductive success of the altruist.

ant garden: Ant nest consisting of epiphytic plants that profit from the association with the ants.

ant mosaics: Spatial patchworks of two or more dominant ant species that have non-overlapping territories.

antennation: Investigation of an encountered ant with the antennae; the interaction allows nestmate recognition at contact or at very short distance.

arboreal ants: Ants that live and forage above the ground in trees and other vegetation.

assemblage: A taxonomic subset of a community.

bait matrix: Substance in which active constituents are delivered to ants. Matrices can be liquid (predominantly water, but sometimes other attractive substances such as sugar or honey) or solid (typically corn grit or fishmeal).

Bergmann's rule: The hypothesis that size of individuals (or for social organisms, colonies) increases with elevation and latitude.

biodiversity: Variation in life on Earth at all levels of biological organization (genetic, species, ecosystem).

biodiversity hotspot: An area of significant biodiversity containing at least 0.5% or 1,500 species of global vascular plants as endemics, and having lost at least 70% of its primary vegetation (see: www.conservation.org). Recently Conservation International has included areas of high irreplaceability but low vulnerability as hotspots.

biogeography: Study of the distribution of biodiversity in space and time, and includes physiological, morphological, and genetic perspectives.

bioregion: See **ecoregion.**

bivouac: Temporary nest structure formed by army and driver ants. The structure consists of a mass of tightly locked individuals that protects the queen and the larvae within it.

brachypterous queen: Queen with short wings that are ineffective for flying.

brood: Immature ant individuals, including eggs, all larval stages and pupae.

budding: A synonym for **dependent colony foundation**, originally meant for polygynous species whose nests remain interconnected; often used interchangeably with the term fission.

carton nest: Nest structures built actively by ants from different substrates such as detritus, earth, plant-fibres or trichomes, or silk collected from spiders.

caste: A group of female individuals distinguished from another group within the colony, by function or morphology. See also **functional caste definition** and **morphological caste definition**.

caste determination: The process by which embryological development of a female ant is determined as either worker or queen; also see **genetic caste determination, environmental caste determination**, and **hybridization-mediated caste determination**.

central place foraging: A means of collecting food in which the forager returns to a central place to deliver food (e.g. in nesting birds or in most social insects).

cheater: A species that takes advantage of one or more benefits exchanged between two mutualists at a cost to one or both of the mutualists.

chemical insignificance: Absence or low quantities of cuticular hydrocarbons that allows acceptance in a social insect colony; characteristic of newly emerged individuals and of some social parasites.

clade: A monophyletic group; a group of biological taxa that includes a single common ancestor and all its descendants.

claustral: A mode of independent colony foundation in which the queen has sufficient metabolic reserves to raise her first brood of workers without a need to forage outside the nest.

cognitive map: A map-like representation of spatial locations stored in the brain.

colony: Eusocial society of cooperating individuals of the same ant species.

colony closure: Inaccessibility of a colony to organisms other than members of the colony.

colony fusion: Merging of two mature colonies into one.

colony odour: The colony specific blend of non-volatile substances (mostly hydrocarbons) found on the cuticle of individuals and shared among all colony members.

colony size: Number of individuals in a colony, mainly depending on the number of workers in the colony.

colony structure: The caste, demographic, genealogical, and spatial make-up of a colony.

community: Interacting, coexisting assemblages of organisms.

co-occurrence analysis: A method for testing for non-random patterns of species occurrences.

correlated random walk: A random walk (an individual turns by a random angle after each step of a given distance), with the added aspect that turning angles are normally distributed around the previous direction of the movement.

crown group ants: Clade composed of the most recent common ancestor of all living ants and their descendants.

cue: A trait that can be used in communication to extract relevant information although it has not evolved for that purpose (see also **signal**).

cuticular hydrocarbons: Lipids found on the cuticle. In social insects, a blend of long-chain hydrocarbons forms the specific colony odour and play role in protecting against insects from desiccation and in communication.

deforestation: Removal of trees (sometimes complete extraction) from forested areas; usually by logging and/or burning.

dependent colony foundation (DCF): Initiation of a new colony by a group of nestmate queen(s) and workers that leave the maternal nest together; accordingly queens never go through a solitary stage; dispersal is on the ground and thus short ranged.

diaspore: A plant dispersal unit consisting of a seed plus any additional tissues.

diploid male vortex: Reduced allelic diversity in a population at the sex-determining locus (e.g. resulting from inbreeding and population size reduction); reduced heterozygosity at the locus increases production of nonviable, diploid males which reduces population growth rate and thus further reduces allelic diversity at the sex determining locus in the population.

discovery–dominance trade-off: The inverse relationship between the ability of a species to discover food and its ability to dominate resources (either behaviorally or numerically).

domain: The geometric area of interest, may be spatial (a mountain range, the globe, a continent) or temporal.

domatia: Plant structures including hollow thorns, stems, and leaf pouches in which colonies of ants reside.

dominance–impoverishment rule: A relationship between ant species richness and dominant species in a community. The fewer ant species in a local community, the more likely it is to be behaviourally dominated by one or two species with large, aggressive colonies. In many studies, however, the direction of this causality has been reversed, and emphasis has been placed on the effect dominants have on species richness, not vice versa.

ecological gradient: A pattern, usually in richness, abundance, or body size, along an ecological axis. Common ecological axes are temperature, elevation, latitude, and net primary productivity.

ecoregion: A large unit of land or water containing geographically distinct communities associated with particular combinations of environmental conditions.

ecosystem engineers: Organisms that directly or indirectly modify ecosystem properties (e.g. the availability of resources for other species) by causing physical changes in biotic or abiotic material.

ecosystem services: Benefits humans derive from ecosystems, including provisioning services such as food and water, regulating services such as regulation of floods, drought, land degradation, and disease, supporting services such as soil formation and nutrient cycling and cultural services such as recreational, spiritual, religious, and other non-material benefits.

elaiosome: Lipid and protein-rich fleshy structures attached to seeds of many plants reliant upon ants for seed dispersal, consumed by many ant species.

endemic species: Species ecologically unique to a specifically defined place and not found elsewhere.

environmental caste determination: Exclusively non-genetic caste determination.

epigaeic ants: ground-active ants; forage primarily on the ground.

epigenetic: Changes in gene expression not involving changes in the underlying nucleotide sequence.

eradication: The intentional extirpation of geographically discrete populations of a species, irrespective of whether other discrete populations still remain within the landscape.

ergatoid queen: Under the functional caste definition, a queen with worker-like external morphology (permanently lacking wings and with a simplified mesosoma and fused sclerites); distinct from dealate queen, which is a winged queen that has shed her wings. A 'multi-purpose' ergatoid queen can either function as reproductive or perform worker-like sterile tasks, and is similar to workers in size. 'Sole-purpose' ergatoid queens function only as reproductives and may be similar to workers in size or extremely dimorphic. Identical with gamergate under the morphological caste definition.

ergatomorphic queen: see **ergatoid queen.**

eusociality: True sociality defined by reproductive division of labour between female castes, cooperative brood care, and the presence of workers of a later generation to the queen(s).

exotic species: species not native to an area of interest.

extent: The domain or scope of an analysis. For example an analysis of New World terrestrial latitudes (North and South America) has a scope of 148° of latitude (ranging from 55°S to ~83°N) or ~16,500 km (see also **grain**).

extinction: The death of every member of a population, an entire species or higher taxon from some area of interest.

extrafloral nectary: A structure on a plant that produces nectar in a location other than within a flower.

facultative polygyny: Polygyny that is not essential for a colony to remain viable. Colonies that display facultative polygyny can switch between monogyny and polygyny.

female calling: The emission, typically of pheromones, by a reproductive female in order to attract males to her.

fission: A synonym for dependent colony foundation, originally meant for monogynous species where colonies reproduce by dividing into two equal parts; often used interchangeably with the term budding.

fitness: The reproductive rate of a genotype.

food bodies: Lipid and protein-rich structures found on some species of plants; believed to have evolved in symbiotic interactions with ants.

formicoid: A well-supported clade that includes three widespread and species-rich ant subfamilies – Dolichoderinae, Formicinae, and Myrmicinae – as well as army ants and relatives (dorylomorphs), bulldog ants (Myrmeciinae), big-eyed arboreal ants (Pseudomyrmecinae), and a few other smaller groups.

functional caste definition: Definition of caste by role in the colony. Under the functional caste definition a queen is a mated individual laying both male- and female-destined eggs, independent of her external morphology, and a worker is an unmated individual that may lay male-destined eggs; compare with **morphological caste definition**.

functional monogyny: Presence of more than one mated female capable of reproduction in a colony, only one of which lays eggs.

fungus garden: The cultivation of basidiomycete fungi as a food source by ants of the Attini tribe.
gamergate: A mated and egg-laying individual with worker-like external morphology; identical with ergatomorphic queen under the functional caste definition.
genetic caste determination: Caste determination encoded in nucleotide sequences; possibly modulated by environmental influences.
Gestalt model: Establishment of a common colony odour by sharing the recognition cues (by trophallaxis and/or allogrooming).
Gondwana: Also called Gondwanaland; southern hemisphere supercontinent in the Palaeozoic Era (248 to 545 Mya), formed with the break up of Pangaea; subsequent splitting resulted in the origins of the modern continents (Antarctica, South America, Africa, Australia-New Guinea) as well as Madagascar, New Zealand, the Arabian peninsula, and the Indian subcontinent.
grain: The resolution of an analysis, can range from 1 m^2 to 100 m^2 for local analyses of ant diversity up to 100 km^2 to 100,000 km^2 for regional or hemispheric scale analyses. Also see '**extent**'.
granary: A nest chamber in seed-harvesting ants in which the ants store plant seeds they have collected.
granivore: An organism that consumes seeds as a main food source.
group selection: Natural selection that works to the advantage of a group of not necessarily related individuals.
gyne: Broad term for a female reproductive; queens are functioning gynes.
habitat disturbance: Any event that removes biomass from a habitat.
habitat specialist: Species that can live and reproduce only in a particular type of habitat.
habitat transformation: Any event that reduces available resources or changes the microclimate or structure of a habitat.
haplodiploid: See **male haploidy**.
haplometrosis: Foundation of a new colony by a single queen.
honeydew: The carbohydrate-rich excreta of hemipterans that feed on plant sap, often collected and fed upon by ants.
hybridization-mediated caste-determination: Genetic caste determination in hybrids fixed in populations, workers bear the genomes of both parental species, but queens that of only one.
hypogaeic ants: Ants that forage and live underground.
inbreeding: Mating between related individuals.
inbreeding depression: Decreased population vitality in terms of growth, survival, or fecundity following inbreeding.
inclusive fitness: The fitness of an individual, taking into account not only that individual's own success in passing on genes to the next generation, but also the success of all its kin, that is, those bearing some portion of the same genotype; see also **kin selection** and **fitness**.
independent colony foundation (ICF): Initiation of a new colony by a solitary queen who raises her first brood of workers without the help of workers from the maternal nest, generally after dispersal flight and mating. See also claustral and non-claustral.
inquilinism: Extreme parasitic state where the parasitic species lacks a worker caste and is thus fully dependent on its host's workers to complete its life cycle.
insect growth regulator: Chemical used in ant baits that halts normal development of insects. They typically affect all stages of development including eggs, larvae, pupae, and adults. In reproductive individuals ovaries are compromised and production of eggs becomes limited or stops altogether. Includes juvenile hormone analogues.
integrated pest management (IPM): Pest control strategy that uses a range of complementary techniques to manage pest populations. It aims to also reduce reliance on chemicals. Techniques used include mechanical and physical devices, genetic, biological, and cultural control, as well as chemical control.
interspecific competition: Competition between and among individuals of different species.
intracolonial relatedness: Degree of average relatedness across all individuals of the colony; extreme values are 0.75 under monogyny–monandry and 0.0 in a spatially extensive supercolony.
intranidal: Within the nest.
intraspecific competition: Competition between individuals of the same species.
invasive: Those ant species that demonstrate ecological, environmental, or economic impacts.
invasive alien species: A species that has been introduced to an area and is invasive; distinct from alien, exotic, introduced, tramp, and other terms that only denote origin.

IUCN Red List of Threatened Species: A list of taxa deemed threatened by IUCN criteria, it provides taxonomic, conservation status, and distribution information (http://www.iucnredlist.org).
juvenile hormone analogue: Synthetic version of a biological juvenile hormone. These hormones regulate many aspects of insect physiology. The juvenile hormone analogues used in ant baits most commonly halt development of workers before becoming adults (at various stages).
keystone species: A species whose impact on its community or ecosystem is disproportionately large relative to its abundance; also a species that regulates local species diversity in lower trophic levels.
kin selection: Natural selection mediated by interactions between relatives. Under Hamilton's formulation, an individual's fitness is divided into personal reproductive success and the effects of the individual on the reproductive success of others weighted by relatedness, a formulation under which reproductive self-sacrifice is selected for if the transmission of genes identical to genes in the altruist by relatives is increased sufficiently.
leaf litter: The layer of decaying leaves, rotting wood, and other organic material that accumulates on the ground, typically under woody vegetation.
macrogyne: The larger queen morph under queen dimorphism.
male aggregation: The mating arrangement whereby males from different nests assemble in a group for queens to join them for the purpose of insemination.
male haploidy: Genetic system in which males are haploid; in Hymenoptera this results from normal males arising from unfertilized eggs and females from fertilized ones. Males are thus a genetic subsample of their mother which results in a relatedness asymmetry in the colony. Generally seen as a synonym for haplodiploidy.
mesosoma: in the Apocrita, the fused thorax and first abdominal segment
metapleural gland: Gland on the alitrunk that is peculiar to ants, although not all ant species possess it. Produces and secretes antibiotics that can prevent the growth of bacteria and fungi on the ants and inside their nest.
microgyne: The smaller queen morph under queen dimorphism.
monandry: Mating of a queen with just one male.
monodomy: The use of a single nest by a colony. Serially monodomous colonies construct more than one nest but live in only one of them at a time.
monogyny: Possessing a single queen. Primary monogyny is monogyny resulting from haplometrosis. Secondary monogyny results from pleometrosis in which only one of the founding queens persists.
monophyletic: See **clade**.
morphological caste definition: Definition of caste under which a queen is morphologically distinguishable from a worker; not applicable to all ant species; to be distinguished from the **Functional caste definition**.
morphospecies: Species that are distinguished solely on the basis of appearance.
multicoloniality: Colony structure characterized by distinct colony and nest boundaries within a population.
multi-level selection: Selection operating at several levels including individual, patriline, matriline, nest, colony, and population.
mutualism: An interaction between two species that is mutually beneficial to the fitness of each. Obligate mutualisms are required for the persistence of a species, facultative mutualisms are not required for the persistence of a species.
myrmechophile: An organism that lives in association with ants.
myrmecochore: A plant that depends in whole or in part upon ants for seed dispersal.
myrmecochory: Ant dispersal of seeds.
myrmecophyte: A plant that permanently hosts a colony of ants in specialized domatia and usually provides food in the form of food bodies or extrafloral nectar.
natural enemy: A predator, parasite, parasitoid, pathogen, or herbivore adversely affecting the species or group of species being described.
nest: Dwelling of a colony.
nest chamber: Cavity within a nest, distinct from tunnels.
nest odour: Mixture of all substances found inside or at vicinity of the nest, including the colony odour, which allows orientation to and recognition of the nest.
nestmate recognition: Recognition of individuals of the same nest sharing a common colony odour allowing altruistic behaviours towards nestmates and discrimination and rejection of non-nestmates.
net diversification: The difference between the rate of extinction and the rate of speciation.
net primary productivity (NPP): A measure of the grams of carbon per unit area per unit time sequestered.

non-claustral: Mode of independent colony foundation where the queen has limited metabolic reserves and needs to forage outside the nest regularly to feed her first brood of workers.

null models: Models that exclude some process of interest in order to consider the pattern expected in its absence. A null model of diversity along gradients is used to examine the pattern of diversity that would be expected in the absence of climatic drivers of diversity.

nutrient balance: Ratio of nutrients (e.g. carbohydrate–protein balance) contained in food. Food that is ingested that diverges from the optimum for an organism requires the excess nutrients to be excreted so that the balance is achieved.

nutritional upgrading: Enhancement of nutritional value of food sources by endosymbiotic bacteria either via recycling of excretions or production of higher quality nutrients from lower quality ones, e.g. production of essential amino acids from non-essential ones.

obligate polygyny: Polygyny that is necessary or essential for the species to ensure full viability and reproduction success of a colony.

oligogyny: A special case of polygyny in which colonies possess a low and limited number of queens that are intolerant of each other and frequently occupy different parts of the nest.

omnivore: A consumer feeding on both plant- and animal-based diets; a mixture of primary and secondary consumption.

parabiosis: Association of two or more ant species in the same nest, involves cooperation of the two colonies but separation of brood.

paraphyletic: A group of biological taxa that contains a common ancestor, but does not contain all descendents of that ancestor.

parataxonomist: A biodiversity collection and inventory specialist, usually recruited from local areas and trained by professional biologists, who typically collects specimens, prepares them and sorts them into morphospecies.

pest risk analysis (PRA): Combination of risk assessment and risk management for a particular pest threat. Pest Risk Analysis aims to directly link the severity of the risk to management objectives and priorities.

pheromone: A chemical substance (or a specific blend of substances) that mediates communication between members of the same species. By contrast, allelomones mediate communication between members of different species (and are further distinguished in allomones, kairomones, and synomones). Pheromones are called releasers if they trigger an immediate behavioural response in the receiver; and primers if they cause physiological changes in the receiver that can eventually result in a behavioural response.

phragmosis: Blocking of the entrance of the nest by a body part, usually the head; usually performed by a morphologically specialized worker subcaste.

phylogenetic diversity: A measure of biodiversity based on the length of evolutionary pathways that connect a given set of taxa (sum of the length of branches of a phylogeny).

phylogeny: The evolutionary development and history of a particular taxonomic group, usually a species or higher taxonomic grouping; constructed using molecular sequencing data and morphological data matrices.

physogastry: Enlargement of the gaster by stretching of the intersegmental membranes, allowing for increased ovarian activity.

pleometrosis: Foundation of a new colony by two or more cooperating queens.

polyandry: Mating of a queen with more than one male.

polydomy: The use of two or more spatially separated nests by one colony.

polyethism: Functional specialization of different workers of the colony leading to a division of labour among workers; may be as a function of age (age polyethism) or permanent, involving specializations of individuals over their entire adult life, sometimes but not always accompanied by morphological adaptation, see also **caste**.

polygyny: Possessing more than one queen, distinct from oligogyny in that several to very many queens mix freely within the nest. Primary polygyny results from pleiometrosis. Secondary polygyny develops from a colony founded by a single queen that later adopts other queens. Serial polygyny is a derived strategy under monogyny whereby after the queen's death she is replaced by a daughter-queen to avoid colony dissolution, resulting in temporary coexistence of worker offspring of different queens.

polymorphism: Substantial differences in the size and/or shape of non-gyne females in the same colony; see also **worker subcastes**.

population viscosity: A measure of the genetic isolation-by-distance effect across colonies of a population.

post-pharyngeal gland: A gland positioned in the head, mainly found in Formicinae, that produces salivary secretions and functions as a reservoir of hydrocarbons that are shared between individuals by trophallaxis.

propaganda pheromone: A volatile substance used by social parasites to influence workers of a host ant colony.

proventriculus: Valve-like muscular and sclerotized structure separating the crop from the midgut. The proventriculus is important in ants feeding largely on liquids since it dams the efflux of liquid from the crop into the midgut and therefore allows storage of large volumes of liquid in the crop.

queen: A mated female reproductively active and able to produce offspring of both sexes; frequently queens differ in their external morphology from workers, mainly by wings, and by larger mesosoma size.

queen dimorphism: Existence of two distinct queen phenotypes in a species.

Rapoport's rule (also Rapoport pattern): The hypothesis that there is a positive relationship between range size and latitude.

rare species: Species with low abundance that are infrequently encountered; may have a widespread or localized distribution.

relatedness asymmetry: The state of one individual being more related to another than the other is to it, e.g., under male haploidy, a father is twice as related to a daughter as she is to him, and sisters are more related to each other than their brothers are to them.

replete: A worker with her crop full of sweet liquids for provision; an example of permanent polyethism, not present in all species.

selective logging: A type of logging where only certain, usually economically important, tree species are removed from forests.

sex allocation: Proportion of investment in either male or female reproductive function, generally measured as the proportion of investment into producing males to queens.

sex determining locus: The genetic locus that governs sex determination whereby heterozygosity results in females but homozygosity in non-viable, diploid males; viable males are haploid. Not yet identified in ants.

sex ratio: Proportion of reproductive individuals of both sexes produced.

signal: A trait that evolves in a signaller to provide information to a receiver in a way that induces a change in the behaviour of the receiver for the benefit of the signaller (see also **cue**).

silvopastoral system: Pasture or grazing systems that include trees.

social parasite: Ant species that live in the nest of another genetically distinct ant species almost exclusively producing sexual offspring while remaining reproductively isolated from their hosts.

soil bioturbation: Movement of soils; caused by ants through the formation of mounds, subterranean galleries and chambers, and the movement of soil particles along the soil profile.

soldier: A distinct worker phenotype that is functionally specialized for the role of colony defence.

speciation: The evolutionary process by which one species forms a new species.

species diversity: Species richness weighted by some measure of abundance such as number of individuals or biomass. Many people use the term species diversity when they are actually referring to species richness.

species richness: Number of different species in a sample, location, region, habitat, ecosystem, or other biological unit.

stable isotope: Any of the different types of atoms of the same chemical element that differ in atomic mass and do not decay.

stable isotope analysis: The measurement of the ratio of two stable isotopes of an element typically to estimate the relative contributions of multiple resources to a consumer or the consumer's trophic position.

stem group ants: All the taxa in a clade preceding a major cladogenesis event. For ants a group containing all organisms more closely related to ants than to any other extant taxa; can refer to extinct taxa that are outside the crown group but that are inferred to be more closely related to ants than to any other living aculeate wasps.

supercolony: An ant colony with multiple queens integrated harmoniously over a large area. Individuals freely mix among nests across spatially separate parts of the colony.

symbiosis: The close and often long-term relationship between different species that is frequently, but not always beneficial to one or both species.

systematics: Study of relationships among taxa through time involving the systematic classification of organisms and the evolutionary relationships among them.

taxon cycle: Sequential stages of expansion and contraction of the ranges of species; often linked with shifts in ecological distribution.

taxonomy (Linnaean): A method of classifying living things; originally devised by Carolus Linnaeus. It uses a combination of a genus name and a single specific characterizing word (species name) to uniquely identify each species.

template: Internal (neural) representation of the colony odour believed to be learned just after emergence (sometimes during the preimaginal period) and constantly updated.

tending: The collective actions of ants collecting honeydew and closely associating with honeydew-producing herbivorous hemipterans.

threatened species: Species that are at risk of extinction due to human activities that directly or indirectly affect their life or their habitat.

trophallaxis: The exchange of either regurgitated or excreted liquid food that occurs among members of an ant colony.

trophic egg: Egg laid for later consumption by another individual in the colony; frequently understood as a term for eggs that also are inviable.

trophic level: A quantitative description of the position a species or group of species occupies in a food chain, with primary producers being the first trophic level, herbivores the second, primary predators the third, and secondary predators the fourth.

trophobiont: see **trophobiosis**.

trophobiosis: A symbiotic association between organisms where food is obtained or provided. The provider of food in the association is referred to as a trophobiont.

umbrella taxon: A taxon for which protection also confers protection on coexisting organisms.

unicoloniality: Colony structure characterized dependent colony foundation, inbreeding, and the lack of distinct colony boundaries within a population.

worker: Female ant, typically incapable of producing offspring, and differing morphologically from the queen by absence of wings and frequently smaller mesosoma size.

worker policing: Destruction of an egg laid by one worker by another worker, to avoid a reduction of relative fitness by the egg-destroyer.

worker subcastes: Extreme case of worker polymorphism in that worker phenotypes differ in a strong deviation from isometry; a morphological adaptation to permanent polyethism.

References

Abbot, P., Grinath, J., Brown, J., Peeden, E., Erickson, D., and Billick, I. (2008) Insect herbivore stoichiometry: the relative importance of hostplant. *Ecological Entomology 33*, 497–502.

Abbott, K.L. (2005) Supercolonies of the invasive yellow crazy ant, *Anoplolepis gracilipes*, on an oceanic island: forager patterns, density and biomass. *Insectes Sociaux 52*, 266–73.

——(2006) Spatial dynamics of supercolonies of the invasive yellow crazy ant, *Anoplolepis gracilipes*, on Christmas Island, Indian Ocean. *Diversity and Distributions 12*, 101–10.

——and Green, P.T. (2007) Collapse of an ant-scale mutualism in a rainforest on Christmas Island. *Oikos 116*, 1238–46.

——Sarty, M., and Lester, P.J. (2006) The ants of Tokelau. *New Zealand Journal of Zoology 33*, 157–64.

—— Greaves, S., Ritchie, P., and Lester, P. (2007) Behaviourally and genetically distinct populations of an invasive ant provide insight into invasion history and impacts on a tropical ant community. *Biological Invasions 9*, 453–63.

Abdelkrim, A., Pascal, M., and Samadi, S. (2007) Establishing causes of eradication failure based on genetics: case study of ship rat eradication in Ste. Anne Archipelago. *Conservation Biology 21*, 719–30.

Abedrabbo, S. (1994) Control of the little fire ant *Wasmannia auropunctata* on Santa Fe Island in the Galápagos Islands. In D.F. Williams, ed. *Exotic Ants: biology, impact, and control of introduced species*, pp. 219–27. Westview Press, Boulder, CO.

Adams, E.S. (1994) Territory defense by the ant *Azteca trigona*: maintenance of an arboreal ant mosaic. *Oecologia 97*, 203–8.

——(2001) Approaches to the study of territory size and shape. *Annual Review of Ecology and Systematics 32*, 277–303.

——and Mesterton-Gibbons, M. (2006) Lanchester's attrition models and fights among social animals. *Behavioral Ecology 14*, 719–23.

Addison, P., and Samways, M.J. (2000) A survey of ants (Hymenoptera: Formicidae) that forage in vineyards in the Western Cape Province, South Africa. *African Entomology 8*, 251–60.

Adis, J. (1982) Eco-entomological observations from the Amazon: III. How do leafcutting ants of inundation-forests survive flooding? *Acta Amazonica 12*, 839–40.

——Marques, M.I., and Wantzen, K.M. (2001) First observations on the survival strategies of terricolous arthropods in the northern Pantanal wetland of Brazil. *Andrias 15*, 127–8.

Adler, F.R., Lebrun, E.G., and Feener, D.H. (2007) Maintaining diversity in an ant community: modeling, extending, and testing the dominance-discovery trade-off. *American Naturalist 169*, 323–33.

Agnew, C.W., Sterling, W.L., and Dean, D.A. (1982) Influence of cotton nectar on red imported fire ants and other predators. *Environmental Entomology 11*, 629–34.

Agosti, D., and Alonso, L.E. (2000) The ALL protocol: a standard protocol for the collection of ground-dwelling ants. In D. Agosti, J.D. Majer, L.E. Alonso, and T.R. Schultz, eds. *Ants: standard methods for measuring and monitoring biodiversity*, pp. 204–6. Smithsonian Institution Press, Washington, DC.

——and Johnson, N.F. (2005) *Antbase*. www.antbase.org.

——Majer, J., Alonso, L.E., and Schultz, T.R. (2000) *Ants: standard methods for measuring and monitoring biodiversity*. Smithsonian Institution Press, Washington, DC.

Agrawal, A.A., and Fordyce, J.A. (2000) Induced indirect defence in a lycaenid-ant association: the regulation of a resource in a mutualism. *Proceedings of the Royal Society of London B 267*, 1857–61.

Akino, T. (2006) Cuticular hydrocarbons of *Formica truncorum* (Hymenoptera: Formicidae): description of new very long chained hydrocarbon components. *Applied Entomology and Zoology 41*, 667–77.

——and Yamaoka, R. (2000) Evidence for volatile and contact signals of nestmate recognition in the black shining ant *Lasius fuliginosus* Latreille (Hymenoptera: Formicidae). *Entomological Science 3*, 1–8.

——(2005a) The chemical basis for trail recognition in *Lasius nipponensis*. *Chemoecology 15*, 13–20.

Akino, T. (2005b) Trail discrimination signal of *Lasius japonicus*. *Chemoecology 15*, 21–30.

——Yamamura, N., Wakamura, S., and Yamaoka, R. (2004) Direct behavioral evidence for hydrocarbons as nestmate recognition cues in *Formica japonica* (Hymenoptera: Formicidae). *Applied Entomology and Zoology 39*, 381–7.

Albrecht, M., and Gotelli, N.J. (2001) Spatial and temporal niche partitioning in grassland ants. *Oecologia 126*, 134–41.

Albuquerque, N.L., and Brandão, C.R.F. (2004) A revision of the Neotropical Solenopsidini ant genus *Oxyepoecus* Santschi, 1926 (Hymenoptera: Formicidae: Myrmicinae). 1. The *Vezenyii* species-group *Papeis Avulsos de Zoologia (São Paulo) 44*, 55–80.

Aleksiev, A.S., Longdon, B., Christmas, M.J., Sendova-Franks, A.B., and Franks, N.R. (2007) Individual choice of building material for nest construction by worker ants and the collective outcome for their colony. *Animal Behaviour 74*, 559–66.

Ali, A.D., and Reagan, T.E. (1985) Vegetation manipulation impact on predator and prey populations in Louisiana sugarcane ecosystems. *Journal of Economic Entomology 78*, 1409–14.

Ali, J.R., and Aitchison, J.C. (2008) Gondwana to Asia: plate tectonics, paleogeography and the biological connectivity of the Indian subcontinent from the Middle Jurassic through latest Eocene (166–35 Ma). *Earth-Science Reviews*, doi:10.1016/j.earscirev.2008.1001.1007.

Allen, A.P., Brown, J.H., and Gillooly, J.F. (2002) Global biodiversity, biochemical kinetics, and the energetic-equivalence rule. *Science 297*, 1545–8.

——Gillooly, J.F., Savage, V.M., and Brown, J.H. (2006) Kinetic effects of temperature on rates of genetic divergence and speciation. *Proceedings of the National Academy of Sciences USA 103*, 9130–5.

Allen, C.R., Epperson, D.M., and Garmestani, A.S. (2004) Red imported fire ant impacts on wildlife: a decade of research. *American Midland Naturalist 152*, 88–103.

——Forys, E.A., Rice, K.G., and Wojcik, D.P. (2001) Effects of fire ants (Hymenoptera: Formicidae) on hatchling turtles and prevalence of fire ants on sea turtle nesting beaches in Florida. *Florida Entomologist 84*, 250–3.

Allen, G.E., and Buren, W.F. (1974) Microsporidian and fungal diseases of *Solenopsis invicta* Buren in Brazil. *Journal of the New York Entomological Society 82*, 125–30.

Alonso, L.E. (2000) Ants as indicators of diversity. In D. Agosti, J. Majer, L.E. Alonso, and T.R. Schultz, eds. *Ants: standard methods for measuring and monitoring biodiversity*, pp. 80–8. Smithsonian Institution Press, Washington, DC.

Alpert, G.D. (2007) A review of the ant genus *Metapone* Forel from Madagascar. *Memoirs of the American Entomological Institute 80*, 8–18.

Altfeld, L., and Stiling, P. (2006) Argentine ants strongly affect some but not all common insects on *Baccharis halimifolia*. *Environmental Entomology 35*, 31–6.

——and Stiling, P. (2009) Effects of aphid-tending Argentine ants, nitrogen enrichment and early-season herbivory on insects hosted by a coastal shrub. *Biological Invasions 11*, 183–91.

Amador, J.A., and Gorres, J.H. (2007) Microbiological characterization of the structures built by earthworms and ants in an agricultural field. *Soil Biology & Biochemistry 39*, 2070–7.

Andersen, A.N. (1991) Parallels between ants and plants: implications for community ecology. In C.R. Huxley and D.F. Cutler, eds. *Ant-plant interactions*, pp. 539–53. Oxford University Press, Oxford.

——(1992) Regulation of momentary diversity by dominant species in exceptionally rich ant communities of the Australian seasonal tropics. *American Naturalist 140*, 401–20.

——(1995) A classification of Australian ant communities, based on functional groups which parallel plant life-forms in relation to stress and disturbance. *Journal of Biogeography 22*, 15–29.

——(1997) Functional groups and patterns of organization in North American ant communities: a comparison with Australia. *Journal of Biogeography 24*, 433–60.

——(2000) A global ecology of rainforest ants: functional groups in relation to environmental stress and disturbance. In D. Agosti, J.D. Majer, L.E. Alonso, and T.R. Schultz, eds. *Ants: standard methods for measuring and monitoring biodiversity*, pp. 25–34. Smithsonian Institution Press, Washington, DC.

——(2002) Common names for Australian ants (Hymenoptera: Formicidae). *Australian Journal of Entomology 41*, 285–93.

——(2003) Ant biodiversity in arid Australia: productivity, species richness and community organisation. *Records of the South Australian Museum Monograph Series 7*, 79–92.

——(2006) A systematic overview of the Australian species of the myrmecine ant genus *Meranoplus* F. Smith, 1853 (Hymenoptera: Formicidae). *Myrmecologische Nachrichten 8*, 157–70.

——(2008) Not enough niches: non-equilibrial processes promoting species coexistence in diverse ant communities. *Austral Ecology 33*, 211–20.

——and Majer, J.D. (2004) Ants show the way Down Under: invertebrates as bioindicators in land management. *Frontiers in Ecology and the Environment 2*, 291–8.

——and Morrison, S.C. (1998) Myrmecochory in Australia's seasonal tropics: effects of disturbance on distance dispersal. *Australian Journal of Ecology 23*, 483–91.

——and Patel, A.D. (1994) Meat ants as dominant members of Australian ant communities: an experimental test of their influence on the foraging success and forager abundance of other species. *Oecologia 98*, 15–24.

——and Yen, A.L. (1985) Immediate effects of fire on ants in the semi-arid mallee region of north-western Australia. *Australian Journal of Ecology 10*, 25–30.

——Blum, M.S., and Jones, T.H. (1991) Venom alkaloids in *Monomorium rothsteini* Forel repel other ants – is this the secret to success by *Monomorium* in Australian ant communities? *Oecologia 88*, 157–60.

——Hertog, T., and Woinarski, J.C.Z. (2006) Long-term fire exclusion and ant community structure in an Australian tropical savanna: congruence with vegetation succession. *Journal of Biogeography 33*, 823–32.

——Hoffmann, B.D., and Somes, J. (2003) Ants as indicators of minesite restoration: community recovery at one of eight rehabilitation sites in central Queensland. *Ecological Management & Restoration 4*, S12-9.

——Parr, C.L., Lowe, L.M., and Müller, W.J. (2007) Contrasting fire-related resilience of ecologically dominant ants in tropical savannas of northern Australia. *Diversity and Distributions 13*, 438–46.

——Fisher, A., Hoffmann, B.D., Read, J.L., and Richards, R. (2004a) The use of terrestrial invertebrates for biodiversity monitoring in Australian rangelands, with particular reference to ants. *Austral Ecology 29*, 87–92.

Andersen, M.C., Adams, H., Hope, B., and Powell, M. (2004b) Risk analysis for invasive species: general framework and research needs. *Risk Analysis 24*, 893–900.

Anderson, C., and Franks, N.R. (2001) Teams in animal societies. *Behavioral Ecology 12*, 534–40.

——and McShea, D.W. (2001a) Individual vs. social complexity, with particular reference to ant colonies. *Biological Reviews 76*, 211–37.

————(2001b) Intermediate-level parts in insect societies: adaptive structures that ants build away from the nest. *Insectes Sociaux 48*, 291–301.

——Boomsma, J.J., and Bartholdi, J.J. (2002) Task partitioning in insect societies: bucket brigades. *Insectes Sociaux 49*, 171–80.

Anderson, K.E., and Munger, J.C. (2003) Effect of temperature on brood relocation in *Pogonomyrmex salinus* (Hymenoptera: Formicidae). *Western North American Naturalist 63*, 122–8.

——Linksvayer, T.A., and Smith, C.R. (2008) The causes and consequences of genetic caste determination in ants (Hymenoptera: Formicidae). *Myrmecological News 11*, 119–32.

André, J.-B., Peeters, C., and Doums, C. (2001) Serial polygyny and colony genetic structure in the monogynous queenless ant *Diacamma cyaneiventre*. *Behavioral Ecology and Sociobiology 50*, 72–80.

————Huet, M., and Doums, C. (2006) Estimating the rate of gamergate turnover in the queenless ant *Diacamma cyaneiventre* using a maximum likelihood model. *Insectes Sociaux 53*, 233–40.

ANeT (2008) http://homepage.mac.com/dorylus/index.html.

Angilletta, M.J.J., Roth, T.C., Wilson, R.S., Niehaus, A.C., and Ribeiro, P.L. (2008) The fast and the fractalous: speed and tortuosity trade off in running ants. *Functional Ecology 22*, 78–83.

Antonialli, W.F.J., Lima, S.M., Andrade, L.H.C., and Suarez, Y.R. (2007) Comparative study of the cuticular hydrocarbon in queens, workers and males of *Ectatomma vizottoi* (Hymenoptera, Formicidae) by Fourier transformation-infrared photoacoustic spectroscopy. *Genetics and Molecular Research 6*, 492–9.

——Suarez, Y.R., Izida, T., Andrade, L.H.C., and Lima, S.M. (2008) Intra- and interspecific variation of cuticular hydrocarbon composition in two *Ectatomma* species (Hymenoptera: Formicidae) based on Fourier transform infrared photoacoustic spectroscopy. *Genetics and Molecular Research 7*, 559–66.

Antony, G., Scanlam, J., Francis, A., Kloessing, K., and Nguyen, Y. (2009) Fire ants revisited – the economics of eradication. *Proceedings of the Australian Agricultural and Resource Economics Society 53rd Annual Conference*, Cairns.

AntWeb (2008) http://www.AntWeb.org/

Archibald, S.B., Cover, S.P., and Moreau, C.S. (2006) Bulldog ants of the Eocene Okanagan Highlands and history of the subfamily (Hymenoptera: Formicidae: Myrmeciinae). *Annals of the Entomological Society of America 99*, 487–523.

Armbrecht, I., and Gallego, M.C. (2007) Testing ant predation on the coffee berry borer in shaded and sun coffee plantations in Colombia. *Entomologia Experimentalis et Applicata 124*, 261–7.

——and Perfecto, I. (2003) Litter-twig dwelling ant species richness and predation potential within a forest fragment and neighboring coffee plantations of contrasting habitat quality in Mexico. *Agriculture Ecosystems & Environment 97*, 107–15.

——Perfecto, I., and Vandermeer, J. (2004) Enigmatic biodiversity correlations: ant diversity responds to diverse resources. *Science 304*, 284–6.

Arnan, X., Rodrigo, A., and Retana, J. (2006) Post-fire recovery of Mediterranean ground ant communities follows vegetation and dryness gradients. *Journal of Biogeography 33*, 1246–58.

Aron, S. (2001) Reproductive strategy: an essential component in the success of incipient colonies of the invasive Argentine ant. *Insectes Sociaux 48*, 25–7.

Atkin, L., and Proctor, J. (1988) Invertebrates in litter and soil on Volcan Barva, Costa Rica. *Journal of Tropical Ecology 4*, 307–10.

Autori, M. (1947) Combate a formiga saúva. *Biológico 13*, 196–9.

Ayre, G.L. (1967) The relationships between food and digestive enzymes in five species of ants (Hymenoptera: Formicidae). *Canadian Entomologist 99*, 408–11.

Azevedo-Ramos, C., de Carvalho Jr., O., and do Amaral, B.D. (2006) Short-term effects of reduced-impact logging on eastern Amazon fauna. *Forest Ecology and Management 232*, 26–35.

Bach, C.E. (1991) Direct and indirect interactions between ants (*Pheidole megacephala*), scales (*Coccus viridis*) and plants (*Pluchea indica*). *Oecologia 87*, 233–9.

Ballinger, A., Lake, P.S., and Mac Nally, R. (2007) Do terrestrial invertebrates experience floodplains as landscape mosaics? Immediate and longer-term effects of flooding on ant assemblages in a floodplain forest. *Oecologia 152*, 227–38.

Banks, W.A., Miles, L.R., and Harlan, D.P. (1983) The effects of insect growth-regulators and their potential as control agents for imported fire ants (Hymenoptera: Formicidae). *Florida Entomologist 66*, 172–81.

Barber, T.C. (1916) The Argentine ant: distribution and control in the United States. *USDA Bureau of Entomology Bulletin 377*, 1–23.

Barbero, F., Thomas, J.A., Bonelli, S., Balletto, E., and Schonrogge, K. (2009) Queen ants make distinctive sounds that are mimicked by a butterfly social parasite. *Science 323*, 782–5.

Baroni-Urbani, C. (1993) The diversity and evolution of recruitment behaviour in ants, with a discussion of the usefulness of parsimony criteria in the reconstruction of evolutionary histories. *Insectes Sociaux 40*, 233–60.

——(1998) The number of castes in ants, where major is smaller than minor and queens wear the shield of the soldiers. *Insectes Sociaux 45*, 315–33.

——and Collingwood, C.A. (1976) A numerical analysis of the distribution of British Formicidae (Hymenoptera, Aculeata). *Verhandlungen der Naturforschenden Gesellschaft (Basel) 85*, 51–91.

————(1977) The zoogeography of ants (Hymenoptera, Formicidae) in northern Europe. *Annales Zoologici Fennici 152*, 1–34.

——and de Andrade, M.L. (1997) Pollen eating, storing, and spitting by ants. *Naturwissenschaften 84*, 256–8.

————(2003) The ant genus *Proceratium* in the extant and fossil record (Hymenoptera: Formicidae). *Museo Regionale di Scienze Naturali Monografie Turin 36*, 1–492.

————(2006) A new *Protanilla* Taylor, 1990 (Hymenoptera: Formicidae: Leptanillinae) from Sri Lanka. *Myrmecologische Nachrichten 8*, 45–7.

——and Wilson, E.O. (1987) The fossil members of the ant tribe Leptomyrmecini (Hymenoptera: Formicidae). *Psyche (Cambridge) 94*, 1–8.

——Bolton, B., and Ward, P.S. (1992) The internal phylogeny of ants (Hymenoptera: Formicidae). *Systematic Entomology 17*, 301–29.

Barrow, L., Parr, C.L., and Kohen, J.L. (2007) Habitat type influences fire resilience of ant assemblages in the semi-arid tropics of Northern Australia. *Journal of Arid Environments 69*, 80–95.

Barzman, M.S., and Daane, K.M. (2001) Host-handling behaviours in parasitoids of the black scale: a case for ant-mediated evolution. *Journal of Animal Ecology 70*, 237–47.

Basset, Y., Novotny, V., Miller, S.E., and Pyle, R. (2000) Quantifying biodiversity: experience with parataxonomists and digital photography in Papua New Guinea and Guyana. *BioScience 50*, 899–908.

——Novotny, V., Miller, S.E., Weiblen, G.D., Missa, O., and Stewart, A.J.A. (2004) Conservation and biological monitoring of tropical forests: the role of parataxonomists. *Journal of Applied Ecology 41*, 163–74.

Basu, P. (1997) Competition hierarchy in the ground foraging ant community in a wet evergreen forest (Western Ghats, India): role of interference behaviour. *Current Science 73*, 173–9.

Beattie, A.J. (1985) *The Evolutionary Ecology of Ant-Plant Mutualisms*. Cambridge University Press, Cambridge.

——(1995) Natural history at the cutting edge. *Ecological Economics 13*, 93–7.

——(2006) The evolution of ant pollination systems. *Botanische Jahrbücher 127*, 43–55.

——and Culver, D.C. (1981) The guild of myrmecochores in the herbaceous flora of West Virginia forests *Ecology 62*, 107–11.

——Turnbull, C., Knox, R.B., and Williams, E.G. (1984) Ant inhibition of pollen function: a possible reason why ant pollination is rare? *American Journal of Botany 71*, 421–6.

————Hough, T., and Knox, R.B. (1986) Antibiotic production: a possible function for the metapleural glands of ants (Hymenoptera: Formicidae). *Annals of the Entomological Society of America 79*, 448–50.

Becerra, J.X.I., and Venable, D.L. (1989) Extrafloral nectaries – a defense against ant-Homoptera mutualisms? *Oikos 55*, 276–80.

Beckers, R., Goss, S., Deneubourg, J.L., and Pasteels, J.M. (1989) Colony size, communication and ant foraging strategy. *Psyche (Cambridge) 96*, 239–56.

Beckmann, R.L., and Stucky, J.M. (1981) Extrafloral nectaries and plantguarding in *Ipomoea pandurata* (L.) G.F. W. Mey. (Convolvulaceae). *American Journal of Botany 68*, 72–9.

Beekman, M., Sumpter, D., and Ratnieks, F. (2001) Phase transition between disordered and ordered foraging in Pharaoh's ants. *Proceedings of the National Academy of Sciences USA 98*, 9703–4.

Beever, E.A., and Herrick, J.E. (2005) Effects of feral horses in Great Basin landscapes on soils and ants: direct and indirect mechanisms. *Journal of Arid Environments 66*, 96–112.

Behar, A., Yuval, B., and Jurkevitch, E. (2005) Enterobacteria-mediated nitrogen fixation in natural populations of the fruit fly *Ceratitis capitata*. *Molecular Ecology 14*, 2637–43.

Beibl, J., Stuart, R.J., Heinze, J., and Foitzik, S. (2005) Six origins of slavery in formicoxenine ants. *Insectes Sociaux 52*, 291–7.

Belshaw, R., and Bolton, B. (1993) The effect of forest disturbance on the leaf-litter ant fauna in Ghana. *Biodiversity and Conservation 2*, 656–66.

Bensasson, D., Zhang, D.X., Hartl, D.L., and Hewitt, G.M. (2001) Mitochondrial pseudogenes: evolution's misplaced witnesses. *Trends Ecology and Evolution 16*, 314–21.

Bentley, B.L. (1976) Plants bearing extrafloral nectaries and the associated ant community: interhabitat differences in the reduction of herbivore damage. *Ecology 57*, 815–20.

Berg, R.Y. (1975) Myrmecochorous plants in Australia and their dispersal by ants. *Australian Journal of Botany 23*, 475–508.

Bergsten, J. (2005) A review of long-branch attraction. *Cladistics 21*, 163–93.

Bergstrom, D.M., Lucieer, A., Kiefer, K., Wasley, J., Belbin, L., Pedersen, T.K., and Chown, S.L. (2009) Indirect effects of invasive species removal devastate World Heritage Island. *Journal of Applied Ecology 46*, 73–81.

Bernasconi, G., and Strassmann, J.E. (1999) Cooperation among unrelated individuals: the ant foundress case. *Trends in Ecology and Evolution 14*, 477–82.

Bernstein, R.A., and Gobbel, M. (1979) Partitioning of space in communities of ants. *Journal of Animal Ecology 48*, 931–42.

Bertness, M.D., and Callaway, R. (1994) Positive interactions in communities. *Trends in Ecology and Evolution 9*, 191–3.

Beshers, S.N., and Fewell, J.H. (2001) Models of division of labor in social insects. *Annual Review of Entomology 46*, 413–40.

Bestelmeyer, B.T. (2000) The trade-off between thermal tolerance and behavioural dominance in a subtropical South American ant community. *Journal of Animal Ecology 69*, 998–1009.

——and Wiens, J.A. (1996) The effects of land use on the structure of ground-foraging ant communities in the Argentine Chaco. *Ecological Applications 6*, 1225–40.

——Agosti, D., and Alonso, L.E. (2000) Field techniques for the study of ground-dwelling ants. In D. Agosti, J.D. Majer, L.E. Alonso, and T.R. Schultz, eds. *Ants: standard methods for measuring and monitoring biodiversity*, pp. 122–44. Smithsonian Institution Press, Washington, DC.

Beye, M., Neumann, P., and Moritz, R.F.A. (1997) Nestmate recognition and the genetic gestalt in the mound- building ant *Formica polyctena*. *Insectes Sociaux 44*, 49–58.

——Neumann, P., Chapuisat, M., Pamilo, P., and Moritz, R.F.A. (1998) Nestmate recognition and the genetic relatedness of nests in the ant *Formica pratensis*. *Behavioral Ecology and Sociobiology 43*, 67–72.

Bickel, T.O., Brühl, C.A., Gadau, J.R., Hölldobler, B., and Linsenmair, K.E. (2006) Influence of habitat fragmentation on the genetic variability in leaf litter ant populations in tropical rainforests of Sabah, Borneo. *Biodiversity and Conservation 15*, 157–75.

Biesmeijer, J.C., and de Vries, H. (2001) Exploration and exploitation of food sources by social insect colonies: a revision of the scout-recruit concept. *Behavioral Ecology and Sociobiology 49*, 89–99.

Bihn, J.H., and Verhaagh, M. (2007) A review of the genus *Mystrium* (Hymenoptera: Formicidae) in the Indo-Australian region. *Zootaxa 1642*, 65–8.

Billen, J., and Buschinger, A. (2000) Morphology and ultrastructure of a specialized bacterial pouch in the digestive tract of *Tetraponera* ants (Formicidae, Pseudomyrmecinae). *Arthropod Structure & Development 29*, 259–66.

Billick, I., Brown, R., and Reithel, J.S. (2005) Importance of fertilization of host plants to ant tending and growth rates in *Glaucopsyche lygdamus* (Lepidoptera: Lycaenidae). *Annals of the Entomological Society of America 98*, 491–5.

——Wiernasz, D.C., and Cole, B.J. (2001) Recruitment in the Harvester ant, *Pogonomyrmex occidentalis*: effects of experimental removal. *Oecologia 129*, 228–33.

Biosecurity New Zealand (2008) Red imported fire ant response update: April 2009, Ministry of Agriculture and Forestry, Auckland.

BirdLife International (2000) *Threatened Birds of the World*. Lynx Edicions Barcelona, Spain.

——(2008) http://www.birdlife.org/.

Bishop, D., and Bristow, C. (2001) Effect of Allegheny mound ant (Hymenoptera: Formicidae) presence on homopteran and predator populations in Michigan jack pine forests. *Annals of the Entomological Society of America 94*, 33–40.

Blancafort, X., and Gomez, C. (2005) Consequences of the Argentine ant, *Linepithema humile* (Mayr), invasion on pollination of *Euphorbia characias* (L.) (Euphorbiaceae). *Acta Oecologia 28*, 49–55.

————(2006) Downfall of pollen carriage by ants after Argentine ant invasion in two Mediterranean *Euphorbia* species. *Vie et Milieu 56*, 1–4.

Blanchard, G.B., Orledge, G.M., Reynolds, S.E., and Franks, N.R. (2000) Division of labour and seasonality in the ant *Leptothorax albipennis*: worker corpulence and its influence on behaviour. *Animal Behaviour 59*, 723–38.

Blanton, C.M., and Ewel, J.J. (1985) Leaf-cutting ant herbivory in successional and agricultural tropical ecosystems. *Ecology 66*, 861–9.

Blomquist, G.J., Tillman, J.A., Mpuru, S., and Seybold, S.J. (1998) The cuticle and cuticular hydrocarbons of insects: structure, function, and biochemistry. In R.K. Vander Meer, M.D. Breed, K. Espelie, and M.L. Winston, eds. *Pheromone communication in social insects. Ants, wasps, bees and termites*, pp. 34–54. Westview Press, Boulder, CO.

Blüthgen, N., and Fiedler, K. (2004a) Competition for composition: lessons from nectar-feeding ant communities. *Ecology 85*, 1479–85.

————(2004b) Preferences for sugars and amino acids and their conditionality in a diverse nectar-feeding ant community. *Journal of Animal Ecology 73*, 155–66.

——and Stork, N.E. (2007) Ant mosaics in a tropical rainforest in Australia and elsewhere: a critical review. *Austral Ecology 32*, 93–104.

——Gebauer, G., and Fiedler, K. (2003) Disentangling a rainforest food web using stable isotopes: dietary diversity in a species-rich ant community. *Oecologia 137*, 426–35.

——Gottsberger, G., and Fiedler, K. (2004a) Sugar and amino acid composition of ant-attended nectar and honeydew sources from an Australian rainforest. *Austral Ecology 29*, 418–29.

——Stork, N.E., and Fiedler, K. (2004b) Bottom-up control and co-occurrence in complex communities: honeydew and nectar determine a rainforest ant mosaic. *Oikos 106*, 344–58.

——Mezger, D., and Linsenmair, K.E. (2006) Ant-hemipteran trophobioses in a Bornean rainforest – diversity, specificity and monopolisation. *Insectes Sociaux 53*, 194–203.

——Menzel, F., Hovestadt, T., Fiala, B., and Blüthgen, N. (2007) Specialization, constraints, and conflicting interests in mutualistic networks. *Current Biology 17*, 341–6.

——Verhaagh, M., Goitía, W., Jaffé, K., Morawetz, W., and Barthlott, W. (2000) How plants shape the ant community in the Amazonian rainforest canopy: the key role of extrafloral nectaries and homopteran honeydew. *Oecologia 125*, 229–40.

Boevé, J.L., and Wäckers, F.L. (2003) Gustatory perception and metabolic utilization of sugars by *Myrmica rubra* ant workers. *Oecologia 136*, 508–14.

Bohlen, P.J. (2006) Biological invasions: linking the aboveground and belowground consequences. *Applied Soil Ecology 32*, 1–5.

Bohm, M., and Mayhew, P.J. (2005) Historical biogeography and the evolution of the latitudinal gradient of species richness in the Papionini (Primata: Cercopithecidae). *Biological Journal of the Linnean Society 85*, 235–46.

Bolger, D.T., Beard, K.H., Suarez, A.V., and Case, T.J. (2008) Increased abundance of native and non-native spiders with habitat fragmentation. *Diversity and Distributions 14*, 655–65.

Bolger, D.T., Suarez, A.V., Crooks, K.R., Morrison, S.A., and Case, T.J. (2000) Arthropods in urban habitat fragments in southern California: area, age, and edge effects. *Ecological Applications 10*, 1230–48.

Bollazzi, M., and Roces, F. (2002) Thermal preference for fungus culturing and brood location by workers of the thatching grass-cutting ant *Acromyrmex heyeri*. *Insectes Sociaux 49*, 153–7.

Bolton, B. (1994) *Identification Guide to the Ant Genera of the World*. Harvard University Press, Cambridge, MA.

——(1995a) *A Catalog of the Ants of the World*. Harvard University Press, Cambridge, MA.

——(1995b) A taxonomic and zoogeographical census of the extant ant taxa (Hymenoptera: Formicidae). *Journal of Natural History 29*, 1037–56.

——(2000) The ant tribe Dacetini. *Memoirs of the American Entomological Institute 65*, 1–1028.

——(2003) Synopsis and classification of Formicidae. *Memoirs of the American Entomological Institute 71*, 1–370.

——(2007) Taxonomy of the dolichoderine ant genus *Technomyrmex* Mayr (Hymenoptera: Formicidae) based on the worker caste. *Contributions of the American Entomological Institute 35*(1), 1–150.

——and Fisher, B.L. (2008a) Afrotropical ants of the ponerine genera *Centromyrmex* Mayr, *Promyopias* Santschi gen. rev. and *Feroponera* gen. n., with a revised key to genera of African Ponerinae (Hymenoptera: Formicidae). *Zootaxa 1929*, 1–37.

———(2008b) The Afrotropical ponerine ant genus *Phrynoponera* Wheeler (Hymenoptera: Formicidae). *Zootaxa 1892*, 35–52.

——and Marsh, A.C. (1989) The Afrotropical thermophilic ant genus *Ocymyrmex* (Hymenoptera: Formicidae). *Journal of Natural History 23*, 1267–308.

——Alpert, G., Ward, P.S., and Naskrecki, P. (2006) *Bolton's catalogue of ants of the world.* Harvard University Press, Cambridge, MA. CD-ROM.

Bond, W.J., and Slingsby, P. (1983) Seed dispersal by ants in shrublands of the Cape Province and its evolutionary implications. *South African Journal of Science 79*, 231–3.

———(1984) Collapse of an ant-plant mutualism: the Argentine ant (*Iridomyrmex humilis*) and myrmecochorous Proteaceae. *Ecology 65*, 1031–7.

Boomsma, J.J., and Franks, N.R. (2006) Social insects: from selfish genes to self organisation and beyond. *Trends in Ecology and Evolution 21*, 303–8.

——Baer, B., and Heinze, J. (2005) The evolution of male traits in social insects. *Annual Review of Entomology 50*, 395–420.

Boomsma, J.J., Schmid-Hempel, P., and Hughes W.O.H. (2005) Life histories and parasite pressure across the major groups of social insects. In M. Fellowes, G. Holoway and J. Rolff, eds. *Insect Evolutionary Ecology*, pp. 139–75 CABI publishing, Wallingford, UK.

——Nielsen, J., Sundström, L., Oldham, N.J., Tentschert, J., Petersen, H.C., and Morgan, E.D. (2003) Informational constraints on optimal sex allocation in ants. *Proceedings of the National Academy of Sciences USA 100*, 8799–804.

Borgesen, L.W. (2000) Nutritional function of replete workers in the pharaoh's ant, *Monomorium pharaonis* (L.). *Insectes Sociaux 47*, 141–6.

Bos, M.M., Tylianakis, J.M., Steffan-Dewenter, I., and Tscharntke, T. (2008) The invasive yellow crazy ant and the decline of forest ant diversity in Indonesian cacao agroforests. *Biological Invasions 10*, 1399–409.

Boswell, G.P., Britton, N.F., and Franks, N.R. (1998) Habitat fragmentation, percolation theory, and the conservation of a keystone species. *Proceedings of the Royal Society of London B 265*, 1921–5.

Bot, A.N.M., Currie, C.R., Hart, A.G., and Boomsma, J.J. (2001a) Waste management in leaf-cutting ants. *Ethology Ecology & Evolution 13*, 225–37.

——Rehner, S.A., and Boomsma, J.J. (2001b) Partial incompatibility between ants and symbiotic fungi in two sympatric species of *Acromyrmex* leaf-cutting ants. *Evolution 55*, 1980–91.

Botes, A., McGeoch, M.A., Robertson, H.G., van Niekerk, A., Davids, H.P., and Chown, S.L. (2006) Ants, altitude and change in the northern Cape Floristic Region. *Journal of Biogeography 33*, 71–90.

Boulay, R., Hefetz, A., Soroker, V., and Lenoir, A. (2000) *Camponotus fellah* colony integration: worker individuality necessitates frequent hydrocarbons exchanges. *Animal Behaviour 59*, 1127–33.

——Cerdá, X., Simon, T., Roldan, M., and Hefetz, A. (2007a) Intraspecific competition in the ant *Camponotus cruentatus*: should we expect the 'dear enemy' effect? *Animal Behaviour 74*, 985–93.

——Hefetz, A., Cerdá, X., Devers, S., Francke, W., Twele, R., and Lenoir, A. (2007b) Production of sexuals in a fission-performing ant: dual effects of queen pheromones and colony size. *Behavioral Ecology and Sociobiology 61*, 1531–41.

Bourke, A.F.G. (2005) Genetics, relatedness and social behaviour in insect societies. In M. Fellowes, G. Holloway, and J. Rolff, eds. *Insect Evolutionary Ecology*, pp. 1–30. CABI publishing, Wallingford, UK.

——and Franks, N.R. (1995) *Social Evolution in Ants*. Princeton University Press, Princeton, NJ.

Bradshaw, J.W.S., Baker, R., and Howse, P.E. (1975) Multicomponent alarm pheromones of the weaver ant. *Nature 258*, 230–1.

Brady, S.G. (2003) Evolution of the army ant syndrome: the origin and long-term evolutionary stasis of a complex of behavioral and reproductive adaptations. *Proceedings of the National Academy of Sciences USA 100*, 6575–9.

——Fisher, B.L., Schultz, T.R., and Ward, P.S. (2006) Evaluating alternative hypotheses for the early evolution and diversification of ants. *Proceedings of the National Academy of Sciences USA 103*, 18172–7.

Braekman, J.C., Daloze, D., Pasteels, J.M., Van Hecke, P., Declercq, J.P., Sinnwell, V., and Francke, W. (1987) Tetraponerine-8, an alkaloidal contact poison in a Neoguinean pseudomyrmecine ant *Tetraponera* sp. *Z. Naturforsch 42*, 627–30.

Brandão, C.R.F. (2000) Major regional and type collections of ants (Formicidae) of the world and sources for the identification of ant species. In D. Agosti, J.D. Majer, L. E. Alonso, and T.R. Schultz, eds. *Ants: standard methods for measuring and monitoring biodiversity*, pp. 172–85. Smithsonian Institution Press, Washington, DC.

——Diniz, J.L.M., and Tomotake, E.M. (1991) *Thaumatomyrmex* strips millipedes for prey: a novel predatory behaviour in ants, and the first case of sympatry in the genus (Hymenoptera: Formicidae). *Insectes Sociaux 38*, 335–44.

———Agosti, D., and Delabie, J.H. (1999) Revision of the Neotropical ant subfamily Leptanilloidinae. *Systematic Entomology 24*, 17–36.

—— Feitosa, R.M., Schmidt, F.A., and de Castro Solar, R.R. (2008) Rediscovery of the putatively extinct ant species

Simopelta minima (Brandão) (Hymenoptera: Formicidae), with a discussion on rarity and conservation status of ant species. *Revista Brasileira de Entomologia 52*, 480–3.

——Diniz, J.L.M., Silva, P.R., Albuquerque, N.L., and Silvestre, R. (2001) The first case of intranidal phragmosis in ants. The ergatoid queen of *Blepharidatta conops* (Formicidae: Myrmicinae) blocks the entrance of the brood chamber. *Insectes Sociaux 48*, 251–8.

Brandstaetter, A.S., Endler, A., and Kleineidam, C.J. (2008) Nestmate recognition in ants is possible without tactile interaction. *Naturwissenschaften 95*, 601–8.

Brandt, M., Heinze, J., Schmitt, T., and Foitzik, S. (2006) Convergent evolution of the Dufour's gland secretion as a propaganda substance in the slave-making ant genera *Protomognathus* and *Harpagoxenus*. *Insectes Sociaux 53*, 291–9.

Braschler, B. (2009) Successfully implementing a citizen-scientist approach to insect monitoring in a resource-poor country. *BioScience 59*, 103–4.

Breed, M.D., Welch, C.K., and Cruz, R.C. (1994) Kin discrimination within honey bee (*Apis mellifera*) colonies: an analysis of the evidence. *Behavioural Processes 33*, 25–40.

Brent, C., Peeters, C., Dietemann, V., Crewe, R., and Vargo, E. (2006) Hormonal correlates of reproductive status in the queenless ponerine ant, *Streblognathus peetersi*. *Journal of Comparative Physiology A 192*, 315–20.

Breton, L.M., and Addicott, J.F. (1992) Density-dependent mutualism in an aphid-ant interaction. *Ecology 73*, 2175–80.

Brew, C.R., O'Dowd, D.J., and Rae, I.D. (1989) Seed dispersal by ants - behavior-releasing compounds in elaiosomes. *Oecologia 80*, 490–7.

Brian, M.V. (1956) Segregation of species of the ant genus *Myrmica*. *Journal of Animal Ecology 25*, 319–37.

——(1978) *Production ecology of ants and termites, IBP 13.* Cambridge University Press, Cambridge.

Briese, D.T. (1983) Different modes of reproductive behaviour (including a description of colony fission) in a species of *Chelaner* (Hymenoptera: Formicidae). *Insectes Sociaux 30*, 308–16.

Bristow, C.M. (1983) Treehoppers transfer parental care to ants: a new benefit of mutualism *Science 220*, 532–3.

——(1991) Are ant-aphid associations a tritrophic interaction? Oleander aphids and Argentine ants. *Oecologia 87*, 514–21.

Broekhuysen, G.J. (1948) The brown house ant (*Pheidole megacephala*, Fabr.). *Union of South Africa Department of Agriculture Bulletin 266*, 1–40.

Bromham, L., and Cardillo, M. (2003) Testing the link between the latitudinal gradient in species richness and rates of molecular evolution. *Journal of Evolutionary Biology 16*, 200–7.

Bronstein, J.L. (1998) The contribution of ant-plant protection studies to our understanding of mutualism. *Biotropica 30*, 150–61.

——Alarcon, R., and Geber, M. (2006) The evolution of plant-insect mutualisms. *New Phytologist 172*, 412–28.

Brooks, T.M., Da Fonseca, G.A.B., and Rodrigues, A.S. (2004) Species, data, and conservation planning. *Conservation Biology 18*, 1682–8.

Brouat, C., Garcia, N., Andary, C., and McKey, D. (2001) Plant lock and ant key: pairwise coevolution of an exclusion filter in an ant-plant mutualism. *Proceedings of the Royal Society of London B 268*, 2131–41.

Brown, J.H., and Lomolino, M.V. (1998) *Biogeography*. Sinaur Associates, Sunderland, MA.

——Gillooly, J.F., Allen, A.P., Savage, V.M., and West, G. B. (2004) Toward a metabolic theory of ecology. *Ecology 85*, 1771–89.

Brown, M.J.F. (1999) Semi-claustral founding and worker behaviour in gynes of *Messor andrei*. *Insectes Sociaux 46*, 194–5.

——and Bonhoeffer, S. (2003) On the evolution of claustral colony founding in ants. *Evolutionary Ecology Research 5*, 305–13.

Brown Jr., W.L. (1954) Remarks on the internal phylogeny and subfamily classification of the family Formicidae. *Insectes Sociaux 1*, 21–31.

——(1960) Contributions toward a reclassification of the Formicidae. III. Tribe Amblyoponini (Hymenoptera). *Bulletin of the Museum of Comparative Zoology 122*, 143–230.

——(1973) A Comparison of the Hylean and Congo-West African rain forest ant faunas. In B.J. Meggers, E.S. Ayensu, and W.D. Duckworth, eds. *Tropical forest ecosystems in Africa and South America: a comparative review*, pp. 161–85. Smithsonian Institution Press, Washington, DC.

——(1975) Contributions toward a reclassification of the Formicidae. V. Ponerinae, tribes Platythyreini, Cerapachyini, Cylindromyrmecini, Acanthostichini, and Aenictogitini. *Search Agriculture (Ithaca NY) 5*(1), 1–115.

——(1980) A remarkable new species of *Proceratium*, with dietary and other notes on the genus (Hymenoptera: Formicidae). *Psyche (Cambridge) 86*, 337–46.

——(2000) Diversity of ants. In D. Agosti, J.D. Majer, L. E. Alonso, and T.R. Schultz, eds. *Ants: standard methods for measuring and monitoring biodiversity*, pp. 45–79. Smithsonian Institution Press, Washington, DC.

——and Kempf, W.W. (1968) *Tatuidris*, a remarkable new genus of Formicidae (Hymenoptera). *Psyche (Cambridge) 74*, 183–90.

——and Wilson, E.O. (1959a) The evolution of the dacetine ants. *The Quarterly Review of Biology 34*, 278–94.

——(1959b) The search for *Nothomyrmecia*. *Western Australian Naturalist 7*, 25–30.

Brühl, C.A., Eltz, T., and Linsenmair, K.E. (2003) Size does matter – effects of tropical rainforest fragmentation on the leaf litter ant community in Sabah, Malaysia. *Biodiversity and Conservation 12*, 1371–89.

——Mohamed, M., and Linsenmair, K.E. (1999) Altitudinal distribution of leaf litter ants along a transect in primary forest on Mount Kinabalu, Sabah, Malaysia. *Journal of Tropical Ecology 15*, 265–7.

Bruno, J.F., Stachowicz, J.J., and Bertness, M.D. (2003) Inclusion of facilitation into ecological theory. *Trends in Ecology and Evolution 18*, 119–25.

Buckley, L.B., and Jetz, W. (2007) Environmental and historical constraints on global patterns of amphibian richness. *Proceedings of the Royal Society of London B 274*, 1167–73.

Buckley, R.C. (1982) Ant-plant interactions: a world review. In R.C. Buckley, ed. *Ant-plant Interactions in Australia*, pp. 111–41. W. Junk, The Hague.

——and Gullan, P. (1991) More aggressive ant species (Hymenoptera: Formicidae) provide better protection for soft scales and mealybugs (Homoptera: Coccidae, Pseudococcidae). *Biotropica 23*, 282–6.

Buczkowski, G., and Bennett, G.W. (2008) Aggressive interactions between the introduced Argentine ant, *Linepithema humile* and the native odorous house ant, *Tapinoma sessile*. *Biological Invasions 10*, 1001–11.

——Kumar, R., Suib, S.L., and Silverman, J. (2005) Diet-related modification of cuticular hydrocarbon profiles of the Argentine ant, *Linepithema humile*, diminishes intercolony aggression. *Journal of Chemical Ecology 31*, 829–43.

——Vargo, E.L., and Silverman, J. (2004) The diminutive supercolony: the Argentine ants of the southeastern United States. *Molecular Ecology 13*, 2235–42.

Burd, M. (2006) Ecological consequences of traffic organisation in ant societies. *Physica A – Statistical Mechanics and Its Applications 372*, 124–31.

——and Howard, J.J. (2005) Central-place foraging continues beyond the nest entrance: the underground performance of leaf-cutting ants. *Animal Behaviour 70*, 737–44.

Buren, W.F., Allen, G.E., Whitcomb, W.H., Lennartz, F.E., and Williams, R.N. (1974) Zoogeography of the imported fire ants. *New York Entomological Society 82*, 113–24.

Buschinger, A. (1990) Evolutionary transitions between types of social parasitism in ants, hypotheses and evidence. In: social insects and the environment. In G.K. Veeresh, B. Mallik, and C.A. Viraktamath, eds. *Social insects and the environment. Proceedings of the 11th International Congress of IUSSI*, pp. 145–6. Oxford & IBH Publishing Co, New Delhi.

——(1991) Host specificity and host selection in parasitic ants. *Proceedings of the 4th ECE/ 13th SIEEC, Gödöllö Hungary*, 682–85.

——(2004) International pet ant trade, increasing risk and danger in Europe - (Hymenoptera, Formicidae). *Aliens 19/20*, 24–6.

——(2005) Experimental evidence for genetically mediated queen polymorphism in the ant species *Myrmecina graminicola* (Hymenoptera: Formicidae). *Entomologia Generalis 27*, 185–200.

——and Maschwitz, U. (1984) Defensive behavior and defensive mechanisms in ants. In H.R. Hermann, ed. *Defensive mechanisms in social insects*, pp. 95–150.

——and Schreiber, M. (2002) Queen polymorphism and queen-morph related facultative polygyny in the ant, *Myrmecina graminicola* (Hymenoptera, Formicidae). *Insectes Sociaux 49*, 344–53.

Buys, B. (1987) Competition for nectar between Argentine ants *Iridomyrmex humilis* and honeybees *Apis mellifera* on black ironbark *Eucalyptus sideroxylon*. *South African Journal of Zoology 22*, 173–4.

Buzas, M.A., Collins, L.S., and Culver, S.J. (2002) Latitudinal difference in biodiversity caused by higher tropical rate of increase. *Proceedings of the National Academy of Sciences USA 99*, 7841–3.

Byrne, M.M. (1994) Ecology of twig-dwelling ants in a wet lowland tropical forest. *Biotropica 26*, 61–72.

Cagniant, H. (2005) Les *Crematogaster* du Maroc (Hym., Formicidae). Clé de détermination et commentaires. *Orsis (Organismes i Sistemes) 20*, 7–12.

Caldera, E.J., Ross, K.G., DeHeer, C.J., and Shoemaker, D.D. (2008) Putative native source of the invasive fire ant *Solenopsis invicta*. *Biological Invasions 10*, 1457–79.

Cammaerts, M.C., and Cammaerts, R. (1998) Marking of nest entrance vicinity in the ant *Pheidole pallidula* (Formicidae, Myrmicinae). *Behavioural Processes 42*, 19–31.

——(2000a) Foraging area in two related *Tetramorium* ant species. *Journal of Insect Behavior 13*, 679–98.

——(2000b) Nest odour in the ant *Myrmica rubra*. *Biologia 55*, 509–23.

Cammeraat, E.L.H., and Risch, A.C. (2008) The impact of ants on mineral soil properties and processes at different spatial scales. *Journal of Applied Entomology 132*, 285–94.

Campomizzi, A. (2008) *Effects of red imported fire ants on songbird nest survival*. Master of Science thesis, Texas A&M University, College Station.

Campos, R.B.F., Schoereder, J.H., and Sperber, C.F. (2007) Small-scale patch dynamics after disturbance in litter ant communities. *Basic and Applied Ecology 8*, 36–43.

Cane, J.S. (1986) Predator deterrence by mandibular gland secretions of bees (Hymenoptera: Apoidea). *Journal of Chemical Ecology 12*, 1295–309.

Cardillo, M. (1999) Latitude and rates of diversification in birds and butterflies. *Proceedings of the Royal Society of London B 266*, 1221–5.

Cardillo, M., Orme, C.D.L., Owens, I.P.F. (2005) Testing for latitudinal bias in diversification rates: an example using New World birds. *Ecology 86*, 2278–87.

Cardinale, B.J., Harvey, C.T., Gross, K., and Ives, A.R. (2003) Biodiversity and biocontrol: emergent impacts of a multi-enemy assemblage on pest suppression and crop yield in an agroecosystem. *Ecology Letters 6*, 857–65.

Carney, S.E., Byerley, M.B., and Holway, D.A. (2003) Invasive Argentine ants (*Linepithema humile*) do not replace native ants as seed dispersers of *Dendromecon rigida* (Papaveraceae) in California, USA. *Oecologia 135*, 576–82.

Carpintero, S., and Reyes-López, J. (2008) The role of competitive dominance in the invasive ability of the Argentine ant (*Linepithema humile*). *Biological Invasions 10*, 25–35.

———and Arias de Reyna, L. (2003) Impact of human dwellings on the distribution of the exotic Argentine ant: a case study in the Donana National Park, Spain. *Biological Conservation 115*, 279–89.

———(2005) Impact of Argentine ants *Linepithema humile* on an arboreal ant community in Doñana National Park, Spain. *Biodiversity and Conservation 14*, 151–63.

—Retana, J., Cerdá, X., Reyes-López, J., and Arias de Reyna, L. (2007) Exploitative strategies of the invasive Argentine ant (*Linepithema humile*) and native ant species in a southern Spanish pine forest. *Environmental Entomology 36*, 1100–11.

Carroll, C.R., and Janzen, D.H. (1973) Ecology of foraging by ants. *Annual Review of Ecology and Systematics 4*, 231–57.

Carson, R. (1962) *Silent Spring*. Houghton Mifflin, Boston, MA.

Carvalho, K.S., and Vasconcelos, H.L. (1999) Forest fragmentation in central Amazonia and its effects on litter-dwelling ants. *Biological Conservation 91*, 151–7.

Carver, M., Gross, G.F., and Woodward, T.E. (1991) Hemiptera. In I.D. Naumann, ed. *The Insects of Australia*. Melbourne University Press, Melbourne.

Case, T.J. (2000) *An Illustrated Guide to Theoretical Ecology*. Oxford University Press, USA.

—and Gilpin, M.E. (1974) Interference competition and niche theory. *Proceedings of the National Academy of Sciences USA 71*, 3073–7.

Causton, C.E., Sevilla, C.R., and Porter, S.D. (2005) Eradication of the little fire ant, *Wasmannia auropunctata* (Roger) (Hymenoptera: Formicidae), from Marchena Island, Galapàgos: on the edge of success? *Florida Entomologist 88*, 159–67.

Cerdá, X. (1998) Critical thermal limits in Mediterranean ant species: trade-offs between mortality risk and foraging performance. *Functional Ecology 12*, 45–55.

—and Retana, J. (1998) Interference interactions and nest usurpation between two subordinate ant species. *Oecologia 113*, 577–83.

———and Cros, S. (1997) Thermal disruption of transitive hierarchies in Mediterranean ant communities. *Journal of Animal Ecology 66*, 363–74.

———and De Haro, A. (1994) Social carrying between nests in polycalic colonies of the monogynous ant, *Cataglyphis iberica* (Hymenoptera: Formicidae). *Sociobiology 23*, 215–31.

———and Manzaneda, A. (1988) The role of competition by dominants and temperature in the foraging of subordinate species in Mediterranean ant communities. *Oecologia 117*, 404–12.

Chacón, P., Bustos, J., Aldana, R.C., and Baena, M.L. (2000) Control de la hormiga loca *Paratrechina fulva* (Hymenoptera: Formicidae), con cebos tóxicos en la Reserva Natural Laguna de Sonso (Valle, Colombia). *Revista Colombiana de Entomología 26*, 151–6.

Chamberlain, S.C., and Holland, J.N. (2009) Quantitative synthesis of context dependency in ant-plant protection mutualisms. *Ecology 90*, 2384–92.

Chang, V.C.S. (1985) Colony revival and notes on rearing and life history of the big-headed ant. *Proceedings of the Hawaiian Entomological Society 25*, 53–8.

Chapuisat, M., Goudet, J., and Keller, L. (1997) Microsatellites reveal high population viscosity and limited dispersal in the ant *Formica paralugubris*. *Evolution 51*, 475–82.

—Bernasconi, G., Hoehn, S., and Reuter, M. (2005) Nestmate recognition in the unicolonial ant *Formica paralugubris*. *Behavioral Ecology 16*, 15–9.

—Oppliger, A., Magliano, P., and Christe, P. (2007) Wood ants use resin to protect themselves against pathogens. *Proceedings of the Royal Society of London B 274*, 2013–7.

Chéron, B., Doums, C., and Monnin, T. (2008) The mechanism of queen selection during queen replacement in a fission-performing ant. In *Proceedings of the 4th European Meeting of the International Union for the Study of Social Insects. La Roche-en-Ardenne, Belgium.*

Cherrett, J.M. (1968) The foraging behaviour of *Atta cephalotes* L. (Hymenoptera: Formicidae), I: Foraging pattern and plant species attacked in tropical rain forest. *Journal of Animal Ecology 37*, 387–403.

—(1972) Some factors involved in the selection of vegetable substrate by *Atta cephalotes* L. (Hymenoptera: Formicidae) in tropical rain forest. *Journal of Animal Ecology 41*, 647–60.

Choe, J.C., and Perlman, D.L. (1997) Social conflict and cooperation among founding queens in ants (Hymenoptera: Formicidae). In J.C. Choe and B.J. Crespi, eds.

The Evolution of Social Behavior in Insects and Arachnids, pp. 392–406. Cambridge University Press, Cambridge.

Chown, S.L., Sinclair, B.J., Leinaas, H.P., and Gaston, K.J. (2004) Hemispheric asymmetries in biodiversity – A serious matter for ecology. *PLoS Biology 2*, e406. doi:10.1371/journal.pbio.0020406.

Christian, C.E. (2001) Consequences of a biological invasion reveal the importance of mutualism for plant communities. *Nature 413*, 635–9.

——and Stanton, M.L. (2004) Cryptic consequences of a dispersal mutualism: seed burial, elaiosome removal and seed-bank dynamics. *Ecology 85*, 1101–10.

Clark, D.B., Guayasamín, C., Pazmiño, O., Donoso, C., Páez de and Villacís, Y. (1982) The tramp ant *Wasmannia auropunctata*: autecology and effects on ant diversity and distribution on Santa Cruz Island, Galapagos. *Biotropica 14*, 196–207.

Clark, J. (1934) Notes on Australian ants, with descriptions of new species and a new genus. *Memoirs of the National Museum of Victoria 8*, 5–20.

Clarke, A., and Gaston, K.J. (2006) Climate, energy and diversity. *Proceedings of the Royal Society of London B 273*, 2257–66.

Clarke, C.M., and Kitching, R.L. (1995) Swimming ants and pitcher plants: a unique ant-plant interaction from Borneo. *Journal of Tropical Ecology 11*, 589–602.

Clémencet, J., Rome, Q., Fédérici, P., and Doums, C. (2007) Aggressions and size-related fecundity of queenless workers in the ant *Cataglyphis cursor*. *Naturwissenschaften 95*, 133–9.

——Viginier, B., and Doums, C. (2005) Hierarchical analysis of population genetic structure in the monogynous ant *Cataglyphis cursor* using microsatellite and mitochondrial DNA markers. *Molecular Ecology 14*, 3735–44.

Clouse, R.M. (2007) The ants of Micronesia (Hymenoptera: Formicidae). *Micronesica 39*, 171–295.

Coenen-Staß, D., Schaarschmidt, B., and Lamprecht, I. (1980) Temperature distribution and calorimetric determination of heat-production in the nest of the wood ant, *Formica polyctena* (Hymenoptera, Formicidae). *Ecology 61*, 238–44.

Colautti, R., and Richardson, D. (2009) Subjectivity and flexibility in invasion terminology: too much of a good thing? *Biological Invasions 11*, 1225–9.

Colby, D., Husseneder, C., and Foil, L. (2008) Microsatellite genotyping of red imported fire ant (Hymenoptera: Formicidae) colonies reveals that most colonies persist in plowed pastures. *Journal of Economic Entomology 101*, 1062–7.

Cole, A.C. (1940) A guide to the ants of Great Smoky Mountains National Park, Tennessee. *The American Midland Naturalist 24*, 1–88.

Cole, B.J. (1983) Assembly of mangrove ant communities: patterns of geographical distribution. *Journal of Animal Ecology 52*, 339–47.

——and Wiernasz, D.C. (2002) Recruitment limitation and population density in the harvester ant. *Pogonomyrmex occidentalis 83*, 1433–42.

Cole, F.R., Medeiros, A.C., Loope, L.L., and Zuehlke, W. W. (1992) Effects of the Argentine ant on arthropod fauna of Hawaiian high-elevation shrubland. *Ecology 73*, 1313–22.

Collett, T.S., and Graham, P. (2004) Animal navigation: path integration, visual landmarks and cognitive maps. *Current Biology 14*, R475-7.

Collingwood, C.A., Pohl, H., Guesten, R., Wranik, W., and van Harten, A. (2004) The ants (Insecta: Hymenoptera: Formicidae) of the Socotra Archipelago. *Fauna of Arabia 20*, 473–95.

Collins, N.M. (1980) The distribution of soil macrofauna on the West Ridge of Gunung (Mount) Mulu, Sarawak. *Oecologia 44*, 262–75.

Colwell, R.K., and Hurtt, G.C. (1994) Nonbiological gradients in species richness and a spurious Rapoport effect. *American Naturalist 144*, 570–95.

——and Lees, D.C. (2000) The mid-domain effect: geometric constraints on the geography of species richness. *Trends in Ecology and Evolution 15*, 70–6.

——Brehm, G., Cardelús, C.L., Gilman, A.C., and Longino, J.T. (2008) Global warming, elevational range shifts, and lowland biotic attrition in the wet tropics. *Science 322*, 258–61.

——Rahbek, C., and Gotelli, N.J. (2004) The mid-domain effect and species richness patterns: what have we learned so far? *American Naturalist 163*, e1–e23.

Commonwealth of Australia (2006) *Threat Abatement Plan to Reduce the Impacts of Tramp Ants on Biodiversity in Australia and its Territories*. Department of the Environment and Heritage, Canberra.

Connell, J.H. (1980) Diversity and the coevolution of competitors, or the ghost of competition past. *Oikos 35*, 131–8.

——(1983) On the prevalence and relative importance of interspecific competition – evidence from field experiments. *American Naturalist 122*, 661–96.

Connor, E.F., and Simberloff, D. (1979) The assembly of species communities: chance or competition? *Ecology 60*, 1132–40.

————(1986) Competition, scientific method, and null models in ecology. *American Scientist 74*, 155–62.

Cook, S.C., and Davidson, D.W. (2006) Nutritional and functional biology of exudate-feeding ants. *Entomologia Experimentalis et Applicata 118*, 1–10.

Coppler, L.B., Murphy, J.F., and Eubanks, M.D. (2007) Red imported fire ants (Hymenoptera: Formicidae) increase the abundance of aphids in tomato. *Florida Entomologist 90*, 419–25.

Copren, K.A., Nelson, L.J., Vargo, E.L., and Haverty, M.I. (2005) Phylogenetic analyses of mtDNA sequences corroborate taxonomic designations based on cuticular hydrocarbons in subterranean termites. *Molecular Phylogenetics and Evolution 35*, 689–700.

Corbara, B., Dejean, A., and Orivel, J. (1999) Ant gardens, a unique epiphyte-ant association. *Année Biologique 38*, 73–89.

Corin, S.E., Abbott, K.L., Ritchie, P.A., and Lester, P.J. (2007a) Large scale unicoloniality: the population and colony structure of the invasive Argentine ant (*Linepithema humile*) in New Zealand. *Insectes Sociaux 54*, 275–82.

——Lester, P.J., Abbott, K.L., and Richie, P.A. (2007b) Inferring historical introduction pathways with mitochondrial DNA: the case of introduced Argentine ants (*Linepithema humile*) into New Zealand. *Diversity and Distributions 13*, 510–8.

Cornelius, M.L., and Grace, J.K. (1997) Influence of brood on the nutritional preferences of the tropical ant species, *Pheidole megacephala* (F.) and *Ochetellus glaber* (Mayr). *Journal of Entomological Science 32*, 421–9.

————and Yates, J.R., III (1996) Acceptability of different sugars and oils to three tropical ant species (Hymen., Formicidae). *Anzeiger für Schädlingskunde Pflanzenschutz Umweltschutz 69*, 41–3.

Costa, A.N., Vasconcelos, H.L., Vieira-Neto, E.H.M., and Bruna, E.M. (2008) Do herbivores exert top-down effects in Neotropical savannas? Estimates of biomass consumption by leaf-cutter ants (*Atta* spp.) in a Brazilian Cerrado site. *Journal of Vegetation Science 19*, 849–54.

Costanza, R., D'Arge, R., de Groot, R., Farber, S., Grasso, M., Hannon, B., Limburg, K., Naeem, S., O'Neill, R.V., Paruelo, J., Raskin, R.G., Sutton, P., and ven den Belt, M. (1997) The value of the world's ecosystem services and natural capital. *Nature 387*, 253–260.

Couzin, I.D., and Franks, N.R. (2003) Self-organized lane formation and optimized traffic flow in army ants. *Proceedings of the Royal Society of London B 270*, 139–46.

Cox, C.B. (2001) The biogeographic regions reconsidered. *Journal of Biogeography 28*, 511–23.

Coyle, F.A. (1966) Defensive behavior and associated morphological features in three species of the ant genus *Paracryptocerus*. *Insectes Sociaux 13*, 93–104.

Cremer, S., Armitage, S.A.O., Schmid-Hempel, P. (2007) Social immunity. *Current Biology 17*, R693-R702.

——Ugelvig, L.V., Drijfhout, F.P., Schlick-Steiner, B.C., Steiner, F.M., Cremer, S., Ugelvig, L.V., Drijfhout, F.P., Schlick-Steiner, B.C., Steiner, F.M., Seifert, B., Hughes, D.P., Schulz, A., Petersen, K.S., Konrad, H., Stauffer, C., Kiran, K., Espadaler, X., d'Ettorre, P., Aktac, N., Eilenberg, J., Jones, G.R., Nash, D.R., Pedersen, J.S., and Boomsma, J.J. (2008) The evolution of invasiveness in garden ants. *PLoS ONE 3*, e3838. doi:10.1371/journal.pone.0003838.

————Lommen, S.T.E., Petersen, K.S., and Pedersen, J.S. (2006) Attack of the invasive garden ant: aggression behaviour of *Lasius neglectus* (Hymenoptera: Formicidae) against native *Lasius* species in Spain. *Myrmecologische Nachrichten 9*, 13–9.

Crist, T.O. (2008) Biodiversity, species interactions, and functional roles of ants (Hymenoptera: Formicidae) in fragmented landscapes: a review. *Myrmecological News 12*, 3–13.

——and Haefner, J.W. (1994) Spatial model of movement and foraging in harvester ants (*Pogonomyrmex*) (II): the roles of environment and seed dispersion. *Journal of Theoretical Biology 166*, 315–23.

Croll, D.A., Maron, J.L., Estes, J.A., Danner, E.M., and Byrd, G.V. (2005) Introduced predators transform subarctic islands from grassland to tundra. *Science 307*, 1959–61.

Crozier, R.H. (1979) Genetics of sociality. In H.R. Hermann, ed. *Social Insects*, pp. 223–86. Academic Press, New York.

——(1986) Genetic clonal recognition abilities in marine invertebrates must be maintained by selection for something else. *Evolution 40*, 1100–1.

——(1987) Genetic aspects of kin recognition: concepts, models, and synthesis. In D.J.C. Fletcher and C.D. Michener, eds. *Kin Recognition in Animals*, pp. 55–73. John Wiley, New York.

——(1989) Kin recognition using innate labels: a central role for piggy-backing? In R.K. Grosberg, D. Hedgecock, and K. Nelson, eds. *Invertebrate Historecognition*, pp. 143–56. Plenum Press, New York and London.

——(2008) Advanced eusociality, kin selection and male haploidy. *Australian Journal of Entomology 47*, 2–8.

——and Dix, M.W. (1979) Analysis of two genetic models for the innate components of colony odor in social Hymenoptera. *Behavioral Ecology and Sociobiology*, *4*, 217–24.

——and Pamilo, P. (1996) *Evolution of Social Insect Colonies. Sex Allocation and Kin Selection.* Oxford University Press, Oxford.

——Jermiin, L.S., and Chiotis, M. (1997) Molecular evidence for a Jurassic origin of ants. *Naturwissenschaften 84*, 22–3.

——Pamilo, P., and Crozier, Y.C. (1984) Relatedness and microgeographic genetic variation in *Rhytidoponera mayri*, an Australian arid-zone ant. *Behavioral Ecology and Sociobiology 15*, 143–50.

Csosz, S., Radchenko, A., and Schulz, A. (2007) Taxonomic revision of the Palaearctic *Tetramorium chefketi* species complex (Hymenoptera: Formicidae). *Zootaxa 1405*, 1–38.

Cuatle, M., Rico-Gray, V., and Diaz-Castelazo, C. (2005) Effects of ant behaviour and presence of extrafloral nectaries on seed dispersal of the Neotropical myrmecochore *Turnera ulmifolia* L. (Turneraceae). *Biological Journal of the Linnaean Society London 86*, 67–77.

Currie, C.R. (2001) A community of ants, fungi, and bacteria: a multilateral approach to studying symbiosis. *Annual Review of Microbiology 55*, 357–80.

——and Stuart, A.E. (2001) Weeding and grooming of pathogens in agriculture by ants. *Proceedings of the Royal Society of London B 268*, 1033–9.

——Bot, A.N.M., and Boomsma, J.J. (2003) Experimental evidence of a tripartite mutualism: bacteria protect ant fungus gardens from specialized parasites. *Oikos 101*, 91–102.

——Poulsen, M., Mendenhall, J., Boomsma, J.J., and Billen, J. (2006) Coevolved crypts and exocrine glands support mutualistic bacteria in fungus-growing ants. *Science 311*, 81–3.

——Scottt J.A., Summerbell, R.C., and Malloch, D. (1999) Fungus-growing ants use antibiotic-producing bacteria to control garden parasites. *Nature 398*, 701–4.

Cushman, J.H. (1991) Host-plant mediation of insect mutualisms: variable outcomes in herbivore-ant interactions. *Oikos 61*, 138–44.

——and Addicott, J.F. (1989) Intraspecific and interspecific competition for mutualists – Ants as a limited and limiting resource for aphids. *Oecologia 79*, 315–21.

——and Whitham, T.G. (1989) Conditional mutualism in a membracid-ant association: temporal, age-specific, and density dependent effects. *Ecology 70*, 1040–7.

————(1991) Competition mediating the outcome of mutualism: protective services of ants as a limiting resource for membracids. *American Naturalist 138*, 851–65.

——Lawton, J.H., and Manly, B.F.J. (1993) Latitudinal patterns in European ant assemblages: variation in species richness and body size. *Oecologia 95*, 30–7.

——Martinsen, G.D., and Mazeroll, A.I. (1988) Density-dependent and size-dependent spacing of ant nests: evidence for intraspecific competition. *Oecologia 77*, 522–5.

——Rashbrook, V.K., and Beattie, A.J. (1994) Assessing benefits to both participants in a Lycaenid-ant association. *Ecology 75*, 1031–41.

Cuvillier-Hot, V., Renault, V., and Peeters, C. (2005) Rapid modification in the olfactory signal of ants following a change in reproductive status. *Naturwissenschaften 92*, 73–7.

——Lenoir, A., Crewe, R., Malosse, C., and Peeters, C. (2004) Fertility signalling and reproductive status in queenless ants. *Animal Behaviour 68*, 1209–19.

d'Ettorre, P., and Heinze, J. (2005) Individual recognition in ant queens. *Current Biology 15*, 2170–4.

——Brunner, E., Wenseleers, T., and Heinze, J. (2004) Knowing your enemies: seasonal dynamics of host-social parasite recognition. *Naturwissenschaften 91*, 594–7.

——Mondy, N., Lenoir, A., and Errard, C. (2002a) Blending in with the crowd: social integration into their host colonies using a flexible signature. *Proceedings of the Royal Society of London B 269*, 1911–8.

——Mora, P., Dibangou, V., Rouland, C., and Errard, C. (2002b) The role of the symbiotic fungus in the digestive metabolism of two species of fungus-growing ants. *Journal of Comparative Physiology B 172*, 169–76.

Daane, K.M., Sime, K.R., Fallon, J., and Cooper, M.L. (2007) Impacts of Argentine ants on mealybugs and their natural enemies in California's coastal vineyards. *Ecological Entomology 32*, 583–96.

Dadd, R.H. (1985) Nutrition: organisms. In G.A. Kerkut and L.I. Gilbert, eds. *Comprehensive Insect Physiology, Biochemistry and Pharmacology*, pp. 313–90. Pergamon Press, Oxford.

Dahbi, A., and Lenoir, A. (1998) Nest separation and the dynamics of the gestalt odor in the polydomous ant *Cataglyphis iberica* (Hymenoptera, Formicidae). *Behavioral Ecology and Sociobiology 42*, 349–55.

————Tinaut, A., Taghizadeh, T., Francke, W., and Hefetz, A. (1996) Chemistry of the postpharyngeal gland secretion and its implication for the phylogeny of Iberian *Cataglyphis* species (Hymenoptera: Formicidae). *Chemoecology 7*, 163–71.

Dalecky, A., Renucci, M., Tirard, A., *et al.* (2007) Changes in composition of cuticular biochemicals of the facultatively polygynous ant *Petalomyrmex phylax* during range expansion in Cameroon with respect to social, spatial and genetic variation. *Molecular Ecology 16*, 3778–91.

Daniels, H., Gottsberger, G., and Fiedler, K. (2005) Nutrient composition of larval nectar secretions from three species of myrmecophilous butterflies. *Journal of Chemical Ecology 31*, 2805–21.

Darwin, C. (1859) *On the Origin of Species by Means of Natural Selection*. W. Clowes and Sons, London.

Dauber, J., and Wolters, V. (2004) Edge effects on ant community structure and species richness in an agricultural landscape. *Biodiversity and Conservation 13*, 901–15.

——Bengtsson, J., and Lenoir, L. (2006a) Evaluating effects of habitat loss and land-use continuity on ant species richness in seminatural grassland remnants. *Conservation Biology 20*, 1150–60.

——Rommeler, A., and Wolters, V. (2006b) The ant *Lasius flavus* alters the viable seed bank in pastures. *European Journal of Soil Biology 42*, S157-63.

——Schroeter, D., and Wolters, V. (2001) Species specific effects of ants on microbial activity and N-availability in the soil of an old-field. *European Journal of Soil Biology 37*, 259–61.

Davic, R.D. (2003) Linking keystone species and functional groups: a new operational definition of the keystone species concept. *Conservation Ecology 7*, r11. http://www.consecol.org/vol17/iss1/resp11/.

Davidson, D.W. (1988) Ecological studies of Neotropical ant gardens. *Ecology 69*, 1138–52.

Davidson, D.W. (1997) The role of resource imbalances in the evolutionary ecology of tropical arboreal ants. *Biological Journal of the Linnaean Society London 61*, 153–81.

——(1998) Resource discovery versus resource domination in ants: a functional mechanism for breaking the trade-off. *Ecological Entomology 23*, 484–90.

——and McKey, D. (1993) The evolutionary ecology of symbiotic ant-plant relationships. *Journal of Hymenoptera Research 2*, 13–83.

——and Morton, S.R. (1981) Myrmecochory in some plants (F. Chenopodiaceae) of the Australian arid zone. *Oecologia 50*, 357–66.

——Cook, S.C., and Snelling, R.R. (2004) Liquid-feeding performances of ants (Formicidae): ecological and evolutionary implications. *Oecologia 139*, 255–66.

——Snelling, R.R., and Longino, J.T. (1989) Competition among ants for myrmecophytes and the significance of plant trichomes. *Biotropica 21*, 64–73.

——Cook, S.C., Snelling, R.R., and Chua, T.H. (2003) Explaining the abundance of ants in lowland tropical rainforest canopies. *Science 300*, 969–72.

——Lessard, J.P., Bernau, C.R., and Cook, S.C. (2007) The tropical ant mosaic in a primary Bornean rain forest. *Biotropica 39*, 468–75.

Davis, N.E., O'Dowd, D.J., Green, P.T., and MacNally, R. (2008) Effects of an alien ant invasion on abundance, behavior, and reproductive success of endemic island birds. *Conservation Biology 22*, 1165–76.

Dawkins, R. (1976) *The Selfish Gene*. Oxford University Press, Oxford.

——(1982) *The Extended Phenotype*. Oxford University Press, Oxford.

de Andrade, M.L., and Baroni-Urbani, C. (1999) Diversity and adaptation in the ant genus *Cephalotes*, past and present. *Stuttgarter Beiträge zur Naturkunde Serie B (Geologie und Paläontologie) 271*, 1–889.

de Bruyn, L.A.L. (1999) Ants as bioindicators of soil function in rural environments. *Agriculture Ecosystems & Environment 74*, 425–41.

De Heer, C.J., and Ross, K.G. (1997) Lack of detectable nepotism in multiple-queen colonies of the fire ant *Solenopsis invicta* (Hymenoptera: Formicidae). *Behavioral Ecology and Sociobiology 40*, 27–33.

de la Fuente, M.A.S., and Marquis, R.J. (1999) The role of ant-tended extrafloral nectaries in the protection and benefit of a Neotropical rainforest tree. *Oecologia 118*, 192–202.

Debout, G., Schatz, B., Elias, M., and McKey, D. (2007) Polydomy in ants: what we know, what we think we know, and what remains to be done. *Biological Journal of the Linnean Society 90*, 319–48.

Debuse, V.J., King, J., and House, A.P.N. (2007) Effect of fragmentation, habitat loss and within-patch habitat characteristics on ant assemblages in semi-arid woodlands of eastern Australia. *Landscape Ecology 22*, 731–45.

Decaëns, T., Asakawa, N., Galvis, J.H., Thomas, R.J., and Amezquita, E. (2002) Surface activity of soil ecosystem engineers and soil structure in contrasted land use systems of Colombia. *European Journal of Soil Biology 38*, 267–71.

Dechaume-Moncharmont, F.X., Dornhaus, A., Houston, A.I., McNamara, J.M., Collins, E.J., and Franks, N.R. (2005) The hidden cost of information in collective foraging. *Proceedings of the Royal Society of London B 272*, 1689–95.

DeFauw, S.L., Vogt, J.T., and Boykin, D.L. (2008) Influence of mound construction by red and hybridized imported fire ants on soil chemical properties and turfgrass in a sod production agroecosystem. *Insectes Sociaux 55*, 301–12.

Dejean, A., and Beugnon, G. (1991) Persistent intercolonial trunkroute-marking in the African weaver ant *Oecophylla longinoda* Latreille (Hymenoptera, Formicidae): tom Thumb's versus Ariadne's orienting strategies. *Ethology 88*, 89–98.

——and Corbara, B. (2003) A review of mosaics of dominant ants in rainforests and plantations. In Y. Basset, V. Novotny, S.E. Miller, and R.L. Kitching, eds. *Arthropods of Tropical Forests: spatio-temporal dynamics and resource use in the canopy*, pp. 341–7. Cambridge University Press, Cambridge.

——and Gibernau, M. (2000) A rainforest ant mosaic: the edge effect (Hymenoptera: Formicidae). *Sociobiology 35*, 385–402.

——Kenne, M., and Moreau, C.S. (2007b) Predatory abilities favour the success of the invasive ant *Pheidole megacephala* in an introduced area. *Journal of Applied Entomology 131*, 625–9.

——Corbara, B., Orivel, J., and Leponce, M. (2007a) Rainforest canopy ants: the implications of territoriality and predatory behavior. *Functional Ecosystems and Communities 1*, 105–20.

——Moreau, C.S., Kenne, M., and Leponce, M. (2008) The raiding success of *Pheidole megacephala* on other ants in both its native and introduced ranges. *Comptes Rendus Biologies 331*, 631–5.

——Le Breton, J., Suzzoni, J.P., Orivel, J., and Saux-Moreau, C. (2005a) Influence of interspecific competition on the recruitment behavior and liquid food transport in the tramp ant species *Pheidole megacephala*. *Naturwissenschaften 92*, 324–7.

——Solano, P.J., Ayroles, J., Corbara, B., and Orivel, J. (2005b) Arboreal ants build a trap to ambush and capture prey. *Nature 434*, 793.

Dekoninck, W. (2007) Acceptance of two native myrmecophilous species, *Platyarthrus hoffmannseggii* (Isopoda:

oniscidea) and *Cyphoderus albinos* (Collembola: Cyphoderidae) by the introduced invasive garden ant *Lasius neglectus* (Hymenoptera: Formicidae) in Belgium. *European Journal of Entomology 104*, 159–61.

Del-Claro, K., and Oliveira, P. (1996) Honeydew flicking by treehoppers provides cues to potential tending ants. *Animal Behaviour 51*, 1071–5.

——and Oliveira, P.S. (2000) Conditional outcomes in a neotropical treehopper-ant association: temporal and species-specific variation in ant protection and homopteran fecundity. *Oecologia 124*, 156–65.

Delabie, J.H.C. (2001) Trophobiosis between Formicidae and Hemiptera (Sternorrhyncha and Auchenorrhyncha): an overview. *Neotropical Entomology 30*, 501–16.

——Agosti, D., and do Nascimento, I. (2000) Litter ant communities of the Brazilian Atlantic rain forest region. In D. Agosti, J. Majer, L. Alonso, and T.R. Schultz, eds. *Sampling Ground-dwelling Ants: case studies from the world's rain forests*. Curtin University of Technology School of Environmental Biology Bulletin No. *18*.

Delaplane, K.S., and LaFage, J.P. (1990) Wood excavations of three species of subterranean termites. *Entomologia Experimentalis et Applicata 54*, 81–7.

Depickère, S., Fresneau, D., Detrain, C., and Deneubourg, J.L. (2004) Marking as a decision factor in the choice of new nesting site in *Lasius niger*. *Insectes Sociaux 51*, 243–6.

Des Granges, J.L. (1979) Organization of a tropical nectar feeding bird guild in a variable environment. *Living Bird 17*, 199–236.

Deslippe, R.J., and Savolainen, R. (1995) Sex investment in a social insect: the proximate role of food. *Ecology 76*, 375–82.

Detrain, C., and Pasteels, J.M. (1992) Caste polyethism and collective defense in the ant, *Pheidole pallidula*: the outcome of quantitative differences in recruitment. *Behavioral Ecology and Sociobiology 29*, 405–12.

——Deneubourg, J.-L., and Pasteels, J.M. (1999) *Information processing in social insects*. Birkhäuser, Basel.

Dettner, K., and Liepert, C. (1994) Chemical mimicry and camouflage. *Annual Review of Entomology 39*, 129–54.

Deutsch, C.A., Tewksbury, J.J., and Huey, R.B. (2008) Impacts of climate warming on terrestrial ectotherms across latitude. *Proceedings of the National Academy of Sciences USA 105*, 6668–72.

Devigne, C., and Detrain, C. (2002) Collective exploitation and area marking in the ant *Lasius niger*. *Insectes Sociaux 49*, 357–62.

DeVries, P.J. (1991) Mutualisms between *Thisbe irenea* butterflies and ants, the role of ant ecology in the evolution of larval-ant associations. *Biological Journal of the Linnean Society 43*, 179–95.

——and Baker, I. (1989) Butterfly exploitation of an ant-plant mutualism: adding insult to herbivory. *Journal of the New York Entomological Society 97*, 332–40.

Deyrup, M. (2003) An updated list of Florida ants (Hymenoptera: Formicidae). *Florida Entomologist 86*, 43–8.

——and Cover, S. (2004) A new species of the ant genus *Leptothorax* from Florida, with a key to the *Leptothorax* of the southeast (Hymenoptera: Formicidae). *Florida Entomologist 87*, 51–9.

Diamond, J.M. (1975) Assembly of species communities. In M.L. Cody and J.M. Diamond, eds. *Ecology and Evolution of Communities*, pp. 342–444. Harvard University Press, Cambridge, MA.

——and Case, T.J. (1986) Overview: introductions, extinctions, exterminations, and invasions. In J.M. Diamond and T.J. Case, eds. *Community Ecology*, pp. 65–79. Harper and Row, New York.

Díaz-Castelazo, C., and Rico-Gray, V. (1998) Numero y variacion estacional de asociaciones hormiga-planta en un bosque montano bajo de Veracruz, Mexico. *Acta Zoologica Mexicana (nueva serie) 73*, 45–55.

————Oliveira, P.S., and Cuatle, M. (2004) Extrafloral nectary-mediated ant-plant interactions in the coastal vegetation of Veracruz, Mexico: richness, occurrence, seasonality, and ant foraging patterns. *Ecoscience 11*, 472–81.

Diaz, R., Knutson, A., and Bernal, J.-S. (2004) Effect of the red imported fire ant on cotton aphid population density and predation of bollworm and beet armyworm eggs. *Journal of Economic Entomology 97*, 222–9.

Dietrich, C.O., Schlick, B., and Steiner, F. (1998) Ameisen bei Hochwasser (Hymenoptera: Formicidae) – Beobachtungen in Ostösterreich im Juli 1997. *Myrmecologische Nachrichten 2*, 35–41.

DiGirolamo, L.A., and Fox, L.R. (2006) The influence of abiotic factors and temporal variation on local invasion patterns of the Argentine ant (*Linepithema humile*). *Biological Invasions 8*, 125–35.

Dijkstra, M.B., and Boomsma, J.J. (2003) *Gnamptogenys hartmani* Wheeler (Ponerinae: Ectatommini): an agro-predator of *Trachymyrmex* and *Sericomyrmex* fungus-growing ants. *Naturwissenschaften 90*, 568–71.

——Nash, D.R., and Boomsma, J.J. (2005) Self-restraint and sterility in workers of *Acromyrmex* and *Atta* leafcutter ants. *Insectes Sociaux 52*, 67–76.

Dill, M., Williams, D.J., and Maschwitz, U. (2002) Herdsmen ants and their mealybug partners. *Abhandlungen der Senckenbergischen Naturforschenden Gesellschaft 557*, 1–373.

Diniz, J.L.M., and Brandão, C.R.F. (1993) Biology and myriapod predation by the neotropical myrmicine ant

Stegomyrmex vizottoi (Hymenoptera: Formicidae). *Insectes Sociaux 40*, 301–11.

Djiéto-Lordon, C., Dejean, A., Gibernau, M., Hossaert-McKey, M., and McKey, D. (2004) Symbiotic mutualism with a community of opportunistic ants: protection, competition, and ant occupancy of the myrmecophyte *Barteria nigritana* (Passifloraceae). *Acta Oecologica 26*, 109–16.

Dlussky, G.M., and Rasnitsyn, A.P. (2003) Ants (Hymenoptera: Formicidae) of Formation Green River and some other Middle Eocene deposits of North America. *Russian Entomological Journal 11*, 411–36.

——Brothers, D.J., and Rasnitsyn, A.P. (2004) The first Late Cretaceous ants (Hymenoptera: Formicidae) from southern Africa, with comments on the origin of the Myrmicinae. *Insect Systematics and Evolution 35*, 1–13.

Domisch, T., Finér, L., and Jurgensen, M.F. (2005) Red wood ant mound densities in managed boreal forests. *Annales Fennici Zoologici 42*, 277–82.

Don, W. (2008) *Ants of New Zealand*. Otago University Press, New Zealand.

Dornhaus, A. (2008) Specialization does not predict individual efficiency in an ant. *PLoS Biology 6*, e285. doi: 10.1371/journal.pbio.0060285.

——and Franks, N.R. (2008) Individual and collective cognition in ants and other insects (Hymenoptera: Formicidae). *Myrmecological News 11*, 215–26.

——Franks, N., Hawkins, R.M., and Shere, H.N.S. (2004) Ants move to improve – colonies of *Leptothorax albipennis* emigrate whenever they find a superior nest site. *Animal Behaviour 67*, 959–63.

——Holley, J.-A., Pook, V.G., Worswick, G., and Franks, N. R. (2008) Why don't all workers work? Colony size and workload during emigrations in the ant *Temnothorax albipennis*. *Behavioral Ecology and Sociobiology 63*, 43–51.

——Klügl, F., Oechslein, C., Puppe, F., and Chittka, L. (2006b) Benefits of recruitment in honey bees: effects of ecology and colony size in an individual-based model. *Behavioral Ecology 17*, 336–44.

——Collins, E.J., Dechaume-Moncharmont, F.X., Houston, A.I., Franks, N.R., and McNamara, J.M. (2006a) Paying for information: partial loads in central place foragers. *Behavioral Ecology and Sociobiology 61*, 151–61.

Douglas, A.E. (1993) The nutritional quality of phloem sap utilized by natural aphid populations. *Ecological Entomology 18*, 31–8.

Doums, C., Cabrera, H., and Peeters, C. (2002) Population genetic structure and male-biased dispersal in the queenless ant *Diacamma cyaneiventre*. *Molecular Ecology 11*, 2251–64.

Drees, B.M., Berger, L.A., Cavazos, R., and Vinson, S.B. (1991) Factors affecting sorghum and corn seed predation by foraging red imported fire ants (Hymenoptera: Formicidae). *Journal of Economic Entomology 84*, 285–9.

Dreier, S., van Zweden, J.S., d'Ettorre, P. (2007) Long-term memory of individual identity in ant queens. *Biology Letters 3*, 459–62.

Dronnet, S., Lohou, C., Christides, J.-P., and Bagnères, A. G. (2006) Cuticular hydrocarbon composition reflects genetic relationship among colonies of the introduced termite *Reticulitermes santonensis* Feytaud. *Journal of Chemical Ecology 32*, 1027–42.

Dudley, R., Byrnes, G., Yanoviak, S.P., Borrell, B., Brown, R., and McGuire, J. (2007) Gliding and the functional origins of flight: biomechanical novelty or necessity? *Annual Review of Ecology Evolution and Systematics 38*, 179–201.

Dugatkin, L.A., and Reeve, H.K. (2000) *Game Theory and Animal Behavior*. Oxford University Press, USA.

Dumpert, K., Maschwitz, U., and Weissflog, A. (2006) Description of five new weaver ant species of *Camponotus* subgenus *Karavaievia* Emery, 1925 (Hymenoptera: Formicidae) from Malaysia and Thailand, with contribution to their biology, especially to colony foundation. *Myrmecologische Nachrichten 8*, 69–82.

Dunn, R.R. (2004) Managing the tropical landscape: a comparison of the effects of logging and forest conversion to agriculture on ants, birds and Lepidoptera. *Forest Ecology and Management 191*, 215–24.

——(2005) Modern insect extinctions, the neglected majority. *Conservation Biology 19*, 1030–6.

——(2008a) *Global Ant Community database*. http://www.antmacroecology.org/.

——(2008b) The trouble with biodiversity. *Seed Magazine, October*, http://seedmagazine.com/news/2008/2010/the_trouble_with_biodiversity.php.

——Colwell, R.K., and Nilsson, C. (2006) The river domain: why are there more species halfway up the river? *Ecography 29*, 251–9.

——Parker, C.R., and Sanders, N.J. (2007c) Temporal patterns of diversity: assessing the biotic and abiotic controls on ant assemblages. *Biological Journal of the Linnean Society 91*, 191–201.

——Sanders, N.J., Fitzpatrick, M.C., Laurent, E.J., Lessard, J.P., Agosti, D., Andersen, A.N., Brühl, C.A, Cerdá, X., Ellison, A.M., Fisher, B.L., Gibb, H., Gotelli, N.J., Gove, A.D., Guénard, B., Janda, M., Kaspari, M., Longino, J.T., Majer, J.D., McGlynn, T.P., Menke, S.B., Parr, C.L., Philpott, S.M., Pfeiffer, M., Retana, J., Suarez, A.V., and Vasconcelos, H.L. (2007d) Global ant (Hymenoptera: Formicidae) biodiversity and biogeography–a new database and its possibilities. *Myrmecological News 10*, 77–83.

——Parker, C.R., Geraghty, M., and Sanders, N.J. (2007b) Reproductive phenologies in a diverse temperate ant fauna. *Ecological Entomology 32*, 135–42.

——Gove, A.D., Barraclough, T.G., Givnish, T.J., and Majer, J.D. (2007a) Convergent evolution of an ant-plant mutualism across plant families, continents, and time. *Evolutionary Ecology Research 9*, 1349–62.

——Agosti, D., Andersen, A.N., Arnan, X., Brühl, C.A, Cerdá, X., Ellison, A.M., Fisher, B.L., Fitzpatrick, M.C., Gibb, H., Gotelli, N.J., Gove, A.D., Guénard, B., Janda, M., Kaspari, M., Laurent, E.J., Lessard, J.P., Longino, J.T., Majer, J.D., McGlynn, T.P., Menke, S.B., Parr, C.L., Philpott, S.M., Pfeiffer, M., Retana, J., Suarez, A.V., Vasconcelos, H.L., Weiser, M.D., and Sanders, N.J. (2009) Climatic drivers of hemispheric asymmetry in global patterns of ant species richness. *Ecology Letters 12*, 324–33.

Dussutour, A., and Simpson, S.J. (2008) Description of a simple synthetic diet for studying nutritional responses in ants. *Insectes Sociaux 55*, 329–33.

——Beshers, S., Deneubourg, J.L., and Fourcassie, V. (2007) Crowding increases foraging efficiency in the leaf-cutting ant *Atta colombica*. *Insectes Sociaux 54*, 158–65.

Dutcher, J.D., Estes, P.M., and Dutcher, M.J. (1999) Interactions in entomology: aphids, aphidophaga and ants in pecan orchards. *Journal of Entomological Science 34*, 40–56.

Dutra, H.P., Freitas, A.V.L., and Oliveira, P.S. (2006) Dual ant attraction in the Neotropical shrub *Urera baccifera* (Urticaceae): the role of ant visitation to pearl bodies and fruits in herbivore deterrence and leaf longevity. *Functional Ecology 20*, 252–60.

Dynesius, M., and Jansson, R. (2000) Evolutionary consequences of changes in species' geographical distributions driven by Milankovitch climate oscillations. *Proceedings of the National Academy of Sciences USA 97*, 9115–20.

Eastwood, R., and Fraser, A.M. (1999) Associations between lycaenid butterflies and ants in Australia. *Australian Journal of Ecology 24*, 503–37.

——Pierce, N.E., Kitching, R.L., and Hughes, J.M. (2006) Do ants enhance diversification in lycaenid butterflies? Phylogeographic evidence from a model myrmecophile, *Jalmenas evagoras*. *Evolution 60*, 315–27.

Edelstein-Keshet, L., Watmough, J., and Ermentrout, G.B. (1995) Trail following in ants: individual properties determine population behaviour. *Behavioral Ecology and Sociobiology 36*, 119–33.

Eguchi, K. (2008) A revision of Northern Vietnamese species of the ant genus *Pheidole* (Insecta: Hymenoptera: Formicidae: Myrmicinae). *Zootaxa 1902*, 1–118.

——Yoshimura, M., and Yamane, S. (2006) The Oriental species of the ant genus *Probolomyrmex* (Insecta: Hymenoptera: Formicidae: Proceratiinae). *Zootaxa 1376*, 1–35.

Ehmer, B., and Hölldobler, B. (1995) Foraging behavior of *Odontomachus bauri* on Barro Colorado Island, Panama. *Psyche (Cambridge) 102*, 215–24.

Eken, G., Bennun, L., and Brooks, T.M. (2004) Key biodiversity areas as site conservation targets. *BioScience 54*, 1110–8.

Elias, M., Rosengren, R., and Sundstrom, L. (2005) Seasonal polydomy and unicoloniality in a polygynous population of the red wood ant *Formica truncorum*. *Behavioral Ecology and Sociobiology 57*, 339–49.

Elmes, G.W., and Petal, J. (1990) Queen number as an adaptable trait: evidence from wild populations of two red ant species (genus *Myrmica*). *Journal of Animal Ecology 59*, 675–90.

——Thomas, J.A., Wardlaw, J.C., Hochberg, M.E., Clarke, R.T., and Simcox, D.J. (1998) The ecology of *Myrmica* ants in relation to the conservation of *Maculinea* butterflies. *Journal of Insect Conservation 2*, 67–78.

Elton, C.S. (1932) Territory among wood ants (*Formica rufa* L.) at Picket Hill. *Journal of Animal Ecology 1*, 69–76.

——(1958) *The Ecology of Invasions by Animals and Plants*. Methuen and Co. Ltd.

Endler, A., Liebig, J., Schmitt, T., Parker, J.E., Jones, G.R., Schreier, P., and Hölldobler, B. (2004) Surface hydrocarbons of queen eggs regulate worker reproduction in a social insect. *Proceedings of the National Academy of Sciences USA 101*, 2945–50.

Engel, M.S., and Grimaldi, D.A. (2005) Primitive new ants in Cretaceous amber from Myanmar, New Jersey, and Canada (Hymenoptera: Formicidae). *American Museum Novitates 3485*, 1–23.

Enquist, B.J., Haskell, J.P., and Tiffney, B.H. (2002) General patterns of taxonomic and biomass partitioning in extant and fossil plant communities. *Nature 419*, 610–3.

EoL (2008) *Encyclopedia of Life*. http://www.eol.org/home.html.

Epperson, D.M., and Heise, C.D. (2003) Nesting and hatchling ecology of gopher tortoises (*Gopherus polyphemus*) in southern Mississippi. *Journal of Herpetology 37*, 315–24.

Erickson, J.M. (1971) The displacement of native ant species by the introduced Argentine ant *Iridomyrmex humilis* Mayr. *Psyche (Cambridge) 78*, 257–66.

Errard, C. (1994a) Development of interpecific recognition behavior in the ants *Manica rubida* and *Formica selysi* (Hymenoptera: Formicidae) reared in mixed-species groups. *Journal of Insect Behaviour 7*, 83–99.

——(1994b) Long-term memory involved in nestmate recognition in ants. *Animal Behaviour 48*, 263–71.

——Hefetz, A., and Jaisson, P. (2006) Social discrimination tuning in ants: template formation and chemical similarity. *Behavioral Ecology and Sociobiology 59*, 9–14.

——Le Guisquet, A.M., Christides, J.P., Mercier, J., Lenoir, A., and Hefetz, A. (2008) Early learning of volatile chemical cues leads to interspecific recognition between two ant species. *Insectes Sociaux 55*, 115–22.

Erthal, M., Silvab, C.P., and Samuelsc, R.I. (2007) Digestive enzymes in larvae of the leaf cutting ant, *Acromyrmex subterraneus* (Hymenoptera: Formicidae: Attini). *Journal of Insect Physiology 53*, 1101–11.

Espadaler, X. (2005) *Monomorium destructor*, la hormiga de Singapur, detectada y detenida en el puerto de Barcelona (Hymenoptera, Formicidae). *Orsis 20*, 27–32.

Espadaler, X. and Rey, S. (2001) Biological constraints and colony founding in the polygynous invasive ant *Lasius neglectus* (Hymenoptera, Formicidae). *Insectes Sociaux 48*, 159–64.

——Tartally, A., Schultz, R., Seifert, B., and Nagy, C. (2007) Regional trends and preliminary results on the local expansion rate in the invasive garden ant, *Lasius neglectus* (Hymenoptera, Formicidae). *Insectes Sociaux 54*, 293–301.

Eubanks, M.D., Blackwell, S.A., Parrish, C.J., DeLamar, Z. D., and Hull-Sanders, H. (2002) Intraguild predation of beneficial arthropods by red imported fire ants in cotton. *Environmental Entomology 31*, 1168–74.

Evans, K.L., and Gaston, K.J. (2005) Can the evolutionary-rates hypothesis explain species-energy relationships? *Functional Ecology 19*, 899–915.

——Warren, P.H., and Gaston, K.J. (2005) Species-energy relationships at the macroecological scale: a review of the mechanisms. *Biological Reviews 80*, 1–25.

Fagan, W.F., Siemann, E., Mitter, C., Denno, R.F., Huberty, A.F., Woods, H.A., Elser, J.J. (2002) Nitrogen in insects: implications for trophic complexity and species diversification. *American Naturalist 160*, 784–802.

FAO (1997) *Framework for pest risk analysis*. https://www.ippc.int.

——(1998) *Guidelines for pest eradication programmes*. https://www.ippc.int.

——(2006) International Standards for Phytosanitary Measures 1–27, Secretariat of the International Plant Protection Convention.

Farji-Brener, A.G. (2001) Why are leaf-cutting ants more common in early secondary forests than in old-growth tropical forests? An evaluation of the palatable forage hypothesis. *Oikos 92*, 167–77.

——and Ghermandi, L. (2008) Leaf-cutting ant nests near roads increase fitness of exotic plant species in natural protected areas. *Proceedings of the Royal Society of London B 275*, 1431–40.

——and Illes, A.E. (2000) Do leaf-cutting ants make 'bottom-up' gaps in neotropical rain forests? A critical review of the evidence. *Ecology Letters 3*, 219–27.

——Corley, J.C., and Bettinelli, J. (2002) The effects of fire on ant communities in north-western Patagonia: the importance of habitat structure and regional context. *Diversity and Distributions 8*, 235–43.

Farley, J. (2008) The role of prices in conserving critical natural capital. *Conservation Biology 22*, 1399–408.

Federle, W., Maschwitz, U., Fiala, B., and Riederer, M. (1997) Slippery ant-plants and skilful climbers: selection and protection of specific ant partners by epicuticular wax blooms in *Macaranga* (Euphorbiaceae). *Oecologia 112*, 217–24.

Fedriani, J.M., Rey, P.J., Garrido, J.L., Guitián, J., Herrera, C.M., Medrano, M., Sánchez-Lafuente, A.M. and Cerdá, X. (2004) Geographical variation in the potential of mice to constrain an ant-seed dispersal mutualism. *Oikos 105*, 181–91.

Feener, D.H. (1981) Competition between ant species: outcome controlled by parasitic flies. *Science 214*, 815–7.

——(2000) Is the assembly of ant communities mediated by parasitoids? *Oikos 90*, 79–88.

——and Moss, K.A.G. (1990) Defense against parasites by hitchhikers in leaf-cutting ants: a quantitative assessment. *Behavioral Ecology and Sociobiology 26*, 17–29.

——and Schupp, E.W. (1998) Effect of treefall gaps on the patchiness and species richness of Neotropical ant assemblages. *Oecologia 116*, 191–201.

——Orr, M.R., Wackford, K.M., Longo, J.M., Benson, W. W., and Gilbert, L.E. (2008) Geographic variation in resource dominance-discovery in Brazilian ant communities. *Ecology 89*, 1824–36.

Feitosa, R.M., and Brandão, C.R.F. (2008) A taxonomic revision of the Neotropical myrmicine ant genus *Lachnomyrmex* Wheeler (Hymenoptera: Formicidae). *Zootaxa 1890*, 1–49.

————and Dietz, B.H. (2007) *Basiceros scambognathus* (Brown, 1949) n. comb., with the first worker and male descriptions, and a revised generic diagnosis (Hymenoptera: Formicidae: Myrmicinae). *Papeis Avulsos de Zoologia (São Paulo) 47*, 15–26.

————and Diniz, J.L.M. (2008) Revisionary studies on the enigmatic Neotropical ant genus *Stegomyrmex* Emery, 1912 (Hymenoptera: Formicidae: Myrmicinae), with the description of two new species. *Journal of Hymenoptera Research 17*, 64–82.

Feldhaar, H., Fiala, B., and Gadau, J. (2005) A shift in colony founding behaviour in the obligate plant-ant *Crematogaster* (*Decacrema*) morphospecies 2. *Insectes Sociaux 52*, 222–30.

——Foitzik, S., and Heinze, J. (2008) Lifelong commitment to the wrong partner: hybridization in ants. *Philosophical Transactions of the Royal Society B 363*, 2891–9.

——Fiala, B., Hashim, R., and Maschwitz, U. (2003) Patterns of the *Crematogaster-Macaranga* association: the ant partner makes the difference. *Insectes Sociaux 50*, 9–19.

——Straka, J., Krischke, M., Berthold, K., Stoll, S., Mueller, M.J. and Gross, R. (2007) Nutritional upgrading for omnivorous carpenter ants by the endosymbiont *Blochmannia*. *BMC Biology 5*, 48. doi:10.1186/1741-7007-5-48.

Fellers, J.H. (1987) Interference and exploitation in a guild of woodland ants. *Ecology 68*, 1466–78.

Felsenstein, J. (1985) Phylogenies and the comparative method. *American Naturalist 125*, 1–15.

Ferguson, B.G., Boucher, D.H., Pizzi, M., and Rivera, C. (1995) Recruitment and decay of a pulse of *Cecropia* in Nicaraguan rain forest damaged by hurricane Joan: relation to mutualism with *Azteca* ants. *Biotropica 27*, 455–60.

Fernández, F. (2003) Revision of the myrmicine ants of the *Adelomyrmex* genus-group (Hymenoptera: Formicidae). *Zootaxa 361*, 1–52.

——(2004) The American species of the myrmicine ant genus *Carebara* Westwood (Hymenoptera: Formicidae. *Caldasia 26*, 191–238.

——(2006) A new species of *Carebara* Westwood (Hymenoptera: Formicidae) and taxonomic notes on the genus. *Revista Colombiana de Entomología 32*, 97–9.

——(2007a) The myrmicine ant genus *Allomerus* Mayr (Hymenoptera: Formicidae). *Caldasia 29*, 159–75.

——(2007b) Two new South American species of *Monomorium* Mayr with taxonomic notes on the genus. *Memoirs of the American Entomological Institute 80*, 128–45.

——and Guerrero, R.J. (2008) *Technomyrmex* (Formicidae: dolichoderinae) in the New World: synopsis and description of a new species. *Revista Colombiana de Entomología 34*, 110–5.

Fernández-Escudero, I., Seppä, P., and Pamilo, P. (2001) Dependent colony founding in the ant *Proformica longiseta*. *Insectes Sociaux 48*, 80–2.

Fernández-Marín, H., Zimmerman, J.K., and Wcislo, W.T. (2007) Fungus garden platforms improve hygiene during nest establishment in *Acromyrmex* ants (Hymenoptera, Formicidae, Attini). *Insectes Sociaux 54*, 64–9.

Ferrier, S., Powell, G.V.N., Richardson, K.S., Manion, G., Overton, J.M., Allnutt, T.F., Cameron, S.E., Mantle, K., Burgess, N.D., Faith, D.R., Lamoreux, J.F., Kier, G., Hijmans, R.J., Funk, V.A., Cassis, G.A., Fisher, B.L., Flemons, P., Lees, D., Lovett, J.C., and Van Rompaey, R.S.A.R. (2004) Mapping more of terrestrial biodiversity for global conservation assessment. *BioScience 54*, 1101–9.

Ferveur, J.F. (2005) Cuticular hydrocarbons: their evolution and roles in *Drosophila* pheromonal communication. *Behavior Genetics 35*, 279–95.

Fewell, J., Harrison, J., Lighton, J., and Breed, M.D. (1996) Foraging energetics of the ant, *Paraponera clavata*. *Oecologia 105*, 419–27.

Fiala, B., Maschwitz, U., Pong, T.Y., and Helbig, A.J. (1989) Studies of a south east Asian ant-plant association: protection of *Macaranga* trees by *Crematogater borneensis*. *Oecologia 79*, 463–70.

Fiedler, K. (1995) Lycaenid butterflies and plants: is myrmecophily associated with particular hostplant preferences? *Ethology Ecology & Evolution 7*, 107–32.

——(2006) Ant-associates of Palaearctic lycaenid butterfly larvae (Hymenoptera: Formicidae; Lepidoptera: Lycaenidae) – a review. *Myrmecologische Nachrichten 9*, 77–87.

——Kuhlmann, F., Schlick-Steiner, B.C., Steiner, F.M., and Gebauer, G. (2007) Stable N-isotope signatures of central European ants – assessing positions in a trophic gradient. *Insectes Sociaux 54*, 393–402.

Fischer, R.C., Ölzant, S.M., Wanek, W., and Mayer, V. (2005) The fate of *Corydalis cava* elaiosomes within an ant colony of *Myrmica rubra*: elaiosomes are preferentially fed to larvae. *Insectes Sociaux 52*, 55–62.

——Richter, A., Hadacek, F., and Mayer, V. (2008) Chemical differences between seeds and elaiosomes indicate an adaptation to nutritional needs of ants. *Oecologia 155*, 539–47.

————Wanek, W., and Mayer, V. (2002) Plants feed ants: food bodies of myrmecophytic *Piper* and their significance for the interaction with *Pheidole bicornis* ants. *Oecologia 133*, 186–92.

Fisher, B.L. (1996) Ant diversity patterns along an elevational gradient in the Réserve Naturelle Intégrale d'Andringitra, Madagascar. *Fieldiana Zoology 85*, 93–108.

——(1997) Biogeography and ecology of the ant fauna of Madagascar (Hymenoptera: Formicidae). *Journal of Natural History 9*, 714–31.

——(1998) Ant diversity patterns along an elevational gradient in the Reserve Speciale d'Anjanaharibe-Sud and on the western Masoala Peninsula, Madagascar. *Zoology 90*, 39–67.

——(1999a) Ant diversity patterns along an elevational gradient in the Réserve Naturelle Intégrale d'Andohahela, Madagascar. *Fieldiana Zoology 94*, 129–47.

——(1999b) Improving inventory efficiency: a case study of leaf litter ant diversity in Madagascar. *Ecological Applications 9*, 714–31.

——(2000) The Malagasy fauna of *Strumigenys*. In B. Bolton, ed. *The Ant Tribe Dacetini. Memoirs of the American Entomological Institute*, pp. 612–710.

——(2003) Ants (Formicidae: Hymenoptera). In S.M. Goodman and J.P. Benstead, eds. *The Natural History of Madagascar*, pp. 811–819. University of Chicago Press.

——(2005) A new species of *Discothyrea* Roger from Mauritius and a new species of *Proceratium* from Madagascar (Hymenoptera: Formicidae). *Proceedings of the California Academy of Sciences (4) 56*, 657–67.

——(2009) Ants. In R.G. Gillespie and D.A. Clague, eds. *Encyclopedia of Islands*, pp. 35–41. University of California Press, Berkeley, CA.

——and Cover, S.P. (2007) *Ants of North America. A guide to genera*. University of California Press, Berkeley, CA.

——and Smith, M.A. (2008) A revision of Malagasy species of *Anochetus* Mayr and *Odontomachus* Latreille (Hymenoptera: Formicidae). *PLoS ONE 3*, e1787, doi:1710.1371/journal.pone.0001787.

Fitzpatrick, M.C., Weltzin, J.F., and Sanders, N.J., and Dunn, R.R. (2007) The biogeography of prediction error: why does the introduced range of the fire ant over-predict its native range? *Global Ecology and Biogeography 16*, 24–33.

Fjerdingstad, E.J., and Crozier, R.H. (2006) The evolution of worker caste diversity in social insects. *American Naturalist 167*, 390–400.

Flatt, T., and Weisser, W.W. (2000) The effects of mutualistic ants on aphid life history traits. *Ecology 81*, 3522–9.

Fleet, R.R., and Young, B.L. (2000) Facultative mutualism between imported fire ants (*Solenopsis invicta*) and a legume (*Senna occidentalis*). *Southwestern Naturalist 45*, 289–98.

Fletcher, D.J.C., and Michener, C.D. (1987) *Kin recognition in animals*. John Wiley and Sons, New York.

Floren, A., and Linsenmair, K.E. (2000) Do ant mosaics exist in pristine lowland rain forests? *Oecologia 123*, 129–37.

——Freking, A., Biehl, M., and Linsenmair, K.E. (2001) Anthropogenic disturbance changes the structure of arboreal tropical ant communities. *Ecography 24*, 547–54.

Fluker, S.S., and Beardsley, J.W. (1970) Sympatric associations of three ants: *Iridomyrmex humilis*, *Pheidole megacephala*, and *Anoplolepis longipes* in Hawaii. *Annals of the Entomological Society of America 63*, 1290–6.

Foitzik, S., and Heinze, J. (1998) Nest site limitation and colony takeover in the ant *Leptothorax nylanderi*. *Behavioral Ecology 9*, 367–75.

Fokuhl, G., Heinze, J., and Poschlod, P. (2007) Colony growth in *Myrmica rubra* with supplementation of myrmecochorous seeds. *Ecological Research 22*, 845–7.

Folgarait, P.J. (1998) Ant biodiversity and its relationship to ecosystem functioning: a review. *Biodiversity and Conservation 7*, 1221–44.

——and Davidson, D.W. (1994) Antiherbivore defenses of myrmecophytic *Cecropia* under different light regimes. *Oikos 71*, 305–20.

————(1995) Myrmecophytic *Cecropia*: antiherbivore defenses under different nutrient treatments. *Oecologia 104*, 189–206.

——and Gilbert, L.E. (1999) Phorid parasitoids affect foraging activity of *Solenopsis richteri* under different availability of food in Argentina. *Ecological Entomology 24*, 163–73.

——Gorosito, N., Pizzio, R., Rossi, J.P., and Fernandez, J. (2007) *Camponotus punctulatus* ant's demography: a temporal study across land-use types and spatial scales. *Insectes Sociaux 54*, 42–52.

Fonseca, C.R. (1993) Nesting space limits colony size of the plant-ant *Pseudomyrmex concolor*. *Oikos 67*, 473–82.

——(1999) Amazonian ant-plant interactions and the nesting space limitation hypothesis. *Journal of Tropical Ecology 15*, 807–25.

Forest, F., Grenyer, R., and Rouget, M., Davies, T.J., Cowling, R.M., Faith, D.P., Balmford, A., Manning, J.C., Proches, S., van der Bank, M., Reeves, G., Hedderson, T.A.J., and Savolainen, V. (2007) Preserving the evolutionary potential of floras in biodiversity hotspots. *Nature 445*, 757–60.

Forestry Commission (2007) http://www.forestresearch.gov.uk/pdf/FCIN090.pdf/$FILE/FCIN090.pdf.

Forte, P.N. (1956) The effectiveness of dieldrin for the control of the Argentine ant in Western Australia. *Journal of the Department of Agriculture, Western Australia 5*, 85.

Foucaud, J., Jourdan, H., Le Breton, J., Loiseau, A., Konghouleux, D., and Estoup, A. (2006) Rare sexual reproduction events in the clonal reproductive system of introduced populations of the little fire ant. *Evolution 60*, 1646–57.

——Fournier, D., Orivel, J., Delabie, J.H.C., Loiseau, A., Le Breton, J., Kergoat, G.J., and Estoup, A. (2007) Sex and clonality in the little fire ant. *Molecular Biology and Evolution 24*, 2465–2473.

Fourcassié, V., and Traniello, J.F.A. (1993) Effects of experience on food-searching behavior in the ant *Formica schaufussi* (Hymenoptera: Formicidae). *Journal of Insect Behavior 6*, 287–99.

Fournier, D., Estoup, A., Orivel, J., Foucaud, J., Jourdan, H., Le Breton, J., and Keller, L. (2005) Clonal reproduction by males and females in the little fire ant. *Nature 435*, 1230–34.

Fowler, H.G., and Roberts, R.B. (1983) Behavioural and developmental effects of some insect growth-regulators on carpenter ants, *Camponotus pennsylvanicus* (Deg) (Hymenoptera: Formicidae). *Journal of Applied Entomology 95*, 507–12.

Fox, C.W., and Czesak, M.E. (2000) Evolutionary ecology of progeny size in arthropods. *Annual Review of Entomology 45*, 341–69.

Francoeur, A. (1997) Ants (Hymenoptera: Formicidae) of the Yukon. *Biological Survey of Canada Monograph Series, 2*, 901–910.

Franks, N.R. (1987) The organization of working teams in social insects. *Trends in Ecology and Evolution 2*, 72–5.

——(1989) Thermoregulation in army ant bivouacs. *Physiological Entomology 14*, 397–404.

——and Partridge, L.W. (1993) Lanchester battles and the evolution of combat in ants. *Animal Behaviour 45*, 197–9.

——Gomez, N., Goss, S., and Deneubourg, J.L. (1991) The blind leading the blind in army ant raid patterns: testing a model of self-organization (Hymenoptera: Formicidae). *Journal of Insect Behavior 4*, 583–607.

——Hooper, J., Webb, C., and Dornhaus, A. (2005) Tomb evaders: house-hunting hygiene in ants. *Biology Letters 1*, 190–2.

——Dornhaus, A., Hitchcock, G., Guillem, R., Hooper, J., and Webb, C. (2007a) Avoidance of conspecific colonies during nest choice by ants. *Animal Behaviour 73*, 525–34.

——Hooper, J.W., Dornhaus, A., Aukett, P.J., Hayward, A. L., and Berghoff, S.M. (2007b) Reconnaissance and latent learning in ants. *Proceedings of the Royal Society of London B 274*, 1505–9.

Frederickson, M., Greene, M.J., Gordon., D.M. (2005) 'Devil's gardens' bedevilled by ants. *Nature 437*, 495–96.

Freeland, J. (1958) Biological and social patterns in the Australian bulldog ants of the genus *Myrmecia*. *Australian Journal of Ecology 6*, 1–18.

Freitas, L., Galetto, L., Bernardello, G., Paoli, A.A.S. (2000) Ant exclusion and reproduction of *Croton sarcopetalus* (Euphorbiaceae). *Flora 195*, 398–402.

Fresneau, D., and Errard, C. (1994) L'identité coloniale et sa "représentation" chez les fourmis. *Intellectica 2*, 91–115.

Friedrich, R., and Philpott, S.M. (2009) Nest-site limitation and nesting resources of ants (Hymenoptera: Formicidae) in urban green spaces. *Environmental Entomology 38*, 600–7.

Frouz, J., and Jilková, V. (2008) The effect of ants on soil properties and processes (Hymenoptera: Formicidae). *Myrmecological News 11*, 191–9.

Gabet, E.J., Reichman, O.J., and Seabloom, E.W. (2003) The effects of bioturbation on soil processes and sediment transport. *Annual Review of Earth and Planetary Science 31*, 249–73.

Gadau, J., Gertsch, P.J., Heinze, J., Pamilo, P., and Hölldobler, B. (1998) Oligogyny by unrelated queens in the carpenter ant, *Camponotus ligniperdus*. *Behavioral Ecology and Sociobiology 44*, 23–33.

Galen, C., and Butchart, B. (2003) Ants in your plants: effects of nectar-thieves on pollen fertility and seed-siring capacity in the alpine wildflower, *Polemonium viscosum*. *Oikos 101*, 521–8.

Gamboa, G.J., Reeve, H.K., and Holmes, W.G. (1991) Conceptual issues and methodology in kin recognition research, a critical discussion. *Ethology 88*, 109–27.

Gammans, N., Bullock, J.M., and Schonrogge, K. (2005) Ant benefits in a seed dispersal mutualism. *Oecologia 146*, 43–9.

Gardner, A., and West, S.A. (2007) Social evolution: the decline and fall of genetic kin recognition. *Current Biology 17*, R810-2.

Gardner, T.A., Barlow, J., and Araujo, I.S., Ávila-Pires, T. C., Bonaldo, A.B., Costa, J.E., Esposito, M.C., Ferreira, L. V., Hawes, J., Hernandez, M.I.M., Hoogmoed, M.S., Leite, R.N., Lo-Man-Hung, N.F., Malcolm, J.R., Martins, M.B., Mestre, L.A.M., Miranda-Santos, R., Overal, W.L., Parry, L., Peters, S.L., Ribeiro Jr, M.A., Da Silva, M.N.F., Da Silva Motta, C., and Peres, C.A. (2008) The cost-effectiveness of biodiversity surveys in tropical forests. *Ecology Letters 11*, 139–50.

Garnas, J.R., Drummond, F.A., and Groden, E. (2007) Intercolony aggression within and among local populations of the invasive ant, *Myrmica rubra* (Hymenoptera: Formicidae), in coastal Maine. *Environmental Entomology 36*, 105–13.

Gaston, K.J. (1996) Species-range-size distributions: patterns, mechanisms and implications. *Trends in Ecology and Evolution 11*, 197–201.

——and Blackburn, T.M. (1997) Age, area and avian diversification. *Biological Journal of the Linnean Society 62*, 239–53.

Gaston, K.J., Gauld, I.D., and Hanson, P. (1998) Rapoport's rule: time for an epitaph? *Trends in Ecology and Evolution 13*, 70–4.

——Davies, R.G., Orme, C.D.L., Olson, V.A., Thomas, G. H., Ding, T.S., Rasmussen, P.C., Lennon, J.J., Bennett, P.M., Owens, I.P.F., and Blackburn, T.M. (2007) Spatial turnover in the global avifauna. *Proceedings of the Royal Society of London B. 274*, 1567–74.

Gaume, L., McKey, D., and Anstett, M.C. (1997) Benefits conferred by "timid" ants: active anti-herbivore protection of the rainforest tree *Leonardoxa africana* by the minute ant *Petalomyrmex phylax*. *Oecologia 112*, 209–16.

————and Terrin, S. (1998) Ant-plant-homopteran mutualism how the third partner affects the interaction between a plant-specialist ant and its myrmecophyte host. *Proceedings of the Royal Society of London B 265*, 560–75.

——Zacharias, M., Grosbois, V., and Borges, R. (2005) The fitness consequences of bearing domatia and having the right ant partner: experiments with protective and non-protective ants in a semi-myrmecophyte. *Oecologia 145*, 76–86.

Gayahan, G.G., and Tschinkel, W.R. (2008) Fire ants, *Solenopsis invicta*, dry and store insect pieces for later use. *Journal of Insect Science 8*, 39, available online: insectscience.org/8.39.

Geraghty, M.J., Dunn, R.R., and Sanders, N.J. (2007) Body size, colony size, and range size in ants (Hymenoptera: Formicidae): are patterns along elevational and latitudinal gradients consistent with Bergmann's rule? *Myrmecological News 10*, 51–8.

Gerlach, J. (2004) Impact of the invasive crazy ant *Anoplolepis gracilipes* on Bird Island, Seychelles. *Journal of Insect Conservation 8*, 15–25.

Ghazoul, J. (2001) Can floral repellents pre-empt potential ant-plant conflicts? *Ecology Letters 4*, 295–9.

Gibb, H. (2005) The effect of a dominant ant, *Iridomyrmex purpureus*, on resource use by ant assemblages depends on microhabitat and resource type. *Austral Ecology 30*, 856–67.

——and Hochuli, D.F. (2003) Colonisation by a dominant ant facilitated by anthropogenic disturbance: effects on ant assemblage composition, biomass and resource use. *Oikos 103*, 469–78.

——and Hochuli, D.F. (2004) Removal experiment reveals limited effects of a behaviorally dominant species on ant assemblages. *Ecology 85*, 648–57.

Gibbons, L., and Simberloff, D. (2005) Interaction of hybrid imported fire ants (*Solenopsis invicta* x *S. richteri*) with native ants at baits in southeastern Tennessee. *Southeastern Naturalist 4*, 303–20.

Giladi, I. (2006) Choosing benefits or partners: a review of the evidence for the evolution of myrmecochory. *Oikos 112*, 481–92.

Giraud, T., Pedersen, J.S., and Keller, L. (2002) Evolution of supercolonies: the Argentine ants of southern Europe. *Proceedings of the National Academy of Sciences USA 99*, 6075–9.

——Blatrix, R., Poteaux, C., Solignac, M., and Jaisson, P. (2000) Population structure and mating biology of the polygynous ponerine ant *Gnamptogenys striatula* in Brazil. *Molecular Ecology 9*, 1835–41.

Giray, T., Giovanetti, M., and West-Eberhard, M.J. (2005) Juvenile hormone, reproduction, and worker behavior in the Neotropical social wasp *Polistes canadensis*. *Proceedings of the National Academy of Sciences USA 102*, 3330–5.

Glaser, F. (2006) Biogeography, diversity and vertical distribution of ants (Hymenoptera: Formicidae) in Vorarlbergt, Austria. *Myrmecological News 8*, 263–70.

Glenn, S., and Holway, D. (2008) Consumption of introduced prey by native predators: Argentine ants and pit-building ant lions. *Biological Invasions 10*, 273–80.

Golden, D.M., and Crist, T.O. (2000) Experimental effects of habitat fragmentation on rove beetles and ants: patch area or edge? *Oikos 90*, 525–38.

Gómez, C., and Oliveras, J. (2003) Can the Argentine ant (*Linepithema humile* Mayr) replace native ants in myrmecochory? *Acta Oecologica 24*, 47–53.

——Pons, P., and Bas, J.-M. (2003) Effects of the Argentine ant *Linepithema humile* on seed dispersal and seedling emergence of *Rhamnus alaternus*. *Ecography 26*, 532–8.

Gómez, L.A., Arcila, A.M., Lastra, L.A., and Chacón, P. (2002) Algunas bases biológicas para el manejo de la hormiga loca *Paratrechina fulva* (Hymenoptera: Formicidae). *Carta Trimestral Cenicaña 24*, 12–3.

Goodisman, M.A.D., and Hahn, D.A. (2005) Breeding system, colony structure, and genetic differentiation in the *Camponotus festinatus* species complex of carpenter ants. *Evolution 59*, 2185–99.

——Kovacs, J.L., and Hunt, B.G. (2008) Functional genetics and genomics in ants (Hymenoptera: Formicidae): the interplay of genes and social life. *Myrmecological News 11*, 107–17.

——Sankovich, K.A., and Kovacs, J.L. (2007) Genetic and morphological variation over space and time in the invasive fire ant *Solenopsis invicta*. *Biological Invasions 9*, 571–84.

Gorb, E., and Gorb, S. (2003) *Seed Dispersal by Ants in a Deciduous Forest Ecosystem. Mechanisms, strategies, adaptations*. Springer-Verlag, New York.

Gordon, D.M. (1987) Group-level dynamics in harvester ants: young colonies and the role of patrolling. *Animal Behaviour 35*, 833–43.

——(1988) Nest-plugging: interference competition in desert ants (*Novomessor cockerelli* and *Pogonomyrmex barbatus*). *Oecologia 75*, 114–8.

——(1996) The organization of work in social insect colonies. *Nature 380*, 121–4.

——(1999) *Ants at Work: how an insect society is organized*. The Free Press, New York.

——(2002) The regulation of foraging activity in red harvester ant colonies. *American Naturalist 159*, 509–18.

——and Kulig, A.W. (1996) Founding, foraging, and fighting: colony size and the spatial distribution of harvester ant nests. *Ecology 77*, 2393–409.

——and Wagner, D. (1997) Neighbourhood density and reproductive potential in harvester ants. *Oecologia 109*, 556–60.

——Holmes, S., and Nacu, S. (2008) The short-term regulation of foraging in harvester ants. *Behavioral Ecology 19*, 217–22.

Gorosito, N.B., Curmi, P., Hallaire, V., Folgarait, P.J., and Lavelle, P.M. (2006) Morphological changes in *Camponotus punctulatus* (Mayr) anthills of different ages. *Geoderma 132*, 249–60.

Gotelli, N.J. (1996) Ant community structure: effects of predatory ant lions. *Ecology 77*, 630–8.

——(2000) Null model analysis of species co-occurrence patterns. *Ecology 81*, 2606–21.

——and Ellison, A.M. (2002) Assembly rules for New England ant assemblages. *Oikos 99*, 591–9.

——and Entsminger, G.L. (2007) *EcoSim: null models software for ecology*. Acquired Intelligence Inc. and Kesey-Bear, Jericho, VT.

——and Graves, G.R. (1996) *Null Models in Ecology*. Smithsonian Institution Press, Washington, DC.

Gotwald Jr., W.H. (1995) *Army Ants: the biology of social predation*. Cornell University Press, Ithaca, NY.

Gotzek, D., and Ross, K. (2007) Genetic regulation of colony social organization in fire ants: an integrative overview. *Quarterly Review of Biology 82*, 201–26.

Gove, A.D., Dunn, R.R., and Majer, J.D. (2007) A keystone ant species promotes seed dispersal in a "diffuse" mutualism. *Oecologia 153*, 687–97.

——Majer, J.D., and Rico-Gray, V. (2005) Methods for conservation outside of formal reserve systems: the case of ants in the seasonally dry tropics of Veracruz, Mexico. *Biological Conservation 126*, 328–38.

Grafen, A. (1990) Do animals really recognize kin? *Animal Behaviour 39*, 42–54.

——(1998) Green beard as death warrant. *Nature 394*, 521–2.

Graham, L.C., Porter, S.D., Periera, R.M., Dorough, H.D., and Kelley, A.T. (2003) Field releases of the decapitating fly *Pseudacteon curvatus* (Diptera: Phoridae) for control of imported fire ants (Hymenoptera: Formicidae) in Alabama, Florida, and Tennessee. *Florida Entomologist 86*, 334–9.

Grangier, J., Orivel, J., Negrini, M., and Dejean, A. (2008) Low intraspecific aggressiveness in two obligate plant-ant species. *Insectes Sociaux 55*, 238–40.

Grasso, D.A., Sledge, M.F., Le Moli, F., Mori, A., and Turillazzi, S. (2005) Nest-area marking with faeces: a chemical signature that allows colony-level recognition in seed-harvesting ants (Hymenoptera, Formicidae). *Insectes Sociaux 52*, 36–44.

Green, P.T., O'Dowd, D.J. (2009) Management of invasive invertebrates: lessons from the management of an invasive alien ant. In M.N. Clout, and P.A. Williams, eds. *Invasive Species Management. A Handbook of Techniques*, pp. 153–71. Oxford University Press, Oxford.

Greene, M.J., and Gordon, D.M. (2003) Cuticular hydrocarbons inform task decision. *Nature 423*, 32.

————(2007a) Interaction rate informs harvester ant task decisions. *Behavioral Ecology 18*, 451–5.

————(2007b) Structural complexity of chemical recognition cues affects the perception of group membership in the ants *Linephithema humile* and *Aphaenogaster cockerelli*. *Journal of Experimental Biology 210*, 897–905.

Greenslade, P.J.M. (1971) Interspecific competition and frequency changes among ants in Solomon Islands coconut plantations. *Journal of Applied Ecology 8*, 323–52.

——(1976) The meat ant *Iridomyrmex purpureus* (Hymenoptera: Formicidae) as a dominant member of ant communities. *Journal of the Australian Entomological Society 15*, 237–40.

——(1978) Ants. In W.A. Lowe, ed. *The physical and biological features of Kunoth Paddock in central Australia*, pp. 109–13. CSIRO Division of Land Resources Technical Paper No. 4, Canberra.

——(1987) Environment and competition as determinants of local geographical distribution of five meat ants, *Iridomyrmex purpureus* and allied species (Hymenoptera; Formicidae). *Australian Journal of Zoology 35*, 259–73.

Gregg, R.E. (1963) *The Ants of Colorado*. The University of Colorado Press, Boulder, CO.

Grimaldi, D., and Agosti, D. (2000) A formicine in New Jersey Cretaceous amber (Hymenoptera: Formicidae) and early evolution of the ants. *Proceedings of the National Academy of Sciences USA 97*, 13678–83.

——and Engel, M.S. (2005) *Evolution of the Insects*. Cambridge University Press, Cambridge.

——Agosti, D., and Carpenter, J.M. (1997) New and rediscovered primitive ants (Hymenoptera: Formicidae) in Cretaceous amber from New Jersey, and their phylogenetic relationships. *American Museum Novitates 3208*, 1–43.

Grime, J.P. (1979) *Plant Strategies, Vegetation Processes, and Ecosystem Properties*. John Wiley, Chichester, UK.

Gris, G., and Cherix, D. (1977) Les grandes colonies de fourmis des bois du Jura (groupe *Formica rufa*). *Mitteilungen der Schweizerischen Entomologischen Gesellschaft 50*, 249–50.

Groden, E., Drummond, F.A., Garnas, J., and Franceour, A. (2005) Distribution of an invasive ant, *Myrmica rubra* (Hymenoptera: Formicidae), in Maine. *Journal of Economic Entomology 98*, 1774–84.

Gronenberg, W. (1995) The fast mandible strike in the trap-jaw ant *Odontomachus* I. Temporal properties and morphological characteristics. *Journal of Comparative Physiology A 176*, 391–8.

——(1996) The trap-jaw mechanisms in the dacetine ants *Daceton armigerum* and *Strumigenys* sp. *Journal of Experimental Biology 199*, 2021–33.

——and Ehmer, B. (1996) The mandible mechanism of the ant genus *Anochetus* (Hymenoptera, Formicidae) and the possible evolution of trap-jaws. *Zoology-Analysis Of Complex Systems 99*, 153–62.

——Tautz, T., and Hölldobler, B. (1993) Fast trap jaws and giant neurons in the ant *Odontomachus*. *Science 262*, 561–3.

——Paul, J., Just, S., and Hölldobler, B. (1997) Mandible muscle fibers in ants: fast or powerful? *Cell and Tissue Research 289*, 347–61.

Grosholz, E.D. (2005) Recent biological invasion may hasten invasional meltdown by accelerating historical introductions. *Proceedings of the National Academy of Sciences USA 102*, 1088–91.

Grotkopp, E., Rejmanek, M., and Rost, T.L. (2002) Towards causal explantion of plant invasiveness: seedling growth and life history strategies of 29 pine (*Pinus*) species. *American Naturalist 159*, 396–419.

Grover, C.D., Dayton, K.C., Menke, S.B., and Holway, D. A. (2008) Effects of aphids on foliar foraging by Argentine ants and the resulting effects on other arthropods. *Ecological Entomology 33*, 101–6.

——Kay, A.D., Monson, J.A., Marsh, T.C., and Holway, D. A. (2007) Linking nutrition and behavioural dominance: carbohydrate scarcity limits aggression and activity in Argentine ants. *Proceedings of the Royal Society of London B 274*, 2951–7.

Guerrieri, F.J., and d'Ettorre, P. (2008) The mandible opening response: quantifiying aggression elicited by chemical cues in ants. *Journal of Experimental Biology 211*, 1109–13.

Guerrero, R.J., and Fernández, F. (2008) A new species of the ant genus *Forelius* (Formicidae: Dolichoderinae) from the dry forest of Colombia. *Zootaxa* 1958, 51–60.

Guitian, P., Medrano, M., and Guitian, J. (2003) Seed dispersal in *Erythronium denscanis* L. (Liliaceae): variation among habitats in a myrmecochorous plant. *Plant Ecology 169*, 171–7.

Gurevitch, J. (2006) Commentary on Simberloff (2006): meltdowns, snowballs and positive feedbacks. *Ecology Letters 9*, 919–21.

Haemig, P.D. (1997) Effects of birds on the intensity of ant rain: a terrestrial form of invertebrate drift. *Animal Behaviour 54*, 89–97.

——(1999) Predation risk alters interactions among species: competition and facilitation between ants and nesting birds in a boreal forest. *Ecology Letters 2*, 178–84.

Haering, R., and Fox, B.J. (1987) Short-term coexistence and long-term competitive displacement of 2 dominant species of *Iridomyrmex:* The successional response of ants to regenerating habitats. *Journal of Animal Ecology 56*, 495–507.

Haines, B.L. (1978) Element and energy flow through colonies of the leaf-cutting ant, *Atta columbica*, in Panama. *Biotropica 10*, 270–7.

Haines, I.H., and Haines, J.B. (1978a) Colony structure, seasonality and food requirements of the crazy ant, *Anoplolepis longipes* (Jerd.), in the Seychelles. *Ecological Entomology 3*, 109–18.

————(1978b) Pest status of the crazy ant, *Anoplolepis longipes* (Jerdon)(Hymenoptera: Formicidae) in the Seychelles. *Bulletin of Entomological Research 68*, 627–38.

Halloy, J., Sempo, G., Caprari, G., Rivault, C., Asadpour, M., Tâche, F., Saïd, I., Durier, V., Canonge, S., Amé, J.M., Detrain, C., Correll, N., Martinolli, A., Mondada, F., Siegwart, R., and Deneubourg, J.-L. (2007) Social intergration of robots into groups of cockroaches to control self-organized choices. *Science 318*, 1155–1158.

Hamilton, W.D. (1963) The evolution of altruistic behavior. *American Naturalist 97*, 354–6.

——(1964) The genetical evolution of social behaviour. I - II. *Journal of Theoretical Biology 7*, 1–52.

——(1987) Discrimination nepotism: expectable, common, overlooked. In D.J.C. Fletcher and C.D. Michener, eds. *Kin Recognition in Animals*, pp. 417–37. John Wiley and Sons, New York.

——and May, R.M. (1977) Dispersal in stable habitats. *Nature 269*, 578–81.

Hammond, R.L., and Keller, L. (2004) Conflict over male parentage in social insects. *PLoS Biology 2*, e248. doi:10.1371/journal.pbio.0020248.

Handel, S.N. (1981) Ants disperse a majority of herbs in a mesic forest community in New York State. *Bulletin of the Torrey Botanical Society 108*, 430–7.

Handler, A.T., Gruner, D.S., Haines, W.P., Lange, M.W., and Kaneshiro, K.Y. (2007) Arthropod surveys on Palmyra Atoll, Line Islands, and insights into the decline of the native tree *Pisonia grandis* (Nyctaginaceae). *Pacific Science 61*, 485–502.

Hanks, L.M., and Sadof, C.S. (1990) The effect of ants on nymphal survivorship of *Coccus viridis* (Homoptera: Coccidae). *Biotropica 22*, 210–3.

Hannonen, M., and Sundström, L. (2003) Worker nepotism among polygynous ants. *Nature 421*, 910.

——Helanterä, H., and Sundström, L. (2004) Habitat age, breeding system and kinship in the ant *Formica fusca*. *Molecular Ecology 13*, 1579–88.

Hansen, M.C., Stehman, S.V, Potapov, P.V, Loveland, T.R., Townshend, J.R.G., DeFries, R.S., Pittman, K. W., Arunarwati, B., Stolle, F., Steininger, M.K., Carroll, M., and DiMiceli, C. (2008) Humid tropical forest clearing from 2000 to 2005 quantified by using multitemporal and multiresolution remotely sensed data. *Proceedings of the National Academy of Sciences USA 105*, 9439–44.

Hanzawa, F.M., Beattie, A.J., and Culver, D.C. (1988) Directed dispersal: demographic analysis of an ant-seed mutualism. *American Naturalist 131*, 1–13.

Harkness, Maroudas (1985) Central place foraging by an ant: a model of searching. *Animal Behaviour 33*, 916–28.

Harley, R. (1991) The greasy pole syndrome. In C.R. Huxley and D.F. Cutler, eds. *Ant-Plant Interactions*, pp. 430–3. Oxford University Press, Oxford.

Harris, M., Knutson, A., Calixto, A., Dean, A., Brooks, L., and Ree, B. (2003) Impact of red imported fire ant on foliar herbivores and natural enemies. *Southwestern Entomologist Supplement 27*, 123–4.

Harris, R.J., Abbott, K.L., Barton, K., Berry, J., Don, W., Gunawardana, D., Lester, P., Rees, J., Stanley, M., Sutherland, A., and Toft, R. (2005) Invasive ant pest risk assessment project for New Zealand. http://www.landcareresearch.co.nz/research/biocons/invertebrates/ants/ant_pest_risk.asp.

Hart, A.G., and Ratnieks, F.L.W. (2001) Task partitioning, division of labour and nest compartmentalization collectively isolate hazardous waste in the leafcutting ant *Atta cephalotes*. *Behavioral Ecology and Sociobiology 49*, 387–92.

Hartley, S., Harris, R., and Lester, P.J. (2006) Quantifying uncertainty in the potential distribution of an invasive species: climate and the Argentine ant. *Ecology Letters 9*, 1068–79.

Hartley, S., and Lester, P.J. (2003) Temperature-dependent development of the Argentine ant, *Linepithema humile* (Mayr) (Hymenoptera: Formicidae): a degree-day model with implications for range limits in New Zealand. *New Zealand Entomologist 26*, 91–100.

Hartmann, A., d'Ettorre, P., Jones, G.R., and Heinze, J. (2005) Fertility signaling – the proximate mechanism of worker policing in a clonal ant. *Naturwissenschaften 92*, 282–6.

Harvey, C.T., and Eubanks, M.D. (2005) Intraguild predation of parasitoids by *Solenopsis invicta*: a non-disruptive interaction. *Entomologia Experimentalis et Applicata 114*, 127–35.

Harvey, P.H., and Pagel, M.D. (1991) *The Comparative Method in Evolutionary Biology*. Oxford University Press, Oxford.

Hasegawa, E. (1993a) Caste specialization in food storage in the dimorphic ant *Colobopsis nipponicus* (Wheeler). *Insectes Sociaux 40*, 261–71.

——(1993b) Nest defense and early production of the major workers in the dimorphic ant *Colobopsis nipponicus* (Wheeler) (Hymenoptera, Formicidae). *Behavioral Ecology and Sociobiology 33*, 73–7.

Hashimoto, Y. (2003) Identification guide to the ant genera of Borneo. In Y. Hashimoto and H. Rahman, eds. *Inventory and Collection. Total protocol for understanding of biodiversity*, pp. 89–162. Research and Education Component, BBEC Programme (Universiti Malaysia Sabah), Universiti Malaysia Sabah, Kota Kinabalu.

Hastings, A. (1980) Disturbance, coexistence, history, and competition for space. *Theoretical Population Biology 18*, 363–73.

Hauser, C.E., and Possingham, H.P. (2008) Experimental or precautionary? Adaptive management over a range of time horizons. *Journal of Applied Ecology 45*, 72–81.

Hawai'i Ant Group (2001) A plan for prevention of establishment of new ant species in Hawai'i with special attention to the red imported fire ant (*Solenopsis invicta*). http://hbs.bishopmuseum.org/ants/Solenopsis/.

——(2007) A plan for prevention of establishment of new ant species in Hawaii with special attention to the red imported fire ant (*Solenopsis invicta*) and little fire ant (*Wasmannia auropunctata*). August 2007 revision. http://hbs.bishopmuseum.org/ants/Solenopsis/.

Hawkins, B.A. (2001) Ecology's oldest pattern? *Trends in Ecology and Evolution 16*, 470.

——Diniz, J.A.F. (2002) The mid-domain effect cannot explain the diversity gradient of Nearctic birds. *Global Ecology and Biogeography 11*, 419–26.

——Diniz, J.A.F., and Weis, A.E. (2005) The mid-domain effect and diversity gradients: is there anything to learn? *American Naturalist 166*, E140-3.

——Albuquerque, F.S., Araujo, M.B., Beck, J., Bini, L.M., Cabrero-Sanudo, F.J., Castro-Parga, I., Diniz, J.A.F., Ferrer-Castan, D., Field, R., Gomez, J.F., Hortal, J., Kerr, J.T., Kitching, I.J., Leon-Cortes, J.L., Lobo, J.M., Montoya, D., Moreno, J.C., Olalla-Tarraga, M.A., Pausas, J.G., Qian, H., Rahbek, C., Rodriguez, M. A., Sanders, N. J. and Williams, P. (2007) A global evaluation of metabolic theory as an explanation for terrestrial species richness gradients. *Ecology 88*, 1877–1888.

Hawkins, C.P., Norris, R.H., Hogue, J.N., and Feminella, J. W. (2000) Development and evaluation of predictive models for measuring the biological integrity of streams. *Ecological Applications 10*, 1456–77.

Hebert, P.D.N., and Gregory, T.R. (2005) The promise of DNA barcoding for taxonomy. *Systematic Biology 54*, 8582–859.

——Cywinska, A., Ball, S.L., deWaard, J.R. (2003) Biological identifications through DNA barcodes. *Proceedings of the Royal Society of London B 270*, 313–21.

Hee, J.J., Holway, D.A., Suarez, A.V., and Case, T.J. (2000) Role of propagule size in the success of incipient colonies of the invasive Argentine ant. *Conservation Biology 14*, 559–63.

Hefetz, A. (2007) The evolution of hydrocarbon pheromone parsimony in ants (Hymenoptera: Formicidae) – interplay of colony odor uniformity and odor idiosyncrasy. A review. *Myrmecological News 10*, 59–68.

——and Blum, M.S. (1978) Biosynthesis and accumulation of formic acid in the poison gland of the carpenter ant *Camponotus pennsylvanicus*. *Science 201*, 454–5.

Heil, M., and McKey, D. (2003) Protective ant-plant interactions as model systems in ecological and evolutionary research. *Annual Review of Ecology Evolution and Systematics 34*, 425–53.

——Rattke, J., and Boland, W. (2005) Postsecretory hydrolysis of nectar sucrose and specialization in ant/plant mutualism. *Science 308*, 560–3.

——Baumann, B., Krüger, R., and Linsenmair, K.E. (2004a) Main nutrient compounds in food bodies of Mexican *Acacia* ant-plants. *Chemoecology 14*, 45–52.

——Fiala, B., Kaiser, W., and Linsenmair, K.E. (1998) Chemical contents of *Macaranga* food bodies: adaptations to their role in ant attraction and nutrition. *Functional Ecology 12*, 117–22.

——Koch, T., Hilpert, A., Fiala, B., Boland, W., and Linsenmair, K.E. (2001) Extrafloral nectar production of the ant-associated plant, *Macaranga tanarius*, is an induced, indirect, defensive response elicited by jasmonic acid. *Proceedings of the National Academy of Sciences USA 98*, 1083–8.

——Greiner, S., Meimberg, H., Krüger, R., Heubl, G., Noyer, J.L., Linsenmair, K.E., and Boland, W. (2004b) Evolutionary change from induced to constitutive expression of an indirect plant resistance. *Nature 430*, 205–208.

——Hilpert, A., Fiala, B., bin Hashim, R., Strohm, E., Zotz, G., and Linsenmair, K. E. (2002) Nutrient allocation of *Macaranga triloba* ant plants to growth, photosynthesis, and indirect defence. *Functional Ecology 16*, 475–83.

Heimpel, G.E., and De Boer, J.G. (2008) Sex determination in the Hymenoptera. *Annual Review of Entomoogyl 53*, 209–30.

Heinze, J. (2008) The demise of the standard ant (Hymenoptera: Formicidae). *Myrmecological News 11*, 9–20.

——and Buschinger, A. (1987) Queen polymorphism in a non-parasitic *Leptothorax* species (Hymenoptera, Formicidae). *Insectes Sociaux 34*, 28–43.

——and Buschinger, A. (1989) Queen polymorphism in *Leptothorax* spec. A: its genetic and ecological background (Hymenoptera: Formicidae). *Insectes Sociaux 36*, 139–55.

——and Hölldobler, B. (1994) Ants in the cold. *Memorabilia Zoologica 48*, 99–108.

——and Tsuji, K. (1995) Ant reproductive strategies. *Research in Population Ecology 37*, 135–49.

——Stengl, B., and Sledge, M.F. (2002b) Worker rank, reproductive status and cuticular hydrocarbon signature in the ant, *Pachycondyla* cf. *inversa*. *Behavioral Ecology and Sociobiology 52*, 59–65.

——Foitzik, S., Hippert, A., and Hölldobler, B. (1996) Apparent dear-enemy phenomenon and environment-based recognition cues in the ant. *Leptothorax nylanderi*. *Ethology 102*, 510–22.

————Kipyatkov, V.E., and Lopatina, E.B. (1998) Latitudinal variation in cold hardiness and body size in the boreal ant species *Leptothorax acervorum* (Hymenoptera: Formicidae). *Entomologia Generalis 22*, 305–12.

——Schrempf, A., Seifert, B., and Tinaut, A. (2002a) Queen morphology and dispersal tactics in the ant, *Cardiocondyla batesii*. *Insectes Sociaux 49*, 129–32.

——Trindl, A., Seifert, B., and Yamauchi, K. (2005) Evolution of male morphology in the ant genus *Cardiocondyla*. *Molecular Phylogenetics and Evolution 37*, 278–88.

——Trunzer, B., Hölldobler, B., and Delabie, J.H.C. (2001) Reproductive skew and queen relatedness in an ant with primary polygyny. *Insectes Sociaux 48*, 149–53.

——Foitzik, S., Fischer, B., Wanke, T., and Kipyatkov, V.E. (2003) The significance of latitudinal variation in body size in a holarctic ant, *Leptothorax acervorum*. *Ecography 26*, 349–55.

——Foitzik, S., Oberstadt, B., Rueppell, O., and Hölldobler, B. (1999) A female caste specialized for the production of unfertilized eggs in the ant *Crematogaster smithi*. *Naturwissenschaften 86*, 93–5.

Heithaus, E.R., and Humes, M. (2003) Variation in communities of seed-dispersing ants in habitats with different disturbance in Knox County, Ohio. *Ohio Journal of Science 103*, 89–97.

Helanterä, H., and Sundström, L. (2007) Worker reproduction in *Formica* ants. *American Naturalist 170*, E14-E25.

Heller, N.E. (2004) Colony structure in introduced and native populations of the invasive Argentine ant, *Linepithema humile*. *Insectes Sociaux 51*, 378–86.

——and Gordon, D.M. (2006) Seasonal spatial dynamics and causes of nest movement in colonies of the invasive Argentine ant (*Linepithema humile*). *Ecological Entomology 31*, 499–510.

——Sanders, N.J., and Gordon, D.M. (2006) Linking temporal and spatial scales in the study of an Argentine ant invasion. *Biological Invasions 8*, 501–7.

——Sanders, N.J., Shors, J.W., and Gordon, D.M. (2008) Rainfall facilitates the spread, and time alters the impact, of the invasive Argentine ant. *Oecologia 155*, 385–95.

Helms Cahan, S., and Vinson, S.B. (2003) Reproductive division of labor between hybrid and nonhybrid offspring in a fire ant hybrid zone. *Evolution 57*, 1562–70.

——Parker, J.D., and Rissing, S.W, Johnson, R.A., Polony, T.S., Weiser, M.D., and Smith, D.R. (2002) Extreme genetic differences between queens and workers in hybridizing *Pogonomyrmex* harvester ants. *Proceedings of the Royal Society of London B 269*, 1871–7.

Helms, K.R., and Vinson, S.B. (2001) Coexistence of native ants with the red imported fire ant, *Solenopsis invicta*. *Southwestern Naturalist 46*, 396–400.

——and Vinson, S.B. (2002) Widespread association of the invasive ant *Solenopsis invicta* with an invasive mealybug. *Ecology 83*, 2425–38.

——and Vinson, S.B. (2003) Apparent facilitation of an invasive mealybug by an invasive ant. *Insectes Sociaux 50*, 403–4.

——and Vinson, S.B. (2008) Plant resources and colony growth in an invasive ant: the importance of honeydew-producing Hemiptera in carbohydrate transfer across trophic levels. *Environmental Entomology 37*, 487–93.

Helmuth, B., Kingsolver, J.G., and Carrington, E. (2005) Biophysics, physiological ecology, and climate change: does mechanism matter? *Annual Review of Physiology 67*, 177–201.

Henderson, G., Wagner, R.O., and Jeanne, R.L. (1989) Prairie ant colony longevity and mound growth. *Psyche (Cambridge) 96*, 257–68.

Henin, J.-M., and Paiva, M. (2004) Interactions between *Orthotomicus erosus* (Woll.) (Col., Scolytidae) and the Argentine ant *Linepithema humile* (Mayr) (Hym., Formicidae). *Journal of Pest Science 77*, 113–7.

Hepper, P.G. (1991) *Kin recognition*. Cambridge University Press, Cambridge.

Herbers, J.M. (1981) Reliability theory and foraging by ants. *Journal of Theoretical Biology 89*, 175–89.

——(1993) Ecological determinants of queen number in ants. In L. Keller, ed. *Queen Number and Sociality in Insects*, pp. 262–93. Oxford University Press, Oxford.

——and Choiniere, E. (1996) Foraging behaviour and colony structure in ants. *Animal Behaviour 51*, 141–53.

——and Johnson, C.A. (2007) Social structure and winter survival in acorn ants. *Oikos 116*, 829–35.

Hernandez, J.V., Goitia, W., Cabrera, A., Lopez, F., and Jaffe, K. (2006) Leaf-cutting ant species (Hymenoptera: *Atta*) differ in the types of cues used to differentiate between self and others. *Animal Behaviour 71*, 945–52.

Herz, H., Hölldobler, B., and Roces, F. (2008) Delayed rejection in a leaf-cutting ant after foraging on plants unsuitable for the symbiotic fungus. *Behavioral Ecology, 19*, 575–82.

Heterick, B. (1997) The interaction between the coastal brown ant, *Pheidole megacephala* (Fabricius), and other invertebrate fauna of Mt Coot-tha (Brisbane, Australia). *Australian Journal of Ecology 22*, 218–21.

——(2006) A revision of the Malagasy ants belonging to genus *Monomorium* Mayr, 1855 (Hymenoptera: Formicidae). *Proceedings of the California Academy of Sciences* (4) *57*, 69–202.

Hickman, J.C. (1974) Pollination by ants: A low-energy system. *Science 184*, 1290–2.

Higashi, S., and Yamauchi, K. (1979) Influence of a supercolonial ant *Formica* (*Formica yessensis* Forel) on the distribution of other ants in Ishikari Coast. *Japanese Journal of Ecology 29*, 257–64.

Hill, M., Holm, K., Vel, T., Shah, N.J., and Matyot, P. (2003) Impact of the introduced yellow crazy ant *Anoplolepis gracilipes* on Bird Island, Seychelles. *Biodiversity and Conservation 12*, 1969–84.

Hill, S.L., and Hoy, M.A. (2003) Interactions between the red imported fire ant *Solenopsis invicta* and the parasitoid *Lipolexis scutellaris* potentially affect classical biological control of the aphid *Toxoptera citricida*. *Biological Control 27*, 11–9.

Hillebrand, H. (2004) On the generality of the latitudinal diversity gradient. *American Naturalist 163*, 192–211.

Hobbs, R.J., Arico, S., Aronson, J., Baron, J.S., Bridgewater, P., Cramer, V.A., Epstein, P.R., Ewel, J.J., Klink, C.A., Lugo, A.E., Norton, D., Ojima, D., Richardson, D.M., Sanderson, E.W., Valladares, F., Vilà, M., Zamora, R., and Zobel, M. (2006) Novel ecosystems: theoretical and management aspects of the new ecological world order. *Global Ecology and Biogeography 15*, 1–7.

Hoffmann, B.D. (2000) Changes in ant species composition and community organisation along grazing gradients in semiarid rangelands of the Northern Territory. *Australian Rangeland Journal 22*, 171–89.

——(2004) *Pest ants and their management on Aboriginal lands in the Northern Territory*. CSIRO, Darwin.

——(2009) Ecological restoration following the local eradication of an invasive ant in northern Australia. *Biological Invasions*, doi 10.1007/s10530-009-9516-2.

——and Andersen, A.N. (2003) Responses of ants to disturbance in Australia, with particular reference to functional groups. *Austral Ecology 28*, 444–64.

——and O'Connor, S. (2004) Eradication of two exotic ants from Kakadu National Park. *Ecological Management and Restoration 5*, 98–105.

——and Parr, C.L. (2008) An invasion revisited: the African big-headed ant (*Pheidole megacephala*) in northern Australia. *Biological Invasions 10*, 1171–81.

——Andersen, A.N., and Hill, G.J.E. (1999) Impact of an introduced ant on native rain forest invertebrates: *Pheidole megacephala* in monsoonal Australia. *Oecologia 120*, 595–604.

Holland, B.R., Penny, D., and Hendy, M.D. (2003) Outgroup misplacement and phylogenetic inaccuracy under a molecular clock: a simulation study. *Systematic Biology 52*, 229–38.

Hölldobler, B. (1974) Home range orientation and territoriality in harvesting ants. *Proceedings of the National Academy of Sciences USA 71*, 3274–7.

——and Wilson, E.O. (1977) The number of queens: an important trait in ant evolution. *Naturwissenschaften 64*, 8–15.

——(1982) Communication, raiding behaviour and prey storage in *Cerapachys* (Hymenoptera: Formicidae). *Psyche (Cambridge) 89*, 3–23.

——(1983) Territorial behavior in the green tree ant (*Oecophylla smaragdina*). *Biotropica 15*, 241–50.

——(1986) Food robbing in ants, a form of interference competition. *Oecologia 69*, 12–5.

——and Carlin, N.F. (1985) Colony founding, queen dominance and oligogyny in the Australian meat ant *Iridomyrmex purpureus*. *Behavioral Ecology and Sociobiology 18*, 45–58.

————(1990) *The Ants*. The Belknap Press of Harvard University Press, Cambridge, MA.

————(1994) *Journey to the Ants: a story of scientific exploration*. The Belknap Press of Harvard University Press, Cambridge, MA.

————(2008) *The Superorganism: the beauty, elegance, and strangeness of insect societies*. W.W. Norton & Company, New York.

Holmes, W.G. (2004) The early-history of Hamilton-based research on kin recognition. *Annales Zoologici Fennici 41*, 691–711.

——and Sherman, P.W. (1983) Kin recognition in animals. *American Scientist, 46–55*.

Holt, S.J. (1955) On the foraging activity of the wood ant. *Journal of Animal Ecology 24*, 1–34.

Holway, D.A. (1995) Distribution of the Argentine ant (*Linepithema humile*) in northern California. *Conservation Biology 9*, 1634–7.

——(1998a) Effect of Argentine ant invasions on ground-dwelling arthropods in northern California riparian woodlands. *Oecologia 116*, 252–8.

——(1998b) Factors governing rate of invasion: a natural experiment using Argentine ants. *Oecologia 115*, 206–12.

——(1999) Competitive mechanisms underlying the displacement of native ants by the invasive Argentine ant. *Ecology 80*, 238–51.

——(2005) Edge effects of an invasive species across a natural ecological boundary. *Biological Conservation 121*, 561–7.

——and Case, T.J. (2000) Mechanisms of dispersed central-place foraging in polydomous colonies of the Argentine ant. *Animal Behaviour 59*, 433–41.

————(2001) Effects of colony-level variation on competitive ability in the invasive Argentine ant. *Animal Behaviour 61*, 1181–92.

——and Suarez, A.V. (2004) Colony-structure variation and interspecific competitive ability in the invasive Argentine ant. *Oecologia 138*, 216–22.

————(2006) Homogenization of ant communities in Mediterranean California: the effects of urbanization and invasion. *Biological Conservation 127*, 319–26.

————and Case, T.J. (1998) Loss of intraspecific aggression in the success of a widespread invasive social insect. *Science 282*, 949–52.

——————(2002b) Role of abiotic factors in governing susceptibility to invasion: a test with Argentine ants. *Ecology 83*, 1610–9.

——Lach, L., Suarez, A.V., Tsutsui, N.D., and Case, T.J. (2002a) The causes and consequences of ant invasions. *Annual Review of Ecology and Systematics 33*, 181–233.

Holzer, B., Kummerli, R., Keller, L., and Chapuisat, M. (2006b) Sham nepotism as a result of intrinsic differences in brood viability in ants. *Proceedings of the Royal Society of London B 273*, 2049–52.

——Chapuisat, M., Kremer, N., Finet, N., and Keller, L. (2006a) Unicoloniality, recognition and genetic differentiation in a native *Formica* ant. *Journal of Evolutionary Biology 19*, 2031–9.

Hook, A.W., and Porter, S.D. (1990) Destruction of harvester ant colonies by invading fire ants in south-central Texas (Hymenoptera: Formicidae). *Southwest Naturalist 35*, 477–8.

Hopper, K.R. (1999) Risk-spreading and bet-hedging in insect population biology. *Annual Review of Entomology 44*, 535–60.

Horstmann, K. (1974) Untersuchungen uber den Nahrungserwerb der Waldameisen (*Formica polyctena* Foerster) im Eichenwald. III. Jahresbilanz. *Oecologia 15*, 187–204.

Horvitz, C.C., and Schemske, D.W. (1984) Effects of ants and an ant-tended herbivore on seed production of a neotropical herb. *Ecology 65*, 1369–78.

————(1986) Seed dispersal of a neotropical myrmecochore: variation in removal rates and dispersal distance. *Biotropica 18*, 319–23.

————(1990) Spatiotemporal variation in insect mutualists of a neotropical herb. *Ecology 71*, 1085–97.

Hosoishi, S., and Ogata, K. (2008) The ant genus *Crematogaster* Lund, subgenus *Physocrema* Forel, in the Indochinese Peninsula (Hymenoptera: Formicidae). *Asian Myrmecology 2*, 1–10.

Howard, D.F., Blum, M.S., Jones, T.H., and Tomalski, M. D. (1982) Behavioral responses to an alkylpyrazine from the mandibular gland of the ant *Wasmannia auropunctata*. *Insectes Sociaux 29*, 369–74.

Howard, J.J. (1988) Leafcutting ant diet selection – relative influence of leaf chemistry and physical features. *Ecology 69*, 250–60.

Howard, K.J. (2006) Three queen morphs with alternative nest-founding behaviors in the ant, *Temnothorax longispinosus*. *Insectes Sociaux 53*, 480–8.

Howse, P.E. (1975) Chemical defenses of ants, termites and other insects: some outstanding questions. In C. Noirot, P. E. Howse, and G. Le Masne, eds. *Pheromones and Defensive Secretions in Social Insects*, p. 23. University Dijon, France.

Huang, H.T., and Yang, P. (1987) The ancient cultured citrus ant. *BioScience 37*, 665–71.

Hubbard, M.D., and Cunningham, W.G. (1977) Orientation of mounds in the ant *Solenopsis invicta* (Hymenoptera, Formicidae, Myrmicinae). *Insectes Sociaux 24*, 3–7.

Hubbell, S.P. (2001) *The Unified Neutral Theory of Biodiversity and Biogeography*. Princeton University Press, Princeton, NJ.

——Foster, R.B., O'Brien, S.T., Harms, K.E., Condit, R., Wechsler, B., Wright, S.J., and Loo de Lao, S. (1999) Light-gap disturbances, recruitment limitation, and tree diversity in a Neotropical forest. *Science 283*, 554–57.

Hudson, R.R., and Coyne, J.A. (2002) Mathematical consequences of the genealogical species concept. *Evolution 56*, 1557–65.

Hughes, D.P., Kronauer, D.J.C., and Boomsma, J.J. (2008) Extended phenotype: nematodes turn ants into bird-dispersed fruits. *Current Biology 18*, 294–5.

Hughes, L. (2000) Biological consequences of global warming: is the signal already apparent? *Trends in Ecology and Evolution 15*, 56–61.

——and Westoby, M. (1992) Capitula on stick insect eggs and elaiosomes on seeds: convergent adaptations for burial by ants *Functional Ecology 6*, 642–8.

———and Jurado, E. (1994) Convergence of elaiosomes and insect prey: evidence from ant foraging behaviour and fatty acid composition. *Functional Ecology 8*, 358–65.

Hughes, W.O.H., and Boomsma, J.J. (2004) Genetic diversity and disease resistance in leaf-cutting ant societies. *Evolution 58*, 1251–60.

———(2006) Does genetic diversity hinder parasite evolution in social insect colonies? *Journal of Evolutionary Biology 19*, 132–43.

Hulme, P.E. (2003) Biological invasions: winning the science battles but losing the conservation war? *Oryx 37*, 178–93.

Human, K.G., and Gordon, D.M. (1996) Exploitation and interference competition between the invasive Argentine ant, *Linepithema humile*, and native ant species. *Oecologia 105*, 405–12.

———(1997) Effects of Argentine ants on invertebrate biodiversity in northern California. *Conservation Biology 11*, 1242–8.

———(1999) Behavioral interactions of the invasive Argentine ant with native ant species. *Insectes Sociaux 46*, 159–63.

——Weiss, S., Weiss, A., Sandler, B., and Gordon, D.M. (1998) Effects of abiotic factors on the distribution and activity of the invasive Argentine ant (Hymenoptera: Formicidae). *Environmental Entomology 27*, 822–33.

Hunt, J.H. (1983) Foraging and morphology in ants: the role of vertebrate predators as agents of natural selection. In P. Jaisson, ed. *Social insects in the tropics.*, pp. 83–104. Presses de l'Université Paris XIII, Paris.

Hunter, M.D., and Price, P.W. (1992) Playing chutes and ladders – heterogeneity and the relative roles of bottom-up and top-down forces in natural communities. *Ecology 73*, 724–32.

Hurlbert, A.H., Ballantyne, F.I., and Powell, S. (2008) Shaking a leg and hot to trot: the effects of body size and temperature on running speed in ants. *Ecological Entomology 33*, 144–54.

Hutchinson, G.E. (1959) Homage to Santa Rosalia or Why are there so many kinds of animals? *American Naturalist 93*, 145–58.

Huxley, C.R. (1991) Ants and plants: a diversity of interactions. In C.R. Huxley and D.F. Cutler, eds. *Ant-Plant Interactions*, pp. 1–11. Oxford University Press, Oxford.

Ichinose, K., Cerda, X., Christides, J.-P., and Lenoir, A. (2005) Detecting nestmate recognition patterns in the fission-performing ant *Aphaenogaster senilis*: a comparison of different indices. *Journal of Insect Behavior 18*, 633–50.

Imai, H.T., Kihara, A., and Kondoh, M., Kubota, M., Kuribayashi, S., Ogata, K., Onoyama, K., Taylor, R.W., Terayama, M., Tsukii, Y., Yoshimura, M., and Ugava, Y. (2003) *Ants of Japan*. Gakken, Tokyo.

——Taylor, R.W., Kubota, M., Ogata, K., and Wada, M.Y. (1990) Notes on the remarkable karyology of the primitive ant *Nothomyrmecia macrops* Clark, and of the related genus *Myrmecia* Fabricius (Hymenoptera: Formicidae). *Psyche (Cambridge) 97*, 133–40.

Ingraham, W., and Foster, S. (2008) The value of ecosystem services provided by the US National Wildlife Refuge System in the contiguous US. *Ecological Economics 67*, 608–18.

Inouye, B.D., and Agrawal, A.A. (2004) Ant mutualists alter the composition and attack rate of the parasitoid community for the gall wasp *Disholcaspis eldoradensis* (Cynipidae). *Ecological Entomology 29*, 692–6.

International Commission on Zoological Nomenclature (1999) *International Code of Zoological Nomenclature, Fourth edition.* International Trust for Zoological Nomenclature, London.

IPCC (2007) Summary for Policymakers. In *Climate Change 2007: the Physical Science Basis. Contribution of Working Group I to the Fourth Assessment Report of the Intergovernmental Panel on Climate Change* Solomon, S., D. Qin, M. Manning, Z. Chen, M. Marquis, K.B. Averyt, M.Tignor, and H.L. Miller, eds. Cambridge University Press, Cambridge and New York.

Isaac, N.J.B., Turvey, S.T., Collen, B., Waterman, C., Baillie, J.E.M. (2007) Mammals on the EDGE: conservation priorities based on threat and phylogeny. *PLoS ONE 2*, e296. doi:10.1371/journal.pone.0000296.

Itino, T., Itioka, T., Hatada, A., and Hamid, A.A. (2001) Effects of food rewards offered by ant—plant *Macaranga* on the colony size of ants. *Ecological Research 16*, 775–86.

Ito, F. (1996) Colony characteristics of the Indonesian myrmicine ant *Myrmecina* sp. (Hymenoptera, Formicidae, Myrmicinae): polygynous reproduction by ergatoid queens. *Annals of the Entomological Society of America 89*, 550–4.

——(1997) Colony composition and morphological caste differentiation between ergatoid queens and workers in the ponerine ant genus *Leptogenys* in the oriental tropics. *Ethology Ecology and Evolution 9*, 335–43.

Ito, Y., Takamine, H., and Yamauchi, K. (1998) Abundance and species diversity of ants in forests of Yanbaru, the Northern Part of Okinawa Honto with special reference to effects of undergrowth removal. *Entomological Science 1*, 347–55.

IUCN (2008) *IUCN Red List of Threatened Species*. www.iucnredlist.org.

Jablonski, D., Roy, K., and Valentine, J.W. (2006) Out of the tropics: evolutionary dynamics of the latitudinal diversity gradient. *Science 314*, 102–6.

Jackson, D.E., Holcombe, M., Ratnieks, F.L.W. (2004) Trail geometry gives polarity to ant foraging networks. *Nature 432*, 907–9.

——Martin, S.J., Ratnieks, F.L.W., and Holcombe, M. (2007) Spatial and temporal variation in pheromone composition of ant foraging trails. *Behavioral Ecology 18*, 444–50.

Jaffé, K. (1993) Surfing ants. *Florida Entomologist 76*, 182–3.

——and Deneubourg, J.L. (1992) On foraging, recruitment systems and optimum number of scouts in eusocial colonies. *Insectes Sociaux 39*, 201–13.

——Caetano, F.H., Sanchez, P., Hernandez, J., Caraballo, L., Vitelli-Flores, J., Monsalve, W., Dorta, B., and Lemoine, V.R. (2001) Sensitivity of ant (*Cephalotes*) colonies and individuals to antibiotics implies feeding symbiosis with gut microorganisms. *Canadian Journal of Zoology 79*, 1120–4.

Jaisson, P. (1991) Kinship and fellowship in ants and social wasps. In P.G. Hepper, ed. *Kin Recognition*, pp. 60–93. Cambridge University Press, Cambridge.

James, F.C. (1970) Geographic size variation in birds and its relation to climate. *Ecology 51*, 361–90.

Jansen, V.A.A., and van Baalen, M. (2006) Altruism through beard chromodynamics. *Nature 440*, 663–6.

Jansson, R., and Dynesius, M. (2002) The fate of clades in a world of recurrent climatic change: Milankovitch oscillations and evolution. *Annual Review of Ecology, Evolution and Systematics 33*, 741–77.

Janzen, D.H. (1966) Coevolution of mutualism between ants and acacias in Central America. *Evolution 20*, 249–75.

——(1967a) Fire, vegetation structure, and the ant x *Acacia* interaction in Central America. *University of Kansas Scientific Bulletin 47*, 315–558.

——(1967b) Why mountain passes are higher in the tropics. *American Naturalist 101*, 233–49.

——(1973) Sweep samples of the tropical foliage insects: effects of seasons, vegetation types, elevation, time of day, and insularity. *Ecology 54*, 687–708.

——(1977) Why don't ants visit flowers? *Biotropica 9*, 252.

——Ataroff, M., Farinas, M., Reyes, S., Rincon, N., Soler, A., Soriano, P., and Vera, M. (1976) Changes in the arthropod community along an elevational transect in the Venezuelan Andes. *Biotropica 8*, 193–203.

——Hallwachs, W., Jimenez, J., and Gamez, R. (1993). The role of the parataxonomists, inventory managers, and taxonomists in Costa Rica's national biodiversity inventory. In Reid WV, Laird SA, Meyer CA, Gamez R, Sittenfeld A, Janzen DH, Gollin MA, Juma C, (eds). *Biodiversity Prospecting: using generic resources for sustainable development*. pp. 223–254. World Resources Institute, Washington, DC.

Jeanne, R.L. (1979) A latitudinal gradient in rates of ant predation *Ecology 60*, 1211–24.

Jetter, K.M., Hamilton, J., and Klotz, J.H. (2002) Red imported fire ants threaten agriculture, wildlife, and homes. *California Agriculture 56*, 26–34.

Jetz, W., and Rahbek, C. (2001) Geometric constraints explain much of the species richness pattern in African birds. *Proceedings of the National Academy of Sciences USA 98*, 5661–6.

——Carbone, C., Fulford, J., and Brown, J. (2004) The scaling of animal space use. *Science 306*, 266–8.

Jiménez, E., Fernández, F., Arias, T.M., and Lozano-Zambrano, F. H. (2008) *Sistemática, Biogeografía y Conservación de las Hormigas Cazadoras de Colombia*. Instituto de Investigación de Recursos Biológicos Alexander von Humboldt, Bogotá, Colombia.

Johnson, H.B. (1975) Plant pubescence: an ecological perspective. *The Botanical Review 41*, 233–58.

Johnson, R.A. (1994) Distribution and natural history of the workerless inquiline ant *Pogonomyrmex anergismus* Cole (Hymenoptera: Formicidae). *Psyche (Cambridge) 101*, 257–62.

——(2002) Semi-claustral colony founding in the seed-harvester ant *Pogonomyrmex californicus*: a comparative analysis of colony founding strategies. *Oecologia 132*, 60–7.

——(2004) Colony founding by pleometrosis in the semi-claustral seed-harvester ant *Pogonomyrmex californicus* (Hymenoptera: Formicidae). *Animal Behavior 68*, 1189–200.

——(2006) Capital and income breeding and the evolution of colony founding strategies in ants. *Insectes Sociaux 53*, 316–22.

——and Ward, P.S. (2002) Biogeography and endemism of ants (Hymenoptera: Formicidae) in Baja California, Mexico: a first overview. *Journal of Biogeography 29*, 1009–26.

Johnson, R.N., Agapow, P.M., and Crozier, R.H. (2003) A tree island approach to inferring phylogeny in the ant subfamily Formicinae, with special reference to the evolution of weaving. *Molecular Phylogenetics and Evolution 29*, 317–30.

Jones, G. (2005) Is the management plan achieving its objectives? In G. Worboys, M. Lockwood, and T. DeLacy, eds. *Protected Area Management Principles and Practice*, pp. 555–7. Oxford University Press, Oxford.

Jouquet, P., Dauber, J., Lagerlöf, J., Lavelle, P., and Lepage, M. (2006) Soil invertebrates as ecosystem engineers: intended and accidental effects on soil and feedback loops. *Applied Soil Ecology 32*, 153–64.

Jourdan, H. (1997) Are serpentine biota free from biological invasions? An example of an ant community from southern New Caledonia. In T. Jaffré, R.D. Reeves, and T. Becquer, eds. *The ecology of ultramafic and metalliferous areas*. Centre ORSTOM, New Caledonia.

Judd, T.M. (2006) Relationship between food stores and foraging behavior in *Pheidole ceres* (Hymenoptera: Formicidae). *Annals of the Entomological Society of America 99*, 398–406.

——and Fasnacht, M.P. (2007) Distribution of micronutrients in social insects: a test in the termite *Reticulitermes flavipes* (Isoptera: Rhinotermitidae) and the ant *Myrmica punctiventris* (Hymenoptera: Formicidae). *Annals of the Entomological Society of America 100*, 893–9.

Junker, R.R., and Blüthgen, N. (2008) Floral scents repel potentially nectar-thieving ants. *Evolutionary Ecology Research 10*, 295–308.

Jutsum, A.R., Saunders, T.S., and Cherrett, J.M. (1979) Intraspecific aggression in the leaf-cutting ant *Acromyrmex octospinosus*. *Animal Behavior 27*, 839–44.

Kalisz, S., Hanzawa, F.M., Tonsor, S.J., Thiede, D.A., and Voigt, S. (1999) Ant-mediated seed dispersal alters pattern of relatedness in a population of *Trillium grandiflorum*. *Ecology 80*, 2620–34.

Kaplan, I., and Eubanks, M.D. (2002) Disruption of cotton aphid (Homoptera: Aphididae): natural enemy dynamics by red imported fire ants (Hymenoptera: Formicidae). *Environmental Entomology 31*, 1175–83.

————(2005) Aphids alter the community-wide impact of fire ants. *Ecology 86*, 1640–9.

Karmaly, K.A. (2004) A new species and a key to species of *Polyrhachis* Smith (Hymenoptera: Formicidae) from India. In K. Rajmohana, K. Sudheer, P. Girish Kumar, and S. Santhosh, eds. *Perspectives on Biosystematics and Biodiversity. Prof. T.C. Narendran commemoration volume*, pp. 539–51. Systematic Entomology Research Scholars Association, Kerala.

Kaspari, M. (1993) Removal of seeds from neotropical frugivore droppings – ant responses to seed number. *Oecologia 95*, 81–8.

——(1996) Litter ant patchiness at the 1-m^2 scale: disturbance dynamics in three Neotropical forests. *Oecologia 107*, 265–73.

——(2000) A primer on ant ecology. In D. Agosti, J.-D. Majer, L.E. Alonso, and T.R. Schultz, eds. *Ants: standard methods for measuring and monitoring biodiversity*, pp. 9–24. Smithsonian Institution Press, Washington, DC.

——(2004) Using the metabolic theory of ecology to predict global patterns of abundance. *Ecology 85*, 1800–2.

——and Majer, J.D. (2000) Using ants to monitor environmental change. In D. Agosti, J.D. Majer, L. Alonso, and T. Shultz, eds. *Ants: standard methods for measuring and monitoring biodiversity*, pp. 89–98. Smithsonian Institution Press, Washington, DC.

——and Vargo, E. (1994) Colony size as a buffer against seasonality: bergmann's rule in social insects. *Bulletin of the Ecological Society of America 75*, 105–6.

——and Weiser, M.D. (1999) The size-grain hypothesis and interspecific scaling in ants. *Functional Ecology 13*, 530–8.

——O'Donnell, S., and Alonso, L. (2000a) Three energy variables predict ant abundance at a geographic scale. *Proceedings of the Royal Society of London B 267*, 485–90.

————and Kercher, J.R. (2000b) Energy, density, and constraints to species richness: ant assemblages along a productivity gradient. *American Naturalist 155*, 280–93.

——Pickering, J., and Windsor, D. (2001b) The reproductive flight phenology of a neotropical ant assemblage. *Ecological Entomology 26*, 245–57.

——Ward, P.S., and Yuan, M. (2004) Energy gradients and the geographic distribution of local ant diversity. *Oecologia 140*, 407–13.

——Yanoviak S.P., and Dudley R. (2008) On the biogeography of salt limitation: a study of ant communities. *Proceedings of the National Academy of Sciences USA 105*, 17848–51.

——Yuan, M., and Alonso, L. (2003) Spatial grain and the causes of regional diversity gradients in ants. *American Naturalist 161*, 459–77.

——Pickering, J., Longino, J.T., and Windsor, D. (2001a) The phenology of a Neotropical ant assemblage: evidence for continuous and overlapping reproduction. *Behavioral Ecology and Sociobiology 50*, 382–90.

Katzav-Gozansky, T., Boulay, R., Ionesko, A., and Hefetz, A. (2008) Nest volatiles as modulators of nestmate recognition in the ant *Camponotus fellah*. *Journal of Insect Physiology 54*, 378–85.

————Vander Meer, R.K., and Hefetz, A. (2004) In-nest environment modulates nestmate recognition in the ant *Camponotus fellah*. *Naturwissenschaften 91*, 186–90.

Katzerke, A., Neumann, P., Pirk, C.W.W., Bliss, P., and Moritz, R.F.A. (2006) Seasonal nestmate recognition in the ant *Formica exsecta*. *Behavioral Ecology and Sociobiology 61*, 143–50.

Kaufmann, E., and Maschwitz, U. (2006) Ant-gardens of tropical Asian rainforests. *Naturwissenschaften 93*, 216–27.

——Malsch, A.K.F., Erle, M., and Maschwitz, U. (2003) Compound nesting of *Strumigenys* sp. (Myrmicinae) and *Diacamma* sp. (Ponerinae), and other nesting symbioses of myrmicine and ponerine ants in Southeast Asia. *Insectes Sociaux 50*, 88–97.

Kay, A. (2002) Applying optimal foraging theory to assess nutrient availability ratios for ants. *Ecology 83*, 1935–44.

——Rostampour, S., and Sterner, R.W. (2006) Ant stoichiometry: elemental homeostasis in stage-structured colonies. *Functional Ecology 20*, 1037–44.

——Scott, S., Schade, J., and Hobbie, S. (2004) Stoichiometric relations in an ant-treehopper mutualism. *Ecology Letters 7*, 1024–8.

Keller, L. (1991) Queen number, mode of colony founding, and queen reproductive success in ants (Hymenoptera Formicidae). *Ethology Ecology and Evolution 3*, 307–16.

——(1997) Indiscriminate altruism: unduly nice parents and siblings. *Trends in Ecology and Evolution 12*, 186–90.

——(1998) Queen lifespan and colony characteristics in ants and termites. *Insectes Sociaux 45*, 235–46.

——(2007) Uncovering the biodiversity of genetic and reproductive systems: time for a more open approach. *American Naturalist 169*, 1–8.

——and Genoud, M. (1997) Extraordinary lifespans in ants: a test of evolutionary theories of ageing. *Nature 389*, 958–60.

——and Passera, L. (1989) Size and fat content of gynes in relation to the mode of colony founding in ants (Hymenoptera; Formicidae). *Oecologia 80*, 236–40.

——and Ross, K.G. (1998) Selfish genes: a green beard in the red fire ant. *Nature 394*, 573–5.

Kellner, K., Trindl, A., Heinze, J., and d'Ettorre, P. (2007) Polygyny and polyandry in small ant societies. *Molecular Ecology 16*, 1263–2369.

Kemel, K., Azevedo-Ramos, C., Moutinho, P., and Malcher, S. (2001) The effects of logging on the ground-foraging ant community in eastern Amazonia. *Studies on Neotropical Fauna and Environment 36*, 215–9.

Kenis, M., Auger-Rozenberg, M.A., Roques, A., Timms, L., Péré, C., Cock, M., Settele, J., Augustin, S., and Lopez-Vaamonde, C. (2009) Ecological effects of invasive alien insects. *Biological Invasions 11*, 21–45.

Kenne, M., and Dejean, A. (1999) Spatial distribution, size and density of nests of *Myrmicaria opaciventris* Emery (Formicidae: Myrmicinae). *Insectes Sociaux 46*, 179–85.

Kerner, A. (1878) *Flowers and their unbidden guests*. C. Kegan Paul & Co., London.

Ketterl, J., Verhaagh, M., and Dietz, B.H. (2004) *Eurhopalothrix depressa* sp. n. (Hymenoptera: Formicidae) from southern Brazil with a key to the Neotropical taxa of the genus. *Studies on the Neotropical Fauna and Environment 39*, 45–8.

Khila, A., and Abouheif, E. (2008) Reproductive constraint is a developmental mechanism that maintains social harmony in ants. *Proceedings of the National Academy of Sciences USA 105*, 17884–9.

Kier, G., Mutke, J., Dinerstein, E., Ricketts, T.H., Kuper, W., Kreft, H., and Barthlott, W. (2005) Global patterns of plant diversity and floristic knowledge. *Journal of Biogeography 32*, 1107–16.

King, J.R., and Porter, S.D. (2007) Body size, colony size, abundance, and ecological impact of exotic ants in Florida's upland ecosystems. *Evolutionary Ecology Research 9*, 757–74.

——and Tschinkel, W.R. (2006) Experimental evidence that the introduced fire ant, *Solenopsis invicta*, does not competitively suppress co-occurring ants in a disturbed habitat. *Journal of Animal Ecology 75*, 1370–8.

——(2008) Experimental evidence that human impacts drive fire ant invasions and ecological change. *Proceedings of the National Academy of Sciences USA 105*, 20339–43.

——Andersen, A.N., and Cutter, A.D. (1998) Ants as bioindicators of habitat disturbance: validation of the functional group model for Australia's humid tropics. *Biodiversity and Conservation 7*, 1627–38.

Kingsland, S.E. (1991) Defining ecology as a science. In L. A. Real and J.H. Brown, eds. *Foundations of Ecology*, pp. 1–13. University of Chicago Press, Chicago.

Kistner, D.H., Chong, K.F., Lee, C.Y. (2007) A new Malaysian myrmecophilous cricket (Orthoptera: Myrmecophilidae). *Sociobiology 50*, 173–82.

Kjar, D.S., and Sunian, T.W. (2007) First records of invasion by the myrmicine Japanese ant *Vollenhovia emeryi* W. M. Wheeler (Hymenoptera: Formicidae) in the United States. *Proceedings of the Entomological Society of Washington 109*, 596–604.

Klein, R.W., Maschwitz, U., and Kovac, D. (1993) Flood-control by ants – a Southeast-Asian bamboo-dwelling *Tetraponera* (Formicidae, Pseudomyrmecinae) bails water from its internode nests. *Insectes Sociaux 40*, 115–8.

Kleineidam, C., and Roces, F. (2000) Carbon dioxide concentrations and nest ventilation in nests of the leaf-cutting ant *Atta vollenweideri*. *Insectes Sociaux 47*, 241–8.

Klingenberg, C., and Verhaagh, M. (2005) Myrmecology in the Internet: possibilities of information gathering. *Beitraege zur Entomologie 55*, 485–98.

Klopfer, P.H., and Macarthur, R.H. (1961) On the causes of tropical species diversity – niche overlap. *American Naturalist 95*, 223–6.

Knaden, M., Tinaut, A., Cerda, X., Wehner, S., and Wehner, R. (2005) Phylogeny of three parapatric species of desert ants, *Cataglyphis bicolor*, *C. viatica* and *C. savignyi*: a comparison of mitochondrial DNA, nuclear DNA and morphological data. *Zoology 108*, 169–77.

Knaden, M., and Wehner, R. (2003) Nest defense and conspecific recognition in the desert ant *Cataglyphis fortis*. *Journal of Insect Behavior 16*, 717–30.

Kneitel, J.M., and Chase, J.M. (2004) Trade-offs in community ecology: linking spatial scales and species coexistence. *Ecology Letters 7*, 69–80.

Knoch, T.R., Faeth, S.H., and Arnott, D.L. (1993) Endophytic fungi alter foraging and dispersal by desert seed-harvesting ants. *Oecologia 95*, 47–473.

Kobayashi, K., Hasegawa, E., and Ohkawara, K. (2008) Clonal reproduction by males of the ant *Vollenhovia emeryi* (Wheeler). *Entomological Science 11*, 167–72.

Koh, L.P., Dunn, R.R., Sodhi, N.S., Colwell, R.S., Proctor, H.S., and Smith, V.S. (2004) Species coextinctions and the biodiversity crisis. *Science 305*, 1632–4.

Kohout, R.J. (1988) A new species of *Polyrhachis* from Papua New-Guinea with a review of the New Guinean and Australian species (Hymenoptera: Formicidae: Formicinae). *Memoirs of the Queensland Museum 25*, 417–28.

——(2006) Review of *Polyrhachis* (*Cyrtomyrma*) Forel (Hymenoptera: Formicidae: Formicinae) of Australia, Borneo, New Guinea and the Solomon Islands with descriptions of new species. *Memoirs of the Queensland Museum 52*, 87–146.

——(2007a) A review of the subgenus *Polyrhachis* (*Campomyrma*) Wheeler from Borneo with descriptions of new species (Hymenoptera: Formicidae: Formicinae). *Asian Myrmecology 1*, 7–17.

——(2007b) Revision of the subgenus *Aulacomyrma* Emery of the genus *Polyrhachis* F. Smith, with descriptions of new species. *Memoirs of the American Entomological Institute 80*, 186–253.

Kohout, R. J. (2008a) A new species of the *Polyrhachis (Myrmhopla) flavoflagellata*-group (Hymenoptera: Formicidae: Formicinae) from Borneo. *Asian Myrmecology 2*, 11–16.

——(2008b) A review of the *Polyrhachis* ants of Sulawesi with keys and descriptions of new species (Hymenoptera: Formicidae: Formicinae). *Memoirs of the Queensland Museum 52*, 255–317.

Koptur, S. (1979) Facultative mutualism between weedy vetches bearing extrafloral nectaries and weedy ants in California. *American Journal of Botany 66*, 1016–20.

——(1992) Extrafloral nectary mediated interactions between insects and plants. In E. Bernays, ed. *Insect-Plant Interactions*, pp. 81–129. CRC Press, Boca Raton.

——and Truong, N. (1998) Facultative ant-plant interactions: nectar sugar preferences of introduced pest ant species in South Florida. *Biotropica 30*, 179–89.

Korb, J., and Heinze, J. (2008) *Ecology of Social Evolution*. Springer, Heidelberg.

Korzukhin, M.D., Porter, S.D., Thompson, L.C., and Wiley, S. (2001) Modeling temperature-dependent range limits of the red imported fire ant (Hymenoptera: Formicidae: *Solenopsis invicta*) in the United States. *Environmental Entomology 30*, 645–55.

Kost, C., Lakatos, T., Böttcher, I., Arendholz, W.-R., Redenbach, M., and Wirth, R. (2007) Non-specific association between filamentous bacteria and fungus-growing ants. *Naturwissenschaften 94*, 821–8.

Kot, M., Lewis, M.A., and van der Driessche, P. (1996) Dispersal data and the spread of invading organisms. *Ecology 77*, 2027–42.

Krebs, C.J. (2000) *Ecology: the experimental analysis of distribution and abundance*. Benjamin Cummings, USA.

Kreft, H., and Jetz, W. (2007) Global patterns and determinants of vascular plant diversity. *Proceedings of the National Academy of Sciences USA 104*, 5925–30.

Kremen, C., Cameron, A., and Moilanen, A. (2008) Aligning conservation priorities across taxa in Madagascar with high-resolution planning tools. *Science 320*, 222–6.

Krieger, M.J.B., and Ross, K.G. (2002) Identification of a major gene regulating complex social behavior. *Science 295*, 328–32.

Krombein, K.V., Norden, B.B., Rickson, M.M., and Rickson, F.R. (1999) Biodiversity of the domatia occupants (ants, wasps, bees, and others) of the Sri Lankan myrmecophyte *Humboldtia laurifolia* Vahl (Fabaceae). *Smithsonian Contributions to Zoology 0*, i–iv, 1–34.

Kronauer, D.J.C. (2008) Recent advances in army ant biology (Hymenoptera: Formicidae). *Myrmecological News 12*, 51–65.

——Johnson, R.A., and Boomsma, J.J. (2007a) The evolution of multiple mating in army ants. *Evolution 61*, 413–22.

——Schöning, C., Vilhelmsen, L.B., and Boomsma, J.J. (2007b) A molecular phylogeny of *Dorylus* army ants provides evidence for multiple evolutionary transitions in foraging niche. *BMC Evolutionary Biology 7*, 56, doi:10.1186/1471-2148-7-56.

Krushelnycky, P.D., and Gillespie, R.G. (2008) Compositional and functional stability of arthropod communities in the face of ant invasions. *Ecological Applications 18*, 1547–62.

——Loope, L.L., and Joe, S.M. (2004) Limiting spread of a unicolonial invasive insect and characterization of seasonal patterns of range expansion. *Biological Invasions 6*, 47–57.

————and Reimer, N.J. (2005b) The ecology, policy, and management of ants in Hawaii. *Proceedings of the Hawaiian Entomological Society 37*, 1–25.

——Joe, S.M., Medeiros, A.C., Daehler, C.C., and Loope, L. L. (2005a) The role of abiotic conditions in shaping the long-term patterns of a high-elevation Argentine ant invasion. *Diversity and Distributions 11*, 319–31.

Kugler, C. (1979) Evolution of the sting apparatus in the myrmicine ants. *Evolution 33*, 117–30.

——(1984) Ecology of the ant *Pogonomyrmex mayri* – foraging and competition. *Biotropica 16*, 227–34.

Kumar, A., and O'Donnell, S. (2009) Elevation and forest clearing effects on foraging differ between surface – and subterranean – foraging army ants (Formicidae: Ecitoninae) *Journal of Animal Ecology 78*, 91–7.

Kümmerli, R., and Keller, L. (2007) Contrasting population genetic structure for workers and queens in the putatively unicolonial ant *Formica exsecta. Molecular Ecology 16*, 4493–503.

Kunin, E., and Gaston, K.J. (1997) *The Biology of Rarity*. Chapman & Hall, London.

Kusnezov, N. (1957) Numbers of species of ants in faunae of different latitudes. *Evolution 11*, 298–9.

Laakso, J., and Setälä, H. (1998) Composition and trophic structure of detrital food web in ant nest mounds of *Formica aquilonia* and in the surrounding forest soil. *Oikos 81*, 266–78.

————(2000) Impacts of wood ants (*Formica aquilonia* Yarr.) on the invertebrate food web of the boreal forest floor. *Annales Zoologici Fennici 37*, 93–100.

Lach, L. (2003) Invasive ants: unwanted partners in ant-plant interactions? *Annals of the Missouri Botanical Garden 90*, 91–108.

——(2005) Interference and exploitation competition of three nectar-thieving invasive ant species. *Insectes Sociaux 52*, 257–62.

——(2007) A mutualism with a native membracid facilitates pollinator displacement by Argentine ants. *Ecology 88*, 1994–2004.

——(2008a) Argentine ants displace floral arthropods in a biodiversity hotspot. *Diversity and Distributions 14*, 281–90.

——(2008b) Floral visitation patterns of two invasive ant species and their effects on other hymenopteran visitors. *Ecological Entomology 33*, 155–60.

——and Thomas, M.L. (2008) Invasive ants in Australia: documented and potential ecological consequences. *Australian Journal of Entomology 47*, 275–88.

——Hobbs, R.J., and Majer, J.D. (2009) Herbivory-induced extrafloral nectar increases native and invasive ant worker survival. *Population Ecology 51*, 237–43.

Lachaud, J.P., Cadena, A., Schatz, B., Pérez-Lachaud, G., and Ibarra-Núñez, G. (1999) Queen dimorphism and reproductive capacity in the ponerine ant, *Ectatomma ruidum* Roger. *Oecologia 120*, 515–23.

Lack, D. (1947) The significance of clutch-size. *Ibis 89*, 302–52.

Lafleur, B., Hooper-Búi, L.M., Mumma, E.P., and Geaghan, J.P. (2005) Soil fertility and plant growth in soils from pine forests and plantations: effect of invasive red imported fire ants *Solenopsis invicta* (Buren). *Pedobiologia 49*, 415–23.

Lake, P.S., and O'Dowd, D.J. (1991) Red crabs in rainforest, Christmas Island: biotic resistance to invasion by an exotic snail. *Oikos 62*, 25–9.

Lambardi, D., Dani, F.R., Turillazzi, S., and Boomsma, J.J. (2007) Chemical mimicry in an incipient leaf-cutting ant social parasite. *Behavioral Ecology and Sociobiology 61*, 843–51.

Lane, D.R., and BassiriRad, H. (2005) Diminishing effects of ant mounds on soil heterogeneity across a chronosequence of prairie restoration sites. *Pedobiologia 49*, 359–66.

Langhammer, P.F., Bakarr, M.I., Bennun, L.A and Brooks, T.M. (2007) *Identification and Gap Analysis of Key Biodiversity Areas: targets for comprehensive protected area systems*. IUCN, Gland, Switzerland and Cambridge, UK.

Langkilde, T. (2009) Invasive fire ants alter behavior and morphology of native lizards. *Ecology 90*, 208–17.

Lanza, J., Vargo, E.L., Pulim, S., and Chang, Y.Z. (1993) Preferences of the fire ants *Solenopsis invicta* and *S. geminata* (Hymenoptera: Formicidae) for amino acid and sugar components of extrafloral nectars. *Environmental Entomology 22*, 411–7.

LaPolla, J.S. (2004) *Acropyga* (Hymenoptera: Formicidae) of the world. *Contributions of the American Entomological Institute 33(4)*, 1–130.

——and Fisher, B.L. (2005) A remarkable new species of *Acropyga* (Hymenoptera: Formicidae) from Gabon, with a key to the Afrotropical species. *Proceedings of the California Academy of Sciences* (4) *56*, 601–5.

Lapolla, J.S., and Sosa-Calvo, J. (2006) Review of the ant genus *Rogeria* (Hymenoptera: Formicidae) in Guyana. *Zootaxa 1330*, 59–68.

Lattke, J.E. (1995) Revision of the ant genus *Gnamptogenys* in the New World (Hymenoptera: Formicidae). *Journal of Hymenoptera Research 4*, 137–93.

——(2004) A taxonomic revision and phylogenetic analysis of the ant genus *Gnamptogenys* Roger in Southeast Asia and Australasia (Hymenoptera: Formicidae: Ponerinae). *University of California Publications in Entomology 122*, 1–266.

——Fernández, F., and Palacio, E.E. (2007) Identification of the species of *Gnamptogenys* Roger in the Americas. *Memoirs of the American Entomological Institute 80*, 254–70.

Laurence, W.F., Lovejoy, T.E., and Vasconcelos, H.L. (2002) Ecosystem decay of Amazonian forest fragments: a 22-year investigation. *Conservation Biology 16*, 605–18.

Leal, I.R., and Oliveira, P.S. (1995) Behavioral ecology of the neotropical termite-hunting ant *Pachycondyla* (=*Termitopone*) *marginata*: colony founding, group-raiding and migratory patterns. *Behavioral Ecology and Sociobiology 37*, 373–83.

Le Breton, J., Chazeau, J., and Jourdan, H. (2003) Immediate impacts of invasion by *Wasmannia auropunctata* (Hymenoptera: Formicidae) on native litter ant fauna in a New Caledonian rainforest. *Austral Ecology 28*, 204–9.

——Delabie, J.H.C., Chazeau, J., Dejean, A., and Jourdan, H. (2004) Experimental evidence of large-scale unicoloniality in the tramp ant *Wasmannia auropunctata* (Roger). *Journal of Insect Behaviour 17*, 263–71.

——Orivel, J., Chazeau, J., and Dejean, A. (2007) Unadapted behaviour of native, dominant ant species during the colonization of an aggressive, invasive ant. *Ecological Research 22*, 107–14.

——Jourdan, H., Chazeau, J., Orivel, J., and Dejean, A. (2005) Niche opportunity and ant invasion: the case of *Wasmannia auropunctata* in a New Caledonian rain forest. *Journal of Tropical Ecology 21*, 93–8.

LeBrun, E.G. (2005) Who is the top dog in ant communities? Resources, parasitoids, and multiple competitive hierarchies. *Oecologia 142*, 643–52.

——and Feener, D.H. (2002) Linked indirect effects in ant-phorid interactions: impacts on ant assemblage structure. *Oecologia 133*, 599–607.

————(2007) When trade-offs interact: balance of terror enforces dominance discovery trade-off in a local ant assemblage. *Journal of Animal Ecology 76*, 58–64.

——Plowes, R.M., and Gilbert, L.E. (2008) Dynamic expansion in recently introduced populations of fire ant parasitiods (Diptera: Phoridae). *Biological Invasions 10*, 989–99.

——Tillberg, C.V., Suarez, A.V., Folgarait, P.J., Smith, C. R., and Holway, D.A. (2007) An experimental study of competition between fire ants and Argentine ants in their native range. *Ecology 88*, 63–75.

Le Conte, Y., and Hefetz, A. (2008) Primer pheromones in social Hymenoptera. *Annual Review of Entomology 53*, 523–42.

Ledoux, A. (1973) A propos du bouturage de societe chez la fourmi *Aphaenogaster senilis* Mayr. *Comptes Rendus Hebdomadaires des Séances de l'Académie des Sciences D 277*, 2199–200.

Lengyel, F., Westerlund, S.A., and Kaib, M. (2007) Juvenile hormone III influences task-specific cuticular hydrocarbon profile changes in the ant *Myrmicaria eumenoides*. *Journal of Chemical Ecology 33*, 167–81.

Lenoir, A., Malosse, C., and Yamaoka, R. (1997) Chemical mimicry between parasites of the genus *Formicoxenus* and their host *Myrmica* (Hymenoptera, Formicidae). *Biochemical Systematics and Ecology 25*, 379–89.

——d'Ettorre, P., Errard, C., and Hefetz, A. (2001) Chemical ecology and social parasitism in ants. *Annual Review of Entomology 46*, 573–99.

——Fresneau, D., Errard, C., and Hefetz, A. (1999) The individuality and the colonial identity in ants: the emergence of the social representation concept. In C. Detrain, J.L. Deneubourg, and J. Pasteels, eds. *Information Processing in Social Insects*, pp. 219–37. Birkhäuser Verlag, Basel.

——Querard, L., Pondicq, N., and Berton, F. (1988) Reproduction and dispersal in the ant *Cataglyphis cursor* (Hymenoptera, Formicidae). *Psyche (Cambridge) 95*, 21–44.

——Nowbahari, E., Quérard, L., Pondicq, N., and Delalande, C. (1990) Habitat exploitation and intercolonial relationships in the ant *Cataglyphis cursor* (Hymenoptera: Formicidae). *Acta Oecologica 11*, 3–18.

Lenoir, J.-C., Schrempf, A., Lenoir, A., Heinze, J., and Mercier, J.L. (2007) Genetic structure and reproductive strategy in the ant *Cardiocondyla elegans*: strictly monogynous nests invaded by unrelated sexuals. *Molecular Ecology 16*, 345–54.

Leonhardt, S.D., Brandstaetter, A.S., and Kleineidam, C.J. (2007) Reformation process of the neuronal template for nestmate-recognition cues in the carpenter ant *Camponotus floridanus*. *Journal of Comparative Physiology A 193*, 993–1000.

Lessard, J.P., and Buddle, C.M. (2005) The effects of urbanization on ant assemblages (Hymenoptera: Formicidae) associated with the Molson Nature Reserve, Quebec. *Canadian Entomologist 137*, 215–25.

——Dunn, R.R., Parker, C.R., and Sanders, N.J. (2007) Rarity and diversity in forest ant assemblages of Great Smoky Mountains National Park. *Southeastern Naturalist Special Issue 1*, 215–28.

Lester, P.J. (2005) Determinants for the successful establishment of exotic ants in New Zealand. *Diversity and Distributions 11*, 279–88.

——and Keall, J.B. (2005) The apparent establishment and subsequent eradication of the Australian giant bulldog ant *Myrmecia brevinoda* Forel (Hymenoptera: Formicidae) in New Zealand. *New Zealand Journal of Zoology 32*, 353–7.

——and Tavite, A. (2004) Long-legged ants, *Anoplolepis gracilipes* (Hymenoptera: Formicidae), have invaded Tokelau, changing composition and dynamics of ant and invertebrate communities. *Pacific Science 58*, 391–401.

——Abbott, K.L., Sarty, M., and Burns, K.C. (2009) Competitive assembly of South Pacific invasive ant communities. *BMC Ecology 9*, doi:10.1186/1472–6785–1189–1183.

Leston, D. (1973) The ant mosaic-tropical tree crops and the limiting of pests and diseases. *Articles and News Summaries 19*, 311–41.

Letourneau, D.K. (1983) Passive aggression: an alternative hypothesis for the *Piper-Pheidole* association. *Oecologia 60*, 122–6.

Letourneau, D.K., Dyer, L.A., and Vega, G. (2004) Indirect effects of a top predator on a rain forest understory plant community *Ecology 85*, 2144–52.

Leung, B., Lodge, D.M., Finnoff, D., Shogren, J.F., Lewis, M.A., and Lamberti, G. (2002) An ounce of prevention or a pound of cure: bioeconomic risk analysis of invasive species. *Proceedings of the Royal Society of London B 269*, 2407–13.

Levia, D.F., and Frost, E.E. (2004) Assessment of climatic suitability for the expansion of *Solenopsis invicta* Buren in Oklahoma using three general circulation models. *Theoretical and Applied Climatology 79*, 23–30.

Levin, S.A. (1992) The problem of pattern and scale in ecology. *Ecology 73*, 1943–67.

Levins, R. (2008) Predacious ants patrol sweet potatoes in Cuba. http:/www.fadr.msu.ru/rodale/agsieve/txt/vol3/1/a6.html.

——and Culver, D. (1971) Regional coexistence of species and competition between rare species (mathematical model/habitable patches). *Proceedings of the National Academy of Sciences USA 68*, 1246–8.

Lewis, T., Cherrett, J.M., Haines, I., Haines, J.B., and Mathias, P.L. (1976) The crazy ant (*Anoplolepis longipes* (Jerd.) (Hymenoptera, Formicidae) in Seychelles, and its chemical control. *Bulletin of Entomological Research 66*, 97–111.

Li, H.W., Medina, F., Vinson, S.B., and Coates, C.J. (2005) Isolation, characterization, and molecular identification of bacteria from the red imported fire ant (*Solenopsis invicta*) midgut. *Journal of Invertebrate Pathology 89*, 203–9.

Liang, D., and Silverman, J. (2000) 'You are what you eat': diet modifies cuticular hydrocarbons and nestmate recognition in the argentine ant, *Linepithema humile*. *Naturwissenschaften 87*, 412–416.

Liautard, C., and Keller, L. (2001) Restricted effective queen dispersal at a microgeographic scale in polygynous populations of the ant *Formica exsecta*. *Evolution 55*, 2484–92.

——Brown, W.D., Helms, K.R., and Keller, L. (2003) Temporal and spatial variations of gyne production in the ant *Formica exsecta*. *Oecologia 136*, 558–64.

Liebert, A.E., and Starks, P.T. (2004) The action component of recognition systems: a focus on the response. *Annales Zoologici Fennici 41*, 747–64.

Liebherr, J.K., and Krushelnycky, P.D. (2007) Unfortunate encounters? Novel interactions of native *Mecyclothorax*, alien *Trechus obtusus* (Coleoptera: Carabidae) and Argentine ant (*Linepithema humile*) (Hymenoptera: Formicidae) across a Hawaiian landscape. *Journal of Insect Conservation 11*, 61–73.

Liebig, J., and Poethke, H.J. (2004) Queen lifespan and colony longevity in the ant *Harpegnathos saltator*. *Ecological Entomology 29*, 203–7.

Liefke, C.W, Dorow, H.O., Hölldobler, B., and Maschwitz, U. (1998) Nesting and food resources of syntopic species of the ant genus *Polyrhachis* (Hymenoptera, Formicidae) in West-Malaysia. *Insectes Sociaux 45*, 411–25.

Lin, C.C., and Wu, W.J. (2003) The ant fauna of Taiwan (Hymenoptera: Formicidae), with keys to subfamilies and genera. *Guoli Taiwan Bowuguan Niankan 46*, 5–69.

Linnaeus, C. (1758) *Systema naturae per regna tria naturae, secundum classes, ordines, genera, species, cum characteribus, differentiis, synonymis, locis. Tomus I. Editio decima, reformata*. L. Salvii, Stockholm.

Litsinger, J.A., Cruz, C.G.D., Canapi, B.L., and Barrion, A. T. (2007) Maize planting time and arthropod abundance in southern Mindanao, Philippines. II. Population dynamics of natural enemies. *International Journal of Pest Management 53*, 161–73.

Little, A.E.F., and Currie, C.R. (2008) Black yeast symbionts compromise the efficiency of antibiotic defenses in fungus-growing ants. *Ecology 89*, 1216–22.

Lockwood, J.L. (1999) Using taxonomy to predict success among introduced avifauna: relative importance of transport and establishment. *Conservation Biology 13*, 560–7.

Lodge, D.M., Williams, S., MacIsaac, H.J., Hayes, K.R., Leung, B., Reichard, S., Mack, R.N., Moyle, P.B., Smith, M.A., Andow, O.A., Carlton, J.T., and McMichael, A. (2006) Biological invasions: recommendations for U.S. policy and management. *Ecological Applications 16*, 2035–54.

Lofgren, C.S., Banks, W.A., and Glancey, B.M. (1975) Biology and control of imported fire ant. *Annual Review of Entomology 20*, 1–30.

Longhurst, C., and Howse, P.E. (1979) Foraging, recruitment and emigration in *Megaponera foetens* (Fab.) (Hymenoptera: Formicidae) from the Nigerian Guinea savanna. *Insectes Sociaux 26*, 204–15.

Longino, J.T. (1989) Geographic variation and community structure in an ant-plant mutualism: *Azteca* and *Cecropia* in Costa Rica. *Biotropica 21*, 126–32.

——(2003) The *Crematogaster* (Hymenoptera, Formicidae, Myrmicinae) of Costa Rica. *Zootaxa 151*, 1–150.

——(2006) A taxonomic review of the genus *Myrmelachista* (Hymenoptera: Formicidae) in Costa Rica. *Zootaxa 1141*, 1–54.

——(2007) A taxonomic review of the genus *Azteca* (Hymenoptera: Formicidae) in Costa Rica and a global revision of the *aurita* group. *Zootaxa 1491*, 1–63.

——(2008) *Leaf Litter Arthropods of Mesoamerica: LLAMA*. http://llama.evergreen.edu/.

——and Fernández, F. (2007) Taxonomic review of the genus *Wasmannia*. *Memoirs of the American Entomological Institute 80*, 271–89.

——Colwell, R.K., and Coddington, J.A. (2002) The ant fauna of a tropical rainforest: estimating species richness three different ways. *Ecology 83*, 689–702.

Losey, J., and Vaughan, M. (2006) The economic value of ecological services provided by insects. *BioScience 56*, 311–23.

Lowe, S. (2000) *100 of the world's worst invasive alien species: a selection from the Global Invasive Species Database* http://www.issg.org/database. The Invasive Species Specialist Group (ISSG) New Zealand.

Lucas, C., Pho, D.B., Fresneau, D., and Jallon, J.M. (2004) Hydrocarbon circulation and colonial signature in *Pachycondyla villosa*. *Journal of Insect Physiology 50*, 595–607.

———Jallon, J.M., and Fresneau, D. (2005) Role of cuticular hydrocarbons in the chemical recognition between ant species in the *Pachycondyla villosa* species complex. *Journal of Insect Physiology 51*, 1148–51.

——Fresneau, D., Kolmer, K., Heinze, J., Delabie, J.H.C., and Pho, D.B. (2002) A multidisciplinary approach to discriminating different taxa in the species complex *Pachycondyla villosa* (Formicidae). *Biological Journal of the Linnean Society 75*, 249–59.

Lude, A., Reich, M., and Plachter, H. (1999) Life strategies of ants in unpredictable floodplain habitats of Alpine rivers (Hymenoptera: Formicidae). *Entomologia Generalis 24*, 75–91.

Lugo, A.E., Farnworth, E.G., Pool, D., Jerez, P., and Kaufman, G. (1973) The impact of the leaf cutter ant, *Atta columbica*, on the energy flow of a tropical rain forest. *Ecology 54*, 1292–301.

Lutz, H. (1986) Eine neue Unterfamilie der Formicidae (Insecta: Hymenoptera) aus dem mittel-eozänen Ölschiefer der 'Grube Messel' bei Darmstadt (Deutschland, S-Hessen). *Senckenbergiana Lethae 67*, 177–218.

——(1990) Systematische und palökologische Untersuchungen an Insekten aus dem Mittel-Eozän der Grube Messel bei Darmstadt. *Courier Forschungsinstitut Senckenberg 124*, 1–165.

——(1992) Giant ants and other rarities: the insect fauna. In S. Schaal and W. Ziegler, eds. *Messel-An Insight into the History of Life and of the Earth*, pp. 53–67. Clarendon Press, Oxford.

Lyu, D.-P. (2006) *Myrmica* in Korea (Hymenptera: Formicidae). *Journal of Asia-Pacific Entomology 9*, 189–202.

Mabelis, A.A. (1979) Wood ant wars: the relationship between aggression and predation in the red wood ant (*Formica polyctena* Forst.). *Netherlands Journal of Zoology 29*, 451–520.

——(2007) Do ants need protecting? *Entomologische Berichten 67*, 145–9.

MacArthur, R.H. (1972) *Geographical Ecology: patterns in the distribution of species*. Harper and Row, New York.

——and Wilson, E.O. (1967) *The Theory of Island Biogeography*. Princeton University Press, Princeton, NJ.

MacGown, J.A., Brown, R.L., Hill, J.G., and Layton, B. (2007) Carpenter ants (Hymenoptera: Formicidae: *Camponotus*) of Mississippi. *Mississippi Agricultural and Forestry Experiment Station Technical Bulletin 1158*, 28.

Macintosh, J.A., Veal, D.A., Beattie, A.J., and Gooley, A.A. (1998) Isolation from an ant *Myrmecia gulosa* of two inducible O-glycosylated proline-rich antibacterial peptides. *Journal of Biological Chemistry 273*, 6139–43.

MacIntyre, R.J., ed. (1985) *Molecular Evolutionary Genetics*. Plenum Press, New York.

Mackay, W.P., and Mackay, E.E. (1984) Why do harvester ants store seeds in their nests? *Sociobiology 9*, 31–47.

——Porter, S., and Gonzalez, D. (1990) A comparison of monogyne and polygyne populations of the tropical fire ant, *Solenopsis geminata* (Hymenoptera: Formicidae) in Mexico. *Journal of the Kansas Entomological Society 63*, 611–5.

——Maes, J.-M., Rojas Fernández, P., and Luna, G. (2004) The ants of North and Central America: the genus *Mycocepurus* (Hymenoptera: Formicidae). *Journal of Insect Science 4*(27), available online: insectscience.org/4.27. 7.

——Rebeles, A., Arredondo, H.C., Rodriguez, A.Z., Gonzales, D.A., and Vinson, S.B. (1991) Impact of the slash and burning of a tropical rain forest on the native ant fauna (Hymenoptera: Formicidae). *Sociobiology 18*, 257–68.

Macom, T.E., and Porter, S.D. (1996) Comparison of polygyne and monogyne red imported fire ant (Hymenoptera: Formicidae) population densities. *Annals of the Entomological Society of America 89*, 535–43.

Magallón, S. (2004) Dating lineages: molecular and paleontological approaches to the temporal framework of clades. *International Journal of Plant Science 165* (4 Suppl), S7-S21.

Mailleux, A.-C., Deneubourg J.-L., Detrain C. (2003) How does colony growth influence communication in ants? *Insectes Sociaux 50*, 24–31.

Majer, J.D. (1972) The ant mosaic in Ghana cocoa farms. *Bulletin of Entomological Research 62*, 151–60.

——(1976) Influence of ants and ant manipulation on cocoa farm fauna. *Journal of Applied Ecology 13*, 157–75.

——(1994) Spread of Argentine ants (*Linepithema humile*), with special reference to Western Australia. In D.F. Williams, ed. *Exotic Ants: biology, impact, and control of introduced species*, pp. 163–73. Westview Press, Boulder, CO.

——and Beeston, G. (1996) The Biodiversity Integrity Index: an illustration using ants in Western Australia. *Conservation Biology 10*, 65–73.

——and Delabie, J.H.C. (1994) Comparison of the ant communities of annually inundated and terra-firme forests at Trombetas in the Brazilian Amazon. *Insectes Sociaux 41*, 343–59.

————(1999) Impact of tree isolation on arboreal and ground ant communities in cleared pasture in the Atlantic rain forest region of Bahia, Brazil. *Insectes Sociaux 46*, 281–90.

————and McKenzie, N.L. (1997) Ant litter fauna of forest, forest edges and adjacent grassland in the Atlantic rain forest region of Bahia, Brazil. *Insectes Sociaux 44*, 255–66.

———and Smith, R.B. (1994) Arboreal ant community patterns in Brazilian cocoa farms. *Biotropica 26*, 73–83.

——Orabi, G., and Bisevac, L. (2007) Ants (Hymenoptera: Formicidae) pass the bioindicator scorecard. *Myrmecological News 10*, 69–76.

——Day, J.E., Kabay, E.D., and Perriman, W.S. (1984) Recolonization by ants in bauxite mines rehabilitated by a number of different methods. *Journal of Applied Ecology 21*, 355–75.

——Kitching, R.L., Heterick, B.E., Hurley, K., and Brennan, K.E.C. (2001) North-South patterns within arboreal ant assemblages from rain forests in eastern Australia. *Biotropica 33*, 643–61.

Mallon, E., and Franks, N. (2000) Ants estimate area using Buffon's needle. *Proceedings of the Royal Society of London B 267*, 765–70.

Maloof, J.E., and Inouye, D.W. (2000) Are nectar robbers cheaters or mutualists? *Ecology 81*, 2651–61.

Malsch, A.K.F., Fiala, B., Maschwitz, U., Mohamed, M., Nais, J., and Linsenmair, K.E. (2008) An analysis of declining ant species richness with increasing elevation at Mount Kinabalu, Sabah, Borneo. *Asian Myrmecology 2*, 33–49.

Manly, B.F.J. (1995) A note on the analysis of species cooccurrences. *Ecology 76*, 1109–15.

Mark, S., and Oleson, J.M. (1996) Importance of elaiosome size to the removal of ant-dispersed seeds *Oecologia 107*, 95–101.

Markin, G.P. (1970a) Food distribution within laboratory colonies of the Argentine ant, *Iridomyrmex humilis* (Mayr). *Insectes Sociaux 17*, 127–58.

——(1970b) The seasonal life cycle of the Argentine ant, *Iridomyrmex humilis* in southern California. *Annals of the Entomological Society of America 635*, 1238–42.

——Oneal, J., and Collins, H.L. (1974) Effects of mirex on general ant fauna of a treated area in Louisiana. *Environmental Entomology 3*, 895–8.

——Dillier, J.H., Hill, S.O., Blum, M.S., and Hermann, H. R. (1971) Nuptial flight and flight ranges of the imported fire ant, *Solenopsis saevissima richteri* (Hymenoptera: Formicidae). *Journal of the Georgia Entomological Society 6*, 145–56.

Marlier, J.F., Quinet, Y., and De Biseau, J.C. (2004) Defensive behaviour and biological activities of the abdominal secretion in the ant *Crematogaster scutellaris* (Hymenoptera: Myrmicinae). *Behavioural Processes 67*, 427–40.

Marr, R. (2003) *Toxic baiting of the yellow crazy ant,* Anoplolepis gracilipes, *on Christmas Island: impacts on non-target litter invertebrates*. Honours thesis, School of Biological Science, Monash University, Melbourne.

Marti, O.G., and Olson, D.M. (2007) Effect of tillage on cotton aphids (Homoptera: Aphididae), pathogenic fungi, and predators in south central Georgia cotton fields. *Journal of Entomological Science 42*, 354–67.

Martin, M.M. (1970) The biochemical basis of the fungus-attine ant symbiosis. *Science 169*, 16–20.

——(1991) The evolution of cellulose digestion in insects. *Philosophical Transactions of the Royal Society B 333*, 281–8.

Martin, S.J., Helanterä, H., and Drijfhout, F.P. (2008a) Evolution of species-specific cuticular hydrocarbon patterns in *Formica* ants. *Biological Journal of the Linnean Society, 95*, 131–40.

——Vitikainen, E., Helanterä, H., and Drijfhout, F.P. (2008b) Chemical basis of nestmate dsicrimination in the ant *Formica exsecta*. *Proceedings of the Royal Society of London B 275*, 1271–8.

Martinez-Ferrer, M.T., Grafton-Cardwell, E.E., and Shorey, H.H. (2003) Disruption of parasitism of the California red scale (Homoptera: Diaspdidae) by three ant species (Hymenoptera: Formicidae). *Biological Control 26*, 279–86.

Martins Jr., J., Solomon, S.E., Mikheyev, A.S., Mueller, U. G., Ortiz, A., and Bacci Jr., M., (2007) Nuclear mitochondrial-like sequences in ants: evidence from *Atta cephalotes* (Formicidae: Attini). *Insect Molecular Biology 16*, 777–84.

Maschwitz, U., and Hänel, H. (1985) The migrating herdsman *Dolichoderus* (*Diabolus*) *cuspidatus*: an ant with a novel mode of life. *Behavioral Ecology and Sociobiology 17*, 171–84.

——and Moog, J. (2000) Communal peeing: a new mode of flood control in ants. *Naturwissenschaften 87*, 563–5.

——and Steghaus-Kovac, S. (1991) Individualism versus cooperation – conflicting hunting and recruiting strategies among tropical Ponerine ants (Hymenpotera, Formicidae). *Naturwissenschaften 78*, 103–13.

——Lenz, S., and Buschinger, A. (1986) Individual specific trails in the ant *Leptothorax affinis* (Formicidae: Myrmicinae). *Experientia 42*, 1173–4.

——Steghaus-Kovac, S., Gaube, R., and Hänel, H. (1989) A South East Asian ponerine ant of the genus *Leptogenys* (Hym., Form.) with army ant habits. *Behavioral Ecology and Sociobiology 24*, 305–16.

Masuko, K. (1986) Larval hemolymph feeding: a nondestructive parental cannibalism in the primitive ant *Amblyopone silvestrii* Wheeler (Hymenoptera: Formicidae). *Behavioral Ecology and Sociobiology 19*, 249–55.

——(1989) Larval hemolymph feeding in the ant *Leptanilla japonica* by use of a specialized duct organ, the "larval hemolymph tap" (Hymenoptera: Formicidae). *Behavioral Ecology and Sociobiology 24*, 127–32.

——(1990) Behavior and ecology of the enigmatic ant *Leptanilla japonica* Baroni Urbani (Hymenoptera: Formicidae: Leptanillinae). *Insectes Sociaux 37*, 31–57.

Mateo, J.M. (2004) Recognition systems and biological organization: the perception component of social recognition. *Annales Zoologici Fennici 41*, 729–46.

Matsuura, K., Tamura, T., Kobayashi, N., and Tatsumi, S. (2007) The antibacterial protein lysozyme identified as the termite egg recognition pheromone. *PLoS ONE 2*, e813. doi:810.1371/journal.pone.0000813.

May, R.M. (1988) How many species are there on earth? *Science 247*, 1441–9.

Mayade, S., Cammaerts, M.C., and Suzzoni, J.P. (1993) Home range marking and territorial marking in *Cataglyphis cursor* (Hymenoptera: Formicidae). *Behavioural Processes 30*, 131–42.

Mayhé-Nunes, A.J., and Brandão, C.R.F. (2005) Revisionary studies on the attine ant genus *Trachymyrmex* Forel. Part 2: the Iheringi group (Hymenoptera: Formicidae). *Sociobiology 45*, 271–305.

———(2006) Revisionary notes on the fungus-growing ant genus *Mycetarotes* Emery (Hymenoptera, Formicidae). *Revista Brasileira de Entomologia 50*, 463–72.

———(2007) Revisionary studies on the attine ant genus *Trachymyrmex* Forel. Part 3: the Jamaicensis group (Hymenoptera: Formicidae). *Zootaxa 1444*, 1–21.

Maynard Smith, J. (1998) *Evolutionary Genetics*. Oxford University Press, Oxford.

Mbanyana, N., and Robertson, H.G. (2008) Review of the ant genus *Nesomyrmex* (Hymenoptera: Formicidae: Myrmicinae) in southern Africa. *African Natural History 4*, 35–55.

McArdle, B. (1990) When are rare species not there? *Oikos 57*, 276–7.

McArthur, A.J. (2007) A key to *Camponotus* Mayr of Australia. *Memoirs of the American Entomological Institute 80*, 290–351.

McCabe, S., Farina, W.M., and Josens, R.B. (2006) Antennation of nectar-receivers encodes colony needs and food-source profitability in the ant *Camponotus mus*. *Insectes Sociaux 53*, 356–61.

McCarthy, M.A., and Possingham, H.P. (2007) Active adaptive management for conservation. *Conservation Biology 21*, 956–63.

McCutchan, J.H., Lewis, W.M., Kendall, C., and McGrath, C.C. (2003) Variation in trophic shift for stable isotope ratios of carbon, nitrogen, and sulfur. *Oikos 102*, 378–90.

McDonald, K.A., and Brown, J.H. (1992) Using montane mammals to model extinctions due to global change. *Conservation Biology 6*, 409–16.

McGlynn, T.P. (1999a) Non-native ants are smaller than related native ants. *American Naturalist 154*, 690–9.

——(1999b) The worldwide transfer of ants: geographic distribution and ecological invasions. *Journal of Biogeography 26*, 535–48.

——(2000) Do Lanchester's laws of combat describe competition in ants? *Behavioral Ecology 11*, 686–90.

——(2006) Ants on the move: resource limitation of a litter-nesting ant community in Costa Rica. *Biotropica 38*, 419–27.

——(2007) Serial monodomy in ants: an antipredator strategy? *Ecological Entomology 32*, 621–6.

——Carr, R.A., Carson, J.H., and Buma, J. (2004) Frequent nest relocation in the ant *Aphaenogaster araneoides*: resources, competition, and natural enemies. *Oikos 106*, 611–21.

McInnes, D.A. (1994) *Comparative ecology and factors affecting the distribution of north Florida fire ants*. Doctoral dissertation, Florida State University, Tallahassee.

McKinney, M. L., and Lockwood, J.L. (1999) Biotic homogenization: a few winners replacing many losers in the next mass extinction. *Trends in Ecology and Evolution 14*, 450–3.

McNamara, J.M., Barta, Z., Wikelski, M., and Houston, A. I. (2008) A theoretical investigation of the effect of latitude on avian life histories. *American Naturalist 172*, 331–45.

McNatty, A., Abbott, K.L., and Lester, P.J. (2009) Invasive ants compete with and modify the trophic ecology of hermit crabs on tropical islands. *Oecologia 160*, 187–94.

McNicol, C. (2006) Surveillance methodologies used within Australia: various methods including visual surveillance and extraordinary detections – above ground and in ground lures. In L.C. Graham, ed. *Proceedings of the Red Imported Fire Ant Conference, March 28–30, 2006, Mobile, AL*, pp. 69–73. Department of Entomology and Plant Pathology, Auburn University, Auburn, AL.

Mehdiabadi, N.J., and Gilbert, L.E. (2002) Colony-level impacts of parasitoid flies on fire ants. *Proceedings of the Royal Society of London B 269*, 1695–9.

Mehlhop, P., and Scott, N.J. (2008) Temporal patterns of seed use and availability in a guild of desert ants. *Ecological Entomology 8*, 69–85.

Melbourne, B.A. (1999) Bias in the effect of habitat structure on pitfall traps: an experimental evaluation. *Australian Journal of Ecology 24*, 228–39.

Menke, S.B., and Holway, D.A. (2006) Abiotic factors control invasion by Argentine ants at the community scale. *Journal of Animal Ecology 75*, 368–76.

——Fisher, R.N., Jetz, W., and Holway, D.A. (2007) Biotic and abiotic controls of Argentine ant invasion success at local and landscape scales. *Ecology 88*, 3164–73.

Menzel, F., Blüthgen, N., and Schmitt, T. (2008) Tropical parabiotic ants: highly unusual cuticular substances and low interspecific discrimination. *Frontiers in Zoology 5*, 16.

Menzel, R., De Marco, R.J., and Greggers, U. (2006) Spatial memory, navigation and dance behaviour in *Apis mellifera*. *Journal of Comparative Physiology a-Neuroethology Sensory Neural and Behavioral Physiology 192*, 889–903.

Mercier, J.L., and Dejean, A. (1996) Ritualized behaviour during competition for food between two Formicinae. *Insectes Sociaux 43*, 17–29.

Meyers, J. (2008) Exotic Texas ant, *Paratrechina* sp. near *pubens*. http://antinstitute.org/PDF/Crazy-Rasberry-Ants.pdf.

Mikheyev, A.S., and Mueller, U.G. (2007) Genetic relationships between native and introduced populations of the little fire ant *Wasmannia auropunctata*. *Diversity and Distributions 13*, 573–9.

——and Tschinkel, W.R. (2004) Nest architecture of the ant *Formica pallidefulva*: structure, costs and rules of excavation. *Insectes Sociaux 51*, 30–6.

——Mueller, U.G., and Abbot, P. (2006) Cryptic sex and many-to-one coevolution in the fungus-growing ant symbiosis. *Proceedings of the National Academy of Sciences USA 103*, 10702–6.

Milford, E.R. (1999) Ant communities in flooded and unflooded riparian forest of the middle Rio Grande. *Southwest Naturalist 44*, 278–86.

Miller, S.E. (2007a) DNA barcoding and the renaissance of taxonomy. *Proceedings of the National Academy of Sciences USA 104*, 4775–6.

Miller, T.E.X. (2007b) Does having multiple partners weaken the benefits of facultative mutualism? A test with cacti and cactus-tending ants. *Oikos 116*, 500–12.

Mittermeier, R.A., Robles Gil, P., Hoffmann, M., Pilgrim, J., Brooks, T., Mittermeier, C.G., Lamoreux, J., and da Fonseca, G.A.B. (2004) *Hotspots Revisited*. CEMEX, Mexico City.

Miyata, H., Shimamura, T., Hirosawa, H., and Higashi, S. (2003) Morphology and phenology of the primitive ponerine army ant *Onychomyrmex hedleyi* (Hymenoptera: Formicidae: Ponerinae) in a highland rainforest of Australia. *Journal of Natural History 37*, 115–25.

Moeller, B.A., Hellgren, E.C., Ruthven III, D.C., Kazmaier, R.T., and Synatzske, D.R. (2005) Temporal differences in activity patterns of male and female Texas horned lizards (*Phrynosoma coronutum*) in southern Texas. *Journal of Herpetology 39*, 336–9.

Möglich, M., and Hölldobler, B. (1974) Social carrying behavior and division of labor during nest moving in ants. *Psyche (Cambridge) 81*, 219–35.

Molet, M., Peeters, C., and Fisher, B.L. (2007a) Permanent loss of wings in queens of the ant *Odontomachus coquereli* from Madagascar. *Insectes Sociaux 54*, 174–82.

——————(2007b) Winged queens replaced by reproductives smaller than workers in *Mystrium* ants. *Naturwissenschaften 94*, 280–7.

——Van Baalen, M., and Peeters, C. (2008) Shift in colonial reproductive strategy associated with a tropical-temperate gradient in *Rhytidoponera* ants. *American Naturalist 172*, 75–87.

——Fisher, B.L., Ito, F., and Peeters, C. (2009) Shift from independent to dependent colony foundation and evolution of 'multi-purpose' ergatoid queens in *Mystrium* ants (subfamily Amblyoponinae). *Biological Journal of the Linnean Society*, in press.

Mondor, E., and Addicott, J. (2007) Do exaptations facilitate mutualistic associations between invasive and native species? *Biological Invasions 9*, 623–8.

Monnin, T., and Peeters, C. (2008) How many gamergates is an ant queen worth? *Naturwissenschaften 95*, 109–16.

——Malosse, C., and Peeters, C. (1998) Solid-phase microextraction and cuticular hydrocarbon differences related to reproductive activity in queenless ant *Dinoponera quadriceps*. *Journal of Chemical Ecology 24*, 473–90.

——Ratnieks, L.W.F., Jones, G.R., and Beard, R. (2002) Pretender punishment induced by chemical signalling in a queenless ant. *Nature 419*, 61–4.

Monteiro, A.F.M., Sujii, E.R., and Morais, H.C. (2008) Chemically based interactions and nutritional ecology of *Labidus praedator* (Formicidae: Ecitoninae) in an agroecosystem adjacent to a gallery forest. *Revista Brasileira de Zoologia 25*, 674–81.

Moog, J., Feldhaar, H., and Maschwitz, U. (2002) On the caulinary domatia of the SE-Asian ant-plant *Zanthoxylum myriacanthum* Wall. ex Hook. f. (Rutaceae) their influence on branch statics, and the protection against herbivory. *Sociobiology 40*, 547–74.

Mooney, K.A. (2007) Tritrophic effects of birds and ants on a canopy food web, tree growth and phytochemistry *Ecology 88*, 2005–14.

——and Agrawal, A.A. (2008) Plant genotype shapes ant-aphid interactions: implications for community structure and indirect plant defense. *American Naturalist 168*, E195-E205.

——and Tillberg, C.V. (2005) Temporal and spatial variation to ant omnivory in pine forests. *Ecology 86*, 1225–35.

——Jones, P., and Agrawal, A.A. (2008) Coexisting congeners: traits of two milkweed feeding aphids and their competitive interactions. *Oikos 117*, 450–8.

Moore, J. (2002) *Parasites and the Behavior of Animals*. Oxford University Press, Oxford.

Morales, M.A., and Beal, A.L.H. (2006) Effects of host plant quality and ant tending for treehopper *Publilia concava*. *Annals of the Entomological Society of America 99*, 545–52.

——and Heithaus, E.R. (1998) Food from seed-dispersal mutualism shifts sex ratios in colonies of the ant *Aphaenogaster rudis*. *Ecology 79*, 734–9.

——Barone, J.L., and Henry, C.S. (2008a) Acoustic alarm signaling facilitates predator protection of treehoppers by mutualist ant bodyguards. *Proceedings of the Royal Society of London B 275*, 1935–41.

——Dodge, G.J., and Inouye, D.W. (2005) A phenological mid-domain effect in flowering diversity. *Oecologia 142*, 83–9.

——Leduc, E.N., and Ness, J.H. (2008b) Complex interactions between populations of native ants, non-native ants, and Japanese knotweed. Ecological Society of America, 93rd Annual Meeting, Milwaukee, WI USA, 3–8 August 2008.

Moreau, C.S., Bell, C.D., Vila, R., Archibald, S.B., and Pierce, N.E. (2006) Phylogeny of the ants: diversification in the age of angiosperms. *Science 312*, 101–4.

Moritz, R.F.A., Lattorff, H.M.G., Neumann, P., Kraus, F.B., Radloff, S.E., and Hepburn, H.R. (2005) Rare royal families in honeybees, *Apis mellifera*. *Naturwissenschaften 92*, 488–91.

Moron, D., Witek, M., and Woyciechowski, M. (2008) Division of labour among workers with different life expectancy in the ant *Myrmica scabrinodis*. *Animal Behaviour 75*, 345–50.

Morris, J.R., and Steigman, K.L. (1993) Effects of polygyne fire ant invasion on native ants of a blackland prairie in Texas. *Southwest Naturalist 38*, 136–40.

Morrison, J.E., Williams, D.F., Oi, D.H., and Potter, K.N. (1997a) Damage to dry crop seed by red imported fire ant (Hymenoptera: Formicidae). *Journal of Economic Entomology 90*, 218–22.

Morrison, L.W. (1996) Community organization in a recently assembled fauna: the case of Polynesian ants. *Oecologia 107*, 243–56.

——(1999) Indirect effects of phorid fly parasitoids on the mechanisms of interspecific competition among ants. *Oecologia 121*, 113–22.

——(2000) Mechanisms of interspecific competition among an invasive and two native fire ants. *Oikos 90*, 238–52.

——(2002a) Island biogeography and metapopulation dynamics of Bahamian ants. *Journal of Biogeography 29*, 387–94.

——(2002b) Long-term impacts of an arthropod community invasion by the imported fire ant, *Solenopsis invicta*. *Ecology 83*, 2337–45.

——and Porter, S.D. (2003) Positive association between densities of the red imported fire ant, *Solenopsis invicta* (Hymenoptera: Formicidae), and generalized ant and arthropod diversity. *Environmental Entomology 32*, 548–54.

————(2005) Testing for population-level impacts of introduced *Pseudacteon tricuspis* flies, phorid parasitoids of *Solenopsis invicta* fire ants. *Biological Control 33*, 9–19.

——Dall'Aglio-Holvorcem, C.G., and Gilbert, L.E. (1997b) Oviposition behaviour and development of Pseudacteon flies (Diptera: Phoridae), parasitoids of *Solenopsis* fire ants (Hymenoptera: Formicidae). *Environmental Entomology 26*, 716–24.

——Korzukhin, M.D., and Porter, S.D. (2005) Predicted range expansion of the invasive fire ant, *Solenopsis invicta*, in the eastern United States based on the VEMAP global warming scenario. *Diversity and Distributions 11*, 199–204.

——Porter, S.D., Daniels, E., and Korzukhin, M.D. (2004) Potential global range expansion of the invasive fire ant, *Solenopsis invicta*. *Biological Invasions 6*, 183–91.

Morrone, J.J. (2002) Biogeographical regions under track and cladistic scrutiny. *Journal of Biogeography 29*, 149–52.

Moutinho, P., Nepstad, D.C., and Davidson, E.A. (2003) Influence of leaf-cutting ant nests on secondary forest growth and soil properties in Amazonia. *Ecology 84*, 1265–76.

Mueller, U.G., Rehner, S.A., and Schultz, T.R. (1998) The evolution of agriculture in ants. *Science 281*, 2034–8.

Müller, M., and Wehner, R. (2007) Wind and sky as compass cues in desert ant navigation. *Naturwissenschaften 94*, 589–94.

Murray, B.D., and Lepschi, B.J. (2004) Are locally rare species abundant elsewhere in their geographic range? *Austral Ecology 29*, 287–93.

Myers, J.H., Simberloff, D., Kuris, A.M., and Carey, J.R. (2000a) Eradication revisited: dealing with exotic species. *Trends in Ecology and Evolution 15*, 316–20.

Myers, N., Mittermeier, R.A., Mittermeier, C.G., da Fonseca, G.A.B., and Kent, J. (2000b) Biodiversity hotspots for conservation priorities. *Nature 403*, 853–8.

Nakamura, A., Catterall, C.P., House, A.P.N., Kitching, R. L., and Burwell, C.J. (2007) The use of ants and other soil and litter arthropods as bio-indicators of the impacts of rainforest clearing and subsequent land use. *Journal of Insect Conservation 11*, 177–86.

Nation, J.L. (2002) *Insect Physiology and Biochemistry*. CRC Press, Boca Raton.

Ndoutoume-Ndong, A., and Mikissa, B. (2007) Influence de la presence de la fourmi *Wasmannia auropunctata* (Roger 1863) (Hymenoptera: Formicidae) sur les autres especes de fourmis dans la reserve de la Lope (centre du Gabon). *Annales de la Societe Entomologique de France 43*, 155–8.

Neece, K.C., and Bartell, D.P. (1981) Insects associated with *Solenopsis* spp. in southeastern Texas, USA. *Southwestern Entomologist 6*, 307–11.

Nel, A., Perrault, G., Perrichot, V., and Néraudeau, D. (2004) The oldest ant in the Lower Cretaceous amber of Charente-Maritime (SW France) (Insecta: Hymenoptera: Formicidae). *Geologica Acta 2*, 23–9.

Nelder, M.P., Paysen, E.S., Zungoli, P.A., and Benson, E.P. (2006) Emergence of the introduced ant *Pachycondyla chinensis* (Formicidae: Ponerinae) as a public health threat in the southeastern United States. *Journal of Medical Entomology 43*, 1094–8.

Nendick, D. (2008) Win-win from offshore container management programme. *Biosecurity 84*, 12.

Ness, J.H. (2003a) *Catalpa bignonioides* alters extrafloral nectar production after herbivory and attracts ant bodyguards. *Oecologia 134*, 210–8.

——(2003b) Contrasting exotic *Solenopsis invicta* and native *Forelius pruinosus* ants as mutualists with *Catalpa bignonioides*, a native plant. *Ecological Entomology 28*, 247–51.

——(2004) Forest edges and fire ants alter the seed shadow of an ant-dispersed plant. *Oecologia 138*, 448–54.

——(2006) A mutualism's indirect costs: the most aggressive plant bodyguards also deter pollinators. *Oikos 113*, 506–14.

——and Bronstein, J.L. (2004) The effects of invasive ants on prospective ant mutualists. *Biological Invasions 6*, 445–61.

——and Morin, D. (2008) Forest edges and landscape history shape interactions between plants, seed-dispersing ants and seed predators. *Biological Conservation 141*, 838–47.

——Morris, W.F., and Bronstein, J.L. (2006) Variation in mutualistic potential among ant species tending extrafloral nectaries of *Ferocactus wislizeni*. *Ecology 87*, 912–21.

————(in press) For ant-protected plants, the best defense is a hungry offense. *Ecology*.

——Bronstein, J.L., Andersen, A.N., and Holland, J.N. (2004) Ant body size predicts dispersal distance of ant-adapted seeds: implications of small-ant invasions. *Ecology 85*, 1244–50.

Nestel, D., and Dickschen, F. (1990) Foraging kinetics of ground ant communities in different Mexican coffee agroecosystems. *Oecologia 84*, 58–63.

Newell, W., and Barber, T.C. (1913) The Argentine ant. *USDA Bureau of Entomology Bulletin 122*.

Newey, P.S., Robson, S.K., and Crozier, R.H. (2008) Near-infrared spectroscopy identifies the colony and nest of origin of weaver ants, *Oecophylla smaragdina*. *Insectes Sociaux 55*, 171–5.

Nichols, B. J., and Sites, R.W. (1991) Ant predators of founder queens of *Solenopsis invicta* (Hymenoptera, Formicidae) in central Texas. *Environmental Entomology 20*, 1024–9.

Nielsen, M.G. (1997) Nesting biology of the mangrove mud-nesting ant *Polyrhachis sokolova* Forel (Hymenoptera, Formicidae) in northern Australia. *Insectes Sociaux 44*, 15–21.

——Christian, K., Henriksen, P.G., and Birkmore, D. (2006) Respiration by mangrove ants *Camponotus anderseni* during nest submersion associated with tidal inundation in Northern Australia. *Physiological Entomology 31*, 120–6.

————and Birkmose, D. (2003) Carbon dioxide concentrations in the nests of the mud-dwelling mangrove ant *Polyrhachis sokolova* Forel (Hymenoptera: Formicidae). *Australian Journal of Entomology 42*, 357–62.

Nonacs, P. (1991) Less growth with more food: how insect-prey availability changes colony demographics in the ant, *Camponotus floridanus*. *Journal of Insect Physiology 37*, 891–8.

——and Dill, L.M. (1990) Mortality risk vs. food quality trade-offs in a common currency – ant patch preferences. *Ecology 71*, 1886–92.

North, R.D., Jackson, C.W., and Howse, P.E. (1997) Evolutionary aspects of ant-fungus interactions in leaf-cutting ants. *Trends in Ecology & Evolution 12*, 386–9.

Norton, G.A., and Mumford, J.D., eds. (1993) *Decision Tools for Pest Management*. CAB International, Wallingford, Oxon, UK.

Nougier, J., Cantagrel, J.M., and Karche, J.P. (1986) The Comores archipelago in the western Indian Ocean: volcanology, geochronology, and geodynamic setting. *Journal of African Earth Science 5*, 135–45.

Novotny, V., Drozd, P., Miller, S.E., Kulfan, M., Janda, M., Basset, Y., and Weiblen, G. D. (2006) Why are there so many species of herbivorous insects in tropical rainforests? *Science 313*, 1115–18.

Nowbahari, E. (2007) Learning of colonial odor in the ant *Cataglyphis niger* (Hymenoptera; Formicidae). *Learning and Behavior 35*, 87–94.

O'Donnell, S., Lattke, J., Powell, S., and Kaspari, M. (2007) Army ants in four forests: geographic variation in raid rates and species composition. *Journal of Animal Ecology 76*, 580–9.

O'Dowd, D.J. (1979) Foliar nectar production and ant activity on a neotropical tree, *Ochroma pyramidale*. *Oecologia 43*, 233–48.

——(1980) Pearl bodies of a neotropical tree, *Ochroma pyramidale*: ecological implications. *American Journal of Botany 67*, 543–9.

——(1982) Pearl bodies as ant food: an ecological role for some leaf emergences of tropical plants. *Biotropica 14*, 40–9.

——Green, P.T., and Lake, P.S. (2003) Invasional 'meltdown' on an oceanic island. *Ecology Letters 6*, 812–7.

Oberrath, R., and Bohning-Gaese, K. (2002) Phenological adaptation of ant-dispersed plants to seasonal variation in ant activity. *Ecology 83*, 1412–20.

Offenberg, J. (2000) Correlated evolution of the association between aphids and ants and the association between aphids and plants with extrafloral nectaries. *Oikos 91*, 146–52.

——(2007) The distribution of weaver ant pheromones on host trees. *Insectes Sociaux 54*, 248–50.

——Nielsen, M.G., MacIntosh, D.J., Havanon, S., and Aksornkoae, S. (2004) Evidence that insect herbivores are deterred by ant pheromones. *Proceedings of the Royal Society of London B (Suppl) 271*, S433-5.

Ogata, K., and Okido, H. (2007) Revision of the ant genus *Perissomyrmex* with notes on the phylogeny of the tribe Myrmecinini. *Memoirs of the American Entomological Institute 80*, 352–69.

Ohashi, M., Finer, L., Domisch, T., Risch, A.C., Jurgensen, M.F., and Niemela, P. (2007) Seasonal and diurnal CO_2 efflux from red wood ant (*Formica aquilonia*) mounds in boreal coniferous forests. *Soil Biology and Biochemistry 39*, 1504–11.

Ohgushi, T. (2008) Herbivore-induced indirect interaction webs on terrestrial plants: the importance of non-trophic, indirect, and facilitative interactions. *Entomologia Experimentalis et Applicata 128*, 217–29.

Ohnishi, H., Imai, H.T., and Yamamoto, M.T. (2004) Molecular phylogenetic analysis of ant subfamily relationship inferred from rDNA sequences. *Genes and Genetic Systems 78*, 419–25.

Oi, D.H., and Drees, B.M. (2009) Fire ant IPM. In E.B. Radcliffe, W.D. Hutchison, and R.E. Cancelado, eds. *Integrated Pest Management: concepts, tactics, strategies, and case studies*, 390–401. Cambridge University Press, Cambridge.

——and Williams, D.F. (2002) Impact of *Thelohania solenopsae* (Microsporidia: Thelohaniidae) on polygyne colonies of red imported fire ants (Hymenoptera: Formicidae). *Journal of Economic Entomology 95*, 558–62.

Oldham, N.J., Morgan, E.D., Agosti, D., and Wehner, R. (1999) Species recognition from postpharyngeal gland contents of ants of the *Cataglyphis bicolor* group. *Journal of Chemical Ecology 25*, 1383–94.

Oldroyd, B.P., and Fewell, J.H. (2007) Genetic diversity promotes homeostasis in insect colonies. *Trends in Ecology and Evolution 22*, 408–13.

Oliveira, P.S., Freitas, A.V.L. (2004) Ant—plant—herbivore interactions in the Neotropical cerrado savanna. *Naturwissenschaften 91*, 557–70.

——Rico-Gray, V., Diaz-Castelazo, C., and Castillo-Guevara, C. (1999) Interaction between ants, extrafloral nectaries and insect herbivores in Neotropical coastal sand dunes: herbivore deterrence by visiting ants increases fruit set in *Opuntia stricta* (Cactaceae). *Functional Ecology 13*, 623–31.

Oliver, T.H., Leather, S.R., and Cook, J.M. (2008) Macroevolutionary patterns in the origin of mutualisms involving ants. *Journal of Evolutionary Biology 21*, 1597–608.

Oliveras, J., Bas, J.M., and Gómez, C. (2005b) Reduction of the ant mandible gap range after biotic homogenization caused by an ant invasion (Hymenoptera: Formicidae). *Sociobiology 45*, 1–10.

————(2007) A shift in seed harvesting by ants following Argentine ant invasion. *Vie et Milieu 57*, 75–81.

——Casellas, D., and Gómez, C. (2005a) Numerical dominance of the Argentine ant vs native ants and consequences on soil resource searching in Mediterranean cork-oak forests (Hymenoptera: Formicidae). *Sociobiology 45*, 643–58.

Olmstead, K.L., and Wood, T.K. (1990) Altitudinal patterns in species richness of neotropical treehopper (Homoptera: Membracidae); the role of ants. *Proceedings of the Entomological Society of Washington 92*, 552–60.

Olson, D.M. (1994) The distribution of leaf litter invertebrates along a Neotropical altitudinal gradient. *Journal of Tropical Ecology 10*, 129–50.

——Dinerstein, E., and Wikramanayake, E.D. (2001) Terrestrial ecoregions of the world: a new map of life on Earth. *BioScience 51*, 933–8.

Olsson, K., and Agren, J. (2002) Latitudinal population differentiation in phenology, life history and flower morphology in the perennial herb *Lythrum salicaria*. *Journal of Evolutionary Biology 15*, 983–96.

Onoyama, K. (1982) Immature stages of the harvester ant *Messor aciculatus* (Hymenoptera, Formicidae). *Kontyu 50*, 324–9.

Oppelt, A., Spitzenpfeil, N., Kroiss, J., and Heinze, J. (2008) The significance of intercolonial variation of cuticular hydrocarbons for inbreeding avoidance in ant sexuals. *Animal Behaviour 76*, 1029–34.

Orians, G.H., and Milewski, A.V. (2007) Ecology of Australia: the effects of nutrient-poor soils and intense fires. *Biological Reviews 82*, 393–423.

——and Pearson, N.E. (1979) On the theory of central place foraging. In D.J. Horn, R.D. Mitchell, and G.R. Stairs, eds. *Analyses of Ecological Systems*, pp. 154–77. Ohio State University Press, Columbus, OH.

Orivel, J., Errard, C., and Dejean, A. (1997) Ant gardens: interspecific recognition in parabiotic ant species. *Behavioral Ecology and Sociobiology 40*, 87–93.

Orr, M.R., De Camargo, R.X., and Benson, W.W. (2003) Interactions between ant species increase arrival rates of an ant parasitoid. *Animal Behaviour 65*, 1187–93.

——Seike, S.H., Benson, W.W., and Gilbert, L.E. (1995) Flies suppress fire ants. *Nature 373*, 292–3.

Orrock, J., and Danielson, B. (2004) Rodents balancing a variety of risks: invasive fire ants and indirect and direct indicators of predation risk. *Oecologia 140*, 662–7.

Oster, G.F., and Wilson, E.O. (1978) *Caste and Ecology in the Social Insects*. Princeton University Press, Princeton, NJ.

Ottonetti, L., Tucci, L., and Santini, G. (2006) Recolonisation patterns of ants in a rehabilitated lignite mine in Central Italy: potential for the use of Mediterranean ants as indicators of restoration processes. *Restoration Ecology 14*, 60–6.

Ouellette, G.D., Fisher, B.L., and Girman, D.J. (2006) Molecular systematics of basal subfamilies of ants using 28S rRNA (Hymenoptera: Formicidae). *Molecular Phylogenetics and Evolution 40*, 359–69.

Ozaki, K., Takashima, S., and Suo, O. (2000) Ant predation suppresses populations of the scale insect *Aulacaspis marina* in natural mangrove forests. *Biotropica 32*, 764–8.

Ozaki, M., Wada-Katsumata, A., Fujikawa, K., Iwasaki, M., Yokohari, F., Satoji, Y., Nisimura, T., and Yamaoka, R. (2005) Ant nestmate and non-nestmate discrimination by a chemosensory sensillum. *Science 309*, 311–314.

Pacheco, J., Herrera, H.W., and Mackay, W. (2007) A new species of thief ant of the genus *Solenopsis* from the Galapagos Islands (Hymenoptera: Formicidae). *Sociobiology 50*, 1075–86.

Pacheco, R., and Vasconcelos, H.L. (2007) Invertebrate conservation in urban areas: ants in the Brazilian Cerrado. *Landscape and Urban Planning 81*, 193–9.

Pacific Invasive Ant Group (2004) *Pacific ant prevention plan*. IUCN/SSC Invasive Species Specialist Group, Auckland, NZ. http://www.issg.org/database/species/reference_files/PAPP.pdf

Page, R.E., and Mitchell, S.D. (1991) Self organization and adaptation in insect societies. *Proceedings of the Biennial Meeting of the Philosophy of Science Association 1990*, University of Chicago Press, 289–98.

Paine, R.T. (1995) A conversation on refining the concept of keystone species. *Conservation Biology 9*, 962–4.

——(1996) Food web complexity and species diversity. *American Naturalist 100*, 65–75.

Palacio, E.E., and Fernández, F. (2003) Claves para las subfamilias y géneros. In F. Fernández, ed. *Introducción a las Hormigas de la Región Neotropical*, pp. 233–60. Instituto de Investigación de Recursos Biológicos Alexander von Humboldt, Bogotá.

Palladini, J.D., Jones, M.G., Sanders, N.J., and Jules, E.S. (2007) The recovery of ant communities in regenerating temperate conifer forests. *Forest Ecology and Management 242*, 619–24.

Palmer, T.M. (2003) Spatial habitat heterogeneity influences competition and coexistence in an African acacia ant guild. *Ecology 84*, 2843–55.

——Stanton, M.L., Young, T.P., Goheen, J.R., Pringle, R. M., and Karban, R. (2008) Breakdown of an ant-plant mutualism follows the loss of large herbivores from an African savanna. *Science 391*, 192–5.

Pamilo, P. (1991) Evolution of colony characteristics in social insects. I. Sex allocation. *American Naturalist 137*, 83–107.

——Crozier, R.H., and Fraser, J. (1985) Internest interactions, nest autonomy, and reproductive specialization in an Australian arid-zone ant, *Rhytidoponera* sp. 12. *Psyche (Cambridge) 92*, 217–36.

——Viljakainen, L., and Vihavainen, A. (2007) Exceptionally high density of NUMTs in the honeybee genome. *Molecular Biology and Evolution 24*, 1340–6.

——Zhu, D.Q., Fortelius, W., Rosengren, R., Seppä, P., and Sundström, L. (2005) Genetic patchwork of network-building wood ant populations. *Annales Zoologici Fennici 42*, 179–87.

Paris, C., Jouquet, P., Lata, J.C., Ranjard, L., and Dauber, J. (2004) Ant-tree aphid mutualism: effects on community structure of soil microorganisms. *XIVth International Colloquium on Soil Zoology and Ecology*, 136.

Parmesan, C., and Yohe, G. (2003) A globally coherent fingerprint of climate change impacts across natural systems. *Nature 421*, 37–42.

Parr, C.L. (2008) Dominant ants can control species richness in a South African savanna. *Journal of Animal Ecology 77*, 1191–8.

——and Andersen, A.N. (2008) Fire resilience of ant assemblages in long-unburnt savanna of northern Australia. *Austral Ecology 33*, 830–838.

————Chastagnol, C., and Duffaud, C. (2007) Savanna fires increase rates and distances of seed dispersal by ants. *Oecologia 151*, 33–41.

——Robertson, H.G., Biggs, H.C., and Chown, S.L. (2004) Response of African savanna ants to long-term fire regimes. *Journal of Applied Ecology 41*, 630–42.

——Sinclair, B.J., Andersen, A.N., Gaston, K.J., and Chown, S.L. (2005) Constraint and competition in assemblages: a cross-continental and modeling approach for ants. *American Naturalist 165*, 481–94.

Parris, L.B., Lamont, M.M., and Carthy, R.R. (2002) Increased incidence of red imported fire ant (Hymenoptera: Formicidae) presence in loggerhead sea turtle (Testudines: Cheloniidae) nests and observations of hatchling mortality. *Florida Entomologist 85*, 514–7.

Pascoe, A. (2003) Red imported fire ant response stood down. *Biosecurity Magazine 45*, 7.

Passera, L. (1994) Characteristics of tramp species. In D.F. Williams, ed. *Exotic Ants: biology, impact, and control of introduced species*, pp. 23–43. Westview Press, Boulder, CO.

——and Aron, S. (2005) *Les fourmis. Comportement, organisation sociale et évolution*. Presses scientifiques du CNRC, Ottawa, Canada.

——and Keller, L. (1990) Loss of mating flight and shift in the pattern of carbohydrate storage in sexuals of ants (Hymenoptera; Formicidae). *Journal of Comparative Physioliolgy B 160*, 207–11.

——Roncin, E., Kaufmann, B., and Keller, L. (1996) Increased soldier production in ant colonies exposed to intraspecific competition. *Nature 379*, 630–1.

Patek, S.N., Baio, J.E., Fisher, B.L., and Suarez, A.V. (2006) Multifunctionality and mechanical origins: ballistic jaw propulsion in trap-jaw ants. *Proceedings of the National Academy of Sciences USA 103*, 12787–92.

Paton, T.R., Humphrey, G.S. and Mitchell, P.B. (1995). *Soils: a new global view*. Yale University Press, New Haven.

Paul, J., and Roces, F. (2003) Fluid intake rates in ants correlate with their feeding habits. *Journal of Insect Physiology 49*, 347–57.

Peakall, R. (1989) The unique pollination of *Leporella fimbriata* (Orchidaceae): pollination by pseudocopulating male ants (*Myrmecia urens*, Formicidae). *Plant Systematics and Evolution 167*, 137–48.

Pearcy, M., Aron, S., Doums, C., and Keller, L. (2004) Conditional use of sex and parthenogenesis for worker and queen production in ants. *Science 306*, 1780–3.

Pearson, R.G., and Dawson, T.P. (2003) Predicting the impacts of climate change on the distribution of species: are bioclimate envelope models useful? *Global Ecology and Biogeography 12*, 361–71.

Peck, S.L., McQuaid, B., and Campbell, C.L. (1998) Using ant species (Hymenoptera: Formicidae) as a biological indicator of agroecosystem condition. *Environmental Entomology 27*, 1102–10.

Pedersen, E.K., Bedford, T.L., Grant, W.E., Vinson, S.B., Martin, J.B., Longnecker, M.T., Barr, C.L., and Drees, B. M. (2003) Effect of red imported fire ants on habitat use by hispid cotton rats (*Sigmodon hispidus*) and northern pygmy mice (*Baiomys taylori*). *Southwestern Naturalist 48*, 419–426.

Pedersen, J.S., and Boomsma, J.J. (1999) Genetic analysis of colony structure in polydomous and polygynous ant populations. *Biological Journal of the Linnean Society 66*, 115–44.

——Krieger, M.J.B., Vogel, V., Giraud, T., and Keller, L. (2006) Native supercolonies of unrelated individuals in the invasive Argentine ant. *Evolution 60*, 782–91.

Peeters, C. (1988) Nestmate discrimination in a ponerine ant (*Rhytidoponera* sp. 12) without a queen caste and with a low intra-nest relatedness. *Insectes Sociaux 35*, 34–46.

——(1991) Ergatoid queens and intercastes in ants: two distinct adult forms which look morphologically intermediate between workers and winged queens. *Insectes Sociaux 38*, 1–15.

——(1997) Morphologically 'primitive' ants: comparative review of social characters, and the importance of queen-worker dimorphism. In J.C. Choe, and B.J. Crespi, eds. *The Evolution of Social Behavior in Insects and Arachnids*, pp. 372–91. Cambridge University Press, Cambridge.

——and Ito, F. (2001) Colony dispersal and the evolution of queen morphology in social Hymenoptera. *Annual Review of Entomology 46*, 601–30.

——and Liebig, J. (2009) Fertility signaling as a general mechanism of regulating reproductive division of labor in ants. In J. Gadau, and J. Fewell, eds. *Organization of Insect Societies: from genome to socio-complexity*, pp. 220–42. Harvard University Press, Cambridge, MA.

——Liebig, J., and Hölldobler, B. (2000) Sexual reproduction by both queens and workers in the ponerine ant *Harpegnathos saltator*. *Insectes Sociaux 47*, 325–32.

Pemberton, R.W. (1988) The abundance of plants bearing extrafloral nectaries in Colorado and Mojave desert communities of southern California. *Madrono 35*, 283–246.

Peñaloza, C., and Farji-Brener, A.G. (2003) The importance of treefall gaps as foraging sites for leaf-cutting ants depends on forest age. *Journal of Tropical Ecology 19*, 603–5.

Peng, R.K., and Christian, K. (2005) Integrated pest management in mango orchards in the Northern Territory of Australia, using the weaver ant, *Oecophylla smaragdina*, (Hymenoptera: Formicidae) as a key element. *International Journal of Pest Management 51*, 149–55.

————and Gibb, K. (1999) The effect of colony isolation of the predacious ant, *Oecophylla smaragdina* (F.) (Hymenoptera: Formicidae), on protection of cashew plantations from insect pests. *International Journal of Pest Management 45*, 189–94.

——————(2004) Implementing ant technology in commercial cashew plantations. *Rural Industries Research and Development Corporation Publication No. W04/088, ISSN 1440–6845*, p. 72.

Penick, C.A., and Tschinkel, W.R. (2008) Thermoregulatory brood transport in the fire ant, *Solenopsis invicta*. *Insectes Sociaux 55*, 176–82.

Pennisi, E. (2009) Evolutionary biology: agreeing to disagree. *Science 323*, 706–8.

Perfecto, I. (1991) Ants (Hymenoptera: Formicidae) as natural control agents of pests in irrigated maize in Nicaragua. *Journal of Economic Entomology 84*, 65–70.

——(1994) Foraging behavior as a determinant of asymmetric competitive interaction between two ant species in a tropical agroecosystem. *Oecologia 98*, 184–92.

——and Castiñeiras, A. (1998) Deployment of the predaceous ants and their conservation in agroecosystems. In P. Barbosa, ed. *Perspectives on the Conservation of Natural Enemies of Pest Species*, pp. 269–89. Academic Press, San Diego.

——and Vandermeer, J. (1993) Distribution and turnover rate of a population of *Atta cephalotes* in a tropical rainforest in Costa Rica. *Biotropica 25*, 316–21.

————(1996) Microclimatic changes and the indirect loss of ant diversity in a tropical agroecosystem. *Oecologia 108*, 577–82.

————(2002) Quality of agroecological matrix in a tropical montane landscape: ants in coffee plantations in southern Mexico. *Conservation Biology 16*, 174–82.

————(2006) The effect of an ant-hemipteran mutualism on the coffee berry borer (*Hypothenemus hampei*) in southern Mexico. *Agriculture Ecosystems & Environment 117*, 218–21.

————(2008) Spatial pattern and ecological process in the coffee agroecosystem. *Ecology 89*, 915–20.

——Armbrecht, I., Philpott, S.M., Soto-Pinto, L., and Dietsch, T.V. (2007) Shaded coffee and the stability of rainforest margins in Latin America. In T. Tscharntke, C. Leuschner, M. Zelle, E. Guhadja, and A. Bidin, eds. *The Stability of Tropical Rainforest Margins: linking ecological, economic and social constraints of land use and conservation*, pp. 227–63. Springer Verlag, Berlin.

Perrichot, V., Lacau, S., Néraudeau, D., and Nel, A. (2008a) Fossil evidence for the early ant evolution. *Naturwissenschaften 95*, 85–90.

——Nel, A., Néraudeau, D., Lacau, S., and Guyot, T. (2008b) New fossil ants in French Cretaceous amber (Hymenoptera: Formicidae). *Naturwissenschaften 95*, 91–7.

Perrings, C. (2005) How to manage biological invasions under globalization. *Trends in Ecology and Evolution 20*, 212–5.

Persad, A.B., and Hoy, M.A. (2004) Predation by *Solenopsis invicta* and *Blatella asahinai* on *Toxoptera citricida* parasitized by *Lysiphlebus testaceipes* and *Lipolexis oregmae* on citrus in Florida. *Biological Control 30*, 531–7.

Peterson, B., and Fry, B. (1987) Stable isotopes in ecosystem studies. *Annual Review of Ecology and Systematics 18*, 293–320.

Petralia, R.S., Sorensen, A.A., and Vinson, S.B. (1980) The labial gland system of larvae of the imported fire ant, *Solenopsis invicta* Buren: ultrastructure and enzyme analysis. *Cell and Tissue Research 206*, 145–56.

Petren, K., and Case, T.J. (1996) An experimental demonstration of exploitation competition in an ongoing invasion. *Ecology 77*, 118–32.

Petrov, I. (2006) The ants of Serbia and Montenegro. *Posebna Izdanja. Srpska Akademija Nauka i Umetnosti 656*, 1–136.

Pfeiffer, M., and Linsenmair, K.E. (2001) Territoriality in the Malaysian giant ant *Camponotus gigas* (Hymenoptera/Formicidae) *Journal of Ethology 19*, 75–85.

Philpott, S.M. (2005a) Changes in arboreal ant populations following pruning of coffee shade-trees in Chiapas, Mexico. *Agroforestry Systems 64*, 219–24.

——(2005b) Trait-mediated effects of parasitic phorid flies (Diptera: Phoridae) on ant (Hymenoptera: Formicidae) competition and resource access in coffee agro-ecosystems. *Environmental Entomology 34*, 1089–94.

——and Armbrecht, I. (2006) Biodiversity in tropical agroforests and the ecological role of ants and ant diversity in predatory function. *Ecological Entomology 31*, 369–77.

——and Foster, P.F. (2005) Nest-site limitation in coffee agroecosystems: artificial nests maintain diversity of arboreal ants. *Ecological Applications 15*, 1478–85.

——Perfecto, I., and Vandermeer, J. (2006) Effects of management intensity and season on arboreal ant diversity and abundance in coffee agroecosystems. *Biodiversity and Conservation 15*, 139–55.

——————(2008a). Effects of predatory ants on lower trophic levels across a gradient of coffee management complexity. *Journal of Animal Ecology 77*, 505–11.

——————(2008b) Behavioral diversity of predatory ants in coffee agroecosystems. *Environmental Entomology 37*, 181–91.

Pianka, E.R. (1966) Latitudinal gradients in species diversity: a review of concepts. *American Naturalist 100*, 33–46.

Pickett, S.T.A., and White, P.S. (1985) *The Ecology of Natural Disturbance as Patch Dynamics*. Academic Press, New York.

Pielou, E.C. (1979) *Biogeography*. John Wiley and Sons, New York.

Pierce, N.E. (1986) Lycaenid butterflies and ants: selection for nitrogen fixing and other protein-rich foods. *American Naturalist 125*, 888–95.

——and Mead, P.S. (1981) Parasitoids as selective agents in the symbiosis between lycaenid butterfly caterpillars and ants. *Science 211*, 1185–7.

——Brady, M.F., Heath, A., Lohman, D.J., Mathew, J., Rand, D.B., and Travassos, M.A. (2002) The ecology and evolution of ant association in the Lycaenidae (Lepidoptera). *Annual Review of Entomology 47*, 733–71.

Pirk, C.W.W., Neumann, P., Moritz, R.F.A., and Pamilo, P. (2001) Intranest relatedness and nestmate recognition in the meadow ant *Formica pratensis* (R). *Behavioral Ecology and Sociobiology 49*, 366–74.

Plentovich, S., Hebshi, A., and Conant, S. (2009) Detrimental effects of two widespread invasive ant species on weight and survival of colonial nesting seabirds in the Hawaiian Islands. *Biological Invasions 11*, 289–98.

Plowes, N.J., and Adams, E.S. (2005) An empirical test of Lanchester's square law: mortality during battles of the fire ant *Solenopsis invicta*. *Proceedings of the Royal Society of London B 272*, 1809–14.

Plowes, R.M., Dunn, J.G., and Gilbert, L.E. (2007) The urban fire ant paradox: native fire ants persist in an urban refuge while invasive fire ants dominate natural habitats. *Biological Invasions 9*, 825–36.

Poinar Jr., G., Lachaud, J.P., Castillo, A., and Infante, F. (2006) Recent and fossil nematode parasites (Nematoda: Mermithidae) of Neotropical ants. *Journal of Invertebrate Pathology 91*, 19–26.

——and Yanoviak, S.P. (2008) *Myrmeconema neotropicum* n. g., n. sp., a new tetradonematid nematode parasitising South American populations of *Cephalotes atratus* (Hymenoptera: Formicidae), with the discovery of an apparent parasite-induced host morph. *Systematic Parasitology 69*, 145–53.

Pontin, A.J. (1960) Field experiments on colony foundation by *Lasius niger* and *L. flavus* (Hym., Formicidae). *Insectes Sociaux 7*, 227–30.

——(1961) Population stabilization and competition between the ants *Lasius flavus* (F.) and *L. niger* (L.). *Journal of Animal Ecology 30*, 47–54.

——(1969) Experimental transplantation of nest-mounds of the ant *Lasius flavus* (F.) in a habitat containing also *L. niger* (L.) and *Myrmica scabrinodis* Nyl. *Journal of Animal Ecology 38*, 747–54.

Porter, S.D. (1988) Impact of temperature on colony growth and developmental rates of the ant *Solenopsis invicta*. *Journal of Insect Physiology 34*, 1127–33.

——and Savignano, D.A. (1990) Invasion of polygyne fire ants decimates native ants and disrupts arthropod community. *Ecology 71*, 2095–106.

——and Tschinkel, W.R. (1986) Adaptive value of nanitic workers in newly founded red imported fire ant colonies (Hymenoptera: Formicidae). *Annals of the Entomological Society of America 79*, 723–6.

——Nogueira de Sá, L.A., and Morrison, L.W. (2004) Establishment and dispersal of the fire ant decapitating fly *Pseudaction tricuspis* in north Florida. *Biological Control 29*, 179–88.

——Williams, D.F., Patterson, R.F., and Fowler, H.G. (1997) Intercontinental differences in the abundance of *Solenopsis* fire ants (Hymenoptera: Formicidae): an escape from natural enemies? *Environmental Entomology 26*, 373–84.

——Vander Meer, R.K., Pesquero, M.A., Campiolo, S., and Fowler, H.G. (1995) *Solenopsis* (Hymenoptera: Formicidae) fire ant reactions to attacks of *Pseudacteon* flies (Diptera: Phoridae) in southeastern Brazil. *Annals of the Entomological Society of America 88*, 570–5.

Poulsen, M., and Boomsma, J. (2005) Mutualistic fungi control crop-diversity in fungus-growing ants. *Science 307*, 741–4.

——and Currie, C.R. (2006) Complexity of insect-fungal associations: exploring the influence of microorganisms on the attine ant-fungus symbiosis. In B. Kostas, ed. *Insect Symbiosis*, pp. 57–77. CRC Press, Hoboken.

——Bot, A.N.M., Currie, C.R., and Boomsma, J.J. (2002) Mutualistic bacteria and a possible trade-off between alternative defence mechanisms in *Acromyrmex* leaf-cutting ants. *Insectes Sociaux 49*, 15–9.

Powell, S. (2008) Ecological specialization and the evolution of a specialized caste in *Cephalotes* ants. *Functional Ecology 22*, 902–911.

——and Clark, E. (2004) Combat between large derived societies: a subterranean army ant established as a predator of mature leaf-cutting ant colonies. *Insectes Sociaux 51*, 342–51.

——and Franks, N.R. (2005) Caste evolution and ecology: a special worker for novel prey. *Proceedings of the Royal Society of London B 272*, 2173–80.

————(2006) Ecology and the evolution of worker morphological diversity: a comparative analysis with *Eciton* army ants. *Functional Ecology 20*, 1105–14.

————(2007) How a few help all: living pothole plugs speed prey delivery in the army ant *Eciton burchelli*. *Animal Behaviour 73*, 1067–76.

——and Tschinkel, W.R. (1999) Ritualized conflict in *Odontomachus brunneus* and the generation of interaction-based task allocation: a new organizational mechanism in ants. *Animal Behaviour 58*, 965–72.

Power, M.E. (1992) Top-down and bottom-up forces in food webs – do plants have primacy? *Ecology 73*, 733–46.

Pusch, K., Seifert, B., Foitzik, S., and Heinze, J. (2006) Distribution and genetic divergence of two parapatric sibling ant species in Central Europe. *Biological Journal of the Linnean Society 88*, 223–34.

Pyne, S.J. (1997) *World fire: the culture of fire on earth*. University of Washington Press, Seattle.

Qian, H. (1998) Large-scale biogeographic patterns of vascular plant richness in North America: an analysis at the generic level. *Journal of Biogeography 25*, 829–36.

——(2008) A latitudinal gradient of beta diversity for exotic vascular plant species in North America. *Diversity and Distributions 14*, 556–60.

Qian, H. and Ricklefs, R.E. (2008) Global concordance in diversity patterns of vascular plants and terrestrial vertebrates. *Ecology Letters 11*, 547–53.

——Badgley, C., and Fox, D.L. (2009) The latitudinal gradient of beta diversity in relation to climate and topography for mammals in North America. *Global Ecology and Biogeography 18*, 111–22.

Quek, S.-P., Davies, S.J., Ashton, P.S., Itino, T., and Pierce, N.E. (2007) The geography of diversification in mutualistic ants: a gene's-eye view into the Neogene history of Sundaland rain forests. *Molecular Ecology 16*, 2045–62.

Queller, D.C. (2000) Relatedness and the fraternal major transitions. *Philosophical Transactions of the Royal Society B 355*, 1647–55.

——and Strassmann, J.E. (2002) The many selves of social insects. *Science 296*, 311–3.

Quinlan, R.J., and Cherrett, J.M. (1979) The role of fungus in the diet of the leaf-cutting ant *Atta cephalotes* (L.). *Ecological Entomology 4*, 151–60.

Quiroz-Robledo, L., and Valenzuela-Gonzalez, J. (1995) A comparison of ground ant communities in a tropical rainforest and adjacent grassland in Los Tuxtlas, Veracruz, Mexico. *Southwestern Entomologist 20*, 203–13.

Rabeling, C., Brown, J.M., and Verhaagh, M. (2008) Newly discovered sister lineage sheds light on early ant evolution. *Proceedings of the National Academy of Sciences USA 105*, 14913–7.

——Cover, S.P., Johnson, R.A., and Mueller, U.G. (2007) A review of the North American species of the fungus-gardening ant genus *Trachymyrmex* (Hymenoptera: Formicidae). *Zootaxa 1664*, 1–53.

Rabinowitz, D. (1981) Seven forms of rarity. In H. Synge, ed. *The Biological Aspects of Rare Plant Conservation*, pp. 205–17. Wiley, New York.

Raczkowski, J.M., and Wenzel, J.W. (2008) Biodiversity studies and their foundation in taxonomic scholarship. *BioScience 57*, 974–9.

Radchenko, A. (2004) A review of the ant genera *Leptothorax* Mayr and *Temnothorax* Mayr (Hymenoptera, Formicidae) of the eastern Palaearctic. *Acta Zoologica Academiae Scientiarum Hungaricae 50*, 109–37.

——(2005a) Monographic revision of the ants (Hymenoptera: Formicidae) of North Korea. *Annales Zoologici (Warsaw) 55*, 127–221.

——(2005b) A review of the ants of the genus *Lasius* Fabricius, 1804, subgenus *Dendrolasius* Ruzsky, 1912 (Hymenoptera: Formicidae) from east Palaearctic. *Annales Zoologici (Warsaw) 55*, 83–94.

Radchenko, A.G., and Elmes, G.W. (2003) A taxonomic revision of the socially parasitic *Myrmica* ants (Hymenoptera: Formicidae) of the Palaearctic region. *Annales Zoologici (Warsaw) 53*, 217–43.

Radnikova, Zh.I. (2003) Distribution patterns of ants in different natural zones and landscapes Kazakhstan and West Siberia along a meridian trend. *EuroAsian Entomological Journal 2*, 235–42.

Rahbek, C. (2005) The role of spatial scale and the perception of large-scale species-richness patterns. *Ecology Letters 8*, 224–39.

——Gotelli, N.J., Colwell, R.K., Entsminger, G.L., Rangel, T., and Graves, G.R. (2007) Predicting continental-scale patterns of bird species richness with spatially explicit models. *Proceedings of the Royal Society of London B 274*, 165–74.

Raine, N.E., Ings, T.C., Dornhaus, A., Saleh, N., and Chittka, L. (2006) Adaptation, genetic drift, pleiotropy, and history in the evolution of bee foraging behavior. *Advances in the Study of Behavior 36*, 305–54.

Raj, M. (2007) *Navigation in the intertidal ant* Polyrhachis sokolova: *the role of path integration and vision*. Honours thesis, School of Marine & Tropical Biology, James Cook University, Townville.

Ramírez, M., Armbrecht, I., and Montoya, J. (2007) Fodder banks as modifiers of arthropod diversity in agricultural landscapes. In P. Leterme, E. Mugueitio, and A. Buldgeen, eds. *Sustainable Pig Production Systems in the Tropics, Based on Tree Foliage*, pp. 139–63, CIPAV: Centre for Research on Sustainable Agricultural Production Systems, Faculte de Sciences Agronomiques de Gembloux, Belgium.

Ramirez, R.A., Thompson, D.C., and Remmenga, M.D. (2006) Influence of low humidity, *Pseudacteon* flies (Diptera: Phoridae), and competition by *Solenopsis xyloni* on *Solenopsis invicta* (Hymenoptera: Formicidae). *Environmental Entomology 35*, 1037–48.

Rangel, T., Diniz-Filho, J.A.F., and Colwell, R.K. (2007) Species richness and evolutionary niche dynamics: a spatial pattern-oriented simulation experiment. *American Naturalist 170*, 602–16.

Rao, A., and Vinson, S.B. (2004) Ability of resident ants to destruct small colonies of *Solenopsis invicta* (Hymenoptera: Formicidae). *Environmental Entomology 33*, 587–98.

Ratchford, J.S., Wittman, S.E., Jules, E.S., Ellison, A.M., Gotelli, N.J., and Sanders, N.J. (2005) The effects of fire, local environment and time on ant assemblages in fens and forests. *Diversity and Distributions 11*, 487–97.

Ratnieks, F.L.W., Foster, K.R., and Wenseleers, T. (2006) Conflict resolution in insect societies. *Annual Review of Entomology 51*, 581–608.

Read, J.L., and Andersen, A.N. (2000) The value of ants as early warning bioindicators: responses to pulsed cattle grazing at an Australian arid zone locality. *Journal of Arid Environments 45*, 231–51.

Redding, D.W., Hartmann, K., Mimotob, A., Bokalb, D., DeVosb, M., and Mooersa, A. (2007) Evolutionarily distinctive species often capture more phylogenetic diversity than expected. *Journal of Theoretical Biology 251*, 606–15.

Reed, J.M. (1996) Using statistical probability to increase confidence of inferring species extinction. *Conservation Biology 10*, 1283–5.

Reeve, H.K. (1989) The evolution of conspecific acceptance thresholds. *American Naturalist 133*, 407–35.

Regan, T.J., McCarthy, M.A., Baxter, P.W.J., F.D. Panetta, F.D., and Possingham, H.P. (2006) Optimal eradication: when to stop looking for an invasive plant. *Ecology Letters 9*, 759–66.

Rejmánek, M., and Richardson, D.M. (1996) What attributes make some plant species more invasive? *Ecology 77*, 1655–61.

Retana, J., and Cerdá, X. (2000) Patterns of diversity and composition of Mediterranean ground ant communities tracking spatial and temporal variability in the thermal environment. *Oecologia 123*, 436–44.

Rettenmeyer, C.W. (1963) Behavioral studies of army ants. *University of Kansas Scientific Bulletin 44*, 281–465.

Rey, S., and Espadaler, X. (2004) Area-wide management of the invasive garden ant *Lasius neglectus* (Hymenoptera: Formicidae) in northeast Spain. *Journal of Agricultural and Urban Entomology 21*, 99–112.

Rheindt, F.E., Gadau, J., Strehl, C.-P., and Hölldobler, B. (2004) Extremely high mating frequency in the Florida harvester ant (*Pogonomyrmex badius*). *Behavioral Ecology and Sociobiology 56*, 472–81.

Ribas, C.R., and Schoereder, J.H. (2002) Are all ant mosaics caused by competition? *Oecologia 131*, 606–11.

Rice, B., and Westoby, M. (1986) Evidence against the hypothesis that ant-dispersed seeds reach nutrient-enriched microsites. *Ecology 67*, 1270–4.

Richard, F.J., Hefetz, A., Christides, J.-P., and Errard, C. (2004) Food influence on colonial recognition and chemical signature between nestmates in the fungus-growing ant *Acromyrmex subterraneus subterraneus*. *Chemoecology 14*, 9–16.

——Poulsen, M., Drijfhout, F.P., Jones, G., and Boomsma, J.J. (2007) Specificity in chemical profiles of workers, brood and mutualistic fungi in *Atta*, *Acromyrmex*, and *Sericomyrmex* fungus-growing ants. *Journal of Chemical Ecology 33*, 2281–92.

Richards, O.W. (1977) Hymenoptera. Introduction and key to families (2nd edition). *Handbook for the Identification of British Insects*, vol. 6, Royal Entomological Society, London.

Richardson, T.O., Sleeman, P.A., McNamara, J.M., Houston, A.I., and Franks, N.R. (2007) Teaching with evaluation in ants. *Current Biology 17*, 1520–6.

Ricklefs, R.E. (2006) Global variation in the diversification rate of passerine birds. *Ecology 87*, 2468–78.

——and Bermingham, E. (2002) The concept of the taxon cycle in biogeography. *Global Ecology and Biogeography 11*, 353–61.

Ricks, B.L., and Vinson, S.B. (1970) Feeding acceptability of certain insects and various water-soluble compounds to two varieties of the imported fire ant. *Journal of Economic Entomology 63*, 145–8.

————(1972) Digestive enzymes of the imported fire ant, *Solenopsis richteri* (Hymenoptera: Formicidae). *Entomologia Experimentalis et Applicata 15*, 329–34.

Rico-Gray, V. (1993) Use of plant-derived food resources by ants in the dry tropical lowlands of coastal Veracruz, Mexico. *Biotropica 25*, 301–15.

——and Oliveira, P.S. (2007) *The Ecology and Evolution of Ant-Plant Interactions*. University of Chicago Press, Chicago.

Rissing, S.W. (1984) Replete caste production and allometry of workers in the honey ant, *Myrmecocystus mexicanus* Wesmael (Hymenoptera, Formicidae). *Journal of the Kansas Entomological Society 57*, 347–50.

Robertson, H.G. (2000) Formicidae (Hymenoptera: Vespoidea). In A.H. Kirk-Spriggs, and E. Marais, eds. *Dâures – biodiversity of the Brandberg Massif, Namibia. Cimbebasia Memoir 9*, pp. 371–82. National Museum of Namibia, Windhoek.

——(2002) Comparison of leaf litter ant communities in woodlands, lowland forests and montane forests of north-eastern Tanzania. *Biodiversity and Conservation 11*, 1637–52.

——and Villet, M.H. (1989) Mating behavior in three species of myrmicine ants (Hymenoptera: Formicidae). *Journal of Natural History 23*, 767–73.

Robinson, E.J.H. (1985) Effects of a juvenile-hormone analog on honey bee foraging behaviour and alarm pheromone production. *Journal of Insect Physiology 31*, 277–82.

——Jackson, D.E., Holcombe, M., Ratnieks, F.L.W. (2005a) 'No entry' signal in ant foraging. *Nature 438*, 442.

Robinson, W., Nicholls, Z., and Collier, P. (2005b) The attractiveness of several commercial ant baits and food attractants to black crazy ants, *Paratrechina longicornis*. University of the Sunshine Coast, Queensland, Technical Report 4088.

Robson, S.K., and Traniello, J.F.A. (2002) Transient division of labor and behavioral specialization in the ant *Formica schaufussi*. *Naturwissenschaften 89*, 128–31.

Roces, F. (2002) Individual complexity and self-organization in foraging by leaf-cutting ants. *Biological Bulletin 202*, 306–13.

Rockwood, L.L., and Hubbell, S.P. (1987) Host-plant selection, diet diversity, and optimal foraging in a tropical leafcutting ant. *Oecologia 74*, 55–61.

Rodrigo, A., and Retana, J. (2006) Post-fire recovery of ant communities in submediterranean *Pinus nigrus* forests. *Ecography 29*, 231–9.

Rodrigues, A.S.L., and Brooks, T.M. (2007) Shortcuts for biodiversity conservation planning: the effectiveness of surrogates. *Annual Review of Ecology, Evolution and Systematics 38*, 713–37.

——and Gaston, K.J. (2002) Maximising phylogenetic diversity in the selection of networks of conservation areas. *Biological Conservation 105*, 103–11.

——Pilgrim, J.D., Lamoreux, J.F., Hoffmann, M., and Brooks, T.M. (2006) The value of the IUCN Red List for conservation. *Trends in Ecology and Evolution 21*, 71–6.

Roff, D.A. (1990) The evolution of flightlessness in insects. *Ecoogical Monographs 60*, 389–421.

——(1992) *The Evolution of Life Histories: theory and analysis.* Chapman & Hall, New York.

Rohde, K. (1996) Rapoport's rule is a local phenomenon and cannot explain latitudinal gradients in species diversity. *Biodiversity Letters 3*, 10–3.

——(1999) Latitudinal gradients in species diversity and Rapoport's rule revisited: a review of recent work and what can parasites teach us about the causes of gradients? *Ecography 22*, 593–613.

Romdal, T.S., and Grytnes, J.A. (2007) An indirect area effect on elevational species richness patterns. *Ecography 30*, 440–8.

Room, P.M. (1971) The relative distributions of ant species in Ghana's cocoa farms. *Journal of Animal Ecology 40*, 735–51.

Roque-Albelo, L., Causton, C.E., and Mieles, A. (2000) The ants of Marchena Island, twelve years after the introduction of the little fire ant *Wasmannia auropunctata. Noticias de Galapápagos 61*, 17–20.

Röschard, J., and Roces, F. (2003) Cutters, carriers and transport chains: distance-dependent foraging strategies in the grass-cutting ant *Atta vollenweideri. Insectes Sociaux 50*, 237–44.

Rosene, W.J. (1958) Whistling cock counts of bobwhite quail on areas treated with insecticide and untreated areas, Decatur Country, Georgia. 12th Annual Conference of the Association of Game and Fish Commissioners, Louisville, KY pp. 14–8.

Rosengaus, R.B., Moustakas, J.E., Calleri, D.V., and Traniello, J.F.A. (2003) Nesting ecology and cuticular microbial loads in dampwood (*Zootermopsis angusticollis*) and drywood termites (*Incisitermes minor, I. schwarzi, Cryptotermes cavifrons*). *Journal of Insect Science 3*, 31, available online: insectscience.org/3.31. 31.

Rosengren, R. (1986) Competition and coexistence in an insular ant community – a manipulation experiment (Hymenoptera: Formicidae). *Annales Zoologici Fennici 23*, 297–302.

——and Pamilo, P. (1983) The evolution of polygyny and polydomy in mound-building *Formica* ants. *Acta Entomologica Fennica 42*, 65–77.

——Sundström, L., and Fortelius, W. (1993) Monogyny and polygyny in *Formica* ants: the results of alternative dispersal tactics. In L. Keller, ed. *Queen Number and Sociality in Insects*, pp. 308–33. Oxford University Press, Oxford.

Rosenzweig, M.L. (1995) *Species Diversity in Space and Time.* Cambridge University Press, Cambridge.

——and Sandlin, E.A. (1997) Species diversity and latitudes: listening to area's signal. *Oikos 80*, 172–6.

Ross, K.G. (1993) The breeding system of the fire ant *Solenopsis invicta*, and its effects on colony genetic structure. *American Naturalist 141*, 554–76.

——(2001) Molecular ecology of social behaviour: analyses of breeding systems and genetic structure. *Molecular Ecology 10*, 265–84.

——and Carpenter, J.M. (1991) Phylogenetic analysis and the evolution of queen number in eusocial Hymenoptera. *Journal of Evolutionary Biology 4*, 117–30.

——and Keller, L. (1998) Genetic control of social organization in an ant. *Proceedings of the National Academy of Sciences USA 95*, 14232–7.

——and Shoemaker, D.D. (2005) Species delimitation in native South American fire ants. *Molecular Ecology 14*, 3419–38.

——Vargo, E.L., and Keller, L. (1996) Social evolution in a new environment: the case of introduced fire ants. *Proceedings of the National Academy of Sciences USA 93*, 3021–5.

——Krieger, M.J.B., Shoemaker, D.D., Vargo, E.L., and Keller, L. (1997) Hierarchical analysis of genetic structure in native fire ant populations: results from three classes of molecular markers. *Genetics 147*, 643–55.

Rosset, H., Schwander, T., and Chapuisat, M. (2007) Nestmate recognition and levels of aggression are not altered by changes in genetic diversity in a socially polymorphic ant. *Animal Behaviour 74*, 951–6.

Roughgarden, J. (1983) Competition and theory in community ecology. *American Naturalist 122*, 583–601.

Roulston, T.H., Buczkowski, G., and Silverman, J. (2003) Nestmate discrimination in ants: effect of bioassay on aggressive behavior. *Insectes Sociaux 50*, 151–9.

Roura-Pascual, N. (2004) Geographic potential of Argentine ants (Mayr) in the face of global climate change. *Proceedings of the Royal Society of London B 571*, 2527–34.

——(2006) Niche differentiation and fine scale projections for Argentine ants based on remotely sensed data. *Ecological Applications 16*, 1832–41.

——and Suarez, A.V. (2008) The utility of species distribution models to predict the spread of invasive ants (Hymenoptera: Formicidae) and to anticipate changes in their ranges in the face of global climate change. *Myrmecological News 11*, 67–77.

Rousset, F., and Roze, D. (2007) Constraints on the origin and maintenance of genetic kin recognition. *Evolution 61*, 2320–30.

Rowles, A.D., and O'Dowd, D.J. (2009) New mutualism for old: indirect disruption and direct facilitation of seed dispersal following Argentine ant invasion. *Oecologia 158*, 709–16.

——(2007) Interference competition by Argentine ants displaces native ants: implications for biotic resistance to invasion. *Biological Invasions 9*, 73–85.

Roy, H.E., Steinkraus, D.C., Eilenberg, J., Hajek, A.E., and Pell, J.K. (2006) Bizarre interactions and endgames: entomopathogenic fungi and their arthropod hosts. *Annual Review of Entomology 51*, 331–57.

Rudgers, J.A., and Strauss, S.Y. (2004) A selection mosaic in the facultative mutualism between ants and wild cotton. *Proceedings of the Royal Society of London B 271*, 2481–8.

Rueppell, O., and Kirkman, R.W. (2005) Extraordinary starvation resistance in *Temnothorax rugatulus* (Hymenoptera, Formicidae) colonies: demography and adaptive behavior. *Insectes Sociaux 52*, 282–90.

Ryder Wilkie, K.T., Mertl, A.L., Traniello, J.F.A. (2007) Biodiversity below ground: probing the subterranean ant fauna of Amazonia. *Naturwissenschaften 94*, 725–31.

Ryti, R.T., and Case, T.J. (1986) Overdispersion of ant colonies – a test of hypotheses. *Oecologia 69*, 446–53.

——(1988) Field experiments on desert ants – testing for competition between colonies. *Ecology 69*, 1993–2003.

——(1992) The role of neighborhood competition in the spacing and diversity of ant communities. *American Naturalist 139*, 355–74.

Saarinen, E.V., and Daniels, J.C. (2006) Miami blue butterfly larvae (Lepidoptera: Lycaenidae) and ants (Hymenoptera: Formicidae): new information on the symbionts of an endangered taxon. *Florida Entomologist 89*, 69–74.

Sagers, C.L., Ginger, S.M., and Evans, R.D. (2000) Carbon and nitrogen isotopes trace nutrient exchange in an ant-plant mutualism. *Oecologia 123*, 582–6.

Sameshima, S., Hasegawa, E., Kitade, O., Minaka, N., and Matsumoto, T. (1999) Phylogenetic comparison of endosymbionts with their host ants based on molecular evidence. *Zoological Science 16*, 993–1000.

Samson, D.A., Rickart, E.A., and Gonzales, P.C. (1997) Ant diversity and abundance along an elevational gradient in the Philippines. *Biotropica 29*, 349–63.

Samways, M.J. (1983) Community structure of ants (Hymenoptera: Formicidae) in a series of habitats associated with citrus. *Journal of Appied Ecology 20*, 833–47.

Sanders, N.J. (2002) Elevational gradients in ant species richness: area, geometry, and Rapoport's rule. *Ecography 25*, 25–32.

——and Gordon, D.M. (2002) Resources and the flexible allocation of work in the desert ant, *Aphaenogaster cockerelli*. *Insectes Sociaux 49*, 371–9.

————(2003) Resource-dependent interactions and the organization of desert ant communities. *Ecology 84*, 1024–31.

——Barton, K.E., and Gordon, D.M. (2001) Long-term dynamics of the distribution of the invasive Argentine ant, *Linepithema humile*, and native ant taxa in northern California. *Oecologia 127*, 123–30.

——Moss, J., and Wagner, D. (2003b) Patterns of ant species richness along elevational gradients in an arid ecosystem. *Global Ecology and Biogeography 12*, 93–102.

——Gotelli, N.J., Heller, N.E., and Gordon, D.M. (2003a) Community disassembly by an invasive species. *Proceedings of the National Academy of Sciences USA 100*, 2474–7.

——Lessard, J.P., Fitzpatrick, M.C., and Dunn, R.R. (2007b) Temperature, but not productivity or geometry, predicts elevational diversity gradients in ants across spatial grains. *Global Ecology and Biogeography 16*, 640–9.

——Crutsinger, G.M., Dunn, R.R., Majer, J.D., Delabie, J. H.C. (2007a) An ant mosaic revisited: dominant ant species disassemble arboreal ant communities but co-occur randomly. *Biotropica 39*, 422–7.

Sanetra, M., and Crozier, R.H. (2001) Polyandry and colony genetic structure in the primitive ant *Nothomyrmecia macrops*. *Journal of Evolutionary Biology 14*, 368–78.

————(2002) Daughters inherit colonies from mothers in the "living-fossil" ant *Nothomyrmecia macrops*. *Naturwissenschaften 89*, 71–4.

————(2003) Patterns of population subdivision and gene flow in the ant *Nothomyrmecia macrops* reflected in microsatellite and mitochondrial DNA markers. *Molecular Ecology 12*, 2281–95.

Santini, G., Tucci, L., Ottonetti, L., and Frizzi, F. (2007) Competition trade-offs in the organisation of a Mediterranean ant assemblage. *Ecological Entomology 32*, 319–26.

Sarnat, E.M. (2006) *Lordomyrma* (Hymenoptera: Formicidae) of the Fiji Islands. *Bishop Museum Occasional Papers 90*, 9–42.

——(2008) A taxonomic revision of the *Pheidole roosevelti*-group (Hymenoptera: Formicidae) in Fiji. *Zootaxa 1767*, 1–36.

Sarty, M., Abbott, K.L., and Lester, P.J. (2006) Habitat complexity facilitates coexistence in a tropical ant community. *Oecologia 149*, 465–73.

——————(2007) Community level impacts of an ant invader and food mediated coexistence. *Insectes Sociaux 54*, 166–73.

Sauer, C., Stackebrandt, E., Gadau, J., Hölldobler, B., and Gross, R. (2000) Systematic relationships and cospeciation of bacterial endosymbionts and their carpenter ant host species: proposal of the new taxon Candidatus *Blochmannia* gen. nov. *International Journal of Systematic and Evolutionary Microbiology 50*, 1877–86.

Saux, C., Fisher, B.L., and Spicer, G.S. (2004) Dracula ant phylogeny as inferred by nuclear 28S rDNA sequences and implications for ant systematics (Hymenoptera: Formicidae: Amblyoponinae). *Molecular Phylogenetics and Evolution 33*, 457–68.

Savolainen, R., and Vepsäläinen, K. (1988) A competition hierarchy among boreal ants: impact on resource partitioning and community structure. *Oikos 51*, 135–55.

————and Wuorenrinne, H. (1989) Ant assemblages in the taiga biome – testing the role of territorial wood ants. *Oecologia 81*, 481–6.

Say-Piau, L., and Chow-Yang, L. (2003) Nestmate recognition and intercolonial aggression in the crazy ant, *Paratrechina longicornis* (Latreille) (Hymenoptera: Formicidae). *Sociobiology 41*, 295–305.

Schaffer, W.M., Jensen, D.B., Hobbs, D.E., Gurevitch, J., Todd, J.R., and Schaffer, M.V. (1979) Competition, foraging energetics, and the cost of sociality in three species of bees. *Ecology 60*, 976–87.

Schatz, B., Lachaud, J., and Beugnon, G. (1999) Spatiotemporal learning by the ant *Ectatomma ruidum*. *Journal of Animal Ecology 202*, 1897–907.

Schemske, D.W. (1980) The evolutionary significance of extrafloral nectar production by *Costus woodsonii* (Zingiberaceae): an experimental analysis of ant protection. *Journal of Ecology 68*, 959–67.

Schilman, P.E., Lighton, J.R.B., and Holway, D.A. (2005) Respiratory and cuticular water loss in insects with continuous gas exchange: comparison across five ant species. *Journal of Insect Physiology 51*, 1295–305.

Schlick-Steiner, B.C., Steiner, F.M., and Steiner, H.M. (2005) Effect of extensification of coppice management on central European ant assemblages (Hymenoptera: Formicidae). *Entomologia Generalis 28*, 23–37.

————Moder, K., Bruckner, A., Fiedler, K., and Christian, E. (2006a) Assessing ant assemblages: pitfall trapping versus nest counting (Hymenoptera, Formicidae). *Insectes Sociaux 53*, 274–81.

————Konrad, H., Seifert, B., Christian, E., Moder, K., Stauffer, C., and Crozier, R. H. (2008) Specificity and transmission mosaic of ant nest-wall fungi. *Proceedings of the National Academy of Sciences USA 105*, 940–943.

————Moder, K., Bruckner, A., Fiedler, K., and Christian, E. (2006b) A multidisciplinary approach reveals cryptic diversity in western Palearctic *Tetramorium* ants (Hymenoptera: Formicidae). *Molecular Phylogenetics and Evolution 40*, 259–73.

Schluter, D. (1984) A variance test for detecting species associations, with some example applications. *Ecology 65*, 998–1005.

Schmid-Hempel, P. (1998) *Parasites in Social Insects*. Princeton University Press, Princeton, NJ.

——Schmid-Hempel, R. (1984) Life duration and turnover of foragers in the ant *Cataglyphis bicolor* (Hymenoptera, Formicidae). *Insectes Sociaux 31*, 345–60.

Schmidt, G.H. (1974) Steuerung der Kastenbildung und Geschlechtsregulation im Waldameisenstaat. In G.H. Schmidt, ed. *Sozialpolymorphismus bei Insekten*, pp. 404–512. Wissenschaftliche Verlagsgesellschaft, Stuttgart.

Schneirla, T.C. (1971) *Army Ants: a study in social organization*. W. H. Freeman & Co., San Francisco.

Schödl, S. (2007) Revision of Australian *Meranoplus*: the *Meranoplus diversus* group. *Memoirs of the American Entomological Institute 80*, 370–424.

Schoener, T.W. (1977) Competition and the niche. In C. Gans, and D.W. Tinkle, eds. *Biology of the Reptilian*, pp. 36–136. Academic Press, New York.

——(1982) The controversy over interspecific competition. *American Scientist 70*, 586–95.

——(1983) Field experiments on interspecific competition. *American Naturalist 122*, 240–85.

——and Adler, G.H. (1991) Greater resolution of distributional complementarities by controlling for habitat affinities – a study with Bahamian lizards and birds. *American Naturalist 137*, 669–92.

Schöning, C., Kinuthia, W., and Franks, N.R. (2005) Evolution of allometries in the worker caste of *Dorylus* army ants. *Oikos 110*, 231–40.

Schooley, R.L., and Wiens, J.A. (2003) Spatial patterns, density dependence, and demography in the harvester ant, *Pogonomyrmex rugosus* in semi-arid grasslands. *Journal of Arid Environments 53*, 183–96.

Schrempf, A., and Heinze, J. (2006) Back to one: consequences of derived monogyny in an ant with polygynous ancestors. *Journal of Evolutionary Biology 20*, 792–9.

Schultz, T., and Brady, S.G. (2008) Major evolutionary transitions in ant agriculture. *Proceedings of the National Academy of Sciences USA 105*, 5435–40.

Schulz, A., and Wagner, T. (2002) Influence of forest type and tree species on canopy ants (Hymenoptera: Formicidae) in Budongo Forest, Uganda. *Oecologia 133*, 224–32.

Schwander, T., Helms Cahan, S., and Keller, L. (2007) Characterization and distribution of *Pogonomyrmex* harvester ant lineages with genetic caste determination. *Molecular Ecology 16*, 367–87.

Seagraves, M.P., and McPherson, R.M. (2006) Monitoring red imported fire ant (Hymenoptera: Formicidae) foraging behavior and impact on foliage-dwelling arthropods on soybean produced under three ant suppression regimes. *Journal of Entomological Science 41*, 374–84.

——and Ruberson, J.R. (2004) Impact of *Solenopsis invicta* Buren suppression on arthropod ground predators and pest species in soybean. *Journal of Entomological Science 39*, 433–43.

Seal, J.N., and Tschinkel, W.R. (2007a) Co-evolution and the superorganism: switching cultivars does not alter the performance of fungus-gardening ant colonies. *Functional Ecology 21*, 988–97.

——(2007b) Complexity in an obligate mutualism: do fungus-gardening ants know what makes their garden grow? *Behavioral Ecology and Sociobiology 61*, 1151–60.

Seaman, R.E., and Marino, P.C. (2003) Influence of mound building and selective seed predation by the red imported fire ant (*Solenopsis invicta*) on an old-field plant assemblage. *Journal of the Torrey Botanical Society 130*, 193–201.

Seifert, B. (2000) Rapid range expansion in *Lasius neglectus* (Hymenoptera: Formicidae) – an Asian invader swamps Europe. *Deutsche Entomologische Zeitschrift 47*, 173–9.

——(2007) *Die Ameisen Mittel- und Nordeuropas*. lutra Verlags- und Vertriebsgesellschaft, Tauer.

Sempo, G., Depickère, S., and Detrain, C. (2006) Spatial organization in a dimorphic ant: caste specificity of clustering patterns and area marking. *Behavioral Ecology 17*, 642–50.

Sendova-Franks, A.B., and Franks, N.R. (1995) Spatial relationships within nests of the ant *Leptothorax unifasciatus* (Latr.) and their implications for the division of labour. *Animal Behaviour 50*, 121–36.

Seppä, P. (2008) Do ants (Hymenoptera: Formicidae) need conservation and does ant conservation need genetics? *Myrmecological News 11*, 161–72.

——Fernández-Escudero, I., Gyllenstrand, N., and Pamilo, P. (2008) Colony fission affects kinship in a social insect. *Behavioral Ecology and Sociobiology 62*, 589–97.

Serfling, R.E. (1952) Historical review of epidemic theory. *Human Biology 24*, 145–66.

Seymour, M. (2007) *Effect of red imported fire ant* (Solenopsis invicta *Buren) on the nesting success of Northern bobwhite* (Colinus virginianus). Masters thesis, Louisiana State University, Baton Rouge.

Sfenthourakis, S., Tzanatos, E., and Giokas, S. (2005) Species co-occurrence: the case of congeneric species and a causal approach to patterns of species association. *Global Ecology and Biogeography 15*, 39–49.

Shaffer, H.B., and Thomson, R.C. (2007) Delimiting species in recent radiations. *Systematic Biology 56*, 896–906.

Shattuck, S.O. (1999) *Australian ants: their biology and identification*. CSIRO Publishing, Collingwood, Victoria.

——(2006) A third species in the rare Australian ant genus *Peronomyrmex* Viehmeyer (Hymenoptera: Formicidae). *Zootaxa 1194*, 49–55.

——(2007) New species of myrmicine ants from Western Australia (Hymenoptera: Formicidae). *Zootaxa 1661*, 47–53.

Shattuck, S.O. (2008a) Australian ants of the genus *Aphaenogaster* (Hymenoptera: Formicidae). *Zootaxa* 1677, 25–45.

——(2008b) Review of the ant genus *Aenictus* (Hymenoptera: Formicidae) in Australia with notes on *A. ceylonicus* (Mayr). *Zootaxa 1926*, 1–19.

——(2008c) Revision of the ant genus *Prionopelta* (Hymenoptera: Formicidae) in the Indo-Pacific region. *Zootaxa 1846*, 21–34.

——and Barnett, N.J. (2006) Australian species of the ant genus *Diacamma* (Hymenoptera: Formicidae). *Myrmecologische Nachrichten 8*, 13–9.

——and Barnett, N.J. (2007) Revision of the ant genus *Mayriella*. *Memoirs of the American Entomological Institute 80*, 437–58.

Sherman, P.W., Reeve, H.K., and Pfenning, D.W. (1997) Recognition systems. In R. Krebs, and N.B. Davies, eds. *Behaviuoral Ecology: an evolutionary approach*, pp. 69–96. Blackwell Scientific, Oxford.

Shigesada, N., and Kawasaki, K. (1997) *Biological Invasions: theory and practice*. Oxford University Press, Oxford.

Shingleton, A.W., Stern, D.L., and Foster, W.A. (2005) The origin of a mutualism: a morphological trait promoting the evolution of ant-aphid mutualisms. *Evolution 59*, 921–6.

Shoemaker, D.D., DeHeer, C.J., Krieger, M.J.B., and Ross, K.G. (2006) Population genetics of the invasive fire ant *Solenopsis invicta* (Hymenoptera: Formicidae) in the United States. *Annals of the Entomological Society of America 99*, 1213–33.

Shrestha, R.K., Stein, T.V., and Clark, J. (2007) Valuing nature-based recreation in public natural areas of the Apalachicola River region, Florida. *Journal of Environmental Management 85*, 977–85.

Shuetrim, A. (2001) *Territoriality, colony structure and spatial dispersion in the intertidal ant* Polyrhachis sokolova *(Formicidae: Formicinae)*. Honours thesis, James Cook University, Cairns.

Shure, D.J., and Phillips, D.L. (1991) Patch size of forest openings and arthropod populations. *Oecologia 86*, 325–34.

Sih, A., Englund, G., and Wooster, D. (1998) Emergent impacts of multiple predators on prey. *Trends in Ecology and Evolution 13*, 350–5.

Silverman, J., and Brightwell, R.J. (2008) The Argentine ant: challenges in managing a unicolonial invasive ant. *Annual Review of Entomology 53*, 231–52.

——and Nsimba, B. (2000) Soil-free collection of Argentine ants (Hymenoptera: Formicidae) based on food-directed brood and queen movement. *Florida Entomologist 83*, 10–6.

Simberloff, D. (2002) Today Tiritiri Matangi, tomorrow the world! Are we aiming too low in invasives control? In C. R. Veitch, and M.N. Clout, eds. *Turning the tide: the eradication of invasive species. Proceedings of the International Conference on Eradication of Island Invasives*. Occasional Paper of the IUCN Species Survival Commission Gland, Switzerland.

——(2003) How much information on population biology is needed to manage introduced species? *Conservation Biology 17*, 83–92.

——(2006) Invasional meltdown 6 years later: important phenomenon, unfortunate metaphor, or both? *Ecology Letters 9*, 912–9.

——(2009) We can eliminate invasions or live with them. Successful management projects. *Biological Invasions 11*, 149–57.

——and Gibbons, L. (2004) Now you see them, now you don't – population crashes of established introduced species. *Biological Invasions 6*, 161–72.

——and von Holle, B. (1999) Positive interactions of nonindigenous species: invasional meltdown? *Biological Invasions 1*, 21–32.

——and Wilson, E.O. (1969) Experimental zoogeography of islands – Colonization of empty islands. *Ecology 50*, 278–96.

Singer, T.L. (1998) Role of hydrocarbons in the recognition systems of insects. *American Zoologist 38*, 394–405.

Smallwood, J. (1982) Nest relocations in ants. *Insectes Sociaux 29*, 138–47.

Smith, A.A., and Haight, K.L. (2008) Army ants as research and collection tools. *Journal of Insect Science 8*, 71, available online: insectscience.org/8.71.

Smith, C.R. (2007) Energy use and allocation in the Florida harvester ant, *Pogonomyrmex badius*: are stored seeds a buffer? *Behavioral Ecology and Sociobiology 61*, 1479–87.

Smith, J.E., Taylor, S.J., Whelan, C.J., Denight, M.L., and Stake, M.M. (2004) Behavioral interactions between fire ants and vertebrate nest predators at two black-capped vireo nests. *Wilson Bulletin 116*, 163–7.

Smith, M.A., Fisher, B.L., and Hebert, P.D.N. (2005) DNA barcoding for effective biodiversity assessment of a hyperdiverse arthropod group: the ants of Madagascar. *Philosophical Transactions of the Royal Society B 360*, 1825–34.

——Woodley, N., Hallwachs, W., Janzen, D.H., and Hebert, P.D.N. (2006) DNA barcodes reveal cryptic host-specificity within the presumed polyphagous members of a genus of parasitoid flies (Diptera: Tachinidae). *Proceedings of the National Academy of Sciences USA 103*, 3657–62.

Smith, M.R. (1934) Ponerine ants of the genus *Euponera* in the United States. *Annals of the Entomological Society of America 27*, 558–64.

Smith, R.A., Mooney, K.A., and Agrawal, A.A. (2008) Coexistence of three specialist aphids on common milkweed, *Asclepias syriaca*. *Ecology 89*, 2187–96.

Smythe, K.D., Bernabo, J.C., Carter, T.B., and Jutro, P.R. (1996) Focusing biodiversity research on the needs of decision makers. *Environmental Management 20*, 865–72.

Snelling, G.C., and Snelling, R.R. (2007) New synonymy, new species, new keys to *Neivamyrmex* army ants of the United States. *Memoirs of the American Entomological Institute 80*, 459–550.

Snelling, R.R. (2006) Taxonomy of the *Camponotus festinatus* complex in the United States of America (Hymenoptera: Formicidae). *Myrmecologische Nachrichten 8*, 83–97.

——(2007) A review of the arboreal Afrotropical ant genus *Axinidris*. *Memoirs of the American Entomological Institute 80*, 551–79.

Soininen, J., Lennon, J.J., and Hillebrand, H. (2007) A multivariate analysis of beta diversity across organisms and environments. *Ecology 88*, 2830–8.

Solow, A.S. (1993) Inferring extinction from sighting data. *Ecology 74*, 962–4.

Srinivasan, M.V., Zhang, S., Altwein, M., and Tautz, J. (2000) Honeybee navigation: nature and calibration of the "Odometer". *Science 287*, 851–3.

Srivastava, D.S., and Lawton, J.H. (1998) Why more productive sites have more species: experimental test of theory using tree-hole communities. *American Naturalist 152*, 510–29.

Stachowicz, J.J. (2001) Mutualism, facilitation, and the structure of ecological communities. *BioScience 51*, 235–46.

Stadler, B., and Dixon, A.F.G. (1998) Costs of ant attendance for aphids. *Journal of Animal Ecology, 67*, 454–9.

————(2005) Ecology and evolution of aphid-ant interactions. *Annual Review of Ecology, Evolution and Systematics 36*, 345–72.

——and Dixon, T. (2008) *Mutualism: ants and their insect partners*. Cambridge University Press, Cambridge.

——Michalzik, B., and Müller, T. (1998) Linking aphid ecology with nutrient fluxes in a coniferous forest. *Ecology 79*, 1514–25.

Stake, M.M., and Cimprich, D.A. (2003) Using video to monitor predation at black-capped vireo nests. *Condor 105*, 348–57.

Staller, E.L., Palmer, W.E., Carroll, J.P., Thornton, R.P., and Sisson, D.C. (2005) Identifying predators at northern bobwhite nests. *Journal of Wildlife Management 69*, 124–32.

Stanaway, M.A., Zalucki, M.P., Gillespie, P.S., and Rodriquez, C.M. (2001) Pest risk assessment of insects in sea cargo containers. *Australian Journal of Entomology 40*, 180–92.

Stanton, M.L., Palmer, T.M., Young, T.P., Evans, A., and Turner, M.L. (1999) Sterilization and canopy modification of a swollen thorn acacia tree by a plant ant. *Nature 401*, 578–81.

Stapley, J.H. (1980) Using the predatory ants, *Oecophylla smaragdina*, to control insect pests of coconuts and cocoa. *Information Circular, South Pacific Commission 85*, 5.

Starks, P.T. (2004) Recognition systems: from components to conservation. *Annales Zoologici Fennici 41*, 689–90.

Steghaus-Kovac, S., Maschwitz., U. (1993) Predation on earwigs – a novel diet specialization within the genus *Leptogenys* (Formicidae, Ponerinae). *Insectes Sociaux 40*, 337–40.

Stein, M., Thorvilson, H.G., and Johnson, J. (1990) Seasonal changes in bait preference of the red imported fire ant, *Solenopsis invicta* Buren (Hymenoptera: Formicidae). *Florida Entomologist 73*, 118–23.

Steiner, F.M., Schlick-Steiner, B.C., Trager, J., Moder, K., Sanetra, M., Christian, E., and Stauffer, C., (2006a) *Tetramorium tsushimae*, a new invasive ant in North America. *Biological Invasions 8*, 117–23.

———Konrad, H. Moder, K., Christian, E., Seifert, B., Crozier, R.H., Stauffer, C., and Buschinger, A. (2006b) No sympatric speciation here: multiple data sources show that the ant *Myrmica microrubra* is not a separate species but an alternate reproductive morph of *Myrmica rubra*. *Journal of Evolutionary Biology 19*, 777–87.

———Moder, K., Stauffer, C., Arthofer, W., Buschinger, A., Espadaler, X., Christian, E., Einfinger, K., Lorbeer, E., Schafellner, C., Ayasse, M., and Crozier, R. H. (2007) Abandoning aggression but maintaining self-nonself discrimination as a first stage in ant supercolony formation. *Current Biology 17*, 1903–7.

———Nikiforov, A., Kalb, R., and Mistrik, R. (2002) Cuticular hydrocarbons of *Tetramorium* ants from Central Europe: analysis of GC-MS data with self-organizing maps (SOM) and implications for systematics. *Journal of Chemical Ecology 28*, 2569–84.

Stevens, G.C. (1989) The latitudinal gradient in geographical range: how so many species coexist in the tropics. *American Naturalist 133*, 240–56.

——(1992) The elevational gradient in altitudinal range: an extension of Rapoport's latitudinal rule to altitude. *American Naturalist 140*, 893–911.

Stevens, R.D., and Willig, M.R. (2002) Geographical ecology at the community level: perspectives on the diversity of new world bats. *Ecology 83*, 545–60.

Stiles, J.H., and Jones, R.H. (2001) Top-down control by the red imported fire ant (*Solenopsis invicta*). *American Midland Naturalist 146*, 171–85.

Stille, M. (1996) Queen/worker thorax volume ratios and nest-founding strategies in ants. *Oecologia 105*, 87–93.

Stoll, S., Gadau, J., Gross, R., and Feldhaar, H. (2007) Bacterial microbiota associated with ants of the genus *Tetraponera*. *Biological Journal of the Linnean Society 90*, 399–412.

Stone, L., and Roberts, A. (1990) The checkerboard score and species distributions. *Oecologia 85*, 74–9.

——(1992) Competitive-exclusion, or species aggregation – an aid in deciding. *Oecologia 91*, 419–24.

Stow, A., Briscoe, D., Gillings, M., Holley, M., Smith, S., Leys, R., Silberbauer, L., Turnbull, C., and Beattie, A. (2007) Antimicrobial defences increase with sociality in bees. *Biology Letters 3*, 422–424.

Stradling, D.J. (1978) Food and feeding habits of ants. In M.V. Brian, ed. *Production Ecology of Ants and Termites*, pp. 81–106. Cambridge University Press, Cambridge.

Straka, J., and Feldhaar, H. (2007) Development of a chemically defined diet for ants. *Insectes Sociaux 54*, 100–4.

Stuart, R.J., Greshambissett, L., and Alloway, T.M. (1993) Queen adoption in the polygynous and polydomous ant, *Leptothorax curvispinosus*. *Behavioral Ecoogy 4*, 276–81.

——Jackson, I.W., and McCoy, C.W. (2003) Predation on neonate larvae of *Diaprepes abbreviatus* (Coleoptera: Curculionidae) in Florida citrus: testing for daily patterns of neonate drop, ant predators and chemical repellency. *Florida Entomologist 86*, 61–72.

Styrsky, J.D., and Eubanks, M.D. (2007) Ecological consequences of interactions between ants and honeydew-producing insects. *Proceedings of the Royal Society of London B 274*, 151–64.

——Kaplan, I., and Eubanks, M.D. (2006) Plant trichomes indirectly enhance tritrophic interactions involving a generalist predator, the red imported fire ant. *Biological Control 36*, 375–284.

Suarez, A.V., and Case, T.J. (2002) Bottom-up effects on persistence of a specialist predator: ant invasions and horned lizards. *Ecological Applications 12*, 291–8.

——and Tsutsui, N.D. (2004) The value of museum collections for research and society. *BioScience 54*, 66–74.

Suarez, A.V., and Case, T.J. (2008) The evolutionary consequences of biological invasions. *Molecular Ecology 17*, 351–60.

——Bolger, D.T., and Case, T.J. (1998) Effect of fragmentation and invasion on native ant communities in coastal Southern California. *Ecology 79*, 2041–56.

——Holway, D.A., and Case, T.J. (2001) Patterns of spread in biological invasions dominated by long-distance jump dispersal: insights from Argentine ants. *Proceedings of the National Academy of Sciences USA 98*, 1095–100.

————and Tsutsui, N.D. (2008) Genetics and behavior of a colonizing species: the invasive Argentine ant. *American Naturalist 172*, S72–84.

————and Ward, P.S. (2005a) The role of opportunity in the unintentional introduction of invasive ants. *Proceedings of the National Academy of Sciences USA 102*, 17032–5.

——Richmond, J.Q., and Case, T.J. (2000) Prey selection in horned lizards following the invasion of Argentine ants in southern California. *Ecological Applications 10*, 711–25.

——Yeh, P., and Case, T.J. (2005b) Impacts of Argentine ants on avian nesting success. *Insectes Sociaux 52*, 378–82.

——Tsutsui, N.D., Holway, D.A., and Case, T.J. (1999) Behavioral and genetic differentiation between native and introduced populations of the Argentine ant. *Biological Invasions 1*, 43–53.

Suchocka, H., Czechowski, W., and Radchenko, A. (2008) Second report on the occurrence of *Camponotus truncatus* (Spinola) (Hymenoptera: Formicidae) in Poland, with a key to the Polish species of the genus *Camponotus* Mayr. *Fragmenta Faunistica (Warsaw) 51*, 9–13.

Sunamura, E., Nishisue, K., Terayama, M., and Tatsuki, S. (2007) Invasion of four Argentine ant supercolonies into Kobe Port, Japan: their distributions and effects on indigenous ants (Hymenoptera: Formicidae). *Sociobiology 50*, 659–74.

Sundström, L. (1993) Foraging responses of *Formica truncorum* (Hymenoptera, Formicidae) – exploiting stable vs spatially and temporally variable resources. *Insectes Sociaux 40*, 147–61.

——(1995) Dispersal polymorphism and physiological condition of males and females in the ant, *Formica truncorum*. *Behavioral Ecology 6*, 132–9.

——(1997) Queen acceptance and nestmate recognition in monogyne and polygyne colonies of the ant *Formica truncorum*. *Animal Behaviour 53*, 499–510.

——Seppä, P., and Pamilo, P. (2005) Genetic population structure and dispersal patterns in *Formica* ants – a review. *Annales Zoologici Fennici 42*, 163–77.

Sutherst, R.W., and Maywald, G. (2005) A climate model for the red imported fire ant, *Solenopsis invicta* Buren (Hymenoptera: Formicidae): implications for invasion of new regions, particularly Oceania. *Environmental Entomology 34*, 317–35.

Tarpy, D.R., Nielsen, R., and Nielsen, D.I. (2004) A scientific note on the revised estimates of effective paternity in *Apis*. *Insectes Sociaux 51*, 203–4.

Tay, W.T., Cook, J.M., Rowe, D.J., and Crozier, R.H. (1997) Migration between nests in the Australian arid-zone ant *Rhytidoponera* sp. 12 revealed by DGGE analyses of mitochondrial DNA. *Molecular Ecology 6*, 403–11.

Taylor, P.H., and Gaines, S.D. (1999) Can Rapoport's rule be rescued? Modeling causes of the latitudinal gradient in species richness. *Ecology* 80, 2474–82.

Taylor, R.W. (1978) *Nothomyrmecia macrops*: a living-fossil ant rediscovered. *Science 201*, 979–85.

Tennant, L.E. (1994) The ecology of *Wasmannia auropunctata* in primary tropical rainforest in Costa Rica and Panama. In D.F. Williams, ed. *Exotic Ants: biology, impact, and control of introduced species*, pp. 80–90. Westview Press, Boulder, CO.

——and Porter, S.D. (1991) Comparison of diets of two fire ant species (Hymenoptera: Formicidae): solid and liquid components. *Journal of Entomological Science 26*, 450–65.

Texas A&M University (2008) *Rasberry crazy ant*. http://urbanentomology.tamu.edu/ants/exotic_tx.cfm.

Thead, L.G., Vogt, J.T., and Streett, D.A. (2005) Dispersal of the fire ant decapitating fly, *Pseudacteon curvatus* (Diptera: Phoridae) in northeast Mississippi. *Florida Entomologist 88*, 214–16.

Thomas, M.L. (2002) Nest site selection and longevity in the ponerine ant *Rhytidoponera metallica* (Hymenoptera, Formicidae). *Insectes Sociaux 49*, 147–52.

——and Holway, D.A. (2005) Condition-specific competition between invasive Argentine ants and Australian *Iridomyrmex*. *Journal of Animal Ecology 74*, 532–42.

——Payne-Makrisa, C.M., Suarez, A.V., Tsutsui, N.D., and Holway, D.A. (2006) When supercolonies collide: territorial aggression in an invasive and unicolonial social insect. *Molecular Ecology 15*, 4303–15.

Thompson, J.N. (1981) Elaiosomes and fleshy fruits: phenology and selection pressures for ant-dispersed seeds. *American Naturalist 117*, 104–8.

Thorne, B.L., and Traniello, J.F.A. (2003) Comparative social biology of basal taxa of ants and termites. *Annual Review of Entomology 48*, 283–306.

Thorvilson, H.G., Cokendolpher, J.C., and Phillips, S.A. (1992) Survival of the red imported fire ant (Hymenoptera: Formicidae) on the Texas high plains. *Environmental Entomology 21*, 964–8.

Thurin, N., and Aron, S. (2008) Seasonal nestmate recognition in the polydomous ant *Plagiolepis pygmaea*. *Animal Behaviour 75*, 1023–30.

Tillberg, C.V., Holway, D.A., LeBrun, E.G., and Suarez, A. V. (2007) Trophic ecology of invasive Argentine ants in their native and introduced ranges. *Proceedings of the National Academy of Sciences USA 104*, 20856–61.

——McCarthy, D.P., Dolezal, A.G., and Suarez, A.V. (2006) Measuring the trophic ecology of ants using stable isotopes. *Insectes Sociaux 53*, 65–9.

Tilley, C.A., and Oldroyd, B.P. (1997) Unequal subfamily proportions among honey bee queen and worker brood. *Animal Behaviour 54*, 1483–90.

Tillman, G., Schomberg, H., Phatak, S., Mullinix, B., Lachnicht, S., Timper, P., and Olson, D. (2004) Influence of cover crops on insect pests and predators in conservation tillage cotton. *Journal of Economic Entomology 97*, 1217–32.

Tilman, D. (1982) *Resource Competition and Community Structure.* Princeton University Press, Princeton, NJ.

Tobin, J.E. (1994) Ants as primary consumers: diet and abundance in the Formicidae. In J.H. Hunt, and C.A. Nalepa, eds. *Nourishment and Evolution in Insect Societies*, pp. 279–308. Westview Press, Boulder, CO.

Todd, B.D., Rothermel, B.B., Reed, R.N., Luhring, T. M., Schlatter, K., Trenkamp, L., and Gibbons, J. W. (2008) Habitat alteration increases invasive fire ant abundance to the detriment of amphibians and reptiles. *Biological Invasions 10*, 539–46.

Torossian, C. (1967) Recherches sur la biologie et l'ethologie de *Dolichoderus quadripunctatus* (L.) (Hym. Form. Dolichoderidae). IV. Étude des possibilités évolutives des colonies avec reine, et des femelles isolées désailées. *Insectes Sociaux 14*, 259–80.

Torres, J.A. (1984) Diversity and distribution of ant communities in Puerto Rico. *Biotropica 16*, 296–303.

Touyama, Y., Ihara, Y., and Ito, F. (2008) Argentine ant infestation affects the abundance of the native myrmecophagic jumping spider *Siler cupreus* Simon in Japan. *Insectes Sociaux 55*, 144–6.

Townsend, C.R., and Hildrew, A.G. (1994) Species traits in relation to a habitat templet for river systems. *Freshwater Biology 31*, 265–75.

Trager, J.C., MacGown, J.A., Trager, M.D. (2007) Revision of the Nearctic endemic *Formica pallidefulva* group. *Memoirs of the American Entomological Institute 80*, 610–36.

Traniello, J.F.A. (1987) Comparative foraging ecology of north temperate ants: the role of worker size and cooperative foraging in prey selection. *Insectes Sociaux 34*, 118–30.

——(1989) Foraging strategies of ants. *Annual Review of Entomology 34*, 191–210.

——and Beshers, S.N. (1991) Maximization of foraging efficiency and resource defense by group retrieval in the ant *Formica schaufussi*. *Behavioral Ecology and Sociobiology 29*, 283–9.

——and Levings, S.C. (1986) Intra- and intercolony patterns of nest dispersion in the ant *Lasius neoniger*: correlations with territoriality and foraging ecology. *Oecologia 69*, 413–9.

Travassos, M.A., and Pierce, N.E. (2000) Acoustics, context and function of vibrational signalling in a lycaenid butterfly—ant mutualism. *Animal Behaviour 60*, 13–26.

Travis, B.V. (1941) Notes on the biology of the fire ant *Solenopsis geminata* (F.) in Florida and Georgia. *Florida Entomologist* 24, 15–22.

Tremper, B.S. (1976) *Distribution of the Argentine ant,* Iridomyrmex humilis *Mayr, in relation to certain native ants of California: ecological, physiological, and behavioral aspects.* Doctoral dissertation, University of California, Berkeley.

Treseder, K.K., Davidson, D.W., and Ehleringer, J.R. (1995) Absorption of ant-provided carbon dioxide and nitrogen by a tropical epiphyte. *Nature 375*, 137–9.

Tripet, F., and Nonacs, P. (2004) Foraging for work and age-based polyethism: the roles of age and previous experience on task choice in ants. *Ethology 110*, 863–77.

Trunzer, B., Heinze, J., and Hölldobler, B. (1998) Cooperative colony founding and experimental primary polygyny in the ponerine ant *Pachycondyla villosa*. *Insectes Sociaux 45*, 267–76.

Tschinkel, W.R. (1988a) Colony growth and the ontogeny of worker polymorphism in the fire ant *S. invicta*. *Behavioral Ecology and Sociobiology 22*, 103–15.

——(1988b) Distribution of the fire ants *Solenopsis invicta* and *S. geminata* (Hymenoptera: Formicidae) in northern Florida in relation to habitat and disturbance. *Annals of the Entomological Society of America 81*, 76–81.

——(1991) Insect sociometry, a field in search of data. *Insectes Sociaux 38*, 77–82.

——(1993) Sociometry and sociogenesis of colonies of the fire ant *Solenopsis invicta* during one annual cycle. *Ecological Monographs 64*, 425–57.

——(1998) Sociometry and sociogenesis of colonies of the harvester ant, *Pogonomyrmex badius*: worker characteristics in relation to colony size and season. *Insectes Sociaux 45*, 385–410.

——(2002) The natural history of the arboreal ant, *Crematogaster ashmaedi*. *Journal of Insect Science 2*, 12, available online: insectscience.org/2.12.

——(2003) Subterranean ant nests: trace fossils past and future? *Palaeogeography, Palaeoclimatology, Palaeoecology 192*, 321–33.

——(2004) The nest architecture of the Florida harvester ant, *Pogonomyrmex badius*. *Journal of Insect Science 4*, 21, available online: insectscience.org/4.21.

——(2006) *The Fire Ants.* The Belknap Press of Harvard University Press, Cambridge, MA.

——and Howard, D.F. (1983) Colony founding by pleometrosis in the fire ant, *Solenopsis invicta*. *Behavioral Ecology and Sociobiology 12*, 103–13.

Tsuji, K. (1996) Queen-male conflict over sperm use in social insects. *Trends in Ecology and Evolution 11*, 490–1.

Tsutsui, N.D. (2004) Scents of self: the expression component of self/non-self recognition systems. *Annales Zoologici Fennici 41*, 713–27.

——Suarez, A.V., and Grosberg, R.K. (2003) Genetic diversity, asymmetrical aggression, and recognition in a widespread invasive species. *Proceedings of the National Academy of Sciences USA 100*, 1078–83.

————Holway, D.A., and Case, T.J. (2000) Reduced genetic variation and the success of an invasive species. *Proceedings of the National Academy of Sciences USA 97*, 5948–53.

————(2001) Relationships among native and introduced populations of the Argentine ant (*Linepithema humile*) and the source of introduced populations. *Molecular Ecology 10*, 2151–61.

Turillazzi, S., Dapporto, L., Pansolli, C., Boulay, R., Dani, F.R., Moneti, G., and Pieraccini, G. (2006) Habitually used hibernation sites of paper wasps are marked with venom and cuticular peptides. *Current Biology 16*, R530–R531.

Turnbull, C.L., and Culver, D.C. (1983) The timing of seed dispersal in *Viola nuttallii*: attraction of dispersers and avoidance of predators. *Oecologia 59*, 360–5.

Ugelvig, L.V., Drijfhout, F.P., Kronauer, D.J.C., Boomsma, J.J., Pedersen, J.S., and Cremer, S. (2008) The introduction history of invasive garden ants in Europe: integrating genetic, chemical and behavioral approaches. *BMC Biology 6*, 11, doi:10.1186/1741-7007-6-11.

UK Biodiversity Action Plan (2007) *Species Action Plan. Red barbed ant, Formica rufibarbis*. http://www.ukbap.org.uk/UKPlans.aspx?ID = 312.

Umphrey, G.J. (2006) Sperm parasitism in ants: selection for interspecific mating and hybridization. *Evolution 87*, 2148–59.

Underwood, A.J. (1986) The analysis of competition by field experiments. In J. Kikkawa, and D.J. Anderson, eds. *Community Ecology: pattern and process*, pp. 240–68. Blackwell Publishing, Melbourne.

Underwood, E.C., and Fisher, B.L. (2006) The role of ants in conservation monitoring: if, when, and how. *Biological Conservation 132*, 166–82.

U.S. Census Bureau, Population Division (2008) World population 1950–2050. http://www.census.gov/ipc/www/idb/worldpopgraph.html.

U.S. Environmental Protection Agency (2000). Pesticide fact sheet: indoxacarb. Office of Prevention, Pesticides, and Toxic Substances. http://www.epa.gov/opprd001/factsheets/indoxacarb.pdf

Valone, T.J., and Kaspari, M. (2005) Interactions between granivorous and omnivorous ants in a desert grassland: results from a long-term experiment. *Ecological Entomology 30*, 116–21.

Van Bael, S.A., Philpott, S.M., Greenberg, R., Bichier, P., Barber, N.A., Mooney, K.A., and Gruner, D.S (2008) Birds as predators in tropical agroforestry systems. *Ecology 89*, 928–934.

van Borm, S., Billen, J., and Boomsma, J.J. (2002) The diversity of microorganisms associated with *Acromyrmex* leafcutter ants. *BMC Evolutionary Biology* 2, 9, doi:10.1186/1471-2148-2-9.

Van Mele, P., and Cuc, N.T.T. (2003) *Ants as Friends: improving your tree crops with weaver ant*. CABI Bioscience, UK.

Van Schagen, J.J., Davis, P.R., and Widmer, M.A. (1994) Ant pests of Western Australia, with particular reference to the Argentine Ant (*Linepithema humile*). In D.F. Williams, ed. *Exotic Ants: biology, impact, and control of introduced species*, pp. 174–80. Westview Press, Boulder, CO.

van Wilgenburg, E., and Elgar, M.A. (2007) Colony structure and spatial distribution of food resources in the polydomous meat ant *Iridomyrmex purpureus*. *Insectes Sociaux 54*, 5–10.

——van Lieshout, E., and Elgar, M.A. (2005) Conflict resolution strategies in meat ants (*Iridomyrmex purpureus*): ritualised displays versus lethal fighting. *Behaviour 142*, 701–16.

——Dang, S., Forti, A.-L., Koumoundouros, T., Ly, A., and Elgar, M. (2007) An absence of aggression between non-nestmates in the bull ant *Myrmecia nigriceps*. *Naturwissenschaften 94*, 787–90.

van Zweden, J.S., Carew, M.E., Henshaw, M.T., Robson, S. K.A., and Crozier, R.H. (2007) Social and genetic structure of a supercolonial weaver ant, *Polyrhachis robsoni*, with dimorphic queens. *Insectes Sociaux 54*, 34–41.

Vander Meer, R.K., and Morel, L. (1998) Nestmate recognition in ants. In R.K. Vander Meer, M.D. Breed, K. Espelie, and M.L. Winston, eds. *Pheromone Communication in Social Insects. Ants, wasps, bees and termites*, pp. 79–103. Westview Press, Boulder, CO.

——Jaffe, K., and Cedeno, A. (1990) *Applied Myrmecology: a world perspective*. Westview Press, Boulder, CO.

Vandermeer, J., and Perfecto, I. (2007) The diverse faces of ecosystem engineers in agroecosystems. In K.M.D. Cuddington, J.E. Byers, A. Hastings, and W.G. Wilson, eds. *Ecosystem Engineers: plants to protists*, pp. 367–86. Elsevier, New York.

——Granzow de la Cerda, I., Boucher, D., Perfecto, I., and Ruiz, J. (2000) Hurricane disturbance and tropical tree species diversity. *Science 290*, 788–91.

Vanderwoude, C., Lobry de Bruyn, L.A., and House, A.P. N. (2000) Response of an open-forest ant community to invasion by the introduced ant, *Pheidole megacephala*. *Austral Ecology 25*, 253–9.

——Elson Harris, M., Hargreaves, J.R., Harris, E.J., and Plowman, K.P. (2003) An overview of the red imported fire ant (*Solenopsis invicta* Buren) eradication plan for Australia. *Records of the South Australian Museum Monograph Series 7*, 11–6.

Vantaux, A., Dejean, A., Dor, A., and Orivel, J. (2007) Parasitism versus mutualism in the ant-garden parabiosis between *Camponotus femoratus* and *Crematogaster levior*. *Insectes Sociaux 54*, 95–9.

Vasconcelos, H.L. (1999a) Effects of forest disturbance on the structure of ground-foraging ant communities in central Amazonia. *Biodiversity and Conservation 8*, 409–20.

——(1999b) Forest fragmentation in central Amazonia and its effects on litter-dwelling ants. *Biological Conservation 91*, 51–7.

——and Cherrett, J.M. (1997) Leaf-cutting ants and early forest succession in central Amazonia: effects of herbivory on tree seedling establishment. *Journal of Tropical Ecology 13*, 357–70.

——and Fowler, H.G. (1990) Foraging and fungal substrate selection by leaf-cutting ants. In R.K. Vander Meer, K. Jaffé, and A. Cedeño, eds. *Applied Myrmecology: a world perspective*, pp. 410–9. Westview Press, Boulder, CO.

——Carvalho, K.S., and Delabie, J.C. (2001) Landscape modifications and ant communities. In R.O. Bierregaard Jr., C. Gascon, T.E. Lovejoy, and R. Mesquita, eds. *Lessons from Amazonia: the ecology and conservation of a fragmented forest*, pp. 199–207. Yale University Press, New Haven.

——Vilhena, J.M.S., and Caliri, G.J.A. (2000) Responses of ants to selective logging of a central Amazonian forest. *Journal of Applied Ecology 37*, 508–14.

——Vieira-Neto, E.H.M., Mundim, F.M., and Bruna, E.M. (2006) Roads alter the colonization dynamics of a keystone herbivore in neotropical savannas. *Biotropica 38*, 661–5.

Vásquez-Bolaños, M. (2007) Una especie nueva del género *Tetramorium* Mayr (Hymenoptera: Formicidae) de Mascota, Jalisco, México. *Dugesiana 14*(2), 93–7.

Vasquez, G.M., and Silverman, J. (2008) Intraspecific aggression and colony fusion in the Argentine ant. *Animal Behaviour 75*, 583–93.

Veeresh, G.K. (1990) Pest ants of India. In R.K. Vander Meer, K. Jaffe, and A. Cedeno, eds. *Applied Myrmecology: a world perspective*, pp. 15–23. Westview Press, Boulder, CO.

Vega, F.E., and Blackwell, M. (2005) Conclusion: symbioses, biocomplexity, and metagenomes. In F. Vega, and M. Blackwell, eds. *Insect-Fungal Association: ecology and evolution*, pp. 293–5. Oxford University Press, New York.

Vepsäläinen, K., and Pisarski, B. (1982) Assembly of island ant communities. *Annales Zoologici Fennici 19*, 327–35.

——Ikonen, H., and Koivula, M.J. (2008) The structure of ant assemblages in an urban area of Helsinki, southern Finland. *Annales Zoologici Fennici 45*, 109–27.

——Savolainen, R., Tiainen, J., and Vilén, J. (2000) Successional changes of ant assemblages: from virgin and ditched bogs to forests. *Annales Zoologici Fennici 37*, 135–49.

Viana, A.M.M., Frézard, A., Malosse, C., Della Lucia, T.M. C., Errard, C., and Lenoir, A. (2001) Colonial recognition of fungus in the fungus-growing ant *Acromyrmex subterraneus subterraneus* (Hymenoptera: Formicidae). *Chemoecology 11*, 29–36.

Vieira-Neto, E.H.M., Mundim, F.M., and Vasconcelos, H. L. (2006) Hitchhiking behaviour in leaf-cutter ants: an experimental evaluation of three hypotheses. *Insectes Sociaux 53*, 326–32.

Villesen, P., Murakami, T., Schultz, T.R., and Boomsma, J. J. (2002) Identifying the transition between single and multiple mating of queens in fungus-growing ants. *Proceedings of the Royal Society of London B 269*, 1541–8.

Villet, M.H. (1991) Colony foundation in *Plectroctena mandibularis* F. Smith, and the evolution of ergatoid queens in *Plectroctena* (Hymenoptera: Formicidae). *Journal of Natural History 25*, 979–83.

Visscher, P.K. (2007) Group decision-making in nest-site selection among social insects. *Annual Review of Entomology 52*, 255–75.

Visser, D., Wright, M.G., and Giliomee, J.H. (1996) The effect of the Argentine ant, *Linepithema humile* (Mayr) (Hymenoptera: Formicidae), on flower-visiting insects of *Protea nitida* Mill. (Proteaceae). *African Entomology 4*, 285–7.

Vogt, J.T., Rice, S.A., and Armstrong, S.A. (2003) Seed preferences of the red imported fire ant (Hymenoptera: Formicidae) in Oklahoma. *Journal of Entomological Science 38*, 696–698.

——Grantham, R.A., Corbett, E., Rice, S.A., and Wright, R. E. (2002) Dietary habits of *Solenopsis invicta* (Hymenoptera: Formicidae) in four Oklahoma habitats. *Environmental Entomology 31*, 47–53.

Völkl, W., Woodring, J., Fisher, M., Lorenz, M.W., and Hoffmann, K.H. (1999) Ant-aphid mutualisms: the impact of honeydew production and honeydew sugar composition on ant-preferences. *Oecologia 118*, 483–91.

von Humboldt, A. (1808) *Ansichten der natur mit wissenschaftlichen Erlauterungen*. J.G. Cotta, Tubingen, Germany.

Vos, P., Hogers, R., and Bleeker, M., Reijans, M., Van de Lee, T., Hornes, M., Frijters, A., Pot, J., Peleman, J., Kuiper, M., and Zabeau, M. (1995) AFLP: a new technique for DNA fingerprinting. *Nucleic Acids Research 23*, 4407–14.

Wäckers, F.L. (2001) A comparison of nectar- and honeydew sugars with respect to their utilization by the hymenopteran parasitoid *Cotesia glomerata*. *Journal of Insect Physiology 47*, 1077–84.

Wagner, D. (1997) The influence of ant nests on *Acacia* seed production, herbivory and soil nutrients. *Journal of Ecology 85*, 83–93.

——and Gordon, D.M. (1999) Colony age, neighborhood density and reproductive potential in harvester ants. *Oecologia 119*, 175–82.

——and Kay, A. (2002) Do extrafloral nectaries distract ants from visiting flowers? An experimental test of an overlooked hypothesis. *Evolutionary Ecology Research 4*, 293–305.

——and Kurina, L. (2003) The influence of ants and water availability on oviposition behaviour and survivorship of a facultatively ant-tended herbivore. *Ecological Entomology 22*, 352–60.

——Jones, J.B., and Gordon, D.M. (2004) Development of harvester ant colonies alters soil chemistry. *Soil Biology & Biochemistry 36*, 794–804.

——Brown, M.J.F., Broun, P., Cuevas, W., Moses, L.E., Chao, D.L., and Gordon, D.M. (1998) Task-related differences in the cuticular hydrocarbon composition of harvester ants, *Pogonomyrmex barbatus*. *Journal of Chemical Ecology 24*, 2021–2037.

——Tissot, M., Cuevas, W., and Gordon, D.M. (2000) Harvester ants utilize cuticular hydrocarbons in nestmate recognition. *Journal of Chemical Ecology 26*, 2245–57.

Waldman, B. (1987) Mechanisms of kin recognition. *Journal of Theoretical Biology 128*, 159–85.

Walin, L., Seppä, P., and Sundström, L. (2001) Reproductive allocation within a polygyne, polydomous colony of the ant *Myrmica rubra*. *Ecological Entomology 26*, 537–46.

Walker, K.L. (2006) Impact of the little fire ant, *Wasmannia auropunctata*, on native forest ants in Gabon. *Biotropica 38*, 666–73.

Wallace, R. (1855) On the law which has regulated the introduction of new species. *Annals and Magazine of Natural History 16*, 184–96.

Waloff, N., and Blackith, R.E. (1962) The growth and distribution of the mounds of *Lasius-flavus* (Fabricius) (Hym, Formiciade) in Silwood Park, Berkshire. *Journal of Animal Ecology 31*, 421–37.

Walsh, P.D., Henschel, P., Abernethy, K.A., Tutin, C.E.G., Telfer, P., and Lahm, S.A. (2004) Logging speeds little red fire ant invasion of Africa. *Biotropica 36*, 637–41.

Walters, A.C. (2006) Invasion of Argentine ants (Hymenoptera: Formicidae) in South Australia: impacts on community composition and abundance of invertebrates in urban parklands. *Austral Ecology 31*, 567–76.

——and Mackay, D.A. (2003a) An experimental study of the relative humidity preference and survival of the Argentine ant, *Linepithema humile* (Hymenoptera, Formicidae): comparisons with a native *Iridomyrmex* species in South Australia. *Insectes Sociaux 50*, 355–60.

————(2003b) The impact of the Argentine ant, *Linepithema humile* (Mayr) on native ants and other invertebrates in South Australia. *Records of the South Australian Museum Monograph Series 7*, 17–24.

————(2004) Comparisons of upper thermal tolerances between the invasive Argentine ant (Hymenoptera: Formicidae) and two native Australian ant species. *Annals of the Entomological Society of America 97*, 971–5.

————(2005) Importance of large colony size for successful invasion by Argentine ants (Hymenoptera: Formicidae): evidence for biotic resistance by native ants. *Austral Ecology 30*, 395–406.

Walters, C.J., and Holling, C.S. (1990) Large-scale management experiments and learning by doing. *Ecology 71*, 2060–8.

Walther, G.R., Post, E., Convey, P., Menzel, A., Parmesan, C., Beebee, T.J.C., Fromentin, J.M., Hoegh-Guldberg, O., and Bairlein, F. (2002) Ecological responses to recent climate change. *Nature 416*, 389–95.

Wandeler, P., Hoeck, P.E.A., and Keller, L.F. (2007) Back to the future: museum specimens in population genetics. *Trends in Ecology and Evolution 22*, 634–42.

Ward, D.F. (2007a) Modelling the potential geographic distribution of invasive ant species in New Zealand. *Biological Invasions 9*, 723–35.

——and Beggs, J. (2007) Coexistence, habitat patterns and the assembly of ant communities in the Yasawa islands, Fiji. *Acta Oecologica 32*, 215–23.

——and Harris, R.J. (2005) Invasibility of native habitats by Argentine ants, *Linepithema humile*, in New Zealand. *New Zealand Journal of Ecology 29*, 215–9.

——and Wetterer, J.K. (2006) Checklist of the ants of Fiji (Hymenoptera: Formicidae). *British Museum Occasional Papers, Fiji Arthropods III 85*, 23–47.

——Beggs, J.R., Clout, M.N., Harris, R.J., and O'Connor, S. (2006) The diversity and origin of exotic ants arriving in New Zealand via human-mediated dispersal. *Diversity and Distributions 12*, 601–9.

Ward, P.S. (1987) Distribution of the introduced Argentine ant (*Iridomyrmex humilis*) in natural habitats of the lower Sacramento Valley and its effects on the indigenous ant fauna. *Hilgardia 55*(2), 1–16.

——(1990) The endangered ants of Mauritius: doomed like the Dodo? *Notes from the Underground 4*, 3–5.

——(1991) Phylogenetic analysis of pseudomyrmecine ants associated with domatia-bearing plants. In C.R. Huxley and D.F. Cutler, eds. *Ant-Plant Interactions*, pp. 335–52. Oxford University Press, Oxford.

——(2000) Broad-scale patterns of diversity in leaf litter ant communities. In D. Agosti, J.D. Majer, L.E. Alonso, and T.R. Schultz, eds. *Ants: standard methods for measuring and monitoring biodiversity*, pp. 99–121. Smithsonian Institution Press, Washington, DC.

——(2001) Taxonomy, phylogeny and biogeography of the ant genus *Tetraponera* (Hymenoptera: Formicidae) in the Oriental and Australian regions. *Invertebrate Taxonomy 15*, 589–665.

——(2006) The ant genus *Tetraponera* in the Afrotropical region: synopsis of species groups and revision of the *T. ambigua*-group (Hymenoptera: Formicidae). *Myrmecologische Nachrichten 8*, 119–30.

——(2007b) The ant genus *Leptanilloides*: discovery of the male and evaluation of phylogenetic relationships based on DNA sequence data. *Memoirs of the American Entomological Institute 80*, 637–49.

——(2007c) Phylogeny, classification, and species-level taxonomy of ants (Hymenoptera: Formicidae). *Zootaxa 1668*, 549–63.

——and Brady, S.G. (2003) Phylogeny and biogeography of the ant subfamily Myrmeciinae (Hymenoptera: Formicidae). *Invertebrate Systematics 17*, 361–86.

——and Downie, D.A. (2005) The ant subfamily Pseudomyrmecinae (Hymenoptera: Formicidae): phylogeny and evolution of big-eyed arboreal ants. *Systematic Entomology 30*, 310–35.

——Brady, S.G., Fisher, B.L., and Schultz, T.R. (2009) Phylogeny and biogeography of dolichoderine ants: effects of data partitioning and relict taxa on historical inference. *Systematic Biology*, in press.

Way, M.J. (1963) Mutualism between ants and honeydew-producing Homoptera. *Annual Review of Entomology 8*, 307–44.

——and Khoo, K.C. (1992) Role of ants in pest management. *Annual Review of Entomology 37*, 479–503.

——Javier, G., and Heong, K.L. (2002) The role of ants, especially the fire ant, *Solenopsis geminata* (Hymenoptera: Formicidae), in the biological control of tropical upland rice pests. *Bulletin of Entomological Research 92*, 431–437.

——Cammell, M.E., Paiva, M.R., and Collingwood, C.A. (1997) Distribution and dynamics of the Argentine ant *Linepithema* (*Iridomyrmex*) *humile* (Mayr) in relation to vegetation, soil conditions, topography and native competitor ants in Portugal. *Insectes Sociaux 44*, 415–33.

Webber, B.L., Moog, J., Curtis, A.S.O., and Woodrow, I.E. (2007) The diversity of ant-plant interactions in the rainforest understorey tree, *Ryparosa* (Achariaceae): food bodies, domatia, prostomata, and hemipteran trophobionts. *Botanical Journal of the Linnean Society 154*, 353–71.

Weber, N.A. (1943) The ants of the Imatong Mountains, Anglo-Egyptian Sudan. *Bulletin of the Museum of Comparative Zoology 93*, 263–389.

Weeks, R.D., Wilson, L.T., Vinson, S.B., and James, W.D. (2004) Flow of carbohydrates, lipids, and protein among colonies of polygyne red imported fire ants, *Solenopsis invicta* (Hymenoptera: Formicidae). *Annals of the Entomological Society of America 97*, 105–10.

Wehner, R. (1987) Spatial organization of foraging behavior in individually searching desert ants, *Cataglyphis* (Sahara desert) and *Ocymyrmex* (Namib desert). *Experientia Supplementum 54*, 15–42.

——and Menzel, R. (1969) Homing in the ant *Cataglyphis bicolor*. *Science 164*, 192–4.

Weir, J.T., and Schluter, D. (2007) The latitudinal gradient in recent speciation and extinction rates of birds and mammals. *Science 315*, 1574–6.

Weiser, M., and Kaspari, M. (2006) Ecological morphospace of new world ants. *Ecological Entomology 31*, 131–42.

Weissflog, A. (2001) *Freinestbau von Ameisen (Hymenoptera: Formicidae) in der Kronenregion feuchttropischer Wälder Südostasiens*. Doctoral thesis, J.W. Goethe Universität, Frankfurt.

Wells, S., Pyle, R.M., Collins, N.M, and IUCN (compilers). 1983. *The IUCN Invertebrate Red Data Book*. International Union for Conservation of Nature, Gland, Switzerland.

Wenseleers, T. (2007) Nepotism absent in insect societies – or is it? *Molecular Ecology 16*, 3063–5.

——and Ratnieks, F.L.W. (2006) Comparative analysis of worker reproduction and policing in eusocial Hymenoptera supports relatedness theory. *American Naturalist 168*, E163–79.

——Billen, J., and Hefetz, A. (2002) Territorial marking in the desert ant *Cataglyphis niger*: does it pay to play bourgeois? *Journal of Insect Behavior 15*, 85–93.

Werner, E.E. (1976) Species interactions in freshwater fish communities. In J. Diamond, and T.J. Case, eds. *Community ecology*, pp. 344–57. Harper and Row, New York.

West, G.B., Brown, J.H., and Enquist, B.J. (1999) The fourth dimension of life: fractal geometry and allometric scaling of organisms. *Science 284*, 167–9.

Western Australian Department of Agriculture. (1988) Argentine ant control annual report for the year ending 30 June 1988.

Westoby, M., Hughes, L., and Rice, B.L. (1991b) Seed dispersal by ants; comparing infertile soils with fertile soils. In C.R. Huxley, and D.F. Cutler, eds. *Ant-Plant Interactions*, pp. 434–47 Oxford University Press, Oxford.

Westoby, M., French, K., Hughes, L., Rice, B., and Rodgerson, L. (1991a) Why do more plant species use ants for dispersal on infertile compared with fertile soils? *Australian Journal of Ecology 16*, 445–55.

Wetterer, J.K. (1995) Forager size and ecology of *Acromyrmex coronatus* and other leaf-cutting ants in Costa Rica. *Oecologia 104*, 409–15.

——(2002) Ants of Tonga. *Pacific Science 56*, 125–35.

——(2006) The vanished plague ants (Hymenoptera: Formicidae) of 19th century Bermuda. *Myrmecologische Nachrichten 8*, 219–24.

——(2007) Biology and impacts of Pacific Island invasive species. 3. The African big-headed ant, *Pheidole megacephala* (Hymenoptera: Formicidae). *Pacific Science 61*, 437–56.

——and Porter, S.D. (2003) The little fire ant, *Wasmannia auropunctata*: distribution, impact, and control. *Sociobiology 42*, 1–41.

——and Wetterer, A.L. (2003) Ants (Hymenoptera: Formicidae) on non-native neotropical ant-acacias (Fabales: Fabaceae) in Florida. *Florida Entomologist 86*, 460–3.

————(2006) A disjunct Argentine ant metacolony in Macaronesia and southwestern Europe. *Biological Invasions 8*, 1123–9.

——Miller, S.E., and Wheeler, D.E. (1999) Ecological dominance by *Paratrechina longicornis* (Hymenoptera: Formicidae), an invasive tramp ant, in Biosphere 2. *Florida Entomologist 82*, 381–8.

——Wood, L.D., Johnson, C., Krahe, H., and Fitchett, S. (2007) predacious ants, beach replenishment, and nest placement by sea turtles. *Environmental Entomology 36*, 1084–91.

——Espadaler, X., Wetterer, A.L., Aguin-Pombo, D., and Franquinho-Aguiar, A.M. (2006) Long-term impact of exotic ants on the native ants of Madeira. *Ecological Entomology 31*, 358–68.

Wheeler, D.E (1996) The role of nourishment in oogenesis. *Annual Review of Entomology 41*, 407–31.

——(1994) Nourishment in ants: patterns in individuals and societies. In J.H. Hunt, and C.A. Nalepa, eds. *Nourishment and Evolution in Insect Societies*, pp. 245–78. Westview Press, Boulder, CO.

——and Martínez, T. (1995) Storage proteins in ants (Hymenoptera: Formicidae). *Journal of Comparative Biochemistry and Physiology B 112*, 15–9.

Wheeler, W.M. (1908) A European ant (*Myrmica levinodis*) introduced into Massachusetts. *Journal of Economic Entomology 1*, 337–9.

——(1910) *Ants*. Columbia University Press, New York.

White, T.C.R. (1993) *The Inadequate Environment: nitrogen and the abundance of animals*. Springer, New York.

White, W.H., Reagan, T.E., Smith Jr., J.W., and Salazar, J.A. (2004) Refuge releases of *Cotesia flavipes* (Hymenoptera: Braconidae) into the Louisiana sugarcane ecosystem. *Environmental Entomology 33*, 627–32.

Whitehouse, M.E.A., and Jaffe, K. (1996) Ant wars: combat strategies, territory and nest defence in the leaf-cutting ant *Atta laevigata*. *Animal Behaviour 51*, 1207–17.

Wiens, J.A. (1989) Spatial scaling in ecology. *Functional Ecology 3*, 385–97.

Wiens, J.J., and Graham, C.H. (2005) Niche conservatism: integrating evolution, ecology, and conservation biology. *Annual Review of Ecology Evolution and Systematics 36*, 519–39.

Wiernasz, D.C., and Cole, B.J. (1995) Spatial-distribution of *Pogonomyrmex occidentalis* recruitment, mortality and overdispersion. *Journal of Animal Ecology 64, 36*, 519–27.

————(2003) Queen size mediates queen survival and colony fitness in harvester ants. *Evolution 57*, 2179–83.

——Perroni, C.L., and Cole, B.J. (2004) Polyandry and fitness in the western harvester ant, *Pogonomyrmex occidentalis*. *Molecular Ecology 13*, 1601–6.

Wild, A.L. (2005) Observations on larval cannibalism and other behaviors in a captive colony of *Amblyopone oregonensis*. *Notes from the Underground 11(1)*, 9.

Wild, A.L. (2007a). A catalogue of the ants of Paraguay. *Zootaxa 1622*, 1–55.

——(2007b) Taxonomic revision of the ant genus *Linepithema* (Hymenoptera: Formicidae). *University of California Publications in Entomology 126*, 1–151 http://repositories.cdlib.org/ucpress/ucpe/vol_126.

——(2009) Evolution of the Neotropical ant genus *Linepithema*. *Systematic Entomology 34*, 49–62.

Wilkinson, E.B., and Feener, D.H. (2007) Habitat complexity modifies ant-parasitoid interactions: implications for community dynamics and the role of disturbance. *Oecologia 152*, 151–61.

Williams, D.F. (1983) The development of toxic baits for the control of the imported fire ant. *Florida Entomologist 66*, 162–72.

——(1994) *Exotic Ants: biology, impact, and control of introduced species*. Westview Press, Boulder, CO.

——and Vail, K.M. (1994) Control of a natural infestation of the pharaoh ant (Hymenoptera: Formicidae) with a corn grit bait of fenoxycarb. *Journal of Economic Entomology 87*, 108–15.

——Collins, H.L., and Oi, D.H. (2001) The red imported fire ant (Hymenoptera: Formicidae): an historical perspective of treatment programs and the development of chemical baits for control. *American Entomologist 47*, 146–59.

——Knue, G.J., Becnel, J.J. (1998) Discovery of *Thelohania solenopsae* from the red imported fire ant, *Solenopsis invicta*, in the United States. *Journal of Invertebrate Pathology 71*, 175–6.

——Oi, D.H., and Knue, G.J. (1999) Infection of red imported fire ant (Hymenoptera: Formicidae) colonies with the entomopathogen *Thelohania solenopsae* (Microsporidia: Thelohaniidae). *Journal of Economic Entomology 92*, 830–6.

——Lofgren, C.S., Banks, W.A., Stringer, W.A., and Plumley, J.K. (1980) Laboratory studies with 9 amidinohydrazones, a promising new class of bait toxicants for control of the red imported fire ants (Hymenopera: Formicidae). *Journal of Economic Entomology 73*, 798–802.

Williamson, M. (1996) *Biological Invasions*. Chapman & Hall, London.

Willig, M.R., and Lyons, S.K. (1998) An analytical model of latitudinal gradients of species richness with an empirical test for marsupials and bats in the New World. *Oikos 81*, 93–8.

——Kaufman, D.M., and Stevens, R.D. (2003) Latitudinal gradients of biodiversity: pattern, process, scale, and synthesis. *Annual Review of Ecology Evolution and Systematics 34*, 273–309.

Willmer, P.G., and Stone, G.N. (1997) How aggressive ant-guards assist seed-set in *Acacia* flowers. *Nature 388*, 165–7.

Wilson, D.S., and Wilson, E.O. (2007) Rethinking the theoretical foundation of sociobiology. *Quarterly Review of Biology 82*, 327–48.

Wilson, E.O. (1951) Variation and adaptation in the imported fire ant. *Evolution 5*, 68–79.

——(1959) Adaptive shift and dispersal in a tropical ant fauna. *Evolution 13*, 122–44.

——(1961) The nature of the taxon cycle in the Melanesian ant fauna. *American Naturalist 95*, 169–93.

——(1971) *The Insect Societies*. The Belknap Press of Harvard University Press, Cambridge, MA.

——(1976) The organization of colony defense in the ant *Pheidole dentata* Mayr (Hymenoptera: Formicidae). *Behavioral Ecology and Sociobiology 1*, 63–81.

——(1980) Caste and division of labor in leaf-cutter ants: I. The overall pattern in *Atta sexdens*. *Behavioral Ecology and Sociobiology 7*, 143–56.

——(1983) Caste and division of labor in leaf-cutter ants (Hymenoptera: Formicidae: *Atta*) IV. Colony ontogeny of *A. cephalotes*. *Behavioral Ecology and Sociobiology 14*, 55–60.

——(1984) Tropical social parasites in the ant genus *Pheidole*, with an analysis of the anatomical parasitic syndrome (Hymenoptera: Formicidae). *Insectes Sociaux 31*, 316–34.

——(1987a) Causes of ecological success: the case of ants. *Journal of Animal Ecology 56*, 1–9.

——(1987b) The little things that run the world (The importance and conservation of invertebrates). *Conservation Biology 1*, 344–6.

——(1988) The biogeography of the West Indian Ants (Hymenoptera: Formicidae). In J.K. Liebherr, ed. *Zoogeography of Caribbean Insects*, pp. 214–30. Cornell University Press, Ithaca.

——(2003) Pheidole *in the New World. A dominant, hyperdiverse ant genus*. Harvard University Press, Cambridge, MA.

——(2005) Early ant plagues in the New World. *Nature 433*, 32.

——(2006) *The Creation*. W.W. Norton and Company, New York.

——and Hölldobler, B. (2005) The rise of the ants: a phylogenetic and ecological explanation. *Proceedings of the National Academy of Sciences USA 102*, 7411–4.

Wilson, E.O., Carpenter, F.M., and Brown Jr., W.L., (1967) The first Mesozoic ants, with the description of a new subfamily. *Psyche (Cambridge) 74*, 1–19.

Wilson, E.O., Eisner, T., Wheeler, G.C., and Wheeler, J. (1956) *Aneuretus simoni* Emery, a major link in ant evolution. *Bulletin of the Museum of Comparative Zoology 115*, 81–99.

Wimp, G.M., and Whitham, T.G. (2001) Biodiversity consequences of predation and host plant hybridization on an aphid-ant mutualism. *Ecology 82*, 440–52.

Winter, U., and Buschinger, A. (1986) Genetically mediated queen polymorphism and caste determination in the slave-making ant, *Harpagoxenus sublaevis* (Hymenoptera: Formicidae). *Entomologia Generalis 11*, 125–37.

Wirth, R., Herz, H., Ryel, R.J., Beyschlag, W., and Hölldobler, B. (2003) *Herbivory of Leaf-Cutting Ants: a case study on* Atta colombica *in the tropical rainforest*. Springer-Verlag, Heidelberg.

Witt, A.B.R., and Giliomee, J.H. (1999) Soil-surface temperatures at which six species of ants (Hymenoptera: Formicidae) are active. *African Entomology 7*, 161–4.

——(2005) Dispersal of elaiosome-bearing seeds of six plant species by native species of ants and the introduced invasive ant, *Linepithema humile* (Mayr) (Hymenoptera: Formicidae) in the Western Cape Province, South Africa. *African Plant Protection 11*, 1–7.

Witte, V., and Maschwitz, U. (2008) Mushroom harvesting ants in the tropical rain forest. *Naturwissenschaften 95*, 1049–54.

Wittlinger, M., Wehner, R., and Wolf, H. (2006) The ant odometer: stepping on stilts and stumps. *Science 312*, 1965–7.

Wittlinger, M., Wehner, R., and Wolf, H. Wehner, R., and Wolf, H. (2007) The desert ant odometer: a stride integrator that accounts for stride length and walking speed. *Journal of Experimental Biology 210*, 198–207.

Wojcik, D.P., and Porter, S.D. (2008) FORMIS: a master bibliography of ant literature. USDA-ARS, CMAVE, Gainesville, FL (http://fm.cits.fcla.edu/cgi-bin/wlv4/DBFM)

——Smittle, B.J., and Cromroy, H.L. (1991) Fire ant myrmecophiles: feeding relationships of *Martinezia dutertrei* and *Euparia castanea* (Coleoptera: Scarabaeidae) with their host ants, *Solenopsis* spp. (Hymenoptera: Formicidae). *Insectes Sociaux 38*, 273–81.

Woodd-Walker, R.S., Ward, P., and Clarke, A. (2002) Large-scale patterns in diversity and community structure of surface water copepods from the Atlantic Ocean. *Marine Ecology Progress Series 236*, 189–203.

Worm, B., Lotze, H.K., Hillebrand, H., Sommer, U. (2002) Consumer versus resource control of species diversity and ecosystem functioning. *Nature 417*, 848–51.

Woyciechowski, M., and Kozlowski, J. (1998) Division of labor by division of risk according to worker life expectancy in the honey bee (*Apis mellifera* L.). *Apidologie 29*, 191–205.

Wright, D.H. (1983) Species-energy theory: an extension of species-area theory. *Oikos 41*, 496.

Wuellner, C.T., and Saunders, J.B. (2003) Circadian and circannual patterns of activity and territory shifts: comparing a native ant (*Solenopsis geminata*, Hymenoptera: Formicidae) with its exotic, invasive congener (*S. invicta*) and its parasitoids (*Pseudacteon* spp., Diptera: Phoridae) at a central Texas site. *Annals of the Entomological Society of America 96*, 54–60.

Wyatt, T.D. (2003) *Pheromones and Animal Behaviour*. Cambridge University Press, Cambridge.

Wyckhuys, K.A.G., and O'Neil, R.J. (2007) Influence of extra-field characteristics to abundance of key natural enemies of *Spodoptera frugiperda* Smith (Lepidoptera: Noctuidae) in subsistence maize production. *International Journal of Pest Management 53*, 89–99.

Wynalda, R.A. (2008) *Nutrient regulation of an exotic, unidentified* Paratrechina *sp. (Hymenoptera: Formicidae) found in Texas*. Master of Science thesis, Texas A&M University, College Station.

Xiong, Y., Chen, J-D., Gu, X-Y., Wu, X-H., Wan, F-H., and Hong, X-Y. (2008) The potential suitability of Jiangsu Province, east China for the invasive red imported fire ant. *Solenopsis invicta. Biological Invasions 10*, 475–81.

Xu, Z.H. (2003) A systematic study on Chinese species of the ant genus *Oligomyrmex* Mayr (Hymenoptera: Formicidae). *Acta Zootaxonomica Sinica 28*, 310–22.

——and Chai, Z.Q. (2004) Systematic study on the ant genus *Tetraponera* F. Smith (Hymenoptera: Formicidae) of China. *Acta Zootaxonomica Sinica 29*, 63–76.

——and Zhou, X.G. (2004) Systematic study on the ant genus *Pyramica* Roger (Hymenoptera, Formicidae) of China. *Acta Zootaxonomica Sinica 29*, 440–50.

Yamaguchi, T. (1995) Intraspecific competition through food robbing in the harvester ant, *Messor aciculatus* (Fr Smith), and its consequences on colony survival. *Insectes Sociaux 42*, 89–101.

——(2004) Influence of urbanization on ant distribution in parks of Tokyo and Chiba City, Japan I. Analysis of ant species richness. *Ecological Research 19*, 209–16.

Yamauchi, K., Furukawa, T., Kinomura, K., Takamine, H., and Tsuji, K. (1991) Secondary polygyny by inbred wingless sexuals in the dolichoderine ant *Technomyrmex albipes*. *Behavioral Ecology and Sociobiology 29*, 313–319.

Yang, A.S. (2006) Seasonality, division of labor, and dynamics of colony-level nutrient storage in the ant *Pheidole morrisi*. *Insectes Sociaux 53*, 456–62.

Yanoviak, S.P., and Dudley, R. (2006) The role of visual cues in directed aerial descent of *Cephalotes atratus* workers (Hymenoptera: Formicidae). *Journal of Experimental Biology 209*, 1777–83.

————and Kaspari, M. (2005) Directed aerial descent in canopy ants. *Nature 433*, 624–6.

——Fisher, B.A., and Alonso, A. (2008a) Directed aerial descent behaviour in African canopy ants (Hymenoptera: Formicidae). *Journal of Insect Behaviour 21*, 164–71.

——Kaspari, M., Dudley, R., and Poinar Jr., G. (2008b) Parasite-induced fruit mimicry in a tropical canopy ant. *American Naturalist 171*, 536–44.

Yao, I., Shibao, H., and Akimato, S. (2000) Costs and benefits of ant attendance to the drepanosiphid aphid *Tuberculatus quercicola*. *Oikos 89*, 3–10.

Ydenberg, R., and Schmid-Hempel, P. (1994) Modelling social insect foraging. *Trends in Ecology and Evolution 9*, 491–3.

Yéo, K., Molet, M., and Peeters, C. (2006) When David and Goliath share a home: compound nesting of *Pyramica* and *Platythyrea* ants. *Insectes Sociaux 53*, 435–8.

Yoshimura, M., and Fisher, B.L. (2007) A revision of male ants of Madagascar (Hymenoptera: Formicidae): key to subfamilies and treatment of the genera of Ponerinae. *Zootaxa 1654*, 21–40.

Young, A.M., and Hermann, H.R. (1980) Notes on foraging of the giant tropical ant *Paraponera clavata* (Hymenoptera: Formicidae: Ponerinae). *Journal of the Kansas Entomological Society 53*, 35–55.

Youngsteadt, E., Nojima, S., Häberlein, C., Schulz, S., and Schal, C. (2008) Seed odor mediates an obligate ant—plant mutualism in Amazonian rainforests. *Proceedings of the National Academy of Sciences USA 105*, 4571–5.

Yu, D.W., and Pierce, N.E. (1998) A castration parasite of an ant-plant mutualism. *Proceedings of the Royal Society of London B 265*, 375–81.

Zapata, F.A., Gaston, K.J., and Chown, S.L. (2003) Mid-domain models of species richness gradients: assumptions, methods and evidence. *Journal of Animal Ecology 72*, 677–90.

————(2005) The mid-domain effect revisited. *American Naturalist 166*, E144–8.

Zappalà, L., Hoy, M.A., and Cave, R.D. (2007) Interactions between the red imported fire ant, the citrus leafminer, and its parasitoid *Ageniaspis citricola* (Hymenoptera: Encyrtidae): laboratory and field evaluations. *Biocontrol Science and Technology 17*, 353–63.

Zavaleta, E.S., and Royval, J.L. (2002) Climate change and the susceptibility of U.S. ecosystems to biological invasions: two cases of expected range expansion. In S.H. Schneider, and T.L. Root, eds. *Wildlife Responses to Climate Change: North American case studies*, pp. 277–341. Island Press, Washington, DC.

Zayed, A., and Packer, L. (2005) Complementary sex determination substantially increases extinction proneness of haplodiploid populations. *Proceedings of the National Academy of Sciences USA 102*, 10742–6.

Zee, J., and Holway, D. (2006) Nest raiding by the invasive Argentine ant on colonies of the harvester ant, *Pogonomyrmex subnitidus*. *Insectes Sociaux 53*, 161–7.

Zelikova, T.J., Dunn, R.R., and Sanders, N.J. (2008) Variation in seed dispersal along an elevational gradient in Great Smoky Mountains National Park. *Acta Oecologica 34*, 155–62.

Zenner-Polania, I. (1994) Impact of *Paratrechina fulva* on other ant species. In D.F. Williams, ed. *Exotic Ants: biology, impact, and control of introduced species*, pp. 121–32. Westview Press, Boulder, CO.

Zera, A.J., and Denno, R.F. (1997) Physiology and ecology of dispersal polymorphism in insects. *Annual Review of Entomology 42*, 207–30.

Zettel, H. (2006) On the ants (Hymenoptera: Formicidae) of the Philippine Islands: I. The genus *Pristomyrmex* Mayr, 1866. *Myrmecologische Nachrichten 8*, 59–68.

Zettler, J.A., Spira, T.P., and Allen, C.R. (2001) Ant-seed mutualisms: can red imported fire ants sour the relationship? *Biological Conservation 101*, 249–53.

——Taylor, M.D., Allen, C.R., and Spira, T.P. (2004) Consequences of forest clear-cuts for native and nonindigenous ants (Hymenoptera: Formicidae). *Annals of the Entomological Society of America 97*, 513–8.

Zhou, H., Chen, J., and Chen, F. (2007) Ant-mediated seed dispersal contributes to the local spatial pattern and genetic structure of *Globba lancangensis* (Zingiberaceae). *Journal of Heredity 98*, 317–24.

Zhou, S.Y., and Xu, Z.H. (2003) Taxonomic study on Chinese members of the ant genus *Strumigenys* F. Smith (Hymenoptera: Formicidae) from the mainland of China. *Acta Zootaxonomica Sinica 28*, 737–40.

——Zhao, S., and Jia, F. (2006) A taxonomic study on the ant genus *Pheidologeton* Mayr (Hymenoptera, Formicidae, Myrmicinae) from China. *Acta Zootaxonomica Sinica 31*, 870–3.

——and Huang, J. (2006) Two new species of the ant genus *Perissomyrmex* Smith (Hymenoptera: Formicidae) from China. *Entomological News 117*, 189–96.

Zimmerman, E.C. (1970) Adaptive radiation in Hawaii with special reference to insects. *Biotropica 2*, 32–8.

Zinck, L., Hora, R.R., Châline, N., and Jaisson, P. (2008) Nestmate recognition in the polydomous and facultative polygynous ant *Ectatomma tuberculatum*. *Entomologia Experimentalis et Applicata 126*, 211–6.

Zoebelein, G. (1956) Der Honigtau als Nahrung der Insekten, Teil I. *Zeitschrift für Angewandte Entomologie 38*, 369–416.

Zoological Society of London. (2007) *ZSL to mount 'operation red ant'*. http://www.zsl.org/zsl-london-zoo/news/zsl-to-mount-operation-red-ant,338,NS.html.

Index

B

The manufacturer's authorised representative in the EU for product safety is Oxford University Press España S.A. of El Parque Empresarial San Fernando de Henares, Avenida de Castilla, 2 - 28830 Madrid (www.oup.es/en or product.safety@oup.com). OUP España S.A. also acts as importer into Spain of products made by the manufacturer.
Printed and bound by CPI Group (UK) Ltd, Croydon, CR0 4YY
14/09/2026
02216284-0018